MTP International Review of Science

Volume 11

Chemical Crystallography

Edited by **J. Monteath Robertson, F.R.S.**
University of Glasgow

Butterworths · London
University Park Press · Baltimore

THE BUTTERWORTH GROUP

ENGLAND
Butterworth & Co (Publishers) Ltd
London: 88 Kingsway, WC2B 6AB

AUSTRALIA
Butterworth & Co (Australia) Ltd
Sydney: 586 Pacific Highway 2067
Melbourne: 343 Little Collins Street, 3000
Brisbane: 240 Queen Street, 4000

NEW ZEALAND
Butterworth & Co (New Zealand) Ltd
Wellington: 26–28 Waring Taylor Street, 1

SOUTH AFRICA
Butterworth & Co (South Africa) (Pty) Ltd
Durban: 152–154 Gale Street

ISBN 0 408 70272 9

UNIVERSITY PARK PRESS

U.S.A. and CANADA
University Park Press Inc
Chamber of Commerce Building
Baltimore, Maryland, 21202

Library of Congress Cataloging in Publication Data

Robertson, John Monteath.
Chemical crystallography.

(Physical chemistry, series one, v. 11) (MTP international review of science)
Includes bibliographies.
1. Crystallography. I. Title.
QD453.2.P58 vol. 11 [QD951] 541′.3′08s [548′.3]
ISBN 0–8391–1025–1 72 2328

Filmset by Photoprint Plates Ltd., Rayleigh, Essex
Printed in England by Redwood Press Ltd., Trowbridge, Wilts
and bound by R. J. Acford Ltd., Chichester, Sussex

MTP International Review of Science

Chemical Crystallography

MTP International Review of Science

Publisher's Note

The MTP International Review of Science is an important new venture in scientific publishing, which we present in association with MTP Medical and Technical Publishing Co. Ltd. and University Park Press, Baltimore. The basic concept of the Review is to provide regular authoritative reviews of entire disciplines. We are starting with chemistry because the problems of literature survey are probably more acute in this subject than in any other. As a matter of policy, the authorship of the MTP Review of Chemistry is international and distinguished; the subject coverage is extensive, systematic and critical; and most important of all, new issues of the Review will be published every two years.

In the MTP Review of Chemistry (Series One), Inorganic, Physical and Organic Chemistry are comprehensively reviewed in 33 text volumes and 3 index volumes, details of which are shown opposite. In general, the reviews cover the period 1967 to 1971. In 1974, it is planned to issue the MTP Review of Chemistry (Series Two), consisting of a similar set of volumes covering the period 1971 to 1973. Series Three is planned for 1976, and so on.

The MTP Review of Chemistry has been conceived within a carefully organised editorial framework. The over-all plan was drawn up, and the volume editors were appointed, by three consultant editors. In turn, each volume editor planned the coverage of his field and appointed authors to write on subjects which were within the area of their own research experience. No geographical restriction was imposed. Hence, the 300 or so contributions to the MTP Review of Chemistry come from many countries of the world and provide an authoritative account of progress in chemistry.

To facilitate rapid production, individual volumes do not have an index. Instead, each chapter has been prefaced with a detailed list of contents, and an index to the 13 volumes of the MTP Review of Physical Chemistry (Series One) will appear, as a separate volume, after publication of the final volume. Similar arrangements will apply to the MTP Review of Organic Chemistry (Series One) and to subsequent series.

Butterworth & Co. (Publishers) Ltd.

Physical Chemistry
Series One
Consultant Editor
A. D. Buckingham
Department of Chemistry
University of Cambridge

Volume titles and Editors

1 **THEORETICAL CHEMISTRY**
Professor W. Byers Brown, *University of Manchester*

2 **MOLECULAR STRUCTURE AND PROPERTIES**
Professor G. Allen, *University of Manchester*

3 **SPECTROSCOPY**
Dr. D. A. Ramsay, F.R.S.C., *National Research Council of Canada*

4 **MAGNETIC RESONANCE**
Professor C. A. McDowell, *University of British Columbia*

5 **MASS SPECTROMETRY**
Professor A. Maccoll, *University College, University of London*

6 **ELECTROCHEMISTRY**
Professor J. O'M Bockris, *University of Pennsylvania*

7 **SURFACE CHEMISTRY AND COLLOIDS**
Professor M. Kerker, *Clarkson College of Technology, New York*

8 **MACROMOLECULAR SCIENCE**
Professor C. E. H. Bawn, F.R.S., *University of Liverpool*

9 **CHEMICAL KINETICS**
Professor J. C. Polanyi, F.R.S., *University of Toronto*

10 **THERMOCHEMISTRY AND THERMODYNAMICS**
Dr. H. A. Skinner, *University of Manchester*

11 **CHEMICAL CRYSTALLOGRAPHY**
Professor J. Monteath Robertson, F.R.S., *University of Glasgow*

12 **ANALYTICAL CHEMISTRY—PART 1**
Professor T. S. West, *Imperial College, University of London*

13 **ANALYTICAL CHEMISTRY—PART 2**
Professor T. S. West, *Imperial College, University of London*

INDEX VOLUME

Inorganic Chemistry
Series One
Consultant Editor
H. J. Emeléus, F.R.S.
Department of Chemistry
University of Cambridge

Volume titles and Editors

1 **MAIN GROUP ELEMENTS—HYDROGEN AND GROUPS I–IV**
Professor M. F. Lappert, *University of Sussex*

2 **MAIN GROUP ELEMENTS—GROUPS V AND VI**
Professor C. C. Addison, F.R.S. and Dr. D. B. Sowerby, *University of Nottingham*

3 **MAIN GROUP ELEMENTS—GROUP VII AND NOBLE GASES**
Professor Viktor Gutmann, *Technical University of Vienna*

4 **ORGANOMETALLIC DERIVATIVES OF THE MAIN GROUP ELEMENTS**
Dr. B. J. Aylett, *Westfield College, University of London*

5 **TRANSITION METALS—PART 1**
Professor D. W. A. Sharp, *University of Glasgow*

6 **TRANSITION METALS—PART 2**
Dr. M. J. Mays, *University of Cambridge*

7 **LANTHANIDES AND ACTINIDES**
Professor K. W. Bagnall, *University of Manchester*

8 **RADIOCHEMISTRY**
Dr. A. G. Maddock, *University of Cambridge*

9 **REACTION MECHANISMS IN INORGANIC CHEMISTRY**
Professor M. L. Tobe, *University College, University of London*

10 **SOLID STATE CHEMISTRY**
Dr. L. E. J. Roberts, *Atomic Energy Research Establishment, Harwell*

INDEX VOLUME

Organic Chemistry
Series One
Consultant Editor
D. H. Hey, F.R.S.
Department of Chemistry
King's College, University of London

Volume titles and Editors

1 **STRUCTURE DETERMINATION IN ORGANIC CHEMISTRY**
Professor W. D. Ollis, F.R.S., *University of Sheffield*

2 **ALIPHATIC COMPOUNDS**
Professor N. B. Chapman, *Hull University*

3 **AROMATIC COMPOUNDS**
Professor H. Zollinger, *Swiss Federal Institute of Technology*

4 **HETEROCYCLIC COMPOUNDS**
Dr. K. Schofield, *University of Exeter*

5 **ALICYCLIC COMPOUNDS**
Professor W. Parker, *University of Stirling*

6 **AMINO ACIDS, PEPTIDES AND RELATED COMPOUNDS**
Professor D. H. Hey, F.R.S. and Dr. D. I. John, *King's College, University of London*

7 **CARBOHYDRATES**
Professor G. O. Aspinall, *University of Trent, Ontario*

8 **STEROIDS**
Dr. W. D. Johns, *G. D. Searle & Co., Chicago*

9 **ALKALOIDS**
Professor K. F. Wiesner, F.R.S., *University of New Brunswick*

10 **FREE RADICAL REACTIONS**
Professor W. A. Waters, F.R.S., *University of Oxford*

INDEX VOLUME

Physical Chemistry
Series One

Consultant Editor
A. D. Buckingham

Consultant Editor's Note

The MTP International Review of Science is designed to provide a comprehensive, critical and continuing survey of progress in research. The difficult problem of keeping up with advances on a reasonably broad front makes the idea of the Review especially appealing, and I was grateful to be given the opportunity of helping to plan it.

This particular 13-volume section is concerned with Physical Chemistry, Chemical Crystallography and Analytical Chemistry. The subdivision of Physical Chemistry adopted is not completely conventional, but it has been designed to reflect current research trends and it is hoped that it will appeal to the reader. Each volume has been edited by a distinguished chemist and has been written by a team of authoritative scientists. Each author has assessed and interpreted research progress in a specialised topic in terms of his own experience. I believe that their efforts have produced very useful and timely accounts of progress in these branches of chemistry, and that the volumes will make a valuable contribution towards the solution of our problem of keeping abreast of progress in research.

It is my pleasure to thank all those who have collaborated in making this venture possible – the volume editors, the chapter authors and the publishers.

Cambridge A. D. Buckingham

Preface

The greater part of chemistry is concerned with structure in some sense, and the science of crystallography gives structure by providing the means for mapping out accurately the positions of atoms in space. The importance of this metrical aspect of structure will be clearly evident to anyone who glances through the various volumes on inorganic, physical and organic chemistry in this series. Many structures and much structural data are described and applied to chemical problems.

This volume is devoted specifically to chemical crystallography, but it was apparent from the outset that only selected aspects of this very large subject could be included. During the four or five years covered by this review there has been an enormous expansion of the literature dealing with chemical crystallography. Hardly a week passes without one or more new structures being described. This explosion of information is due in large measure to the increasing availability of large and very efficient computers, automatic diffractometers, and so on, which speed up the collection and handling of data and remove the computational burden. An average structure can now be solved in weeks instead of months, and even the very large structures of biological molecules may sometimes be solved in months instead of years. With this speeding up, accuracy has also been increased by a large factor, and so the results become much more significant.

All this has created great difficulties in compiling this volume, and many chapters have had to be drastically shortened and some omitted. No attempt has been made to describe methods of analysis. The heavy atom and isomorphous replacement methods for phase determination are still probably the most important, and indeed they are the only methods so far available for the determination of large protein and enzyme structures. A feature of recent work, however, has been the increasing power and application of the so-called direct methods, which depend on mathematical relations between the structure factors and do not require a heavy atom derivative. A chapter is devoted to recent applications of this method.

Crystallography is a big subject, but we hope this volume will give some indication of the generality of the method and of the more important advances that are being made in its chemical applications.

Glasgow

J. M. Robertson

Contents

1
Hydrogen Bonding, with Special Reference to O··H··O

J. C. SPEAKMAN
University of Glasgow

'One of the difficulties attending investigations of the hydrogen bond is that reliable determinations of the position of the hydrogen atom are difficult to make.'
(J. M. Robertson, *Organic Crystals and Molecules*, 1953)

1.1 INTRODUCTION AND BACKGROUND

1.1.1 Origins

Hydrogen bonding is used to describe a situation resulting from the process

$$A—H \quad + \quad B \quad \rightarrow \quad A—H\cdots B,$$

whereby the entities AH and B become attached to one another. It is a weak form of valency bonding, with an energy usually less than 10% of that associated with ordinary covalency. A and B, whether they be isolated atoms or parts of larger molecules, must be electronegative. Relatively strong hydrogen bonding is largely confined to cases where A and B are atoms of the three most electronegative elements – fluorine, oxygen and nitrogen. Though weak bonding sometimes occurs with chlorine, or bromine, or sulphur – even with carbon, when it is attached to very electronegative atoms – the interaction is then feeble and hard to distinguish from the general dispersion forces that operate between contiguous molecules.

This Review is mainly concerned with systems in which A and B are both oxygen atoms.

From another, though related, point of view AH must be acidic; B must be basic. Within limits, increasing acid and basic strength favours hydrogen bonding. But, if the strengths become too great, an electrovalency will result:

$$A—H \quad + \quad B \quad \rightarrow \quad A—H\cdots B \quad \rightarrow \quad \bar{A} \quad + \quad H—\overset{+}{B}$$

Though the name, hydrogen bond, is more recent, the phenomenon was first explicitly recognised in the first decade of this century. For the roots of the concept we need to go back at least another century to the recognition of water as a remarkably unorthodox chemical compound. The simplest illustration of this is the boiling point, which is strangely high for material of molecular weight 18. No other material with a molecular weight less than 20 is not gaseous at ordinary temperatures. When physical chemistry began to develop, molecular association was recognised as the general cause of the abnormal behaviour of water in particular, and of other liquids in a lesser degree. These were usually compounds whose molecules included hydroxyl or amino groups: ROH, RNH_2, etc. Association could be correlated with the presence of such polar groups in the molecule.

The abnormality disappeared when the hydrogen atoms of the OH or NH_2 groups were replaced by a methyl group, for example. When the hydrogen atoms of water are so replaced, to yield dimethyl ether, the boiling point drops by 124 °C, despite more than a doubling of molecular weight. Many observations of this sort implied that the hydrogen atom is an essential link in the machinery of molecular association. By about 1907 certain chemists were bold enough to write formulae which carried the disturbing implication that hydrogen could behave as if it were bivalent: for instance, F—H···F or H_2O···H—O—H. The meaning of the dotted line was obscure, though not any more so – at the time – than that of the ordinary chemical bonds. Then in 1920, Latimer and Rodebush wrote the electronic formula (1):

```
  H
  ..       ..
: O : H : O : H
  ..       ..
  H   (1)
```

The hydrogen bond was explicit, though this formula could hardly have been taken quite literally even in 1920.

1.1.2 A note on the mechanism of hydrogen bonding

The causes of hydrogen bonding came within range of speculation, at the atomic level, after 1927. The first type of explanation, which we may term 'chemical', arose from a formal resemblance between the holding of two atoms or groups together by a hydrogen atom and coordination through a metal atom. Just as the complex $[Co(NH_3)_6]^{3+}$ was attributed to dative covalency ($H_3N \rightarrow Co$), so a dimerised water molecule might be formulated as

```
H
 \
  O → H—O—H
 /
H
```

This was quickly rejected because the 1s orbital of hydrogen cannot accommodate two pairs of electrons*.

Proponents of a chemical mechanism then invoked the concept of resonance which Pauling had popularised. The hydrogen bond between (say) two water molecules might owe its stability to hybridisation between two electronic formulations (2a) and (2b):

```
H                     H
 \                     \
  O:  H—O—H             O———H:O—H
 /                     /
H                     H

      (2a)                  (2b)
```

This idea fell into disfavour when it was pointed out that form (b), as we have drawn it, would have a much higher energy than (a). Resonance stabilisation

*The contingency amendment, that the donated electron pair went into a second-shell orbital, was summarily rejected because this would be at too high an energy level. But Dr. Brian Webster opines that this objection is not necessarily so valid as it appeared to be in 1935.

would be significant only if the proton were centrally placed between the two oxygen atoms, or nearly so. This is certainly not true of most hydrogen bonds. However, in this Review we shall give prominence to some crystals which do have symmetrical, or quasi-symmetrical, O···H···O bonds. For such bonds, the resonance mechanism – or its more modern counterpart – may be important (see Section 1.7.1).

The second type of explanation, which may be labelled 'physical', attributes hydrogen bonding to a straightforward, classical electrostatic force. In the A—H B system, AH is necessarily dipolar in the sense symbolised below; and B, for similar reasons, carries a negative charge – a full charge if it is an anion (as in HF_2^-), a partial charge if it is part of a larger molecule. When in the mutual orientation

$$\overset{\delta-}{A}—\overset{\delta+}{H} \quad \overset{\delta-}{B}$$

an attractive force operates between A and B. One advantage of this scheme was that it lent itself to simple quantitative calculations. Provided secondary effects were neglected, quite elementary calculations gave bonding energies of correct order of magnitude, at any rate for weak hydrogen bonds.

The theory of hydrogen bonding is surveyed in the standard monographs by Pimentel and McClellan[1] and by Hamilton and Ibers[2]. From a specialist position there is a recent review by Bratož[3]; and Murrell[4] has written a more elementary account for the general reader.

The molecular interactions we classify as hydrogen bonds cover a wide range, from the minimal, or notional, attraction in (say) N—H···Cl, to the strongest bonding in [F—H—F]$^-$. Each exists in a molecular system that is stable in its own right. As Murrell has emphasised, it is artificial to divide the total molecular bonding energy amongst separate bonds, each with a bonding energy resolvable into components due to distinct forces. It is therefore dangerous to think we have clarified the situation when we say that the weak N—H···Cl bond and the very strong F—H—F differ merely in the relative participation in them of electrostatic and delocalisation forces. Nevertheless it comes naturally to many chemists to do just this.

1.1.3 Infrared spectra and hydrogen bonding

In this Review we are concerned with hydrogen bonding as it may be studied by crystal-diffraction methods. However, other methods require brief mention, one such being infrared spectroscopy.

An unperturbed hydroxyl group in a gaseous molecule is associated with a stretching frequency, which appears as a sharp absorption band centred near 3700 cm^{-1} (1.1×10^{14} Hz). The spectra of solutions of hydroxylic compounds in an inert solvent show a similar peak. When the hydroxyl enters into hydrogen bonding, this feature changes notably. The peak moves to lower frequency; it becomes less sharp; it also becomes broader, with enhancement of its integrated intensity. These changes are progressive with increasing strength of the hydrogen bonding. For instance, there is a smooth, inverse correlation between frequency and strength as indicated by the overall shortening of the O—H···O distance[1,2]. It can be followed down to frequencies of about 2000 cm^{-1}.

To get access to a range of measurable hydrogen bonds, we need to study crystalline solids. The interpretation of the spectra of solids is much more difficult than for gases. But it is clear that shortening of the bond is associated not only with a lowering of frequency, but also with more profound changes of spectrum in very strong bonds. This subject has been extensively studied

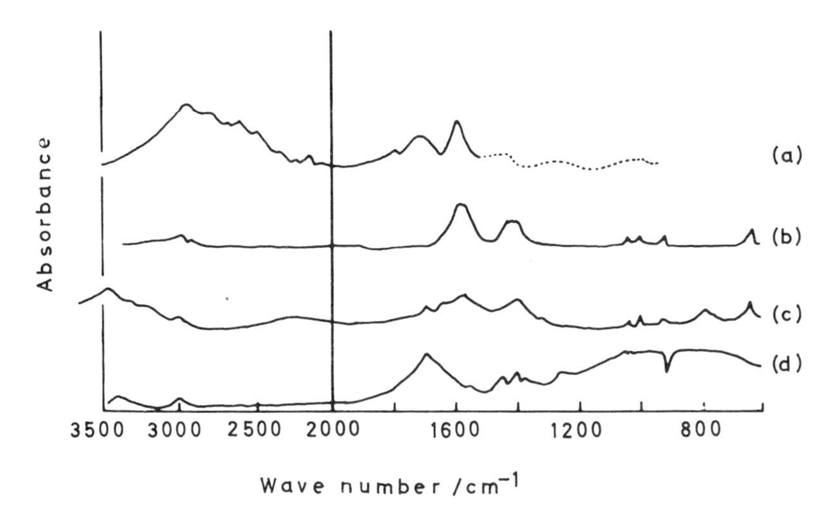

Figure 1.1 Infrared spectra of acetic acid and some salts: (a) crystalline acid; (b) sodium acetate; (c) sodium acetate trihydrate; (d) sodium hydrogen diacetate

by many authors, and reference may be made to recent articles by Hadzi[5] and Sheppard[6]. An example is shown in Figure 1.1, which is concerned with acetic acid and its sodium salts. The first spectrum (a) is that of solid acetic acid, in which the molecules are linked into infinite chains by moderately strong hydrogen bonds (3).

$$\cdots HO—\underset{|}{\overset{CH_3}{C}}=O\cdots HO—\overset{CH_3}{\underset{|}{C}}=O\cdots$$

(3)

The large peak ~3000 cm^{-1} is attributable to the O—H stretching. The ~700 cm^{-1} lowering of frequency is typical. Figure 1.1(b) gives the spectrum of the anhydrous sodium salt, (c) that of its trihydrate. In (b) there is no peak attributable to OH: in (c) there is a broad peak near 3500 cm^{-1}, due to hydrogen-bonded water molecules. Figure 1.1(d) shows the spectrum of the *acid salt*, $NaH(CH_3CO_2)_2$, to which we shall refer later. This is characteristic of the spectra given by systems (particularly, but not necessarily, crystalline) with very strong O···H···O bonds. They have been classified by Hadži[5] as Type (ii) spectra. Sheppard has described them as 'remarkable spectra by any standards'. Their principal peculiarities are two: there is no peak unambiguously attributable to O—H stretching, though the peak near 1700 cm^{-1} in this case might possibly involve interaction of this mode with the C—O stretching; instead, there is a vast region of general absorption culminating near 900 cm^{-1}, sometimes known as the 'D band'. These features of Type (ii) spectra are diagnostic of very strong hydrogen bonding, and have led to successful predictions of unknown crystal structures[7].

1.2 CRYSTAL-STRUCTURE ANALYSIS AND HYDROGEN BONDING

Some of the earliest direct evidence of hydrogen bonding was derived from the x-ray analysis of crystals[8,9]. It is still an important method. In principle it enables us to assign relative positions to all the atoms; when applied to a molecular crystal, it reveals hydrogen bonds *in situ*, whether between different molecules or between different parts of the same molecule.

1.2.1 x-Ray study of hydrogen bonds

x-Rays are scattered by the electron-density condensation around an atomic nucleus. Atoms therefore show up more prominently the greater the atomic number. Because the electron density associated with a hydrogen atom is low, hydrogen atoms are more difficult to find. In early x-ray work they were not located at all, and the evidence for hydrogen bonding was to that extent incomplete, though convincing enough. The typical experimental finding was a pair of oxygen atoms too far apart for them to be directly linked by covalency, yet too close together for them to be non-bonded. The Pauling value for the van der Waals radius of oxygen is 1.40 Å*, the covalent radius 0.66 Å (though the O—O distance in the hydrogen peroxide molecule is 1.50 Å). Thus any pair of atoms separated by a distance in the range 1.5–2.8 Å would be suspected of some form of liaison weaker than normal covalency. In practice distances significantly shorter than 2.4 Å are unknown. Hydrogen bonding would then be indicated by O···O distances between 2.4 and 2.8 Å.

The upper limit needs some increase to meet the following consideration. If there is to be any bonding, a hydrogen atom must be attached to one of the oxygens. Supposing this to be – as it must – nearly on the O···O line, we must add to the van der Waals radius of the acceptor oxygen atom the O—H distance at the donor, plus the van der Waals radius of the hydrogen; these amount, respectively, to $\sim$1.0 and $\sim$0.9 Å, giving a total O···O distance of 3.3 Å for the upper limit. (However this has been questioned by Bellamy[10].)

Application of this criterion to the results of x-ray work revealed many examples of O—H···O bonding, overall distances being mainly in the range 2.5–3.2 Å. Bonds with O···O > 2.75 Å came to be called 'long', whilst those between 2.5 and 2.6_5 Å were 'short'. We now have plenty of examples of bonds with O···O between 2.4 and 2.5 Å. These have had to be termed

*As is still the almost universal practice amongst crystallographers, we express interatomic distances in ångstroms: 1 Å = 100 pm.

'very short' hydrogen bonds. They are of particular concern for us in this Review.

At the level of x-ray structure analysis implied above, participation of the hydrogen atom was deduced by inference. As there was always chemical evidence for the presence of hydrogen, and in a stereochemically reasonable position, the inference was strong. Improvements in x-ray methods substantiated the inference. In a modern analysis, based on more accurate intensity measurements, the small electron-density peak due to hydrogen

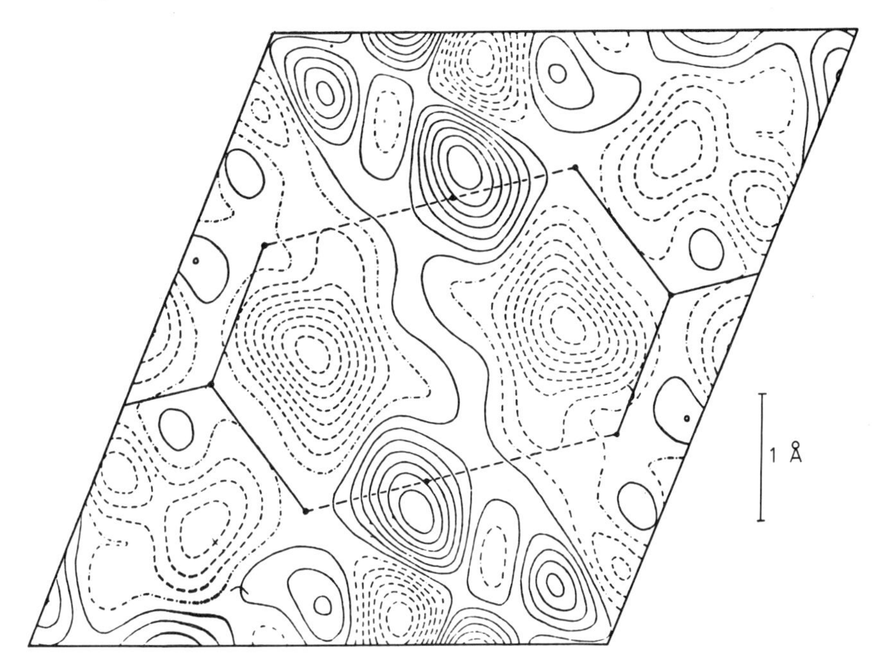

Figure 1.2 Electron-density 'difference' synthesis of the region between two carboxyl groups, based on the x-ray study of β-succinic acid (Negative contour lines are broken. The positions of the carbon and oxygen atoms of the carboxyl groups are indicated)
(From Robertson, J. M.[9], by permission of the Royal Society)

can normally be detected, at any rate in crystals without elements of high atomic number. Such peaks show up better in the electron-density 'difference synthesis'–a Fourier series whose coefficients are $(F_o - F'_c)$, where F_o is the observed structure amplitude (suitably phased) and F'_c is the structure factor calculated for all the heavier atoms, but omitting hydrogens. An example is given in Figure 1.2. With the mountains taken out, the foothills show up more prominently.

1.2.2 The study of hydrogen bonds by neutron diffraction

Neutrons are scattered by the atomic nuclei. Though the scattering powers for 'thermal' neutrons ($\lambda \sim 1.0$ Å) show a general increase with rising atomic number of the scatterer, the increase is gradual and often over-ridden by special effects with particular nuclides. Table 1.1 compares some neutron-scattering factors with (average) values for x-ray scattering. For convenience, different kinds of units are used. With x-rays it is natural to base the scattering power on the number of electrons in the atom. The scattering effect of an atomic nucleus is expressed as an amplitude, which is of the order of the nuclear diameter ($\sim 10^{-12}$ cm). Thus, the amplitude for ^{12}C is 0.66×10^{-12} cm. It has recently become conventional to use a unit 10 times smaller than 10^{-12} cm. In SI this is defined as 10^{-15} m, and known as the fermi. The

Table 1.1 Comparison of x-ray and neutron scattering factors

	H	D	C	O	F	Cl	Br
x-Rays (electrons) at $(\sin\theta)/\lambda = 0.3$	0.25		2.5	4.1	5.2	9.4	23.8
Neutrons (fermis)	−3.8	+6.5	+6.6	+5.8	+5.5	+9.9	+6.7

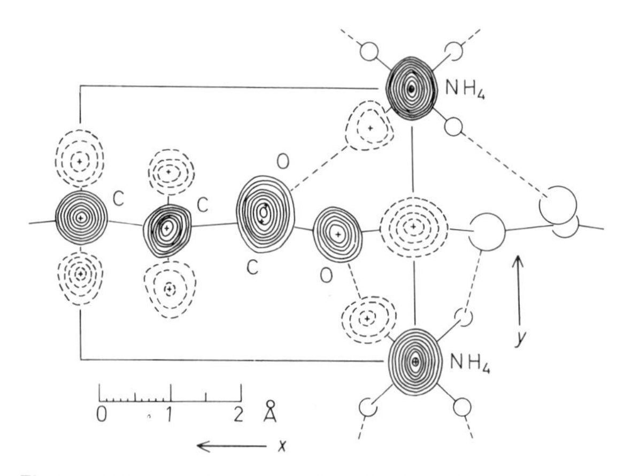

Figure 1.3 Neutron-scattering density in ammonium hydrogen glutarate, seen in the *c* axial projection (The region covered includes rather more than half of a hydrogen–glutarate residue, $O_2C{\cdot}CH_2{\cdot}CH_2{\cdot}CH_2{\cdot}CO_2$, linked at each end to a similar residue by an O···H···O bond across a twofold axis, and two independent ammonium ions, which form N—H···O bonds to neighbouring oxygen atoms. Negative contours are represented by broken lines. See reference 11.)

negative amplitude for H, and a few other nuclides, merely signifies that the act of scattering involves an unorthodox change of phase of the neutron-wave. It results in negative Fourier peaks for hydrogen atoms.

We see that in x-ray analysis a hydrogen (or deuterium) atom is unfavourably recorded, compared with oxygen, by a factor of about 16. With neutrons. ordinary hydrogen is inferior to oxygen only by a factor of <2; whilst, if we use a deuteriated crystal the hydrogen is superior. There is an additional effect that makes deuteriation technically beneficial. Ordinary hydrogen nuclei produce a high background intensity because protons may scatter neutrons incoherently. This is avoided with deuterons. Figure 1.3 reproduces a neutron-scattering Fourier projection in which the positions of protons round two ammonium ions, each hydrogen-bonded to four oxygen neighbours, are clearly seen, as well as other hydrogen atoms[11].

1.2.3 Errors in hydrogen positions determined by x-rays

For reasons implied in Section 1.2.2, hydrogen atoms are much more precisely located by neutrons. In analyses of comparable quality, standard deviations of hydrogen positions based on x-ray intensities will be about ten times larger than those based on neutron intensities. (Against this neutron diffraction suffers two disadvantages: it is much more expensive, and it requires much larger crystals, which are often difficult to procure.)

The positions found by x-rays do not agree with those found by neutrons. In almost all cases A—H distances appear to be shorter by x-rays, and the difference is probably greater the more electronegative is A. This discrepancy is discussed by Hamilton and Ibers[2]. Recent analyses of glycollic acid, $CH_2OH{\cdot}CO_2H$, by x-ray[12] and neutron diffraction[13] illustrate this phenomenon particularly well. Some results are compared in Table 1.2*.

Table 1.2 Comparison of O—H distances (Å) in glycollic acid as found by analysis with x-rays (X) and neutrons (N)

Bond	1	2	3	4
X	0.934(25)	0.898(25)	0.819(26)	0.834(26)
N	1.003(2)	1.001(2)	0.971(2)	0.970(3)
Δ	−0.07	−0.11	−0.15	−0.14

These are typical differences; they are significant despite the higher uncertainty of the x-ray distances.

When we speak of the 'position of a hydrogen atom', we normally imply the position of the proton†. This is what neutron analysis finds. Associated with the proton there must be a local maximum of electron density. This maximum must coincide with the proton, or very nearly so. Allegedly, x-ray analysis finds it. But this may not be true, for the following reason. Because the hydrogen atom is near to its covalent, and electronegative, partner, the local electron-density cloud will be far from symmetrical. The model used for any atom, in structure refinement based on x-ray data, is a spherical electron-density condensation. (Any more sophisticated model is

*Here, and elsewhere, we give standard deviations in parentheses. Thus 0.934(25 is short for the more conventional 0.934 ± 0.025. It is perhaps worthwhile to remind the reader that this is a measure of precision, not of accuracy. With some reservations, we may suppose the result unlikely to be in error by twice 0.025, and the true result very unlikely to be outside the limits 1.009–0.859. Not sensational news, in this case.

†More precisely, some sort of *mean* position. The differences between various sorts are not important in the context.

very difficult to handle.) The refinement procedure tries to fit this model to the non-spherical cloud implicit in the observational data. For a hydrogen it finds a 'best fit' when the sphere is displaced from the density maximum in a direction towards the more electronegative atom. The error (for we consider the x-ray result to be in error) will be greater the higher the electronegativity of the atom to which the hydrogen is covalently bonded.

1.3 THE GEOMETRY OF O—H···O BONDING

Conditions favourable to O—H···O bonding have been discussed by several authors, and their conclusions are summarised by Hamilton and Ibers[2]. The R—O—H angle is usually a little less than tetrahedral. This molecule will form a hydrogen bond with another oxygen atom, only if the O—H group points towards – or nearly towards – the acceptor atom. As hydrogen-bonding energies are not usually large enough to cause large changes in valency angles, this favourable orientation has to be achieved mainly by torsional adjustments. Exact alignment of O, H and O is not, of course, essential; but in strong hydrogen bonds O···H···O is always greater than (say) 160 degrees*. It is also advantageous if the conformational situation at the acceptor atom is such as to allow us to suppose that this atom presents a lone pair of electrons towards the donor.

In weaker hydrogen bonds these geometrical considerations become less important. In marginal cases the notion of a hydrogen bond ceases to have significance. The hydrogen atom of a peripheral hydroxyl group has to be somewhere. When circumstances do not allow it to form a palpable hydrogen bond, it may still minimise the total energy a little if it lies more or less in the direction of some rather distant negative atom.

Many, if not perhaps all, examples of the 'bifurcated' hydrogen bond can be explained away by an extension of the argument sketched in the previous paragraph. A proton may find a 'point of rest' – a point of minimum energy – more or less equidistant from two neighbouring acceptor atoms.

1.3.1 The Nakamoto–Margoshes–Rundle curve

In a very weak O—H···O liaison, the covalent bonding O—H is little affected, as is evidenced by the smallness of any change in the stretching frequency. The O—H will be more perturbed in a stronger liaison. It is intuitively reasonable to suppose that the O—H distance increases as O···O diminishes overall; and there is plenty of evidence to support this idea. The manner in which these distances may be inversely related is of interest.

A direct experiment in which the O—H distance is observed as a second oxygen atom approaches is impracticable. We have to depend on neutron-

*Histograms showing the frequency of occurrence of various O···H···O angles have been interpreted as showing a small preference for angles rather less than 180 degrees. It seems to the Author that this is based on a fallacy: if angles of 170, 180 and 190 degrees occurred with equal frequency, this could be misinterpreted to mean that 170 degrees was twice as common as 180 degrees.

diffraction measurements of the geometry at the hydrogen bonds in a series of different crystalline materials. For reasonable comparability the various hydrogen bonds should be in chemically similar situations. But this is impossible if we wish to cover a wide range of O···O distances. Consequently, a graphical plot of O—H *v.* O···O will yield a scatter of points, due partly to experimental errors, but – even if these errors could be eliminated – also to inherent differences in the O—H···O systems being compared. One of the first such graphs to be published was that of Nakamoto, Margoshes and Rundle (NMR)[14]. A later version, based on fuller experimental material, was given by Pimentel and McClellan[1]. We reproduce their (averaged) curve in Figure 1.4, the broken line. At its right-hand side, the curve has been

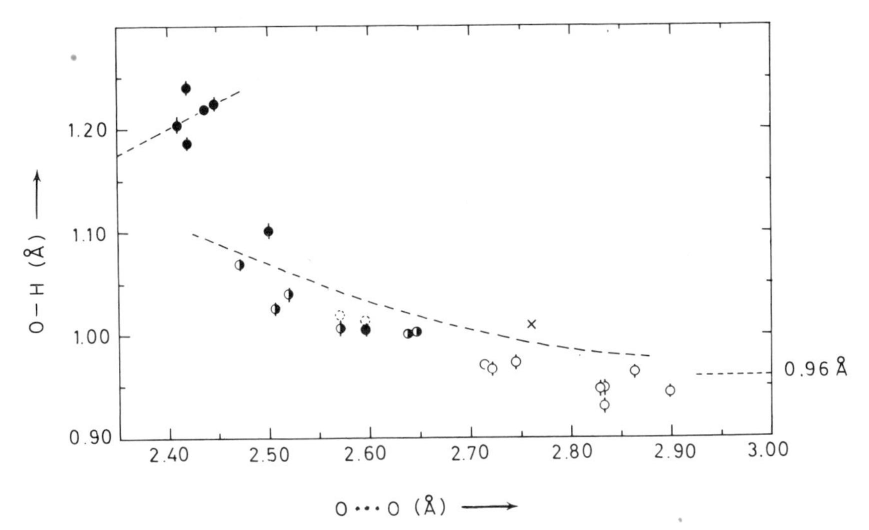

Figure 1.4 The Nakamoto–Margoshes–Rundle curve of Pimentel and McClellan[1] (broken lines), and some points from recent neutron analyses (The circles represent O···H···O bonds, and the vertical line in each the standard deviation of the O—H distance. The filled circles are for systems in which both donor and acceptor are carboxyl groups (or groups of similarly strong acidity); half-filled circles systems in which carboxyl is the donor; open circles systems in which it is the acceptor. The broken circles, in two cases, show the effect of applying an estimated correction for libration. The cross is for the O—D···O bonds in 'heavy' ice.)

drawn to converge on to a horizontal line at O—H = 0.96 Å – the bond-length in a gaseous water molecule. At the left-hand side, we show the straight line of slope 0.5 corresponding to truly symmetrical O···H···O bonds, supposing such bonds to exist. Similar graphs have been plotted for other hydrogen-bonded systems, such as F—H···F, N—H···O or O—H···N[1].

For O···O > 2.6 Å the general trend of the NMR curve is well defined. At shorter distances the trend becomes less certain, and more interesting. Reliable experimental results are few in their region. Possibly we are approaching a catastrophic situation, where the O—H···O bond – rather suddenly, perhaps discontinuously – changes character and becomes O···H···O. In this Review we are particularly concerned with these short bonds.

In Figure 1.4 we have also plotted some recent results from fairly precise neutron-diffraction work (marked by circles). We have restricted the points to bonds where the O···H···O angle is greater than 160 degrees, and to systems in which a carboxyl group (or one of similar acidity) is involved as donor, or acceptor, or both. Though our sample is not large, we notice that it does not make a good fit with the Pimentel and McClellan version of the NMR curve. The points suggest that the curve ought to be pulled down in its central and right-hand parts, and to rise much more steeply to the left. The difference may be partly due to our restrictions on the choice of hydrogen bonds.

The O—H distances plotted are the values uncorrected for the effects of librational motion. The correction is small, and difficult to assess, between hydrogen-bonded molecules (see Section 1.6.4). If it had been applied, it would increase O—H, though not enough to reach the curve.

In heavy ice[15] the O···O distance is 2.76 Å and the mean O—D 1.00(1) Å if corrected for libration. This result is represented in Figure 1.4 by a cross, which fits the curve better.

1.3.2 The potential-energy diagram for O··H··O bonds

Relevant to our discussion is a graph representing the variation of potential energy as the proton moves along a straight line between two fixed oxygen atoms. Some conceivable situations are described by the curves of Figure 1.5. The first (a) could correspond to a long bond between a hydroxyl group and an uncomplicated keto group. The (mean) position of the proton lies at the lower minimum in the normal state: —O—H···O=C< ; at the higher minimum in the less stable tautomeric alternative: —$\bar{\text{O}}$···H—$\overset{+}{\text{O}}$=C< .
A potential-energy curve of this type obtains in most O—H···O bonds. The second curve (b) corresponds to a situation where the bond, though still rather long, has exact, or virtual, symmetry. A familiar example, with virtual symmetry, is ice. There are alternative positions for the proton; diffraction methods find a statistical 'half-proton' in each*.

As the overall O···O distance decreases, the energy barrier separating the double minima will diminish in height, as shown in (c), and ultimately disappear, leaving a single-minimum curve, which might be of the parabolic shape indicated by curve (e), or of the flattened shape (d). (There is now theoretical support for the sequence (c), (d) and (e) from quantum-mechanical calculations on simple model systems[16].) Curves (d) and (e) represent genuinely symmetrical O···H···O bonds, and correspond to the line of 0.5 slope at the left-hand side of Figure 1.4.

Curve (e) would apply to an ideally symmetrical bond, in which the

---

*Such symmetry of the potential-energy curve in ice would be valid only in a restricted sense. Were a single proton to move to the other site, without sympathetic movements elsewhere in the crystal domain, it would find itself in an energetically less favoured situation.

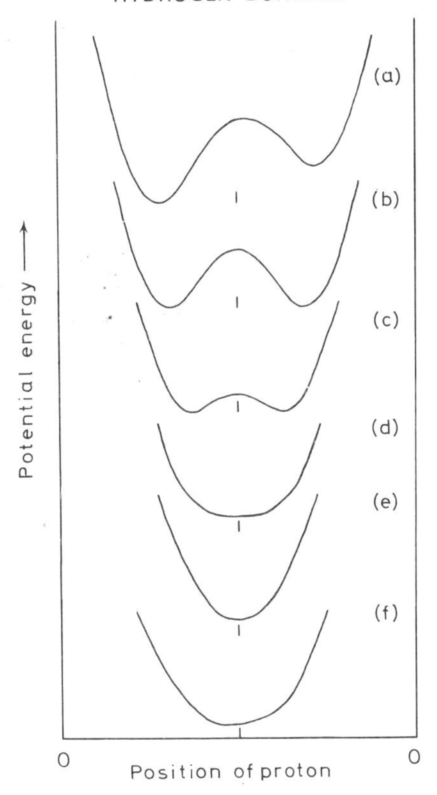

Figure 1.5 The variation of potential energy as a proton moves along the line between two oxygen atoms (The various curves correspond to different, hypothetical, situations.)

proton would vibrate harmonically about the mid-point. Somewhere between (c) and (e) a complication arises when the energy barrier is only about as high as the vibrational zero-point level. Such a situation may result in an effective potential-energy curve of the flattened shape drawn in (d); with the trough no longer parabolic, the vibration of the proton along the bond – an asymmetric stretching – will be anharmonic.

1.4 CRYSTALLOGRAPHICALLY SYMMETRICAL HYDROGEN BONDS

1.4.1 Acid salts of monobasic acids

The great majority of hydrogen bonds that have been studied by crystal-diffraction methods do not possess any symmetry by virtue of their crystallographic situation. An important, and long-recognised, exception is the

F—H—F bond in the bifluorides of some alkali metals. We shall return to this later (Section 1.5). It was only in 1949 that analogously symmetrical O—H—O bonds were first reported[17, 18]. They were in structures where the two participating oxygen atoms are related by a crystallographic symmetry element, such as a centre of inversion. Granted that a hydrogen bond exists between them (which must be so with O···O ≈ 2.5 Å), the hydrogen atom can (without finesse) only be placed at the mid-point.

A typical example, which has been extensively studied[19–24], is potassium hydrogen bisphenylacetate, KHX_2, where HX = phenylacetic acid, $C_6H_5{\cdot}CH_2{\cdot}CO_2H$. According to elementary theory, this monobasic acid

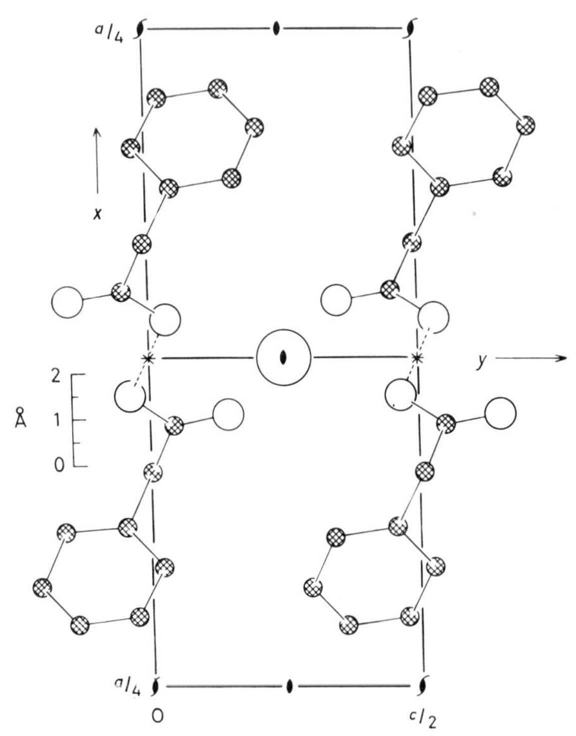

Figure 1.6 The crystal structure of potassium hydrogen bisphenylacetate, in its *b*-axial projection[19] (The larger open circles represent K^+ ions, the smaller, oxygen atoms; the cross-hatched circles, carbon. The short O···H···O bond lies across a centre of symmetry denoted by an asterisk.)

should form only the one series of neutral salts (e.g. KX). In fact, acid salts of monocarboxylic acids are common, a number of examples (accidentally discovered) being recorded in the literature[25], and many more being preparable.

Potassium hydrogen bisphenylacetate crystallises in the monoclinic space group *I*2/*a* (No. 15). As No. 15 is an eighth-order space group, and as the unit cell contains four KHX_2 units, we have the implications that the asymmetric unit is only half of KHX_2, that the potassium ion and the acidic hydrogen atom are in positions of special symmetry and that the two

X units of the formula are crystallographically equivalent. x-Ray structure analysis led to the results represented in Figure 1.6. The two phenylacetate residues are related by a centre of symmetry, at which the hydrogen atom is effectively situated: the $[X\cdots H\cdots X]^-$ unit is thus centrosymmetric, with $O\cdots H\cdots O = 2.443(4)$ Å. This is a very short bond.

Acid salts of stoichiometric formula MHX_2, where M is a univalent cation, whose crystal structures turn out to be of this symmetrical character, have been classified as Type *A*[26, 27]. Their structures are more properly formulated as $M^+ (XHX)^-$. (In contrast, some other acid salts have crystal structures in which X^- and HX can be distinguished, so that $M^+X^-\cdot HX$ would be a more appropriate formula. These are classified as Type *B*.)

Study of Type *A* acid salts of a number of monocarboxylic acids, by x-ray and neutron diffraction, shows them all to contain very short $O\cdots H\cdots O$ bonds. Results from the structure analyses that have been carried out with moderate precision are collected in Table 1.3. Duly weighted according to the standard deviation of each measurement, the results lead to the mean value $O\cdots O = 2.447(2)$ Å. Although there is no *a priori* reason for supposing that all these hydrogen bonds have exactly the same length, the agreement is impressive. All C—O···O angles are in the range 110–115 degrees.

Table 1.3 Summary of inter-carboxyl hydrogen bonding in some Type *A* acid salts, MHX_2, of monocarboxylic acids

((N) indicates a neutron study; standard deviations are in parentheses.)

HX	M	*Symmetry of bond*	O···H···O (Å)	*Reference*
Phenylacetic	K	$\bar{1}$	2.443(4)	24
p-OH-benzoic	K(hydrate)	$\bar{1}$	2.458(6)	62
Acetic	Na	2	2.444(10)	63
Cinnamic	NH_4	$\bar{1}$	2.51(3)	64
p-Cl-benzoic	K	$\bar{1}$	2.457(13)	65
Trifluoroacetic	Cs	$\bar{1}$	2.38(3)	66
	K	$\bar{1}$	{ 2.435(7) 2.437(4) (N)	66 11
	K	$\bar{1}$	{ 2.455(5) 2.448(4) (N)	17 49
Aspirin	Rb	$\bar{1}$	2.48(2)	68
Anisic	K	2	2.476(18)	69

1.4.2 Acid salts of symmetrical dicarboxylic acids

That a dibasic acid, H_2Y, should form an acid salt (MHY) is in accordance with the book of classical chemistry. What is unexpected is that the 'half-salts' of many symmetrical dicarboxylic acids are found, by structure analysis, to have their carboxyl groups crystallographically equivalent. Potassium hydrogen malonate, for example[28], does not have a crystal structure corresponding to the classical formula $K^+\ ^-O_2C\cdot CH_2\cdot CO_2H$. Instead, there are infinite hydrogen–anion chains (4),

$$\cdots\underset{*}{H}\cdots\overset{\frac{1}{2}-}{O_2}C\cdot HCH\cdot\overset{\frac{1}{2}-}{CO_2}\cdots H\cdots\overset{\frac{1}{2}-}{O_2}C\cdot H\underset{\uparrow}{C}H\cdot\overset{\frac{1}{2}-}{CO_2}\cdots\underset{*}{H}\cdots,$$

(4)

where the asterisks and arrows stand, respectively, for centres of symmetry at the hydrogen atoms and for axes of twofold symmetry passing through the methylene carbon atoms. This scheme presents a complete analogy with that in Type *A* acid salts of monocarboxylic acids, as can be seen if we represent the anion structure of potassium hydrogen bisphenylacetate (Figure 1.6) as in (5).

$$\cdots\underset{*}{H}\cdots\overset{\frac{1}{2}-}{O_2}C\cdot CH_2\cdot C_6H_5\ \underset{\uparrow}{}\ H_5C_6\cdot CH_2\cdot\overset{\frac{1}{2}-}{CO_2}\cdots\underset{*}{H}\cdots\overset{\frac{1}{2}-}{O_2}C\cdot CH_2\cdot C_6H_5\ \underset{\uparrow}{}$$

(5)

In potassium hydrogen malonate, whose crystal structure is shown in Figure 1.7, the symmetrical O···H···O bond is very short (see Table 1.4). We classify salts of dibasic acids with this structural pattern as Type A_2 [27].

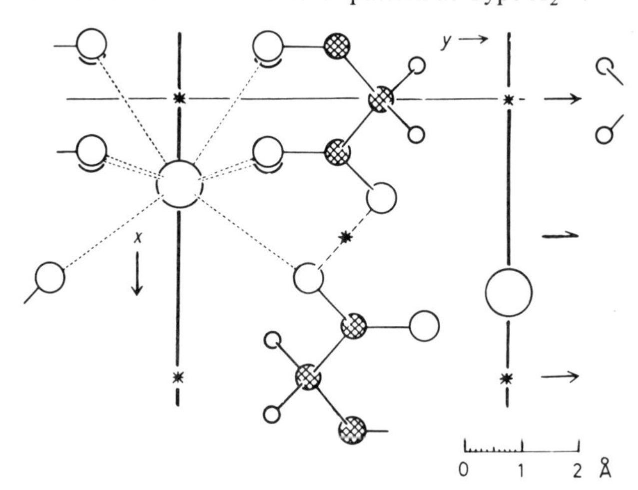

Figure 1.7 The crystal structure of potassium hydrogen malonate, in its *a*-axial projection[27, 28] (The smallest open circles represent methylenic hydrogen atoms. Other conventions are explained in the caption to Figure 1.6.)

In Table 1.4 we summarise the results at present available for the hydrogen bonds in Type A_2 acid salts. The weighted average length is O···O = 2.450(2) Å. All C—O···O angles are in the range 110–115 degrees.

The hydrogen bonds collected in Tables 1.3 and 1.4 constitute a remarkable set. Apart from some chelated systems which we shall mention later, they represent the shortest O···H···O bonds that have been measured with any precision. The great majority of them lie across crystallographic centres of symmetry, so that the O···H···O angle is necessarily (identically) 180 degrees. For the exceptions, in which the bond is symmetrical about a twofold axis, this is not necessary, though the proton is unlikely to be far from the mid-point of the O···O line. (A small deviation, in this sense, may be seen by inspection of Figure 1.3, but it is not statistically significant.)

Acid salts of Types *A* and A_2, as well as crystals with similarly symmetrical hydrogen bonds, always have anomolous infrared spectra of the sort described in Section 1.1.3 (Hadži's Type (ii)) and illustrated by Figure 1.1(d).

Table 1.4 Summary of inter-carboxyl hydrogen bonds in some Type A_2 acid salts, MHY, of dicarboxylic acids

((N) indicates a neutron study; standard deviations are in parentheses.)

H_2Y	M	*Symmetry of bond*	O···H···O (Å)	*Reference*
Malonic	K	$\bar{1}$	2.459(5)	28
			2.468(6) (N)	27
Succinic	K	2	2.446(4)	55
			2.444(10) (N)	
Glutaric	Rb	2/*m*	2.40(2)	70
	NH_4	2	2.460(7)	70
	K	2/*m*	2.445(3)	11
Oxalic	N_2H_5	$\bar{1}$	2.450(4)	71
			2.448(7) (N)	72

1.4.3 Some chelated O··H··O bonds

The acid potassium salt of maleic acid (6) has been of special interest since 1952 when the suggestion was made (partly on spectroscopic grounds, partly because the space group implied unexpected symmetry) that its hydrogen–anion had the symmetrical structure (8) rather than a structure corresponding to its 'classical' formula (7). This was confirmed by x-ray[29] and neutron[30] structure analyses which showed the hydrogen–anion to lie across a crystallographic mirror-plane (*m* in formula (8)).

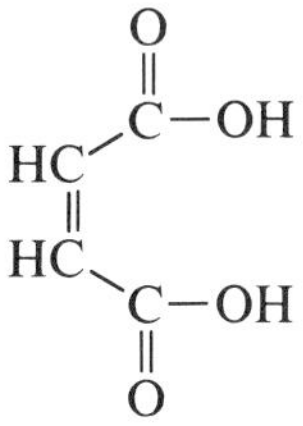

```
      O                  O                  O
      ||                 ||                 ||
     C—OH               C—O⁻               C—O½⁻
    /                  /                  /    :
  HC                 HC                 HC     :
  ||                 ||              ···||·····H···m
  HC                 HC                 HC     :
    \                  \                  \    :
     C—OH               C—OH               C—O½⁻
      ||                 ||                 ||
      O                  O                  O
     (6)                (7)                (8)
```

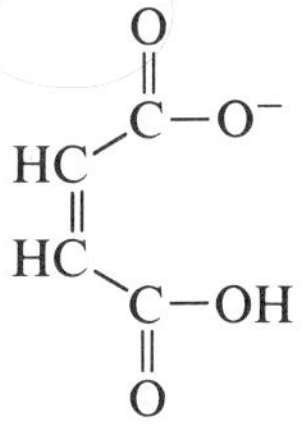

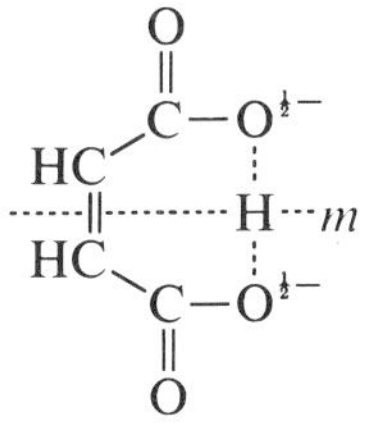

This is essentially a Type A_2 situation, requiring the O···H···O bond to be symmetrical. It is a special case in that the linked carboxyl groups are part of the same molecule; instead of infinite hydrogen-bonded chains, we have a hydrogen chelate. The O···O distance is very short, 2.437 Å.

Since technical difficulties (see Section 1.5) hamper the accurate study of hydrogen bonds that lie across a crystallographic symmetry element, Ellison and Levy[31] made a precise study of the corresponding chloromaleate (represented by formulae (6)–(8) with one ethylenic hydrogen atom replaced by chlorine). This substitution removes the possibility of strict *m*-symmetry. However, the chlorine atom excepted, the hydrogen–chloromaleate anion approaches virtual mirror-symmetry closely. Though the acidic proton is not required to be equidistant from the two oxygen atoms, the two O—H distances (1.206(5) and 1.199(5) Å) do not differ significantly. The O···H···O angle is 175 degrees. The overall O···O distance is even shorter than in the unsubstituted anion; it is 2.403(3) Å, or 2.41 Å after correction for libration (see Section 1.6.4). (This was a neutron-diffraction study.)

The seven atoms of the ring in formula (8) are virtually co-planar. Were the covalent bond lengths and bond angles within this ring normal, the distance between the oxygen atoms of the O···H···O bond would be only ~1.5 Å. To force them apart so as to raise O···O to ~2.4 Å imposes severe strains on the angles within the ring, as Darlow has shown[32]. The compressional stress on the hydrogen bond in this chelated situation must surely be partly responsible for its extreme shortness. A further aspect of a chelate is that transition from hydrogen bonding to electrovalency (a catastrophe described in Section 1.1.1), by complete transference of the proton, is prevented. Hence any tendency towards very strong bonding is allowed free rein.

Special significance attaches to a recent neutron study[33] of a nickel coordination compound, $[NiL_2]^+Cl^-\cdot H_2O_2$ (where L is a bidentate ligand), which includes the chelated system formulated in (9).

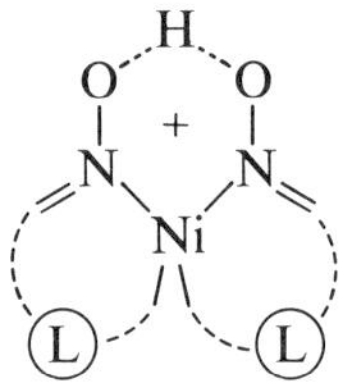

(9)

The right- and left-hand sides of the cation are chemically equivalent, and, if it were possible to measure this entity in the gaseous state the hydrogen bond would presumably turn out to be symmetrical and the proton equidistant from the two oxygen atoms. However, no special crystal symmetry is involved so that the environment of the cation is unsymmetrical. The analysis reveals unquestionable deviations from symmetry in the hydrogen bond: the O—H distances are 1.187(5) and 1.242(5) Å. The proton is more than 0.025 Å from the mid-point. It is also off the O···O line, for O···H···O = 170(0.3) degrees. With O···O = 2.419(3) Å, we again have an extremely short bond in a chelate. The lack of symmetry within the bond must be an environmental effect, which we shall discuss later.

1.4.4 Some basic salts and the $H_5O_2^+$ ion

Certain monoacidic bases form crystalline basic salts, analogous to Type *A* acid salts, MHX_2. For instance, acetamide ($CH_3\cdot CO\cdot NH_2 = B$) yields a 'hemihydrochloride', stoichiometrically $B{\cdot}\frac{1}{2}HCl$, but better formulated as $[BHB]^+Cl^-$. (The completeness of the analogy is emphasised if we write the acid salts as $[XHX]^-M^+$.) This basic salt crystallises with two molecules in a cell belonging to the fourth-order space group $P2_1/c$ (No.14), implying a centrosymmetric cation. The preliminary report of an x-ray study—published as long ago as 1957—showed that the basic function of the acetamide molecule was sited at the oxygen atom, with a very short O···H···O bond, across a centre of symmetry, between the acetamide residues. Though a fuller account of this work does not seem to have been published, the result was confirmed by a neutron analysis, briefly reported[34], which finds O···H···O = 2.418(11) Å. The infrared spectrum is of Type (ii)[5].

On the basis of its spectrum, Hadži[35] predicted a similar symmetrical >N—O···H···O—N< system in the hemihydrobromide of α-picoline-*N*-oxide (10). This prompted an x-ray study, a preliminary report of which finds O···O = 2.5 Å [36].

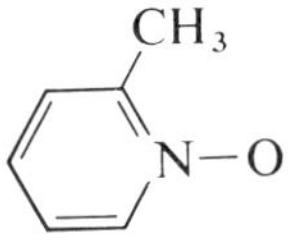

CH_3

N—O

(10)

Recently, Dunlop[37] has made an x-ray study (now complete) and a neutron study (not yet complete) of another salt of this same base (10). The stoichiometry of the crystals corresponds to the formula ($B{\cdot}HCl{\cdot}H_2O$) for a hydrated neutral salt. Structure analysis reveals the presence of the ions $[B_2H]^+$ and $[HCl_2]^-$, both centrosymmetric. (This may be described as a disproportionation, which has made a neutral salt revert in part to a basic salt and in part of an acid salt.) The O···H···O distance in the cation is 2.40(2) Å. The spectrum is almost identical, apart from a feature attributable to the hydration, with that of Hadži's 'hemihydrobromide'[2].

Thus we have strong—if not yet fully documented—evidence for very short hydrogen bonds in basic cations.

The 'hydrogen ion' of aqueous solution is certainly a hydrated proton. The simplest possibility is H_3O^+, and there are now plenty of examples of this entity in crystals, a recent case being $HBr{\cdot}H_2O$, which is really H_3O^+ Br^- [38]. Another possibility is the doubly hydrated proton, $H_5O_2^+$. There is evidence that this entity may exist in concentrated hydrochloric acid solutions: x-ray diffraction, from the liquid, shows a spacing of ~2.6 Å [39]. The first x-ray analysis to detect $H_5O_2^+$ in a crystal was reported in 1952 [40]. A coordination compound, supposed to be described by the formula $[Co(en)_2Cl_2]Cl{\cdot}HCl{\cdot}2H_2O$ (where en = ethylenediamine, and the cation has the *trans* configuration), was found to have a structure corresponding to $[Co(en)_2Cl_2]^+H_5O_2^+2Cl^-$. At that date the demonstration of hydrogen bonding was necessarily based on inference, as the hydrogen atoms were not located. It depended on finding two oxygen atoms related by, and only 1.33 Å from, a crystallographic centre of symmetry. A O···O separation of 2.66 Å inescapably indicated a hydrogen bond. The situation is what we have classed as a Type *A* structure. This hydrogen bond must be supposed, on chemical grounds, to involve the acidic hydrogen atom.

This supposition has now been confirmed by a neutron analysis (Williams)[41]. The proton was duly found at the centre of the bond, with O···O more accurately determined as 2.50(3) Å.

The ion $H_5O_2^+$ has been identified in a number of other crystals by x-ray or neutron diffraction. In all cases the O···H···O distance is very short, though there is not always strict crystallographic symmetry. (The system is then a pseudo-Type *A* structure.) We may instance $HCl{\cdot}2H_2O$, which, according to a low-temperature x-ray study[42], contains the $H_2O\cdots\overset{+}{H}\cdots OH_2$ ion in a 'gauche' conformation, linked by O—H···Cl bonds to four chloride ions. The O···H···O distance is 2.417(7) Å, which increases to 2.42 or 2.44 Å if corrected, in different ways, for libration. Similar results have been obtained for the hydrates of HBr [38].

The hydrogen bond between neutral water molecules (see Section 1.8) has O—H···O ≈ 2.75 Å. The contraction in $H_5O_2^+$ is noteworthy, and may well mean a change in the character of the bonding. We have cited similarly short bonds in the B_2H^+ ions of basic salts. In an obvious way, the contraction may be attributed to the overall positive charge. From another point of view, H_3O^+ is a very strong acid, as also is BH^+ – the acid conjugate to a weak base such as α-picoline-*N*-oxide. High acidity favours donation of the proton to form a strong hydrogen bond.

1.5 THE BIFLUORIDE ANION

The shortest and strongest hydrogen bond known occurs in salts of the bifluoride anion, $[FHF]^-$. This relatively simple system has been extensively studied[2,43]. We need to describe it here because it constitutes a model against whose background we can discuss very short O···H···O bonds which necessarily occur in systems that are more complicated*. The F···H···F system is generally conceded to be symmetrical. A major question asked (but not fully answered) in this Review is: are there any O···H···O bonds structurally equivalent to F···H···F?

At an early stage, x-ray work showed both KHF_2 and $NaHF_2$ to have crystal structures that would now be classed as Type *A* (see Section 1.4.1). The former has four KHF_2 units in a cell belonging to the eighth-order space group *I*4/*mcm*; the latter, three $NaHF_2$ in a cell of the sixth-order $R\bar{3}m$. If the space groups have been correctly assigned, the $[F\cdots H\cdots F]^-$ ion lies across a centre of symmetry in each crystal. Recent x-ray and neutron analyses find F···F distances of 2.277(6) Å (KHF_2)[44] and 2.264(3) Å ($NaHF_2$)[45], which are ~0.4 Å shorter than double the van der Waals, or ionic, radius of fluorine. The bonding energy is estimated to be 40 kcal mol^{-1} (~170 kJ mol^{-1}).

McGaw and Ibers, whose paper[45] also reported a neutron study of $NaDF_2$ (in which F···F = 2.265(7) Å), discuss their results in detail, and give special consideration to the symmetry problem. Certainly $[F\cdots H\cdots F]^-$ is very nearly symmetrical, in the naïve sense of the adjective. What might

---

*The short, but perhaps not very short, bonds in $HCrO_2$, $HCoO_2$ and their deuteriates are perhaps exceptions, but crystal structure work[2,46] on them is difficult.

still be uncertain is a decision between two models: first, the proton (or deuteron) vibrates, along the F···F direction, about a single potential-energy minimum at the bond centre, and with a r.m.s. amplitude of (say) 0.14 Å; or secondly, it vibrates, with a smaller amplitude (say 0.11 Å) about one or other of two minima less than 0.1 Å on either side of the centre, these alternative sites being randomly occupied. McGaw and Ibers demonstrated that either hypothesis would equally well explain the experimentally measured intensities of the neutron reflexions. It is therefore impossible to distinguish between these models by diffraction measurements alone.

However, by an argument based on spectroscopic considerations (which we shall outline in the next section), they concluded that the F···H···F bond in these bifluorides is correctly represented by the first model, with the proton (or deuteron) executing, probably, anharmonic vibrations along the bond. The potential-energy curve would be of the type sketched in Figure 1.5(d).

1.6 THE ANALYSIS OF ATOMIC VIBRATIONS IN CRYSTALS

1.6.1 General

In primitive x-ray work, the investigator was well satisfied if he could find satisfactory coordinates to represent the (mean) positions of the atoms. Following improvements in the methods of measuring intensities and of computing, he can now gain detailed information about the anisotropic vibrations of the atoms. Indeed these vibrations must be taken into consideration before he can achieve the highest accuracy in the determination of molecular geometry.

The vibration of an atom in a crystal is customarily described in terms of six parameters—the components of a symmetric 3×3 tensor. The anisotropic, and supposedly harmonic, vibration is thus characterised by a centrosymmetric ellipsoid; three parameters define its principal axes, three its orientation. (More sophisticated methods, which allow for anharmonicity and other possible complications, have been formulated[46, 47], but they introduce so much additional complexity into the calculations that they have been applied only to a few, very simple, crystals.)

The vibrational figure for any atom in a molecular crystal may be analysed into components due to three sources of vibrational motion. First, there will be vibration of the atom within its molecule. This is usually regarded as being of minor importance in a soft (e.g. organic) crystal. Secondly, the molecule may then be treated as a rigid body, which can vibrate in two ways—by translational movement, and by libration (i.e. torsional oscillation about its centre of mass). The mean-square amplitudes, from each type of vibration, are additive to yield the total mean-square amplitude characteristic of each atom.

The model upon which structure refinement is based assumes the atom to be spherical: a spherical mass of electron density, of defined radial fall-off, for x-ray analysis; a much smaller spherical nucleus (virtually a point), for neutron-diffraction work. The least-squares procedure then adjusts three positional and (in general) six vibrational parameters for each atom so as to achieve the best fit with the actual shape of the atom implied by the observed intensity measurements. Vibrational parameters derived from neutron work nearly always correspond to smaller amplitudes than those from x-ray work. They also seem to be more reliable. This is because the point-atom model is a better approximation to the neutron-scattering nucleus than is the spherical electron cloud to the experimental electron-density function, which is liable to have been distorted in its outer regions by bonding or lone-pair formation. (By a careful combination of accurate x-ray and neutron-diffraction measurements on the same crystal, Coppens and others[48], using the so-called $(F_x - F_N)$-synthesis, have produced density maps which convincingly record bonding and lone-pair electrons in some simple molecules.) These considerations enhance the value of neutron diffraction in the study of hydrogen bonding, beyond the advantages described in Section 1.2.2.

1.6.2 The bifluoride anion

Detailed analysis of vibrations in a molecular crystal is wholly impracticable. However, in relatively simple cases such as this, useful approximation may be made. In sodium bifluoride, the mean-square amplitudes along the F···H···F bond are 0.0145 Å^2 for F and 0.0245 Å^2 for H, as deduced from the neutron study. The difference, 0.0100 Å^2, represents the vibration of the proton relative to the fluorine atoms, supposed stationary. Assuming a linear, symmetrical F···H···F ion, and using the spectroscopically measured symmetric and asymmetric frequencies of HF_2^-, McGaw and Ibers[45] made an approximate calculation of this longitudinal mean-square amplitude. Their result agreed well with the above estimate based on their neutron-diffraction analysis. A similar agreement was found for DF_2^-. They conclude that this is 'the most powerful evidence that the F—H—F and F—D—F bonds are symmetric'.

For vibration perpendicular to the F—H—F bond, the difference between mean-square amplitudes is much smaller: 0.0032 Å*. Thus, the root-mean-square amplitude of the proton is 0.06 Å across the bond, compared with 0.10 Å along it. This implies that F—H—F is 'stiffer' for bending about its centre than for asymmetric stretching, which is the reverse of the usual trend and indicative of the profound changes which occur when F—H enters into so strong a hydrogen bond.

1.6.3 Potassium hydrogen di-trifluoroacetate

Macdonald[11] has recently completed neutron-diffraction studies of this Type *A* acid salt, as well as of its deuteriate (see Table 1.3). He has made a

*In DF_2^- the transverse difference is negative. This cannot be taken literally; it is zero within the limits of precision.

similar analysis of the vibration in the very short O···H···O and O···D···O bonds. His findings are set out in Table 1.5, which shows the mean-square amplitude (U), parallel to the hydrogen bond, and in two directions ($\perp$ and $\perp'$) mutually perpendicular, and both transverse, to the bond. The standard deviations prove the anisotropy of Δ to be very significant. Again, these bonds are more easily stretched than bent.

Table 1.5 Vibrational parameters in potassium hydrogen di-trifluoroacetate and potassium deuterium di-trifluoroacetate from neutron-diffraction analysis

(U is the mean-square amplitude, in Å^2, with standard deviations of the differences, Δ, in parentheses: $\|$ indicates vibration along the O···H···O bond; $\perp$ and $\perp'$ in perpendicular directions across the bond.)

	H	O	Δ	D	O	Δ
$U(\|)$	0.063	0.034	0.029(3)	0.063	0.036	0.027(2)
$U(\perp)$	0.054	0.049	0.005(3)	0.061	0.058	0.003(2)
$U(\perp')$	–	–	0.009(3)	–	–	0.005(2)

The infrared spectra of these acid salts are not well enough interpreted for us to attempt a calculation of amplitudes from the frequencies.

The results of the neutron study of potassium hydrogen disaspirinate[49] have been used in a similar vibrational analysis, and with similar conclusions.

1.6.4 Librational correction of bond lengths: a note

When libration of a molecule is considerable, atomic coordinates derived from diffraction analysis need correction. The effect of applying a correction is always to move an atom further away from the centre of libration of the molecule. For simple organic molecules intramolecular distances are increased by amounts up to 0.02 Å. In this Review we have, as far as possible, cited bond lengths without corrections. This is because, in crystals where the molecules are hydrogen bonded to their neighbours, the corrections are small (though by no means necessarily negligible[50]) and perhaps hard to apply. Furthermore, the application of such correction leads, formally, to a *reduction* of *intermolecular* O—H···O bond lengths. As we may be alleged to have some special interest in very short hydrogen bonds, it seems appropriately conservative to eschew librational corrections in this context.

1.7 PROBLEMS OF VERY SHORT O··H··O BONDS

1.7.1 The symmetry problem

We now return to the very short O··H··O bonds in Type *A* structures, and to the question whether their symmetry is genuine. Disorder of the proton between alternative sites grossly displaced from the mid-point (say by 0.2 Å) can be ruled out in all cases where a neutron, or a precise x-ray, study has been made. The question reduces to that asked by Ibers in his work on the bifluorides: a single potential-energy minimum, or two minima less than 0.1 Å on either side of the centre? The answer is the same: we cannot decide between these alternatives even by neutron diffraction – still less by x-rays. However – as happened with HF_2^- – other considerations may lead us to favour the genuinely symmetrical model.

The trend of the NMR curve (Figure 1.4) encourages confidence that, in the very short bonds, the proton cannot be far from the centre. In Type *A* structures the total environment of the bond is symmetrical. In terms of the 'resonance' picture (see Section 1.1.2) the alternative bond representations (11a) and (11b) are identical. This will make any 'resonance' contribution

—O—H :O—
(11a)

and

—O: H—O—
(11b)

to the bonding energy more important than in an unsymmetrical bond. The stronger and shorter the bond, the more likely is the proton to be at the centre. Thus symmetry of the site would favour symmetry of the bond.

In HF_2^-, the F···F distance is reduced below the sum of the radii of the fluorine atoms by about 0.40 Å. The van der Waals radius of oxygen being 1.4 Å, a nearly equal reduction has been achieved in the bonds listed in Tables 1.3 and 1.4.

Only three Type *A* crystals with unchelated O···H···O bonds have been precisely studied with neutrons: potassium hydrogen diaspirinate by Sequeira, Berkebile and Hamilton[49] and the normal and deuteriated forms of potassium hydrogen di-trifluoroacetate[11]. In their discussion of the diaspirinate, Hamilton and his co-workers report their vibrational analysis, which they relate to Ellison and Levy's work on the chloromaleate anion[30]. We may apply their argument to the results, for the fluoroacetates, listed in Table 1.5. The root-mean-square amplitude of the proton and deuteron along the bond are, respectively, 0.17 and 0.16 Å. If there were disorder between sites 0.1 Å on either side of the centre, then 0.010 $Å^2$ would have to be subtracted from the total mean-square amplitude of 0.029 $Å^2$ in the case of the proton. The residual 0.019 $Å^2$ would correspond to a root-mean-square amplitude, about either site, of 0.13 Å. But such an amplitude would carry the proton well across the middle into the domain of the other potential-energy minimum. At this level, the difference between single-minimum and disordered double-minimum models loses its significance.

We may agree with Hamilton's conclusion: 'the motion (of the proton) must obviously be extremely anharmonic and can best be described in terms of a broad, flat anharmonic potential, possibly with a bump in the centre no more than 200 cm^{-1} in height but also possibly completely flat'.

This picture of a flat, broad minimum potential is represented by the curve of Figure 1.5(d). It finds confirmation in the structure (formula (9)) described at the end of Section 1.4.3. Quite a small lack of symmetry in the crystalline environment of this hydrogen bond might reasonably be expected to tilt the potential-energy minimum from the form shown in (d) to that in (f). With a broad, anharmonic, potential – but not so much with a narrower parabolic curve – such a tilt would displace the position of the minimum considerably.

1.7.2 The isotope effect

Since the early work of Robertson and Ubbelohde[51], it has been realised that useful information on the nature of hydrogen bonding might be obtained by studying changes of the bond length resulting from deuteriation. The results of such studies have so far been disappointing[2, 52], and this is because the effects are usually rather small – and therefore need very precise analyses of both ordinary and deuteriated crystals. Further, the effects may depend on the overall length of the bond. There is some reason for supposing that a large isotope effect ($\sim$0.07 Å) may occur in bonds with O···O close to 2.5 Å.

However Rundle[53] suggested that a short hydrogen bond, with a single potential well, ought to have zero isotope effect. This is what was found by McGaw and Ibers (see Section 1.5). Macdonald's results for the acid trifluoroacetates[46] are: O···H···O = 2.437(4) Å and O···D···O = 2.437(3) Å. The absence of isotope effect in these bonds supports a belief in their 'genuine' symmetry.

1.7.3 The Kroon–Kanters–McAdam effect

A careful study of potassium hydrogen mesotartrate has been made by Peerdeman, Kroon and Kanters[54], based on a large number of x-ray intensity measurements at low temperature. This structure is rather complex, but it includes two independent O···H···O bonds lying across centres of symmetry and in a Type *A* situation such as we have described in Sections 1.4.1 and 1.4.2. The O···O distances are 2.48 and 2.46 Å. At the end of their refinement the authors carried out a 'difference synthesis' with coefficients [*F* (observed) – *F*(calculated for all the atoms except the two hydrogens involved in these bonds)]. Such a synthesis ought to reveal the electron density associated with these hydrogen atoms. The outcome for one O···H···O bond was the remarkable map (kindly sent early in 1969 by Drs. Kroon and Kanters) reproduced in Figure 1.8. The second of the O···H···O bonds in the mesotartrate also showed double peaks, though less remarkable than those in Figure 1.8.

Independently, McAdam at Glasgow had already computed a similar map from his own room-temperature x-ray analysis of potassium hydrogen succinate[55]. This map* shows a similar, though less elegant, effect in a similarly very short, crystallographically symmetrical hydrogen bond (O···O = 2.446(4) Å). I venture to call this phenomenon the Kroon–Kanters–McAdam (KKM) effect.

---

*This map is reproduced in Reference 27.

It is very common for spurious electron density to show up at special positions, and especially at points of symmetry, such as occur at the mid-points of the O···H···O bonds in these Type *A* crystals. This complicates the interpretation of 'difference syntheses' in such regions. Errors in the data will produce effects here; and, of course, the symmetry is built into the

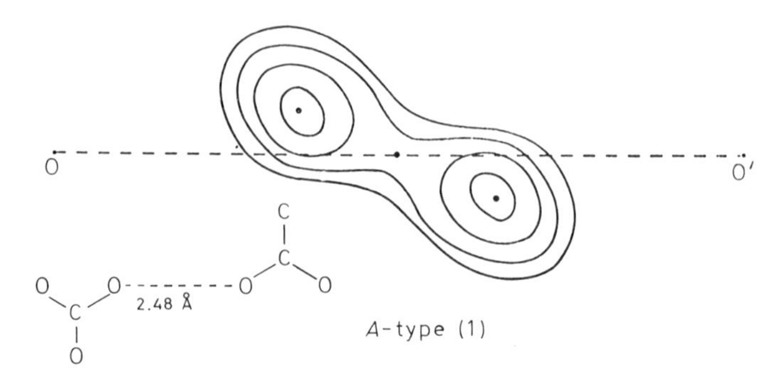

Figure 1.8 The KKM effect (An electron-density 'difference' synthesis in the region of a short hydrogen bond in potassium hydrogen mesotartrate: drawing supplied by Peerdeman, Kanters and Kroon[54].)

calculation of electron density. We have subsequently observed traces of the KKM effect in some other Type *A* crystals, though not in all. It is noteworthy that the effect appears in unusually precise analyses. It is not to be dismissed as due to crude errors.

Taken at its face value, Figure 1.8 means that the hydrogen atom in these very short bonds is not centrally placed at all, but that it occurs randomly in two possible sites, just as in ice. Measurements on Figure 1.8 are then consistent with a situation sketched in (12).

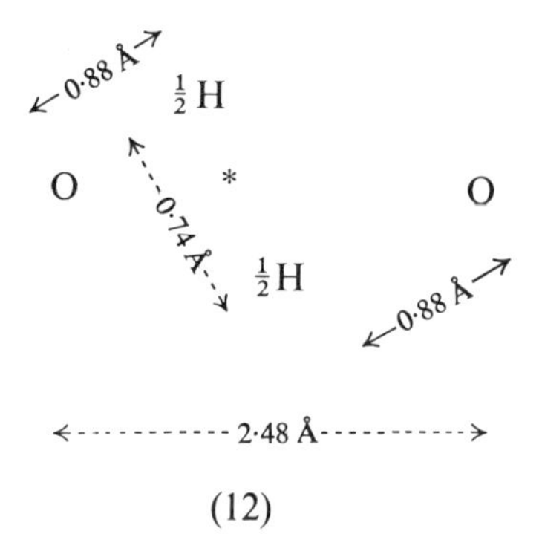

(12)

We do not believe this interpretation is tenable. It would be greatly at variance with the trend of the NMR curve (Figure 1.4). A neutron-diffraction study of the mesotartrate has not yet been made. However, for the acid succinate a neutron study, though based on a limited amount of data, unambiguously indicated a normal situation, with the proton effectively centred at the centre of the bond (McAdam[55]).

Agreement on an interpretation of the KKM effect has not yet been achieved. We may notice, however, that there is a resemblance between the misplacement of the hydrogen atom in Figure 1.2 and the apparent misplacements in Figure 1.8. It should be added that this hydrogen bond, which gave the particularly remarkable example of the KKM effect, is between carboxyl groups with a unique conformation; and that the interpretation of the fine details of 'difference' maps is now becoming sophisticated (e.g. Hirshfeld[55a]).

1.8 THE POLYMORPHS OF ICE

The water molecule, in the gaseous state, has $2m(C_{2v})$ symmetry, O—H distances of 0.96 Å, and the H—O—H angle 105 degrees. The spectrum and other properties of ice make it certain that the molecule is not greatly changed.

The structure of ordinary (hexagonal) ice – now usually specified as ice Ih – has been a classical problem in crystal chemistry[56]. As far as the oxygen atoms are concerned, the crystal structure presents no difficulty. Each oxygen atom is environed by four others at distances of ~2.76 Å and in directions nearly, though not identically, tetrahedral. These O···O contacts must represent hydrogen bonds. The structure is 'open'; the second-nearest oxygen neighbour is at ~4.8 Å which is in accord with the unusually low density, 0.92 g/ml^{-1}.

The difficulty was to place double the number of hydrogen atoms in this oxygen scheme so as to compose discrete H_2O molecules, and at the same time to comply with the overall crystal symmetry. The basic explanation was given by Pauling[51]. He postulated a disordered arrangement of protons, yielding an averaged structure with two statistical half-protons along each O···O bond at sites ~1.0 Å from each oxygen atom. The immediate evidence for this hypothesis was that it accounted for the experimental 'residual' entropy of ice. This hypothesis was fully vindicated by Peterson and Levy's neutron-diffraction analysis of 'heavy ice' (D_2O, Ih)[15], which found scattering density, equivalent to half a deuteron, at two positions, 1.01 Å from each oxygen atom, along each of the crystallographically independent O···O directions. An electron-diffraction study of the cubic form (H_2O, Ic) comes to the equivalent conclusion[58].

Eight other polymorphic forms of ice are now recognised*. Structural study of these materials might be supposed impracticable as they are thermodynamically stable only under pressure of the order, 10^4 atmospheres. However, though such pressures are needed for the formation of the polymorphs, once formed they may be 'quenched': provided the polymorph is first cooled to liquid nitrogen temperature, it will continue to exist in metastable equilibrium after the pressure has been released. (This will cause a dilation, with an increase of all interatomic distances amounting to several per cent.) Crystal-structure analysis may then be performed by the well-developed

*A general account of these forms may be found in a recent monograph by Eisenberg, D. and Kauzmann, W. (1969), *The Structure and Properties of Water* (Oxford: Clarendon Press).

techniques for low-temperature work. In this way, by the use of either single crystals, or more often a polycrystalline mass, the structures of these other forms of ice were elucidated by Kamb and his co-workers[59]. This x-ray work has recently been supplemented by neutron diffraction, which locates the hydrogen atoms. With neutrons, it is advantageous to use deuteriated ices; the D_2O structures seem to be almost identical with those of the H_2O crystals.

In all ice structures we have the same theme: a water molecule – little changed from the molecule in the vapour or in ordinary ice Ih – is linked by weak hydrogen bonds to four neighbouring molecules in nearly tetrahedral directions. In the high-pressure forms, the hydrogen-bonding scheme is distorted (from that in ice Ih) so as to curtail the amount of empty space. Whilst the O—H···O distances do not change much, the distances between second-nearest neighbour oxygen atoms are notably reduced. The distortion takes various forms, which will be exemplified shortly.

Most of the ice polymorphs are disordered with respect to the proton positions, as occurs in ice Ih. One exception is ice II, which has its protons ordered so as to constitute identifiable 'permanent' water molecules. Though most of the other forms are disordered, they tend to become more ordered at lower temperatures.

In one case, ice III, the ordering is so marked that the low-temperature version is considered to be an independent polymorph, with a transition at ~173 K. It has been denoted ice IX. The crystal structure is sketched in Figure 1.9. This is a tetragonal (pseudo-cubic) crystal, with $a = b \approx c = 6.75$ Å, and twelve H_2O (or D_2O) molecules in the cell[59, 60]. The oxygen atoms are of two sorts, O(1) and O(2) – the latter being distinguished by hatching in Figure 1.9. These O(2) water molecules occupy special fourfold positions of the space group, and lie on dyad axes. The eight O(1) molecules are in general positions. Such molecules are linked by hydrogen bonds, with O(1)···O(1) = 2.71 Å, into infinite chains which spiral round fourfold screw axes parallel to *c*. Different spirals are intricately cross-linked by O(2) molecules, which engage in two non-equivalent hydrogen bonds: O(1)—D···O(2) = 2.70 Å, and O(2)—D···O(1) = 2.89 Å. There are various non-bonded contacts with O···O down to 3.68 Å.

The neutron study of ice IX fixes the positions of the deuterium atoms as shown in Figure 1.9, and so defines the two water molecules. For $D_2O(1)$, O—D = 0.99(6) and 0.93(4) Å, and D—O—D = 101(4) degrees; for $D_2O(2)$, the O—D distances, equivalent by crystal symmetry, are 1.00(4) Å; and D—O—D = 90(4) degrees. The authors[59] do not regard the dimensions of the two sorts of molecule as being significantly different; nor, perhaps, are they very significantly different from those of the gaseous water molecule.

This ordered structure is indeed conditioned by the requirement that the molecule should not be much distorted. Were the deuterons to be moved in the sense O–D···O $\rightarrow$ O···D–O the resulting D_2O molecules would be very distorted, or their deuterons would have to lie well off the O···O lines. (However, ice III, which takes over from IX at higher temperatures, does have some degree of disordering of hydrogen.)

The much more tightly packed structures of ice VII and VIII, with densities ~1.6 g ml^{-1}, are based on a cubic body-centred lattice of oxygen atoms, with two molecules per cell. An oxygen atom at the cell centre is environed by eight others at the corners, with O···O = 2.95 Å (at atmospheric pressure). As there are only four hydrogen atoms per cell, only four of these eight contacts can be hydrogen bonded. In ice VII there is disorder, each O···O contact having, in total, a statistical half-hydrogen. Ice VIII has more

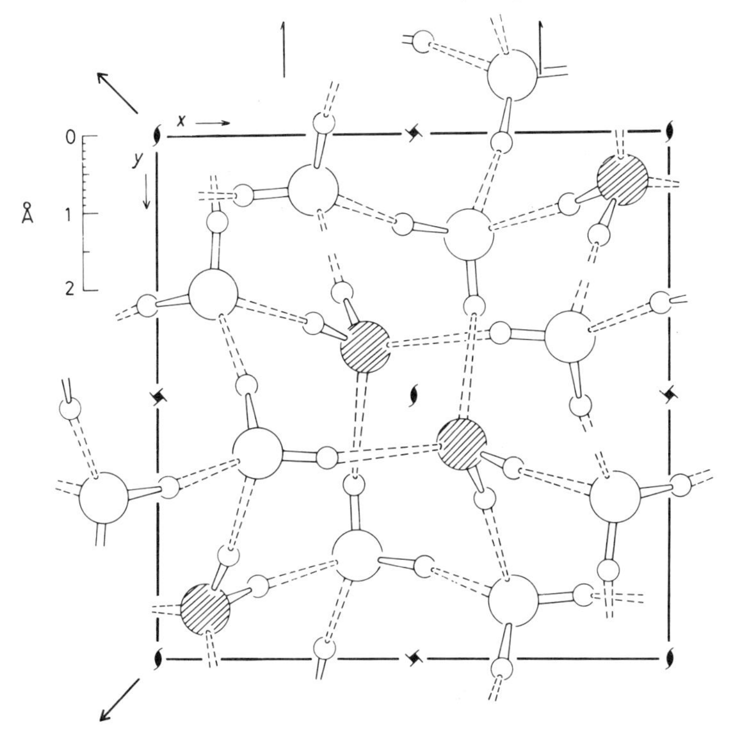

Figure 1.9 The crystal structure of 'heavy' ice – IX. (The larger circles, open or hatched, represent the two non-equivalent oxygen atoms; the smaller circles, deuterium.
(Based on Rabideau, S. W. *et al.*[60])

ordered hydrogen atoms: they are localised along four of the eight O···O contacts in any individual cell; these four contacts are directed towards alternate corners, so as to give four tetrahedrally orientated hydrogen bonds about each oxygen atom. In this way, the crystal structure becomes a clathrate, composed of two equivalent interpenetrating lattices – each akin to the diamond lattice. A side-effect is that the cubic cell becomes slightly distorted, in consequence of which the non-bonded O···O contacts become a little shorter than the O—H···O contacts[61].

1.9 CONCLUSION

The very short O···H···O bonds, lying across symmetry elements in acid salts and some other crystals, and with the O···O distance less than 2.46 Å, deserve serious consideration as genuinely symmetrical hydrogen bonds. The evidence for this is particularly strong in the acid potassium salts of trifluoroacetic, maleic, malonic, succinic and glutaric acids and of aspirin. In these six cases at any rate, the hydrogen atom moves in a symmetrical, single-minimum potential well, probably of the form symbolised in Figure 1.5(d).

Acknowledgements

Mr. William Cook drew most of the diagrams.

I have profited from discussions with Dr. J. K. Tyler, Dr. M. Currie, and Mr. R. S. Dunlop.

Note added in proof

The work on potassium hydrogen di-trifluoroacetate, which is included in Ref. 11, is to be published as Macdonald, A. L., Speakman, J. C. and Hadži, D. (1972). *J. C. S. Perkin II*, in the press. Pimentel and McClellan have published a valuable supplement to Ref. 1. (1971). *Ann. Rev. Phys. Chem.*, **22,** 347.

References

1. Pimentel, G. C. and McClellan, A. L. (1960). *The Hydrogen Bond.* (San Fransisco and London: Freeman)
2. Hamilton, W. C. and Ibers, J. A. (1968). *Hydrogen Bonding in Solids.* (New York: Wiley)
3. Bratož, S. (1967). *Advan. Quantum Chem.*, **3,** 208
4. Murrell, J. N. (1969). *Chem. in Brit.*, **5,** 107
5. Hadži, D. (1965). *Pure Appl. Chem.*, **11,** 435
6. Claydon, M. F. and Sheppard, N. (1969). *Chem. Commun.*, 1421
7. Hadži, D. and Novak, A. (1960). *Proc. Chem. Soc.*, 241
8. Zachariasen, W. H. (1934). *Z. Krist.*, **89,** 442
9. Robertson, J. M. (1940). *Trans. Faraday Soc.*, **36,** 913; Robertson, J. M. and Woodward, I. (1936). *J. Chem. Soc.*, 1817
10. Bellamy, L. J. and Pace, R. A. (1969). *Spectrochim. Acta*, **25A,** 319
11. Macdonald, A. L. (1971). Personal communication; work to be reported in *J. Chem. Soc.*
12. Pijper, W. P. (1971). *Acta Crystallogr.*, **B27,** 344
13. Ellison, R. D., Johnson, C. K. and Levy, H. A. (1971). *Acta Crystallogr.*, **B27,** 333
14. Nakamoto, K., Margoshes, M. and Rundle, R. E. (1955). *J. Chem. Phys.*, **77,** 6480
15. Peterson, S. W. and Levy, H. A. (1957). *Acta Crystallogr.*, **10,** 70
16. Schuster, P. (1970). *Theoret. Chim. Acta*, **19,** 212
17. Speakman, J. C. (1948). *Nature (London)*, **162,** 695
18. Brown, C. J., Peiser, H. S. and Turner-Jones, A. (1949). *Acta Crystallogr.*, **2,** 167
19. Speakman, J. C. (1949). *J. Chem. Soc.*, 3357
20. Davies, M. and Thomas, J. O. (1951). *J. Chem. Soc.*, 2858
21. Hadži, D. and Novak, A. (1955). *Nuovo Cimento*, **2,** 715
22. Bacon, G. E. and Curry, N. A. (1957). *Acta Crystallogr.*, **10,** 524
23. Albert, N. and Badger, R. M. (1958). *J. Chem. Phys.*, **29,** 1193
24. Manojlović, Lj. and Speakman, J. C. (1968). *Acta Crystallogr.*, **B24,** 323
25. See, for example, Gerhardt, C. F. (1853). *Annalen*, **87,** 57, 149
26. Shrivastava, H. N. and Speakman, J. C. (1961). *J. Chem. Soc.*, 1151

27. Currie, M. and Speakman, J. C. (1970). *J. Chem. Soc. A*, 1923
28. Sime, J. G., Speakman, J. C. and Parthasarathy, R. (1970). *J. Chem. Soc. A*, 1919
29. Darlow, S. F. and Cochran, W. (1961). *Acta Crystallogr.*, **14**, 1250
30. Peterson, S. W. and Levy, H. A. (1958). *J. Chem. Phys.*, **29**, 948
31. Ellison, R. D. and Levy, H. A. (1965). *Acta Crystallogr.*, **18**, 260
32. Darlow, S. F. (1961). *Acta Crystallogr.*, **14**, 1257
33. Schlemper, E. O., Hamilton, W. C. and La Placa, S. J. (1971). *J. Chem. Phys.*, **54**, 3990
34. Peterson, S. W. and Worsham, J. E., Jr. (1959). *U.S. Atomic Energy Commission, ORNL, 2782*, 69
35. Hadži, D. (1962). *J. Chem. Soc.*, 5128
36. Mills, H. H. and Speakman, J. C. (1963). *Proc. Chem. Soc.*, 216
37. Dunlop, R. S. (1971). Personal communication. Work to be published in *J. Chem. Soc.*
38. See, for example, Lundgren, J.-O. (1970). *Acta Crystallogr.*, **B26**, 1893
39. Lee, S. C. and Kaplow, R. (1970). *Science*, **167**, 477
40. Nakahara, A., Saito, Y. and Kuroyo, H. (1952). *Bull. Chem. Soc., Jap.*, **25**, 331
41. Williams, J. M. (1967). *Inorg. Nucl. Chem. Lett.*, **3**, 297
42. Lundgren, J.-O. and Olovsson, I. (1967). *Acta Crystallogr.*, **23**, 966; see also Follner, H. (1970). *Acta Crystallogr.*, **B26**, 1544
43. Blinc, R. (1958). *Nature (London)*, **182**, 1016
44. Ibers, J. A. (1964). *J. Chem. Phys.*, **40**, 402; **41**, 25
45. McGaw, B. L. and Ibers, J. A. (1963). *J. Chem. Phys.*, **39**, 2677
46. Hamilton, W. C. and Ibers, J. A. (1963). *Acta Crystallogr.*, **16**, 1209; see also Reference 2
47. Willis, B. T. M. and Johnson, C. K. (1970), in Willis, B. T. M., Ed., *Thermal Neutron Diffraction.* (Oxford: Clarendon Press)
48. See, for example, Coppens, P. (1967). *Science*, **158**, 1577; Coppens, P. and Vos, A. (1971). *Acta Crystallogr.*, **B27**, 146
49. Sequeira, A., Berkebile, C. A. and Hamilton, W. C. (1967–8). *J. Mol. Structure*, **1**, 283
50. Coppens, P. and Sabine, T. M. (1969). *Acta Crystallogr.*, **B25**, 2442
51. See, for example, Robertson, J. M. and Ubbelohde, A. R. (1939). *Proc. Roy. Soc. (London)*, **A185**, 448
52. Delaplane, R. G. and Ibers, J. A. (1969). *Acta Crystallogr.*, **B25**, 2423
53. Rundle, R. E. (1964). *J. Physique*, **25**, 187; see also Singh, T. R. and Wood, J. L. (1968). *J. Chem. Phys.*, **48**, 4567; (1969), **50**, 3572
54. Kroon, J., Kanters, J. A. and Peerdeman, A. F. (1971). *Nature (London), Phys. Sci.*, **229**, 120; see also (1971), **232**, 107
55. McAdam, A., Currie, M. and Speakman, J. C. (1971). *J. Chem. Soc. A*, 1994
55a. Hirshfeld, F. L. (1971). *Acta Crystallogr.*, **B27**, 769
56. Bernal, J. D. and Fowler, R. H. (1933). *J. Chem. Phys.*, **1**, 515
57. Pauling, L. (1935). *J. Amer. Chem. Soc.*, **57**, 2680
58. Honjo, G. and Shimaoka, K. (1957). *Acta Crystallogr.*, **10**, 710
59. See, for example, Kamb, B. (1965). *J. Chem. Phys.*, **43**, 3967
60. Rabideau, S. W., Finch, E. D., Arnold, G. P. and Bowman, A. L. (1969). *J. Chem. Phys.*, **49**, 2514
61. Kamb, B. (1969). *Acta Crystallogr.*, **A25**, S117; see also papers to be published by Kamb, B., Hamilton, W. C., La Placa, S. J. and Prakash, A. (1971). *J. Chem. Phys.*, **55**, 1934
62. Manojlović, Lj. (1968). *Acta Crystallogr.*, **B24**, 326
63. Speakman, J. C. and Mills, H. H. (1961). *J. Chem. Soc.*, 1164
64. Bryan, R. F., Mills, H. H. and Speakman, J. C. (1963). *J. Chem. Soc.*, 4350
65. Mills, H. H. and Speakman, J. C. (1963). *J. Chem. Soc.*, 4355
66. Golić, Lj. and Speakman, J. C. (1965). *J. Chem. Soc.*, 2530
67. Manojlović, Lj. and Speakman, J. C. (1967). *J. Chem. Soc. A*, 971
68. Grimvall, S. and Wengelin, R. F. (1967). *J. Chem. Soc. A*, 968
69. McGregor, D. R. and Speakman, J. C. (1968). *J. Chem. Soc. A*, 2106
70. Macdonald, A. L. and Speakman, J. C. (1971). *J. Cryst. Molec. Structure*, **1**, 189
71. Ahmed, N. A. K., Liminga, R. and Olovsson, I. (1968). *Acta Chem. Scand.*, **22**, 88
72. Nilsson, A., Liminga, R. and Olovsson, I. (1968). *Acta Chem. Scand.*, **22**, 719

2
Structural Chemistry of Transition Metal Complexes: (1) 5-Coordination, (2) Nitrosyl Complexes

BERTRAM A. FRENZ and JAMES A. IBERS
Northwestern University, Evanston, Illinois

2.1 INTRODUCTION

A complete review of recent developments in the field of structural inorganic chemistry is an immense and probably impossible task. The field has grown to such proportions that it is even difficult to keep abreast of the many reviews that continuously appear in journals and books. To be of service to the reader we offer in Table 2.1 a list of recent reviews. Reviews of transition metal complexes that include some discussion of structural work by x-ray diffraction are listed in the table for the period 1969 to mid-1971.

In this review, we single out two topics: 5-coordinate transition metal complexes and transition metal nitrosyl complexes. Both of these will serve to illustrate that our knowledge of chemical bonding is still primitive and so progress in structural chemistry must still be made through intercomparisons of closely related structures.

Structural data play a major role in the development of our understanding of chemical bonding. In coordination chemistry those structures which consist of 5 unidentate ligands about a central transition metal atom play an especially crucial role, as the trigonal-bipyramid and tetragonal-pyramid occur commonly in 5-coordination and the principles that govern one configuration being favoured over another are still not well understood. Since the last reviews of 5-coordinate structures[42–44] new information has increased to a point where a review seems in order.

With the confirmation of an M—N—O bond angle of 120 degrees in transition metal complexes[45] the field of transition metal nitrosyls has taken on new importance in relation to ideas about chemical bonding. Since this area has not, to our knowledge, been reviewed we offer here a complete review of structures of transition metal nitrosyl complexes.

Table 2.1 Recent reviews of x-ray diffraction studies

Title	*Reference*
The transition elements	1
The absolute configuration of transition metal complexes	2
Optical activity and molecular dissymmetry in coordination chemistry	3
A survey of recent x-ray structural studies of organometallic compounds	4
Transition metal carbonyls and organometallic complexes	5
The Lewis basicity of transition metals	6
Stereochemically non-rigid structures	7
Cis and *trans* effects in cobalt(III) complexes	8

Title	*Reference*
Spectra of 3d 5-coordinate complexes	9
The stereochemistry of 5-coordinate nickel(II) and cobalt(II) complexes	10
Organometallic structures: transition metals annual survey covering the year 1968	11
Vibrational spectra and metal–metal bonds	12
Iron, ruthenium and osmium: annual survey covering the year 1968	13
Cobalt, rhodium and iridium: annual survey covering the year 1968	14
Nickel, palladium and platinum: annual survey covering the year 1968	15
Crystal structures of complexes of platinum group metals	16
Organoplatinum(IV) compounds	17
Complexes of platinum(II) with unsaturated hydrocarbons	18
The formation, structure and reactions of binuclear complexes of cobalt	19
The electronic properties and stereochemistry of mononuclear complexes of the copper(II) ion	20
Copper(II) ammonia complexes	21
Transition metal oxo complexes	22
Olefin complexes of the transition metals	23
Mono-olefin and acetylene complexes of Ni, Pd and Pt	24
Transition metal complexes of azulene and related ligands	25
Transition metal clusters with π-acid ligands	26
Structural studies on transition metal complexes containing σ-bonded carbon atoms	27
Metal–metal bonds and covalent atomic radii of transition metals in their π-complexes and polynuclear carbonyls	28
The chemistry of transition metal carbonyls: structural considerations	29
Some aspects of the chemistry of polynuclear metal carbonyl compounds	30
Structure, bonding, and reactivity of (stable) transition metal carbonyl carben complexes	31
The reactions of hydrazine with transition metal complexes	32
The coordination of ambidentate ligands	33
Schiff-base metal complexes as ligands	34
Molecular structures of transition metal hydride complexes	35
Recent developments in the inorganic chemistry of metal complex hydrides	36
Structural studies on some complex species with bridged hydrogens	37
The coordination chemistry of cyclic phosphines and arsines	38
Stereochemistry of tartrato(4)-bridged binuclear complexes	39
Divalent transition metal β-keto-enolate complexes as Lewis acids	40
Structure and bonding in inorganic derivatives of β-diketones	41
The chemistry of the dithioacid and 1,1-dithiolate complexes	159
Zirconium and hafnium chemistry	160
Structural studies of iron–sulphur proteins	161
The structure of porphyrins and metalloporphyrins	162

2.2 5-COORDINATION

Since the earlier reviews[42–44] on 5-coordination, new examples of this once rare coordination number have abounded in the literature. Many examples of the two ideal geometries, trigonal-bipyramid and square- or tetragonal-pyramid (Figure 2.1), are now known, as well as a host of intermediate configurations. Only monomeric transition metal complexes with unidentate ligands will be reviewed in detail. The assignment of the coordination number to a metal in a dimer or polymer system is usually equivocal. Multidentate ligands often impose severe steric restrictions on the geometries of

Table 2.2 Summary of structures of 5-coordinate transition metal complexes known from diffraction studies

		Number of published structures											
		d^0	d^1	d^2	d^3	d^4	d^5	d^6	d^7	d^8	d^9	d^{10}	*Totals*
Only unidentate ligands	TBP*	0	2	1	1	0	1	2	3	45	3	2	60
	TP†	4	0	1	0	9	0	9	1	6	2	0	32
Multidentate ligands	TBP*	1	1	1	0	0	2	1	3	13	8	7	37
	TP†	0	4	2	0	2	5	1	6	19	19	2	60
													189

*TBP = trigonal-bipyramid.
†TP = tetragonal-pyramid.

5-coordinate complexes. As shown in Table 2.2 complexes with multidentate ligands tend to favour a tetragonal-pyramidal structure over a trigonal-bipyramid. The reverse trend is currently observed for complexes with unidentate ligands: the trigonal-bipyramid is favoured over the tetragonal-bipyramid. These trends probably reflect fashions in research, rather than a fundamental principle.

In this review, complexes with five identical ligands will be examined first.

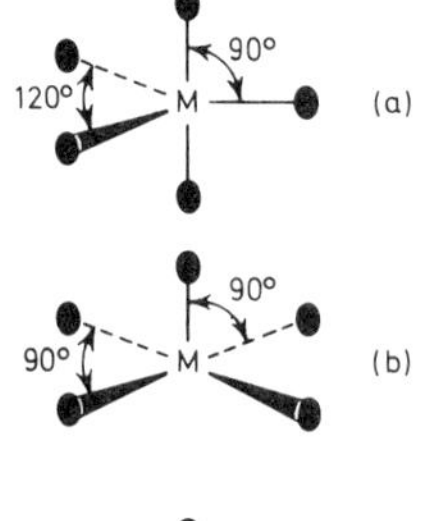

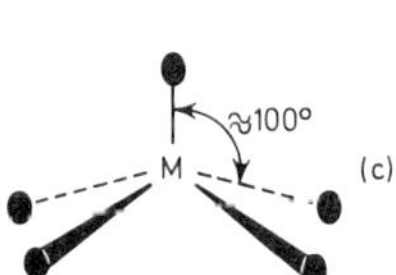

Figure 2.1 5-Coordinate geometries: (a) trigonal-bipyramid, (b) square- or tetragonal-pyramid, (c) 'most stable' square pyramid

Then recent structures will be reviewed in order of increasing number of outer-shell d electrons of the metal. A list of these structures is given in Table 2.3. Finally, a separate section is devoted to phosphine ligands.

Table 2.3 Recent diffraction studies of monomeric 5-coordinate transition metal complexes with unidentate ligands*

Type	*Compound*	d^x	*Structure*	*Axial ligand(s)*	*Figure*	*Reference*
ML_5						
	$Nb(NMe_2)_5$	d^0	TP		3	51
	$Nb(NC_5H_{10})_5$	d^0	TP		3	51
	$[MnCl_5]^{2-}$	d^4	TP			55
	$[Fe(N_3)_5]^{2-}$	d^5	TBP		2	49

Table 2.3 *(continued)*

Type	*Compound*	d^x	*Structure*	*Axial ligand(s)*	*Figure*	*Reference*
	$[Co(C_6H_7NO)_5]^{2+}$	d^7	TBP			50
	$[Mn(CO)_5]^-$	d^8	TBP		7	65
	$Fe(CO)_5$†	d^8	TBP			64
	$[Ni(CN)_5]^{3-}$	d^8	TP		5	54
			TBP		4	52–54
	$Ni[P(OCH)_3(CH_2)_3]_5^{2+}$	d^8	TBP			48
	$[Ni(OAsMe_3)_5]^{2+}$	d^8	TP		6	56
	$[CuCl_5]^{3-}$	d^9	TBP			46, 47
	$[CdCl_5]^{3-}$	d^{10}	TBP			47
ML_4L'						
	$WSCl_4$	d^0	TP	S	8	66
	$WSBr_4$	d^0	TP	S		66
	$Mn(NO)(CO)_4$	d^8	TBP	C, C	25	145
	$[Fe(CN)(CO)_4]^-$	d^8	TBP	CN, CO		111
	$Fe(CO)_4(PHPh_2)$	d^8	TBP	C, P		112
	$Fe(CO)_4(Ph_2C_4Ph_2)$	d^8	TBP	CO, CO		113
	$Fe(CO)_4[C_2H_2(CO_2H)_2]$	d^8	TBP	CO, CO		115
	$Co(SiH_3)(CO)_4$†	d^8	TBP	C, Si		116
	$Co(SiF_3)(CO)_4$	d^8	TBP	C, Si	15	117
	$Co(SiCl_3)(CO)_4$	d^8	TBP	C, Si		118
	$Co(AuPPh_3)(CO)_4$	d^8	TBP	C, Au		119
	$Co[Ag(As_3C_{17}H_{23})](CO)_4$	d^8	TBP	C, Ag		119
	$CoH(PF_3)_4$	d^8	TBP	H, P		124
	$CoH[PPh(OEt)_2]_4$	d^8	TBP	H, P	16	123
	$RhH(PPh_3)_4$	d^8	TBP	—		125
$ML_3L'_2$						
	$TiBr_3(NMe_3)_2$	d^1	TBP	N, N		70
	$VCl_3(NMe_3)_2$	d^2	TBP	N, N		74
	$CrCl_3(NMe_3)_2$	d^3	TBP	N, N		76
	$CoBr_2(PHPh_2)_3$	d^7	TBP	P, P		101
	$NiBr_3(PMePh_2)_2$	d^7	TBP	P, P	14	100
	$NiI_2(PHPh_2)_3$	d^8	TBP	P, P		101
	$Ni(CN)_2(PMe_2Ph)_3$	d^8	TBP	C, C		132
	$Ni(CN)_2[PPh(OEt)_2]_3$	d^8	TBP	C, C		131
	$Ni(CN)_2[PMe(C_{12}H_8)]_3$	d^8	TBP	C, C		133
			TP	C		133
	$Ni(CN)_2[PEt(C_{12}H_8)]_3$	d^8	TBP	C, C		133
	$Os(CO)_3(PPh_3)_2$	d^8	TBP	P, P		134
	$[Ir(CO)_3(PMe_2Ph)_2]^+$	d^8	TBP	P, P		154
	$CuBr_3Cl_2$	d^9	TBP	Cl, Cl		147
	$CuCl_2(C_5H_8N_2)_3$	d^9	TBP	N, N	18	148
	$Cu(O_2COPh)_2(H_2O)_3$	d^9	TP	OH_2		149
$ML_3L'L''$						
	$MoCl(\pi\text{-}C_5H_5)(CO)_3$‡	d^4	TP	$\pi\text{-}C_5H_5$		79
	$MoEt(\pi\text{-}C_5H_5)(CO)_3$	d^4	TP	$\pi\text{-}C_5H_5$		78
	$Mo(C_3F_7)(\pi\text{-}C_5H_5)(CO)_3$	d^4	TP	$\pi\text{-}C_5H_5$		75
	$Mo(AuPPh_3)(\pi\text{-}C_5H_5)(CO)_3$	d^4	TP	$\pi\text{-}C_5H_5$		80
	$RuHCl(PPh_3)_3$	d^6	TBP	P, P		98
	$Mn(NO)(CO)_3(PPh_3)$	d^8	TBP	C, P	25	144
	$Co(CF_2CF_2H)(CO)_3(PPh_3)$	d^8	TBP	C, P		121
	$Co(GePh_3)(CO)_3(PPh_3)$	d^8	TBP	Ge, P		122
	$CoH(N_2)(PPh_3)_3$	d^8	TBP	H, N		127
	$NiCl_3(C_7H_{15}N_2)(H_2O)$	d^8	TBP	O, N	17	146
	$RuH(NO)(PPh_3)_3$	d^8	TBP	H, N		129

Table 2.3 *(continued)*

Type	*Compound*	d^x	*Structure*	*Axial ligand(s)*	*Figure*	*Reference*
	$RhH(PPh_3)_3(AsPh_3)$	d^8	TBP	—		126
	$[IrH(NO)(PPh_3)_3]^+$	d^8	TBP	H, N	22	130
$ML_2L'_2L''$						
	$VOCl_2(NMe_3)_2$	d^1	TBP	N, N		69
	$VOCl_2[OC(NMe_2)_2]_2$	d^1	TP	O	9	68
	$ReNCl_2(PPh_3)_2$	d^2	TP	N	10	73
	$[RuCl(NO)_2(PPh_3)_2]^+$	d^6	TP	N	23	92
	$RhI_2Me(PPh_3)_2$	d^6	TP	Me		96
	$RhHCl_2[P(C_3H_7)_2(CMe_3)]_2$	d^6	TP	H	12	95
	$IrCl_2(NO)(PPh_3)_2$	d^6	TP	N	22	93
	$[Os(OH)(NO)_2(PPh_3)_2]^+$	d^6	TP	N		91
	$Mn(NO)(CO)_2(PPh_3)_2$	d^8	TBP	P, P	25	143
	$IrCl(CO)_2(PPh_3)_2$	d^8	TBP	P, P		135
	$IrH(CO)_2(PPh_3)_2$	d^8	TBP	H, P		136
$ML_2L'L''L'''$						
	$FeH(SiCl_3)_2(\pi\text{-}C_5H_5)(CO)$‡	d^4?	TP	$\pi\text{-}C_5H_5$		84
	$MoI(\pi\text{-}C_5H_5)(CO)_2(PPh_3)$	d^4	TP	$\pi\text{-}C_5H_5$		82
	$Mo(COMe)(\pi\text{-}C_5H_5)(CO)_2(PPh_3)$	d^4	TP	$\pi\text{-}C_5H_5$		81
	$Mo(NCO)(\pi\text{-}C_5H_5)(CO)(PPh_3)_2$	d^4	TP	$\pi\text{-}C_5H_5$		83
	$RhHCl(SiCl_3)(PPh_3)_2$	d^6	TBP	P, P	13	99
	$[IrCl(NO)(CO)(PPh_3)_2]^+$	d^6	TP	N	22	45
	$[IrI(NO)(CO)(PPh_3)_2]^+$	d^6	TP	N	22	90
	$IrI(NO)Me(PPh_3)_2$	d^6	TP	N	22	94
	$RuCl(NO)(O_2)(PPh_3)_2$	d^8	TBP	P, P		155
	$RhCl(CO)(SO_2)(PPh_3)_2$	d^8	TP	S		156
	$IrCl(O_2)(CO)(PPh_3)_2$	d^8	TBP	P, P		141
	$IrCl(O_2)(CO)(PEtPh_2)_2$	d^8	TBP	P, P		157
	$IrBr(O_2)(CO)(PPh_3)_2$	d^8	TBP	P, P		158
	$IrH(CO)(PPh_3)_2[C_2H_2(CN)_2]$	d^8	TBP	H, CO		137
	$IrBr(CO)(PPh_3)_2[C_2(CN)_4]$	d^8	TBP	Br, CO		138, 139
	$Ir(C_6HN_4)(CO)(PPh_3)_2[C_2(CN)_4]$	d^8	TBP	N, CO		140

*Structures reported prior to 1966 which are listed in reference 43 are not re-listed here. TBP = trigonal-bipyramid; TP = tetragonal-pyramid. Distorted structures are assigned the geometry that they most closely resemble. TBP/TP implies that the structure can be interpreted in terms of either of the regular geometries. The following abbreviations are also used in the tables and the text: Me = CH_3, Et = C_2H_5 and Ph = C_6H_5.

†Studied by electron diffraction.

‡(π-C_5H_5) is considered here as a unidentate ligand.

2.2.1 ML_5 structures

2.2.1.1 Distortions in bond angles

Because of their simplicity ML_5 structures are of great importance to our understanding of chemical bonding. The structures of one dozen ML_5 complexes have been reported, although structural details are not available for some. In general these structures can be viewed as trigonal-bipyramidal with varying degrees of distortion. Table 2.4 compares the observed bond angles with the two ideal geometries. As illustrated, the trigonal-bipyramid and tetragonal-pyramid differ in the position of L_2 and L_3; for the former the angle L_2—M—L_3 is 120 degrees, for the latter the angle is 180 degrees.

Table 2.4 Bond angles (in degrees) in ML_5 complexes

TBP   TP

Compound	*Av. of* $L_2—M—L_4$, $L_2—M—L_5$, $L_3—M—L_4$, $L_3—M—L_5$	*Av. of* $L_1—M—L_4$, $L_1—M—L_5$	$L_4—M—L_5$	*Av. of* $L_1—M—L_2$, $L_1—M—L_3$	$L_2—M—L_3$	*Reference*
ideal TBP	90	90	180	120	120	
$[CuCl_5]^{3-}$	90	90	180	120	120	46
$Ni[P(OCH)_3(CH_2)_3]_5^{2+}$	90(1)*	91(1)	178	118(1)	123(1)	48
$[Mn(CO)_5]^-$	90(1)	91(1)	179(1)	118(1)	123(1)	65
$[Fe(N_3)_5]^{2-}$	89(1)	93(1)	178(2)	118(1)	124(2)	49
$[Mn(CO)_5]^-$	90(1)	91(1)	178(1)	118(1)	125(1)	65
$[Co(CNCH_3)_5]^+$	90(1)	89(1)	180(0)	116(1)	128(1)	61
$[Co(C_6H_7NO)_5]^{2+}$	89(6)	93(2)	173(1)	115(1)	130(1)	50
$[Ni(CN)_5]^{3-}$ TBP	89(2)	93(1)	173(1)	109(3)	141(1)	54
$Nb(NC_5H_{10})_5$	87(2)	99(1)	161	109(2)	141	51
$Nb(NMe_2)_5$	87(1)	102(1)	157	109(1)	141	51
$[Ni(CN)_5]^{3-}$ TP	88(1)	100(1)	160(1)	100(5)	159(1)	54
$[MnCl_5]^{2-}$	89(1)	95(4)	170(1)	99(1)	161(1)	55
most stable TP	88	100	160	100	160	57
ideal TP	90	90	180	90	180	

*In this and following tables, the standard deviations in terms of the least significant digits are given in parentheses. For average values the estimated standard deviation of the mean is reported.

The compounds are listed in order of increasing distortion from trigonal-bipyramid to tetragonal-pyramid as judged by the $L_2—M—L_3$ angle. Thus $[CuCl_5]^{3-}$ [46, 259], $Ni[P(OCH)_3(CH_2)_3]_5^{2+}$ [48], $[Fe(N_3)_5]^{2-}$ [49] (Figure 2.2), and $[Mn(CO)_5]^{-}$ [65] are regular trigonal-bipyramids and $[Co(C_6H_7NO)_5]^{2-}$ [50] is a somewhat distorted trigonal-bipyramid. The two niobium

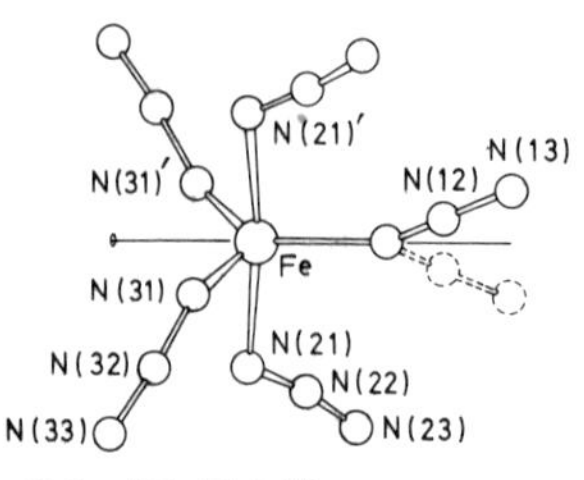

Figure 2.2 $[Fe(N_3)_5]^{2-}$
(From Drummond, J. and Wood, J. S.[49], by permission of The Chemical Society)

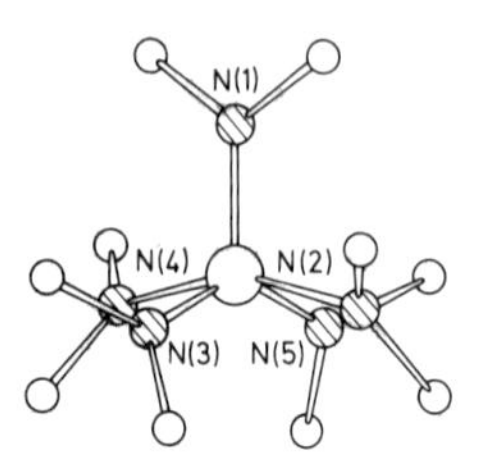

Figure 2.3 $Nb(NMe_2)_5$
(From Heath, C. and Hursthouse, M. B.[51], by permission of The Chemical Society)

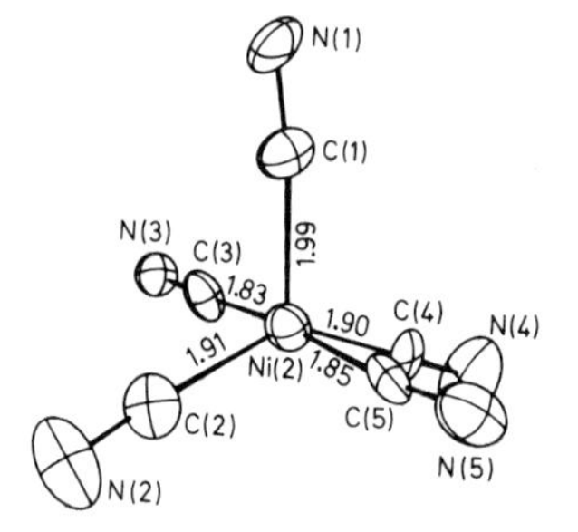

Figure 2.4 Trigonal-bipyramidal form of $[Ni(CN)_5]^{3-}$
(From Raymond, K. N. *et al.*[54], by permission of The American Chemical Society)

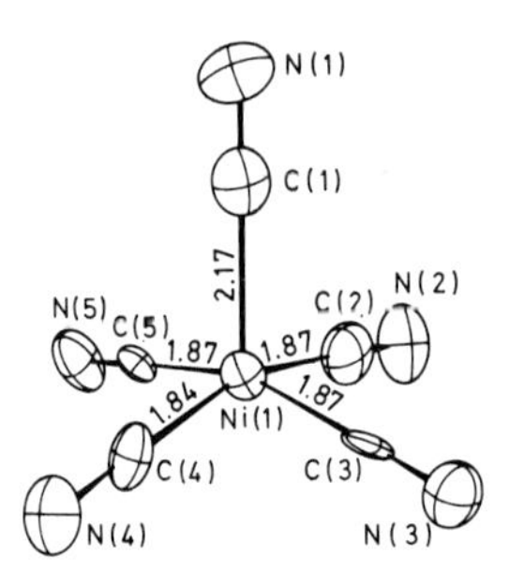

Figure 2.5 Tetragonal-pyramidal form of $[Ni(CN)_5]^{3-}$
(From Raymond, K. N. *et al.*[54], by permission of The American Chemical Society)

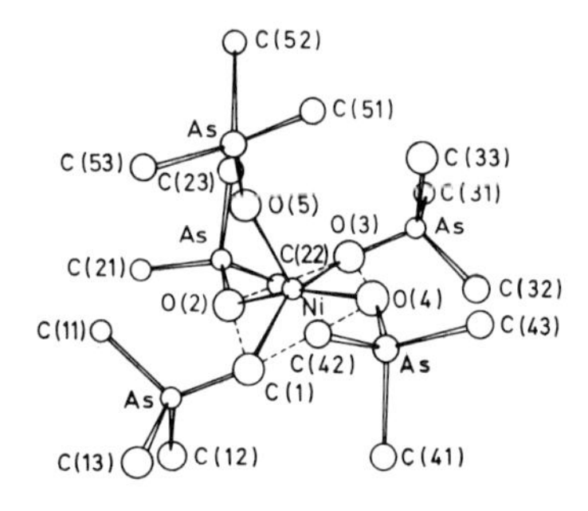

Figure 2.6 $[Ni(OAsMe_3)_5]^{2+}$
(From Hunter, S. H. *et al.*[56], by permission of The Chemical Society)

d^0 complexes[51] (Figure 2.3) and one form of $[Ni(CN)_5]^{3-}$ [52–54] (Figure 2.4) are significantly distorted from trigonal-bipyramidal geometry; $[MnCl_5]^{2-}$ [55] and the other form of $[Ni(CN)_5]^{3-}$ [54] (Figure 2.5) can more accurately be described as tetragonal-pyramids. In addition to these entries, the limited data available indicate that $[Ni(OAsMe_3)_5]^{2+}$ [56] (Figure 2.6) is a tetragonal-pyramid and $[CdCl_5]^{3-}$ is a trigonal-bipyramid with D_{3h} symmetry[47].

In the most stable form of the tetragonal-pyramid[57] the metal atom lies above the plane of the basal atoms such that the apex-to-base angles are *c.* 100 degrees according to Gillespie[57] or *c.* 104 degrees according to Zemann[58] (Figure 2.1). As shown in Table 2.4, $[MnCl_5]^{3-}$, one form of $[Ni(CN)_5]^{3-}$ and the niobium compounds fit this geometry within experimental error.

2.2.1.2 *Distortions in bond distances*

Table 2.5 compares axial and equatorial bond lengths for trigonal-bipyramidal ML_5 complexes. The differences in these bond lengths, especially in the d^8 and d^9 systems, were predicted by Gillespie[59]. Because of the repulsion between the electron pairs, the axial M—L bonds are expected to be longer than the equatorial bonds. This is generally observed for non-transition metal complexes, e.g. PF_5 (axial = 1.577 Å, equatorial = 1.534 Å)[60]. For transition metal complexes the repulsion between valency-shell electron-pairs and d electrons must also be considered. For d^7 (low spin), d^8 and

Table 2.5 Axial and equatorial M—L bond lengths of trigonal-bipyramidal complexes with five identical ligands

Compound	d^x	*Equatorial/Å*	*Axial/Å*	*Equatorial −axial/Å*	*Reference*
$[Fe(N_3)_5]^{2-}$	d^5 (high spin)	1.967	2.041(15)	−0.074	49
$[Co(C_6H_7NO)_5]^{2+}$	d^7 (high spin)	1.975(8)	2.098(4)	−0.123(9)	50
$[Mn(CO)_5]^-$	d^8	1.798(15)	1.820(14)	−0.022(21)	65
$Fe(CO)_5$ x-ray	d^8	1.795(20)	1.795(20)	0.00(3)	62
e.d.*		1.832(5)	1.820(6)	0.012(8)	63
e.d.*		1.833(4)	1.806(5)	0.027(7)	64
$[Co(CNCH_3)_5]^+$	d^8	1.88(2)	1.84(2)	0.04(3)	61
$[Ni\{P(OCH)_3(CH_2)_3\}_5]^{2+}$	d^8	2.190(10)	2.144(15)	0.046(18)	48
$[Ni(CN)_5]^{3-}$	d^8	1.94	1.839(9)	0.10	54
$[CuCl_5]^{3-}$	d^9	2.391(1)	2.296(1)	0.095(1)	46
$[CdCl_5]^{3-}$	d^{10}	2.564(5)	2.527(5)	0.037(7)	47

*Electron diffraction, trigonal-bipyramidal geometry assumed.

d^9 systems the d electron cloud is an oblate ellipsoid because the d_{z^2} orbital is either empty or only half filled. Thus the d electron–ligand repulsions, and hence the bond distances, are greater in the equatorial plane than in the axial direction. The actual difference in axial and equatorial M—L distances will depend on the relative importance of d electron–ligand repulsions versus ligand–ligand repulsions. As shown in Table 2.5, for d^8 systems the difference in M—L distances (i.e., equatorial minus axial) increases in the following series: $Mn^{-I} < Fe^0 < Co^I < Ni^{II}$. On going from Mn to Ni the nuclear charge of the metal increases while the number of electrons remains constant. Thus the electrons are held more tightly and the metallic radius becomes progressively smaller. Concomitant with this is a shorter metal–ligand bond distance and an increase in d electron–ligand repulsion. Because of the effects of the d electron–ligand repulsion, the equatorial bonds are significantly longer than the axial bonds in $[Ni(CN)_5]^{3-}$ [54]. For $[Co(CNMe)_5]^{+}$ [61] and $Fe(CO)_5$ [62–64] the differences are less. In $[Mn(CO)_5]^{-}$ the d electron–ligand repulsions play a minor role compared with the ligand–ligand repulsions, and the axial bonds do not differ significantly from the equatorial bonds (Figure 2.7) [65].

The arguments used above for d^8 systems can be applied to d^9 complexes. Thus, as expected, the equatorial bond is significantly longer than the axial bond in $[CuCl_5]^{3-}$ [46, 47]. Gillespie's explanation can be extended to account for the high spin d^5 and d^7 complexes; the d shell for a high spin d^7 system is prolate in shape as it lacks one electron in the d_{z^2} orbital and two electrons in the *xy* plane. Hence d electron–ligand repulsions would increase the axial

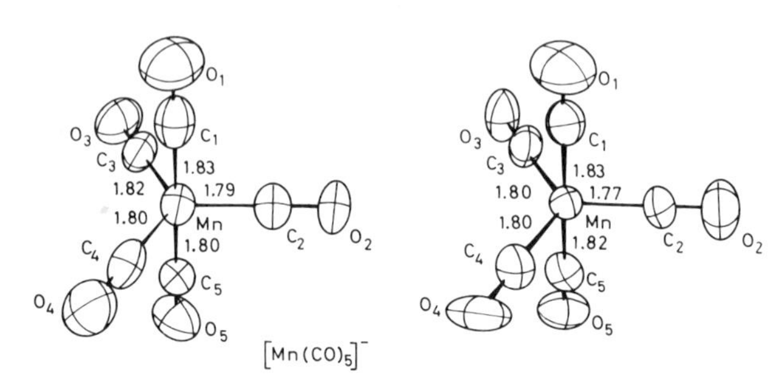

Figure 2.7 Two independent molecules in the asymmetric unit of $[Mn(CO)_5]^{-}$ [65]

bond length. In the pentakis(*N*-picoline oxide)cobalt(II) example there is additional axial elongation arising from strong ligand–ligand repulsions of the relatively bulky ligands[50]. Thus the highly significant difference of 0.123(9) Å in bond lengths appears reasonable. For a spin-free d^5 shell the electrons are symmetrically distributed in the five d orbitals, giving rise to a spherical d shell. Only ligand–ligand repulsions are important and the observed[49] bond lengths for $[Fe(N_3)_5]^{2-}$ are in accord with this. The same argument should apply to a d^{10} complex. Thus it is surprising that the equatorial Cd—Cl bond is moderately longer than the axial bond in $[CdCl_5]^{3-}$ [47].

2.2.2 d^0 Complexes

The niobium d^0 complexes[51] have been briefly discussed above. The structure (Figure 2.3) has been described as a distorted tetragonal-pyramid with a shortened axial Nb—N bond. The axial bonds for $Nb(NMe_2)_5$ and $Nb(NC_5H_{10})_5$ are 1.977(17) and 1.986(13) Å compared with the average basal bonds of 2.042(15) and 2.047(16) Å, respectively. As shown in Table 2.4 the complexes could also be described as distorted trigonal-bipyramids with a unique equatorial bond that is shorter than the other four bonds. The short bond has been attributed to p_π–d_π bonding.

The structures[66] of $WSCl_4$ and $WSBr_4$ may be viewed either as tetragonal pyramids with sulphur at the apex or as dimers formed through weak W—X

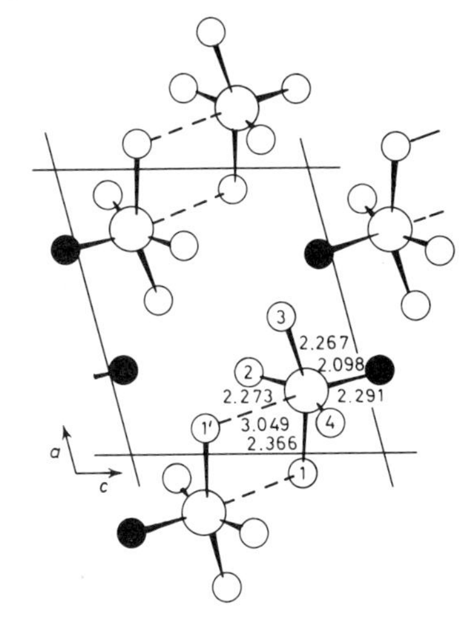

Figure 2.8 The contents of the unit cell of $WSCl_4$
(From Drew, M. G. B. and Mandyczewsky, R.[66], by permission of The Chemical Society)

linkages (X = Cl, Br) as shown in Figure 2.8. The structures contrast with those of $WOCl_4$ and $WOBr_4$ which involve polymeric —W—O—W— bridges along fourfold axes[67].

2.2.3 d¹ Complexes

The structure of $VOCl_2[OC(NMe_2)_2]_2$ is a tetragonal-pyramid with the four atoms forming the base of the pyramid coplanar to within ± 0.015 Å [68]. The vanadium atom is raised out of the plane as shown in Figure 2.9. The V—Cl distance of 2.340(5) Å is significantly longer than the corresponding bond length of 2.250(5) Å in $VOCl_2(NMe_3)_2$ [69]. Structures of $VOCl_2(NMe_3)_2$

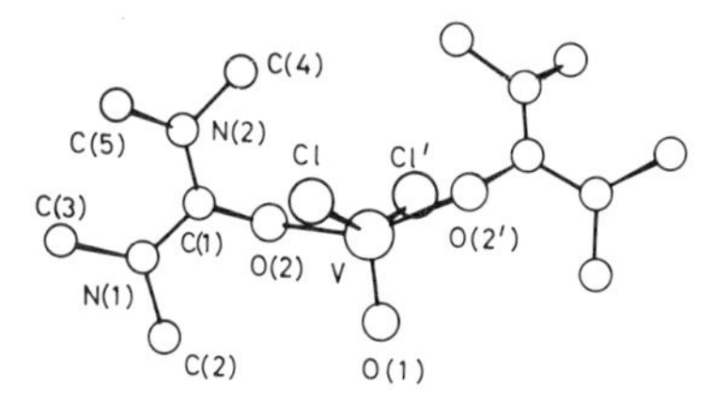

Figure 2.9 $VOCl_2[OC(NMe_2)_2]_2$
(From Coetzer, J.[68], by permission of the International Union of Crystallography)

and $TiBr_3(NMe_3)_2$ [70] are trigonal-bipyramids with axial trimethylamine groups.

Several oxovanadium(IV) structures with multidentate ligands are known, and there has been a recent communication on the structure of bis-(2-methyl-8-quinolinolato)oxovanadium(IV) [71], and a report on the structure of *N,N′*-ethylenebis(acetylacetoneiminato)oxovanadium(IV)[72].

2.2.4 d² Complexes

The only examples of this class of structures are $ReNCl_2(PPh_3)_2$ [73] and $VCl_3(NMe_3)_2$ [74]; the latter structure is discussed in the next section. The Re structure, shown in Figure 2.10, is intermediate between the idealised geometries. The distorted angles are: P—Re—P, 163 degrees; P—Re—N, 98 degrees; P—Re—Cl, 87 degrees av; Cl—Re—N, 110 degrees; Cl—Re—Cl, 141 degrees. An imposed crystallographic twofold axis lies along the Re—N vector. The Re—Cl and Re—N distances are significantly shorter than those in the 6-coordinate $ReNCl_2(PEt_2Ph)_3$ complex[75].

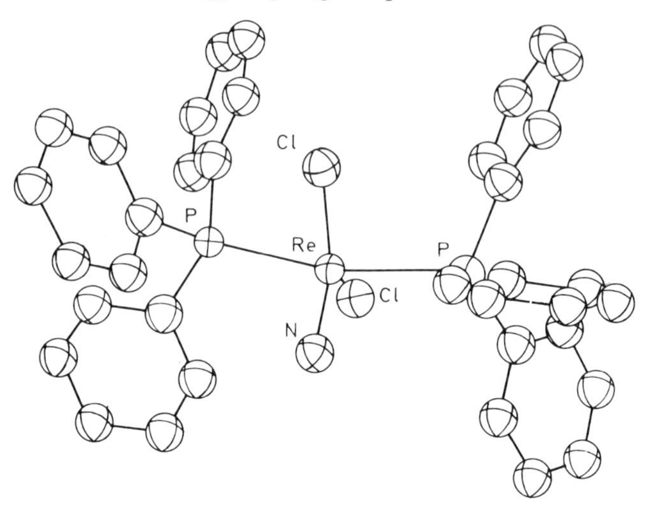

Figure 2.10 $ReNCl_2(PPh_3)_2$
(From Doedens, R. J. and Ibers, J. A.[73], by permission of The American Chemical Society)

2.2.5 d³ Complexes

The only structural determination of a 5-coordinate d^3 complex is that of $CrCl_3(NMe_3)_2$ [76]. Like the bromo-titanium and chloro-vanadium complexes mentioned above this structure consists of a trigonal plane of halogen atoms and axial NMe_3 ligands; crystallographic C_s symmetry is imposed on each molecule. The equatorial Cl—Cr—Cl bond angles are significantly distorted from the ideal 120 degrees. For the Ti and V complexes the corresponding angles show smaller distortions. The angles are as follows: Ti, 121.3(2), 117.5(4) degrees; V, 121.0(1), 118.1(2) degrees; Cr, 124.2(2), 111.6(2) degrees. (There are two angles of the first value in each case because of the mirror symmetry.) For the titanium d^1 and chromium d^3 complexes the departures from idealised D_{3h} symmetry have been interpreted in terms of a Jahn–Teller distortion. The angular distortions in the vanadium d^2 complex were attributed to crystal-packing forces.

2.2.6 d⁴ Complexes

The structure of $MnCl_5^{2-}$ has been reported[55] to be square-pyramidal with distortions towards trigonal-bipyramidal. The Mn atom lies 0.37 Å above the basal plane. The apical Mn—Cl bond of 2.583(5) Å is considerably longer than the basal bonds of 2.241(8), 2.328(8), and 2.336(8) Å.

A series of molybdenum(II) complexes can be considered to be 5-coordinate if the π-cyclopentadienyl ligand is classed as a unidentate ligand. These complexes are of the general formula $Mo(\pi\text{-}C_5H_5)(CO)_2XY$, where X = CO and Y = C_3F_7 [77], Et [78], Cl [79], $AuPPh_3$ [80] or X = PPh_3 and Y = COMe [81], I [82]. The structures of $Mo(NCO)(\pi\text{-}C_5H_5)(CO)(PPh_3)_2$ [83] and $MoCl(\pi\text{-}C_5H_5)(CO)(PPh_2CH_2CH_2PPh_2)$ [82] also have been studied. All of these structures may be described as having a distorted tetragonal-pyramidal shape with the $\pi\text{-}C_5H_5$ group at the apex of the pyramid.

Although the formal oxidation state of the Fe atom in the compound $FeH(SiCl_3)_2(\pi\text{-}C_5H_5)(CO)$, is open to question it will be included in this section. The structure is tetragonal-pyramidal with the cyclopentadienyl ring (considered here as a unidentate ligand) centred at the apex[84]. The Fe atom lies significantly above the basal plane as indicated by the apex-to-base angles of 119.4, 118.1 and 125.8 degrees for Si, Si′, and C, respectively. The hydrogen ligand was not located but can be accommodated *trans* to the carbonyl group without unreasonably short non-bonded interactions.

2.2.7 d^5 Complexes

The structure[49] of $[Fe(N_3)_5]^{2-}$ was discussed above and is shown in Figure 2.2. Recent structural determinations of d^5 complexes with multidentate ligands include $[Fe\{S_2C_2(CF_3)_2\}_2(OPPh_3)]^-$ [85], FeCl(*N*-n-propylsalicylaldiminato)$_2$ [86], $FeCl(O_2C_2Me_2CH)_2$ [87], $FeCl(S_2CNEt_2)_2$ [88] and $Fe(NO)[S_2C_2(CN)_2]_2$ [253].

The isostructural series of high-spin complexes $M^{II}Br[N(CH_2CH_2NMe_2)_3]^+$ with M = Mn, Fe, Co, Ni, Cu and Zn [89] has been studied and is included here as an example of structural trends in multidentate complexes. The general structure is shown in Figure 2.11; Table 2.6 compares axial and

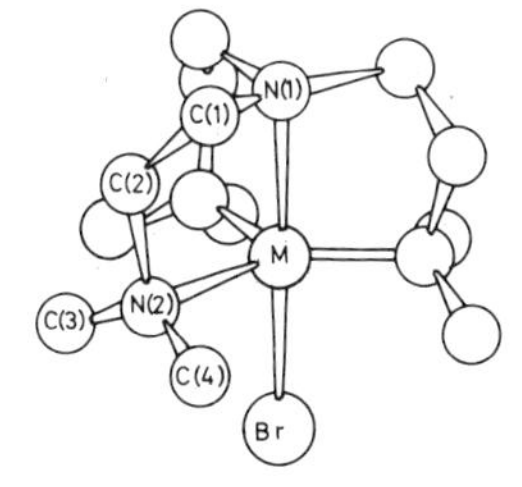

Figure 2.11 $MBr[N(CH_2CH_2NMe_2)_3]^+$ (From Di Vaira, M. and Orioli, P. L.[89], by permission of the International Union of Crystallography)

equatorial M—N distances. The d^6, d^7, d^9 and d^{10} systems are in agreement with that expected on the basis of the shape of the d electron cloud. However, the high-spin d^5 and d^8 systems would be expected to have spherical d shells so that d electron–ligand repulsions would be non-differentiating; ligand–ligand repulsions would favour longer axial bonds relative to equatorial bonds. Unfortunately the errors on the bond distances are

Table 2.6 Metal–nitrogen distances in high-spin $M^{II}Br[N(C_2H_4NMe_2)_3]$ complexes

(From Di Vaira, M. *et al.*[89])

M	d^x	*Axial*/Å	*Equatorial*/Å	*Equatorial –axial*/Å
Mn	d^5	2.19(3)	2.27(2)	0.08(4)
Fe	d^6	2.21(1)	2.15(1)	−0·06(1)
Co	d^7	2.15(2)	2.08(2)	−0.07(3)
Ni	d^8	2.10(1)	2.13(1)	0.03(1)
Cu	d^9	2.07(1)	2.14(1)	0.07(1)
Zn	d^{10}	2.19(2)	2.11(2)	−0.08(3)

sufficiently large that the small differences in bond distances cannot be meaningfully discussed.

2.2.8 d^6 Complexes

All of the complexes in this class can be described as tetragonal-pyramidal, although some are greatly distorted structures. The tetragonal-pyramids are distorted such that the metal atom lies above the basal plane. For example, the metal atom-to-plane distance is 0.23 Å in $[IrI(NO)(CO)(PPh_3)_2]^+$ [90], 0.30 Å in $[Os(OH)(NO)_2(PPh_3)_2]^+$ [91] and 0.39 Å in $[RuCl(NO)_2(PPh_3)_2]^+$ [92].

For the six nitrosyl complexes in this class[45, 90–94] the M—N—O angles are close to 120 degrees. (In the dinitrosyl complexes only one nitrosyl group is bent.) If these ligands are given the formal oxidation state of −1 then the

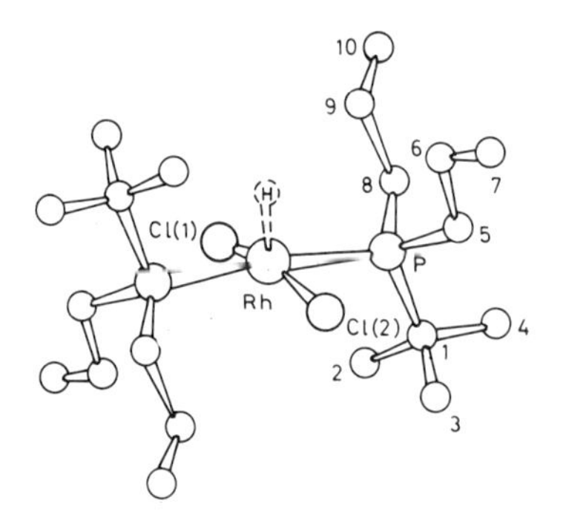

Figure 2.12 $RhHCl_2[P(CMe_3)(C_3H_7)_2]_2$ (From Masters, C. *et al.*[95], by permission of The Chemical Society)

d^6 nitrosyl complexes are tetragonal-pyramids with bent nitrosyl groups at the apices of the pyramids.

Among the other complexes, $RhHCl_2[P(CMe_3)(C_3H_7)_2]_2$ [95] (Figure 2.12) and $RhI_2Me(PPh_3)_2$ [96] are tetragonal-pyramids with H and Me ligands in axial positions, respectively. The complexes $RuCl_2(PPh_3)_3$[97], $RuHCl(PPh_3)_3$[98] and $RhHCl(SiCl_3)(PPh_3)_2$ [99] resemble one another, although they deviate substantially from either of the ideal geometries. The complexes can be described as trigonal-bipyramids with axial phosphine groups. The distortion is evident in the P—M—P angles of 156.4(2), 153.1(2), and 161.7(1) degrees, respectively. A tetragonal-pyramidal description is equally inadequate. However in this case it appears that a sixth coordination site is occupied by a phenyl α-hydrogen atom at distances of 2.59, 2.85 and 2.79 Å from Ru and Rh, so that these complexes may equally well be considered octahedral (Figure 2.13).

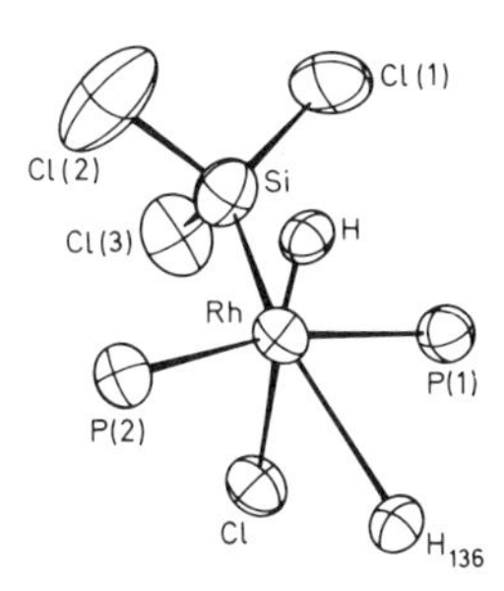

Figure 2.13 $RhHCl(SiCl_3)(PPh_3)_2$ (From Muir, K. W. and Ibers, J. A.[99], by permission of The American Chemical Society)

2.2.9 d^7 Complexes

Bond angles and distances for pentakis(2-picoline *N*-oxide)cobalt(II)[50] are given in Tables 2.4 and 2.5. The distortion from trigonal-bipyramidal geometry is very similar to that in the Ni^{III} complex, $NiBr_3(PMe_2Ph)_2$[100]. In each, one of the angles in the equatorial plane is significantly larger than the other two. The bond opposite this enlarged angle is longer than the other two equatorial bonds (Figure 2.14). The distortions could be ascribed to Jahn–Teller effects for a d^7 electronic configuration; however, similar distortions have been observed for several Ni^{II} d^8 complexes where Jahn–Teller effects cannot be invoked.

The Ni^{III} complex crystallises with $NiBr_2(PMe_2Ph)_2$ and provides an unusual example of an x-ray structural determination of a compound that

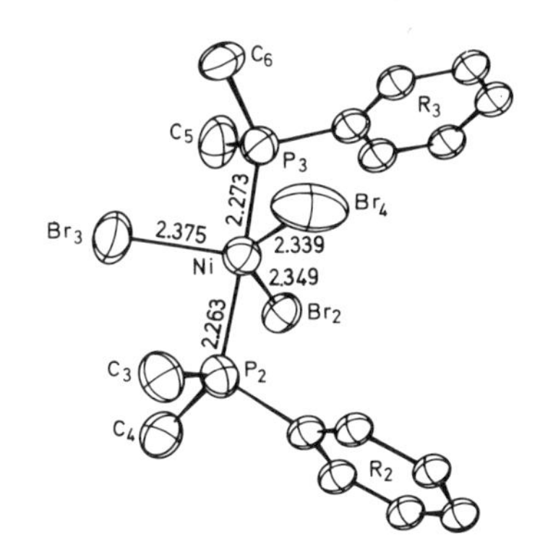

Figure 2.14 $NiBr_3(PMe_2Ph)_2$ (From Meek, D. W. *et al.*[100], by permission of The American Chemical Society)

contains simultaneously two different oxidation states of nickel, two coordination numbers and two unrelated stereochemistries. The complex $CoBr_2(PHPh_2)_3$ has a distorted trigonal-bipyramidal structure[101] with axial phosphine groups and an equatorial plane consisting of the third $PHPh_2$ group and the two Br ligands. The Co complex and the analogous

$NiI_2(PHPh_2)_3$ compound[101] are the only examples of trigonal-bipyramidal $ML_3L'_2$ structures where the three L ligands and the two L′ ligands do not occupy equatorial and axial positions, respectively. Recent structural determinations of d^7 complexes involving multidentate ligands include Fe(NO) $(S_2CNR_2)_2$, R = Me[102] and Et[103], and seven Co^{II} complexes[104–110]. Orioli has recently reviewed the stereochemistry of 5-coordinate Ni^{II} and Co^{II} complexes[10]; these complexes will not be considered in detail here.

2.2.10 d^8 Complexes

As indicated in Table 2.2, by far the most common d shell for which 5-coordinate transition metal structures have been reported is the d^8 system. The great majority of the structures with unidentate ligands are trigonal-bipyramidal, although those involving multidentate ligands are evenly divided between the two geometric forms.

2.2.10.1 ML_5 *structures*

Of the ML_5 complexes three are trigonal-bipyramidal ($[Mn(CO)_5]^-$ [65], $Fe(CO)_5$ [64] and $Ni[P(OCH)_3(CH_2)_3]_5^{2+}$ [48]) and one is square-pyramidal, $[Ni(OAsMe_3)_5]^{2+}$ [56]. $[Ni(CN)_5]^{3-}$ is known to exhibit both geometries[52–54]. In fact, in $[Cr(NH_2C_2H_4NH_2)_3][Ni(CN)_5]$ [54] both forms appear in the same unit cell; these are shown in Figures 2.4 and 2.5. This structure illustrates the small energy difference that must exist between the two ideal geometries of 5-coordination.

2.2.10.2 $FeL(CO)_4$ *structures*

Five complexes of the form $FeL(CO)_4$ with L = CN [111], $PHPh_2$ [112], $Ph_2C_4Ph_2$ [113], H_2CCHCN [114] and $C_2H_2(CO_2H)_2$ [115] have been reported; all are trigonal-bipyramidal. When L is acrylonitrile the axial Fe—C bond length of 1.99 Å is significantly longer than the equatorial bond of 1.77 Å. This is reversed for $Fe(CN)(CO)_4$ where the average equatorial Fe—CO bond of 1.768(14) Å is longer than the axial bond of 1.723(8) Å. In the other complexes the Fe—C bond lengths do not differ significantly.

2.2.10.3 $CoL(CO)_4$ *structures*

The structures of $Co(SiX_3)(CO)_4$ have been studied by electron diffraction (X = H)[116] and x-ray diffraction (X = F [117] and Cl [118]) and found to be trigonal-bipyramidal with the silyl group in an axial position (Figure 2.15). The Co—Si distances are 2.381(7), 2.226(5) and 2.254(3) Å for X = H, F, Cl, respectively. These distances are somewhat shorter than those predicted from sums of covalent radii; this could be interpreted in terms of $(d \rightarrow d)\pi$ bonding between Co and Si. In both halogen derivatives, the axial Co—C bonds are slightly longer than the equatorial bonds and the equatorial carbonyl groups are displaced towards the silyl group.

Trigonal-bipyramidal structures have been reported[119] for $CoL(CO)_4$, L = $AuPPh_3$ and $Ag(As_3C_{17}H_{23})$. As in the silyl complexes the L group is axial and the equatorial CO groups are displaced towards the L group. The average L—Co—C(equatorial) angles for each of the four derivatives are as follows: $SiCl_3$, 85.2; SiF_3, 85.4; $AuPPh_3$, 78; $Ag(As_3C_{17}H_{23})$, 78 degrees.

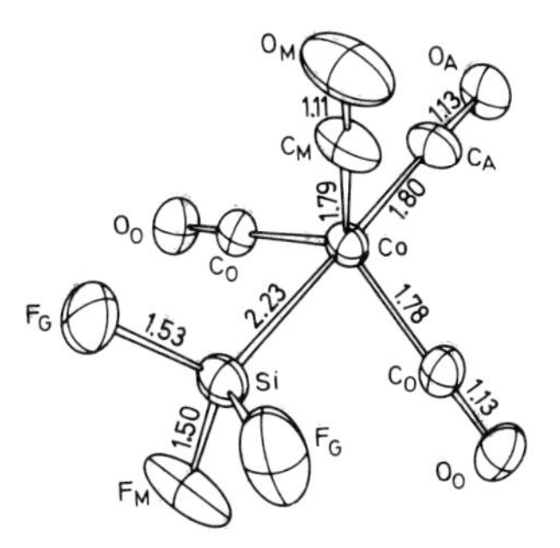

Figure 2.15 $Co(SiF_3)(CO)_4$
(From Emerson, K. *et al.*[117], by permission of The American Chemical Society)

Similar distortions have been observed in Mn complexes, e.g. $Mn_2(CO)_{10}$ [120]. An explanation, based on reduced d electron repulsions as a result of rehybridisation, has been given[119] to account for these distortions.

The structures of $CoL(CO)_3(PPh_3)$, L = CF_2CF_2H [121] and $GePh_3$ [122] have been reported. The carbonyl groups lie in the equatorial plane of the nearly ideal trigonal-bipyramids. The Co—Ge bond length of 2.34(2) Å indicates some degree of multiple bonding.

2.2.10.4 **MHP_4** *and* **$MLHP_3$** *structures*

The complexes of the form MHP_4, where P = phosphine, show greatly distorted trigonal-bipyramidal structures. The equatorial phosphine groups are displaced towards the presumed axial position of the hydride ligand such that the metal atom lies below the plane of the P atoms by 0.49 Å in

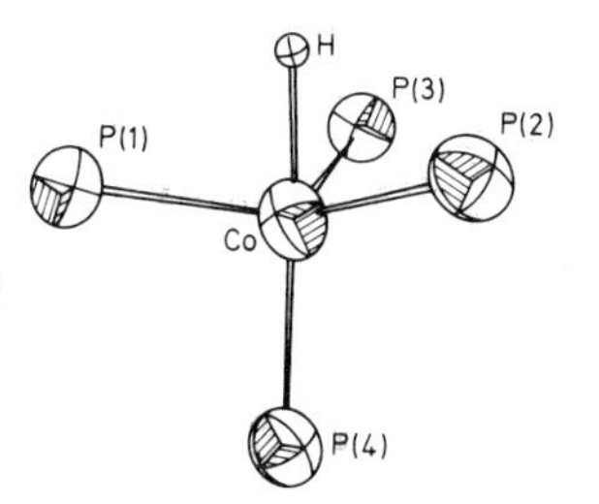

Figure 2.16 $CoH[PPh(OEt)_2]_4$
(From Titus, D. D. *et al.*[123], by permission of The Chemical Society)

$CoH[PPh(OEt)_2]_4$ [123] (Figure 2.16), 0.59 Å in $CoH(PF_3)_4$ [124], and 0.70 Å in $RhH(PPh_3)_4$ [125]. In the latter compound the P atoms closely approach a tetrahedral arrangement around the Rh atom. A similar structure is found for $RhH(PPh_3)_3(AsPh_3)$ [126]. The distortion has been attributed to the minimal steric requirements of the hydride ligand compared with the bulky phosphine groups[35].

Four hydride complexes of the form $MLH(PPh_3)_3$, where ML = $Co(N_2)$[127], Rh(CO)[128], Ru(NO)[129] and $Ir(NO)^+$[130] have been studied. The equatorial phosphine groups are displaced towards the axial hydrogen ligand as shown in Figure 2.22 for $[IrH(NO)(PPh_3)_3]^+$. The metal lies below the plane of the P atoms by 0.30, 0.36, 0.55, and 0.51 Å for Co, Rh, Ru and Ir, respectively[35].

2.2.10.5 $Ni(CN)_2L_3$ structures

The geometry of the d^8 $ML_3L'_2$ complexes is that of a trigonal-bipyramid with equatorial L groups and axial L′ groups (with the exception of $NiI_2(PHPh_2)_3$ as pointed out above). The structures of $Ni(CN)_2L_3$, L = CN[54], $PPh(OEt)_2$[131], PMe_2Ph[132], $PMe(C_{12}H_8)$[133] and $PEt(C_{12}H_8)$[133] are distorted such that one equatorial bond angle is larger than the other two, and the Ni—L bond opposite this angle is longer than the other Ni—L bonds as shown in Table 2.7. This distortion is not observed for $Os(CO)_3(PPh_3)_2$[134] since the molecule is crystallographically required to possess D_3 symmetry. The distortion is, in effect, a movement towards a tetragonal-pyramidal structure. The apical bond of a tetragonal-pyramid is expected to be longer than the basal bonds[57]; this may be related to the observed lengthening of the bond opposite the enlarged angle in the nickel structures.

2.2.10.6 Derivatives of $IrCl(CO)(PPh_3)_2$

The structures of a large number of derivatives of $IrCl(CO)(PPh_3)_2$ have been investigated; these are shown in Table 2.8 with bond angles and distances. For complexes containing NO and SO_2 as ligands a tetragonal-pyramidal geometry is observed; in all other cases the structure is trigonal-bipyramidal with PPh_3 groups axial, equatorial, or both. In $IrH(CO)_2(PPh_3)_2$ a small displacement of equatorial ligands towards the hydride ligand is observed[136]. In the olefin complexes the C=C bond lies in the equatorial plane; the bulkiness of the ligand probably is responsible for the decreased P—Ir—P angles of 110–114 degrees[137–140]. In the tetragonal-pyramidal structures the Ir atom lies above the basal plane. This is evident in the nonlinear P—Ir—P and X—Ir—CO angles.

2.2.10.7 Other d^8 structures

The $Mn(NO)(CO)_{4-x}(PPh_3)_x$ (x = 0–2)[143–145] series of compounds has been studied and will be discussed in the nitrosyl section of this review.

The structure of the zwitterion $NiCl_3(C_7H_{15}N_2)(H_2O)$ is illustrated in Figure 2.17[146]. The magnetic moment of this complex is 3.7 BM, indicating a high-spin d^8 system. The structure is distorted as indicated by the axial O—Ni—N bond angle of 171.4(3) degrees.

The very large number of structural reports of d^8 complexes with multidentate ligands cannot be discussed adequately here. A checklist of the more recent determinations is given in Table 2.9.

Table 2.7 Distortions in the trigonal plane of $Ni(CN)_2L_3$ structures

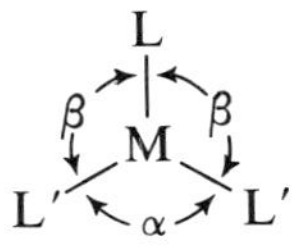

L	α/degrees	*Av* β/*degrees*	$\alpha-\beta$/degrees	M—L/Å	*Av* M—L′/Å	(M—L)–(M—L′)/Å	*Reference*
CN*	159.3	100.3	59	2.17(1)	1.85(2)	0.32(2)	54
CN*	141.2	109.5	32	1.99(1)	1.91(1)	0.08(1)	54
$PPh(OEt)_2$	133.5	113.3	20	2.289(5)	2.20(1)	0.09(1)	131
PMe_2Ph	127.0	116.5	11	2.261(3)	2.223(3)	0.038(4)	132
$PMe(C_{12}H_8)$*	161.2	99.4	62	2.321(2)	2.179(2)	0.142(3)	133
$PMe(C_{12}H_8)$*	132.8	112.6	20	2.293(4)	2.224(4)	0.069(6)	133
$PEt(C_{12}H_8)$	126.4	116.8	10	2.267(5)	2.246(6)	0.021(8)	133

*Two isomers exist for this complex. For L = CN both isomers are in the same unit cell; for L = $PMe(C_{12}H_8)$ the isomers are found in different crystals with different colours.

Table 2.8 Bond distances and angles for a series of $IrX(CO)(PPh_3)_2L$ complexes

Compound (P = PPh_3)	*Structure*	X	L	Ir—P/Å	Ir—X/Å	Ir—L/Å	Ir—CO/Å	P—Ir—P/degrees	P—Ir—X/degrees	P—Ir—L/degrees	P—Ir—CO/degrees	X—Ir—L/degrees	X—Ir—CO/degrees	L—Ir—CO/degrees	*Reference*
$IrCl(CO)_2P_2$		Cl	CO	2.341(7) 2.322(7)	2.37	2.04	2.04	176.8(2)	89.6 90.4	92.2 90.4	88.4 91.7	118.2	120.9	120.9	135
$IrH(CO)_2P_2$		H	CO	2.375(2) 2.377(2)	1.66(20)	1.834(9) 1.868(9)	1.834(9) 1.868(9)	101.4(1)	171	116.0(6)	114.2(4), 93.8(3) 116.0(6), 95.3(4)	—	—	125.9(7)	136
$IrH(CO)P_2[C_2H_2(CN)_2]$*		H	$C_2H_2(CN)_2$†	2.317(3)	—	2.110(9)	1.98(2)	114.1(1)	—	103.1(3) 142.8(3)	87.9(6)	—	—	89.3(8)	137, 138
$IrBr(CO)P_2[C_2(CN)_4]$		Br	$C_2(CN)_4$†	2.397(3) 2.402(3)	2.508(2)	2.146(11) 2.151(11)	—	110.4(1)	—	144.6(3), 103.6(3) 146.1(2), 105.0(3)	—	—	—	—	138, 139
$Ir(C_6N_4H)(CO)P_2[C_2(CN)_4]$		C_6N_4H	$C_2(CN)_4$†	2.397(2) 2.387(2)	2.024(8)	2.155(9) 2.176(10)	1.849(10)	111.6(1)	84.8(2) 92.5(2)	102.8(2), 104.3(2) 143.8(2), 145.5(2)	86.4(3) 94.2(3)	88.6(3) 89.1(3)	178.1(3)	—	140

$IrCl(O_2)(CO)P_2$	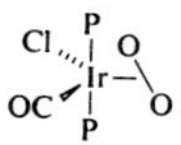	Cl	O_2‡	2.38(1) 2.36(1)	2.38(2)§ 2.42(2)	2.09(3) 2.04(3)	2.38(2)§ 2.42(2)	172.8(5)	90.9(5) 93.0(5)	85.4(7), 90.6(7) 89.2(9), 84.1(9)	90.9(5) 93.0(5)	116(1), 107.4(9) 152.4(8), 144(1)	100.1(6)	116(1), 107.4(9) 152.4(8), 144(1)	141
$[IrCl(NO)(CO)P_2]^+$	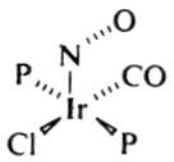	Cl	NO	2.407(3) 2.408(3)	2.343(3)	1.97(1)	1.86(1)	175.7(1)	86.8(1) 88.9(1)	90.1(3) 91.3(3)	91.2(3) 92.7(3)	101.3(3)	161.3(3)	97.4(5)	45
$[IrI(NO)(CO)P_2]^+$	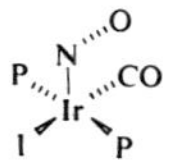	I	NO	2.35(1) 2.37(1)	2.666(3)	1.89(3)	1.70(5)	168.2(3)	86.1(3) 87.6(3)	94(1) 97(1)	90(1) 91(1)	101(1)	158(1)	101(2)	90
$IrI(NO)(Me)P_2$	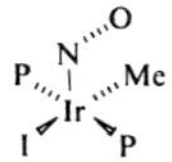	I	NO	2.348(3)	2.726(2)	1.91(2)	2.05(4)‖	169.2(2)	88.3(1) 89.4(1)	88(1) 103(1)	88(1)‖ 89(1)	102(1) 102(1)	151(1)‖	107(2)‖	94
$IrCl_2(NO)P_2$	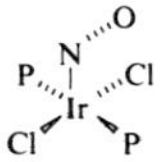	Cl	NO	2.367(2)	2.348(2)	1.94(2)	—	170.2(1)	90.0(1) 88.1(1)	101(1) 89(10)	—	100(1) 103(1)	—	—	93
$IrCl(SO_2)(CO)P_2$	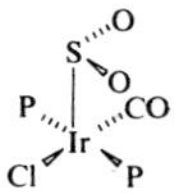	Cl	SO_2	2.359(9) 2.328(8)	2.37(1)	2.49(1)	1.96(4)	169.5(4)	87.3(4)	92.6(4) 97.8(4)	93(1)	97.4(3)	173(1)	90(1)	142

*A crystallographic twofold axis passes through the Ir atom, bisecting the C—C bond. The H and CO ligands are disordered.
†Bond angles and distances pertain to each of the carbon atoms of the olefinic bond; rather than to the centre of the bond.
‡Bond angles and distances pertain to each of the oxygen atoms, rather than the centre of the bond.
§The Cl and CO ligands are disordered and were refined essentially as two Cl atoms.
‖Refers to methyl group.

Table 2.9 5-Coordinate structures of d^8-complexes with multidentate ligands*

Compound†	*Structure*	*Reference*
Bidentate		
$Co(NO)(S_2CNMe_2)_2$	TP	163
$Ni[S_2P(OMe)_2]_2(Me_2phen)$	TP	164
$[NiCl(atsc)_2]^+$	TBP	165
$[Ru(NO)(diphos)_2]^+$	TBP	129
$[Rh(O_2)(diphos)_2]^+$	TBP	166
$RhCl(C_4H_6)_2$	TP	167
$RhCl(SbPh_3)_2[C_4(CF_3)_4]$	TBP	168
$[Ir(O_2)(diphos)_2]^+$	TBP	166, 169
$[Ir(S_2)(diphos)_2]^+$	TBP	170
$[Ir(CO)(diphos)_2]^+$	TBP	254
Tridentate		
$NiI_2(DSP)$	TP	171
$NiBr_2(Me_2dpma)$	TBP	172
$NiBr_2[MeOC_6H_4CHN(CH_2)_8NR_2]$	TBP/TP	173
$NiBr_2(Tas)$	TBP	174
$NiBr_2(bda)$	TP	175
$Ni(NCS)_2(N_3As)$	TP	176
$[Ni(diars)(triars)]^{2+}$	TP	177
$Ni(Cl\text{-}SALen\text{-}NEt_2)_2$	TP	178
Quadradentate		
$[Ni(CN)(TAP)]^+$	TBP	179
$[NiCl(3,22,3\text{-}tet)]^+$	TP	180
$[NiI(tda)]^+$	TBP	181
$[NiBr(Me_6tren)]^+$	TBP	89
$[NiBr(C_{15}H_{22}N_4)]^+$	TP	182
$[Ni(NCS)(bdma)]^+$	TP	183
$[NiI(bdma)]^+$	TP	184
$[NiCl(TSP)]^+$	TBP	185
$Ni(dacoDA)(H_2O)$	TP	186
$[PdCl(TPAS)]^+$	TP	187
Quinquedentate		
$Ni(SalMedpt)$	TBP	188
$Ni(C_{20}H_{23}N_3O_2)$	TBP	189
$[Ni(tpen)]^{2+}$	TP	190

*See Table 2.2, footnote *.

†The following abbreviations are used in the table: atsc = acetonethiosemicarbazone; bda = $HN^*(C_2H_4P^*Ph)_2$; bdma = $MeO^*C_2H_4N^*(C_2H_4P^*Ph_2)_2$; Cl-SALen-NEt$_2$ = *N*-β-diethylaminoethyl-5-chlorosalicylaldimine; dacoDA = 1,5-diazocyclo-octane-*N,N'*-diacetate; diars = $C_6H_4(As^*Me_2)_2$; diphos = $Ph_2P^*C_2H_4P^*Ph_2$; DSP = $PhP^*(C_6H_4S^*Me)_2$; Me$_2$phen = 2,9-dimethyl-1,10-phenanthroline; Me$_2$dpma = $HN^*[CH_2(N^*C_5H_3)Me]_2$; Me$_6$tren = $N^*(C_2H_4N^*Me_2)_3$; N$_3$As = $Ph_2AsC_2H_4N^*(C_2H_4N^*Et_2)_2$; SalMedpt = $MeN^*(C_3N_6N^*CHC_6H_4O^*)_2$; TAP = $P^*(C_3H_6As^*Me_2)_3$; Tas = $MeAs^*(C_3H_6As^*Me_2)$; tda = $N^*(C_2H_4P^*Ph_2)_3$; TPAS = $Me(MeAs^*Ph)_4Me$; tpen = $(C_5H_4N^*)C_2H_4N^*HC_2H_4N^*[C_2H_4(C_5H_4N^*)]_2$; triars = $MeAs^*(C_6H_4As^*Me_2)_2$; TSP = $P^*(C_5H_4S^*Me)_3$; 3,22,3-tet = *N,N'*-di(3-aminopropyl)piperazine. The asterisk denotes the atom bonded to the metal.

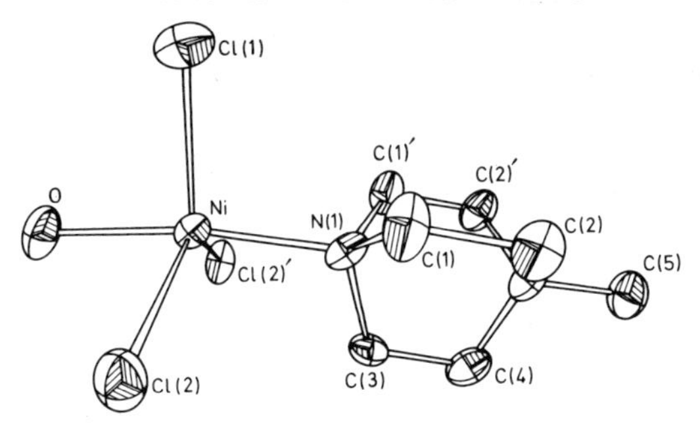

Figure 2.17 $NiCl_3(C_7H_{15}N_2)(H_2O)$
(From Ross, F. K. and Stucky, G. D.[146], by permission of The American Chemical Society)

2.2.11 d⁹ Complexes

The structures of $[CuCl_5]^{3-}$ [46, 47] and $[CuBr_3Cl_2]^{3-}$ [147] are perfect trigonal-bipyramids since both possess crystallographic point symmetry 32 (D_3). In $[CuCl_5]^{3-}$ the equatorial Cu—Cl bond of 2.391(1) Å is significantly longer than the axial bond of 2.296(1) Å [46]. Replacement of chlorine by bromine increases ligand–ligand repulsions; hence the axial Cu—Cl bond increases to 2.375(2) Å in $[CuBr_3Cl_2]^{3-}$ [260]. A similar lengthening of equatorial bonds by 0.14 Å relative to axial bonds has been observed[148] for $CuCl_2(C_5H_8N_2)_3$. The

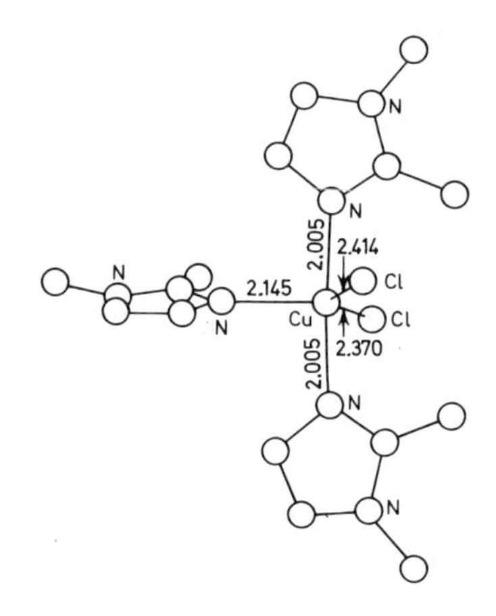

Figure 2.18 $CuCl_2(C_5H_8N_2)_3$
(From Huq, F. and Skapski, A. C.[148], by permission of The Chemical Society)

equatorial plane is distorted such that the Cl—Cu—Cl angle is 138 degrees and the Cl—Cu—N angles are 118 and 104 degrees, where the smaller angle borders on the methylated side of the 1,2-dimethylimidazole ring (Figure 2.18).

The structure of $Cu(O_2COPh)_2(H_2O)_3$ is tetragonal-pyramidal with an axial oxygen atom belonging to a water molecule; identical ligands in the basal plane are mutually *trans*[149]. Although only a few d⁹ complexes with unidentate ligands have been studied, a large number of copper structures involving multidentate ligands have been reported. Recent structures are tabulated in Table 2.10.

Table 2.10 5-Coordinate structures of d^9 complexes with multidentate ligands*

Compound†	*Structure*	*Reference*
Bidentate		
$CuCl_2(Me_2phen)(H_2O)$	TBP	191
Cu(L-isoleucinato)$_2$(H_2O)	TP	192
Cu(L-serinato)$_2$	TP	193
Cu(DL-lactato)$_2$(H_2O)	TP	194
$[Cu(dach)_2(H_2O)]^{2+}$	TP	195
$[Cu(dach)_2(NO_3)]^+$	TP	195
$Cu(OC_6H_4CHNPh)_2(C_5H_4N)$	TP	196
$Cu(C_8H_6O_3)_2(C_5H_4NMe)$	TP	197
$[Cu(SCN)(tn)_2]^+$	TBP	198
Tridentate		
Cu(gly–gly)($C_3N_2H_4$)(H_2O)	TP	199
Cu(gly–glyH)($C_3N_2H_4$)(H_2O)	TP	199
$CuBr(N_3)(Et_4dien)$	TBP	200
Quadradentate		
$[CuBr(bpe)]^+$	TBP/TP	201
$[CuBr(Me_6tren)]^+$	TBP	89
$[Cu(NCS)(tetramine)]^+$	TBP	202
$[CuCl\{N_6P_6(NMe_2)_{12}\}]^+$	TP	203
$Cu(ppp)(H_2O)$	TP	204

*See Table 2.2, footnote *.

†The following abbreviations are used in the table: bpe = $(C_5H_4N^*)CHN^*C_2H_4N^*CH(C_5H_4N^*)$; dach = 1,4-diazocycloheptane; Et_4dien = $HN^*(C_2H_4N^*Et_2)_2$; Gly–Gly = $H_2N^*CH_2CON^*CH_2CO^*_2$; Gly–GlyH = $H_2N^*CH_2CON^*HCH_2CO^*_2$; ($Me_2phen$) = 2,9-dimethyl-1,10-phenenthroline; Me_6tren = $N^*(C_2H_4N^*Me_2)_3$; ppp = phenylalaninepyridoxylidine-5-phosphate; tetramine = $(C_5H_4N^*)CHN^*C_3H_6N^*CH(C_5H_4N^*)$; tn = $H_2N^*C_3H_6N^*H_2$. The asterisk denotes the atom bonded to the metal.

2.2.12 d^{10} Complexes

The only recent example of a structural determination of a d^{10} transition metal complex with unidentate ligands is $[CdCl_5]^{3-}$ [47]. The crystallographic site symmetry of the Cd atom is 32, imposing perfect trigonal-bipyramidal geometry on the molecule. The equatorial Cd—Cl bonds are 0.037(7) Å longer than the axial bonds. This is unexpected since $[CdCl_5]^{3-}$ should be structurally similar to PF_5 where ligand–ligand repulsions cause longer axial bonds relative to equatorial bonds[60].

Several zinc complexes with quadradentate ligands have recently been reported: $ZnBr[MeN(CH_2CH_2NMe_2)_2]^+$ [150]; $ZnX[N(CH_2CH_2NH_2)_3]^+$, X = Cl [151] and NCS [152]; and $ZnI[H_2NC_2H_4NHC_2H_4NHC_2H_4NH_2]$ [153]. The first three structures are trigonal-bipyramidal; the last is square-pyramidal with an axial I atom. In all, the quadradentate ligands are attached through the nitrogen atoms to the zinc atom.

2.2.13 Phosphine ligands

The phosphine ligand is the most common one found in structures of unidentate 5-coordinate complexes. Because phosphines are generally very bulky, they have an important effect on the geometry of the complex. The following rules hold for 5-coordinate complexes:

(a) Trigonal-bipyramids.
 (i) One or two phosphine groups: P atom(s) is axial (9 examples).
 (ii) Three phosphine groups: P atoms are equatorial (8 examples).
 (iii) Four phosphine groups: P atoms form a pseudo-tetrahedral arrangement (3 examples).
(b) Tetragonal-pyramids: P atom(s) is basal (15 examples).

The following exceptions are known:

(a) Pseudo-octahedral structures.
 (i) Ir complexes with ligands $(NC)_2C{=}C(CN)_2$ [138–140] and (NC)HC=CH(CN) [137, 138].
 (ii) $RuCl_2(PPh_3)_3$ [97], $RuHCl(PPh_3)_3$ [98] and $RhHCl(SiCl_3)(PPh_3)_2$ [99] with phenyl α-hydrogens occupying the sixth coordination positions.
(b) $CoBr_2(PHPh_2)_3$ and $NiI_2(PHPh_2)_3$ [101] where halogen atoms are equatorial.
(c) $IrH(CO)_2(PPh_3)_2$ [136] where one P atom is axial and one is equatorial.

2.3 NITROSYL COMPLEXES

The structural properties of the nitrosyl group coordinated to transition metals are well documented in the earlier literature and have been the subject of a comprehensive review[205]. In general, nitrosyl complexes show linear M—N—O bond angles. Recently, a new mode of coordination, involving an M—N—O bond angle of *c.* 120 degrees, has been confirmed[45]. This report has sparked activity in several laboratories towards enlarging the available data on M—N—O linkages and theorising on the bonding forces responsible for the wide variety of observed bond angles.

In order to systematise the many complexes whose structures have been investigated, the compounds have been divided into three groups on the basis of M—N—O bond angles: (I) linear M—N—O linkages, (II) intermediate bond angles, and (III) M—N—O bond angles near 120 degrees. Table 2.11 lists most of the known complexes containing a nitrosyl ligand whose structures have been determined by x-ray or neutron diffraction techniques. In addition, the following structures also have been determined, but the reported results are either incomplete or inappropriate for inclusion in Table 2.11: $[Co(NO)_2Cl]_2$ [206], $[Co(NO)_2Cl]_n$ [207], $[Co(NO)_2Br]_n$ [206], $Ru_3(NO)_2(CO)_{10}$ [208], $Mn_3(NO)_4(\pi\text{-}C_5H_5)_3$ [209], $Co(NO)(CO)_2(PPh_3)$ [210] and $[Mo(OH)(CO)_2(NO)]_4$ [211]. The structures of $Fe(NO)_2(CO)_2$ and $Co(NO)(CO)_3$ have been determined[212] by electron diffraction (M—N—O = 180 degrees assumed, Fe—N = 1.77 ± 0.02 Å, Co—N = 1.76 ± 0.03 Å), and the structure of $Ni(NO)(\pi\text{-}C_5H_5)$ has been studied by microwave techniques (Ni—N—O = 180 degrees, Ni—N = 1.676 ± 0.020 Å)[213].

Table 2.11 Nitrosyl complexes studied by x-ray or neutron diffraction

*Compound**	M—N/Å	N—O/Å	M—N—O/ *degrees*	*Reference*
Group I†				
$Fe(NO)[S_2C_2(CN)_2]_2$	1.56	1.06	168(6)	253
$Ni(NO)[MeC(CH_2PEt_2)_3]$	1.58	1.20	180	234
$[Co(NO)_2I]_n$	1.61(4)	1.17(5)	171(4)	227
$[Fe(CN)_5(NO)]^{2-}$	1.63(2)	1.13(2)	178(1)	218
$Fe(NO)_2[Ph_2PC{=}CPPh(CF_2)_2CF_2]$	1.65(1)	1.18(1)	177.4(7)	235
$Cr_2(NO)_3(\pi\text{-}C_5H_5)NH_2$	1.66(2)	1.20(2)	172(3)	236
$[Fe_4(NO)_7S_3]^-$	1.66(5)	1.20(6)	168(5)	237
$[V(CN)_5(NO)]^{3-}$	1.66(4)	1.29(5)	171(3)	238
$[Mn(CN)_5(NO)]^{3-}$	1.66(1)	1.21(2)	174(1)	217
$Ir(NO)(PPh_3)_3$	1.67(2)	—	180(0)	239
$[Fe(NO)_2SEt]_2$	1.67(1)	1.17(2)	167(4)	240
$[IrH(NO)(PPh_3)_3]^+$	1.68(3)	1.21(3)	175(3)	130
$Fe(NO)(S_2CNEt_2)_2$	1.69(4)	1.16(5)	174(4)	103
$[RuCl_5(NO)]^{2-}$	1.738(2)	1.131(3)	177(1)	221
$Fe(NO)(S_2CNMe_2)_2$	1.71(2)	1.02(2)	173(2)	102
$[Cr(CN)_5(NO)]^{3-}$	1.71(1)	1.21(1)	176(1)	220
$CrCl(NO)_2(\pi\text{-}C_5H_5)$	1.71(1)	1.14(2)	169(3)	226
$[Os(OH)(NO)_2(PPh_3)_2]^+$	1.71(4)	1.25(5)	—	91
$Ru(NO)(S_2CNEt_2)_3$	1.72	1.17	170	242
$Mn(NO)(CO)_2(PPh_3)_2$	1.73(1)	1.18(1)	178(1)	143
$Mo(NO)[S_2CN(C_4H_9)_2]_3$	1.73(1)	1.15(1)	173	243
$[Ru(NO)(Ph_2PC_2H_4PPh_2)_2]^+$	1.74(1)	1.20(1)	174(1)	129
$RuCl(NO)_2(PPh_3)_2$	1.74(2)	1.16(2)	180(2)	92
$[Ru(NO_2)_4(OH)(NO)]^{2-}$	1.748(4)	1.127(7)	180.0(6)	244
$Mo(NO)(\pi\text{-}C_5H_5)_3$	1.751(3)	1.207(4)	179.2(2)	245
$Ir_2O(NO)_2(PPh_3)_2$	1.76(2)	1.12(2)	177(2)	246
$Mn(NO)(CO)_3(PPh_3)$	1.78(2)	1.15(1)	178(1)	144
$OsCl_2(HgCl)(NO)(PPh_3)_2$	1.79(4)	1.03(6)	178(2)	247
$Mn(NO)(CO)_4$	1.80(1)	1.15(2)	180(0)	145
$RuH(NO)(PPh_3)_3$	1.80(1)	1.18(1)	176(1)	129
$[Mo(CN)_5(NO)]^{4-}$	1.95(3)	1.23(4)	175(3)	219
Average	1.71	1.17	175	
Group II‡				
cis-$[Cr(NO)(\pi\text{-}C_5H_5)NMe_2]$	1.63(1)	1.22(2)	169(1)	248
trans-$[Cr(NO)(\pi\text{-}C_5H_5)NMe_2]$	1.63(2)	1.23(2)	169(1)	248
$[Cr(NO)(\pi\text{-}C_5H_5)(SPh)]_2$	1.662(7)	1.19(1)	169.9(7)	228
$[Co(NO)_2(NO_2)]_n$	1.67	1.12	167(2)	249
$[Fe(NO)_2I]_2$	1.67(4)	1.15(6)	161(3)	227
$Ni(N_3)(NO)(PPh_3)_2$	1.686(7)	1.164(8)	152.7(7)	229
$Cr(NO)_2(NCO)(\pi\text{-}C_5H_5)$	1.72(1)	1.16(1)	171.0(6)	248
$[Co(NO)_2Cl]_2$	1.73(3)	1.12(5)	166(3)	207
$[Ir(NO)_2(PPh_3)_2]^+$	1.77(1)	1.21(1)	163(1)	230
$[Ru(NO)(NH_3)_5]^{3+}$	1.80(1)	1.11(2)	167(1)	257
$[RuCl(NO)_2(PPh_3)_2]^+$	1.86(2)	1.17(2)	136(2)	92
$[RuCl_4(OH)(NO)]^{2-}$	2.04(5)	1.13(7)	153	250
$[Ru(OH)(NO)(NH_3)_4]^{2+}$	2.07	1.14	150	258

Table 2.11 *(continued)*

*Compound**	M—N/Å	N—O/Å	M—N—O/ *degrees*	*Reference*
Group III§				
$Co(NO)(S_2CNMe_2)_2$	1.70	1.1	127	163
$[CoCl(NO)(HNC_2H_4NH)_2]^+$	1.82(1)	1.04(2)	124(1)	223
$[Co(NO)(NH_3)_5]^{2+}$	1.871(6)	1.154(7)	119(1)	222
$[IrI(NO)(CO)(PPh_3)_2]^+$	1.89(3)	1.17(4)	125(3)	90
$IrI(NO)(Me)(PPh_3)_2$	1.91(2)	1.23(2)	120(2)	94
$IrCl_2(NO)(PPh_3)_2$	1.94(2)	1.03(2)	123(2)	93
$[IrCl(NO)(CO)(PPh_3)_2]^+$	1.97(1)	1.16(1)	124(1)	45
$[Os(OH)(NO)_2(PPh_3)_2]^+$	1.98(5)	1.12(6)	128(2)	91
Average	1.89	1.13	124	

*The following abbreviations are used: Ph = C_6H_5, Et = C_2H_5, Me ≐ CH_3.
†Includes those compounds with M—N—O angles in the range 174–180 degrees, plus those whose angles are not significantly different from 180 degrees, i.e. the observed angles differ from 180 degrees by less than three standard deviations.
‡Includes those compounds with M—N—O angles significantly different from 180 and 120 degrees.
§Includes those compounds with M—N—O angles in the range 120–128 degrees.

2.3.1 Structures with linear M—N—O angles

Complexes in Group I are characterised by (i) linear M—N—O linkages, (ii) short M—N bond lengths, and (iii) little or no *trans* influence of the NO group. In these complexes NO acts as a σ electron donor (Lewis base) toward transition metals and therefore is formally considered to be NO^+. The nitrogen atom is sp hybridised, donating its lone pair of electrons to the metal and accepting electrons into its unoccupied π^* orbitals from the occupied metal d orbitals. The following Lewis dot structure is appropriate:

$$M::N::\ddot{O}:$$

This cumulated double bond system is analogous to allene, $>C{=}C{=}C<$ and a linear M—N—O bond angle is expected. The multiple bond character of the M—N bond is in agreement with the observed short M—N distances.

Table 2.12 Selected metal–nitrogen bond lengths

Compound	*Nitrogen ligand*	*Distance/Å*	*Reference**
Average of Group I complexes	NO	1.71	—
Average of Group III complexes	NO	1.89	—
$ReNCl_2(PPh_3)_2$	N^{3-}	1.602(9)	73
$ReCl_3(NMe)(PEt_2Ph)_2$	NMe^{2-}	1.685(11)	255
$ReCl_3(NC_6H_4COMe)(PEt_2Ph)_2$	$[NC_6H_4COMe]^{2-}$	1.690(5)	256
$ReCl_3(NC_6H_4OMe)(PEt_2Ph)_2$	$[NC_6H_4OMe]^{2-}$	1.709(4)	256
$ReCl(N_2)(PMe_2Ph)_4$	N_2	1.97(2)	241
$CoH(N_2)(PPh_3)_3$	N_2	1.80(3)	127
$[Co(NO_2)(NH_3)_5]^{2+}$	NO_2^-	1.92(2)	251
$[Co(N_3)(NH_3)_5]^{2+}$	N_3^-	1.943(5)	252
$[Co(en)_3]^{3+}$	$NH_2—CH_2—CH_2—NH_2$	1.964(4)	220
$[Co(NH_3)_6]^{3+}$	NH_3	1.968(11)	214

*The Re^V and Co^{III} radii are 1.38 and 1.22 Å, respectively (see references 215 and 216).

Examples of M—N bond lengths for various nitrogen ligands are listed in Table 2.12. The average M—N bond length for NO^+ ligands is 1.71 Å. This is significantly shorter than the formal single bond length of 1.97 Å in $[Co(NH_3)_6]^{3+}$ [214]. The nitrosyl N single bond radius is considerably longer than the nitrido and imido N bond radii, when the 0.16 Å difference in Co and Re metallic radii[215, 216] is taken into consideration (Table 2.12).

The NO ligand in Group I complexes shows little or no *trans* influence. For the octahedral $[M(CN)_5(NO)]^{n-}$ complexes, with M = Mn, Fe, and Mo, the M—C distances are not significantly different in each compound[217–219]. For M = Cr the M—C bond *trans* to the NO group is 2.075(14) Å compared with the *cis* M—C bond of 2.033(14) Å [220]; this difference of 0.042(20) is barely significant. In $[RuCl_5(NO)]^{2-}$ the Ru—Cl bond lengths are essentially the same[221].

2.3.2 Structures with 120 degree M—N—O angles

The characteristics of Group III nitrosyl complexes are strikingly different from those of Group I described above. Group III complexes have (i) M—N—O bond angles near 120 degrees, (ii) longer M—N bond lengths, and (iii) a large *trans* influence of the NO group. Here the nitrosyl ligand is acting as a σ electron acceptor (Lewis acid). Hence these Group III complexes are formally considered to contain NO^- ligands. A Lewis dot structure for NO^- coordinated to a transition metal would be:

M:N⋰Ö: (with lone-pair dots on N and O)

which is analogous to propene,

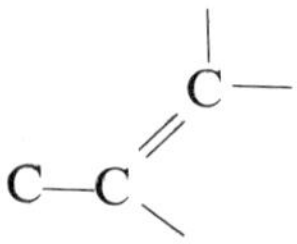

C—C=C (propene skeletal diagram)

An M—N—O angle of 120 degrees would be anticipated, since the N atom is sp^2 hybridised. As the electron dot structure indicates, the M—N bond is formally a single bond and the observed bond distances are longer than those of the Group I nitrosyls. However, the M—N bond still has some multiple bond character as shown in Table 2.12.

The *trans* influence of NO^- appears to be considerably greater than that of NO^+. As shown in Figure 2.19, in $[Co(NO)(NH_3)_5]^{2+}$ the Co—N bond of 2.220(4) Å, *trans* to the NO group, is 0.239(7) Å longer than the average of the *cis* Co—N bonds (1.981(6) Å) [222]. Also, in $[CoCl(NO)(H_2NCH_2CH_2NH_2)_2]^+$ the Co—Cl bond *trans* to the nitrosyl ligand is 0.31 Å longer than normally expected[223]. The different *trans* influencing abilities of NO^+ and NO^- are consistent with the suggestion, based on calculated overlap integrals[224], that strong σ-donor ligands and weak π-acceptor ligands (e.g., NO^-) will exert a *trans* bond weakening effect[223].

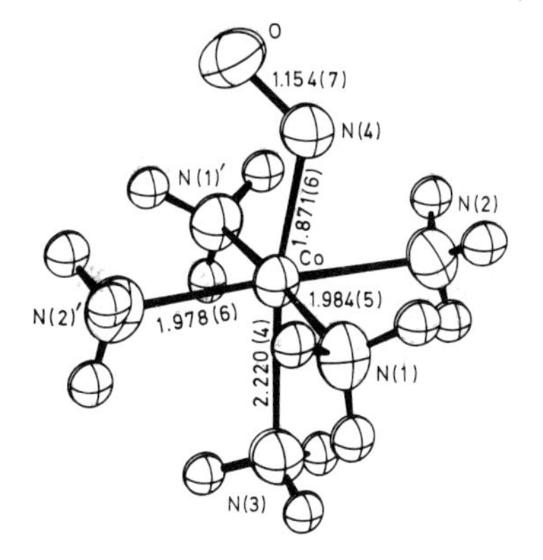

Figure 2.19 $[Co(NO)(NH_3)_5]^{2+}$
(From Pratt, C. S. *et al.*[222], by permission of The Chemical Society)

2.3.3 Structures with intermediate M—N—O angles

The complexes in Group II have M—N—O angles that are significantly different from 180 and 120 degrees. Several reasons have been given for the observed bond angles, but in general the subject is not well understood.

Kettle has shown[225] that for the $M(CO)_3$ group the two sets of $\pi^*(CO)$ orbitals transform independently. Thus there is no symmetry requirement

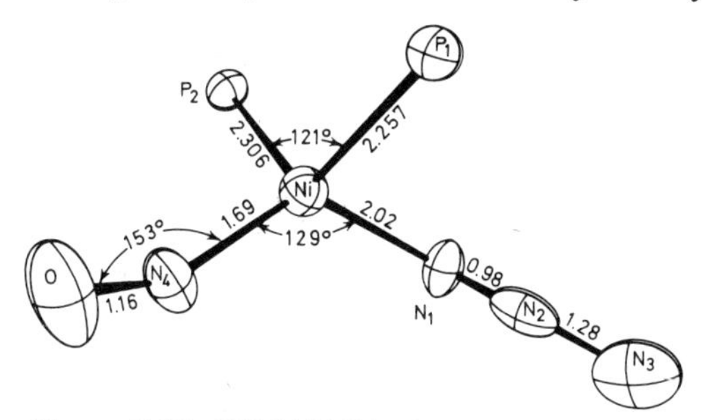

Figure 2.20 $Ni(N_3)(NO)(PPh_3)_2$
(Reproduced by permission of Professor J. H. Enemark[229], The University of Arizona, Tucson, Arizona, USA)

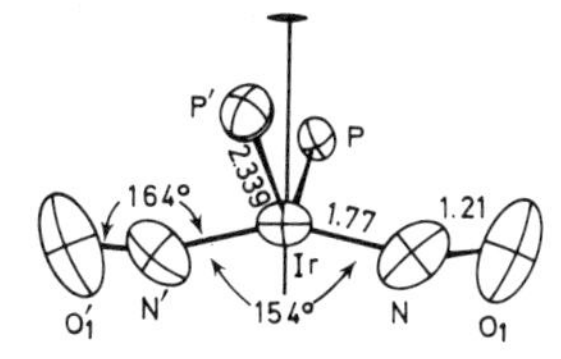

Figure 2.21 $Ir(NO)_2(PPh_3)_2{}^+$
(From Mingos, D. M. P. and Ibers, J. A.[230], by permission of The American Chemical Society)

that they interact equally with metal orbitals. A difference in the occupation of the two π^* orbitals will lead to a non-linear M—C—O bond angle and a deviation of ≈ 5 degrees from linearity often is observed for these angles. A similar argument has been applied to non-linear M—N—O linkages in $CrCl(NO)_2(\pi\text{-}C_5H_5)$ [226] and $[Fe(NO)_2I]_2$ [227]. Steric effects have also been suggested to be responsible for the observed distortions[227, 228]. Recently[229] Enemark has begun an investigation of 4-coordinate metal nitrosyls and has applied Kettle's arguments to these complexes. Both $Ni(N_3)(NO)(PPh_3)_2$ [229] and $[Ir(NO)_2(PPh_3)_2]^+$ [230] exhibit grossly distorted tetrahedral structures as shown in Figures 2.20 and 2.21. More structural work on 4-coordinate nitrosyl complexes will be required before a clear understanding of their geometries is obtained.

2.3.4 Other bonding theories

The bonding of the nitrosyl group to transition metals has been described in terms of molecular orbital theory by many authors[93, 130, 231, 232]. In a discussion[93, 130] of iridium complexes the versatility of the NO group has been associated with the relative energies of the nitrosyl $\pi^*(NO)$ orbitals and the metal d_{z^2} orbital. For complexes with linear M—N—O linkages the electronic configuration of lowest energy is $(d_{xy})^2(d_{xz})^2(d_{yz})^2(\pi^*(NO))^0(d_{z^2})^2$ (formally Ir^I and NO^+). When the electronic configuration of lowest energy is $(d_{xy})^2(d_{xz})^2(d_{yz})^2(\pi^*(NO))^2$ then the nitrosyl ligand will coordinate with an M—N—O angle of ≈ 120 degrees (formally Ir^{III} and NO^-). Mingos has pointed out[233] that in these complexes the direct product of the ground and first excited states will have $\pi \times \sigma = \pi$ symmetry. Thus the π-bending mode is symmetry allowed and gives rise to a bent M—N—O bond. Although a bond angle of 120 degrees is not specified by symmetry arguments the experimental results seem to indicate a sharp separation of 180 and 120 degree linkages in the Ir system (compare $[IrH(NO)(PPh_3)_3]^+$ with the Ir compounds in Group III of Table 2.11).

2.3.5 5-Coordinate nitrosyl complexes

Most of the recent structural work on complexes with bent M—N—O linkages has been on 5-coordinate compounds; these compounds are discussed here in a separate section and are compared with other 5-coordinate nitrosyl structures.

As reviewed in Section 2.2, 5-coordinate transition metal complexes exhibit a diverse range of structures; the nitrosyl complexes clearly illustrate this diversity. All five of the Group III 5-coordinate complexes are tetragonal-pyramidal with the NO group in the axial position. That this establishes a general rule is open to question since four of the five complexes are closely related iridium systems and would be expected to have similar structures (Figure 2.22). Bond distances and angles for some of these structures are compared in Table 2.8.

A tetragonal-pyramid was found[92] for $[RuCl(NO)_2(PPh_3)_2]^+$. The two NO ligands are in axial and equatorial positions as shown in Figure 2.23. The Ru—N—O linkages are strikingly different; bond angles are 179.5(18) and 136.0(16) degrees indicating a formal NO^+ type ligand in the basal position and a formal NO^- type ligand at the apex. A similar structure was found[91] for $[Os(OH)(NO)_2(PPh_3)_2]^+$. However the apical Os—N—O bond

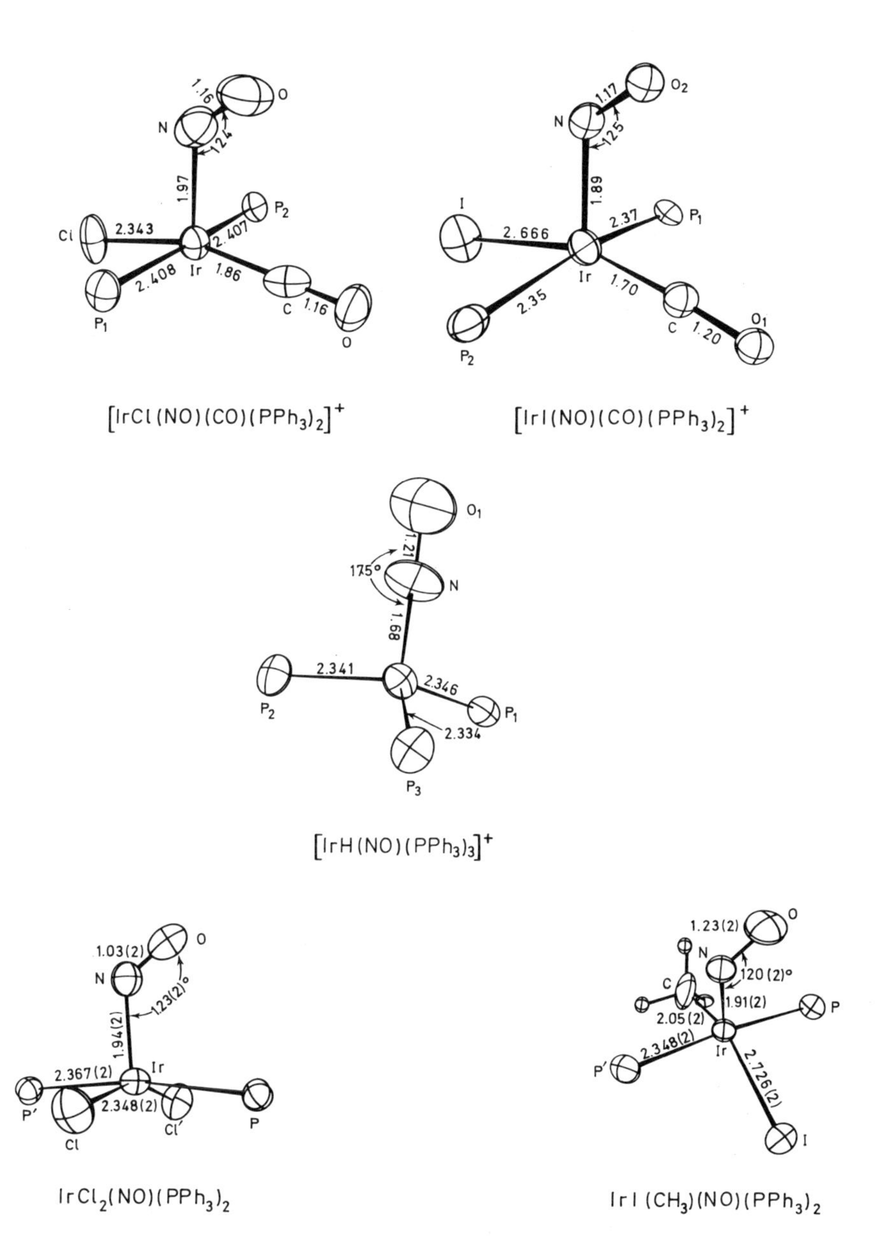

Figure 2.22 Iridium nitrosyl structures
(From Hodgson, D. J. *et al.*[45], Hodgson, D. J. and Ibers, J. A.[90], Mingos, D. M. P. and Ibers, J. A.[93, 130], Mingos, D. M. P. *et al.*[94], by permission of The American Chemical Society)

angle of 127.5(20) degrees is closer than the Ru complex to the 120 degree angles found in the other Group III complexes. The Os atom lies 0.30 Å above the base of the tetragonal pyramid.

5-coordinate nitrosyl complexes of Group I exhibit a wide variety of geometries. The structures of $Fe(NO)(S_2CNR_2)_2$, R = Me [102] and Et [103], and $Fe(NO)[S_2C_2(CN)_2]_2$ [253] are square-pyramidal with an axial NO group. The

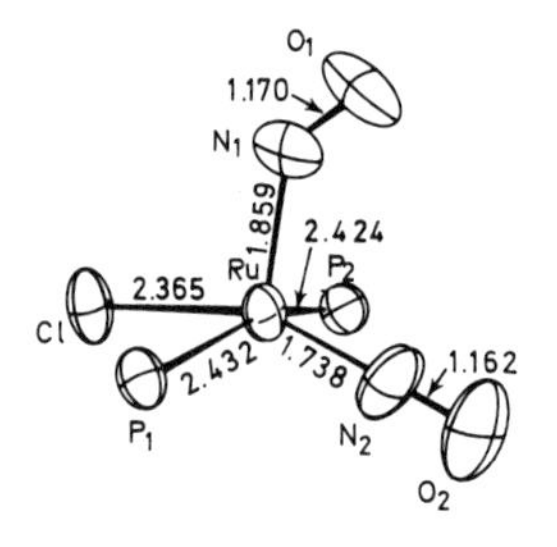

Figure 2.23 $[RuCl(NO)_2(PPh_3)_2]^+$
(From Pierpont, C. G. *et al.*[92], by permission of The American Chemical Society)

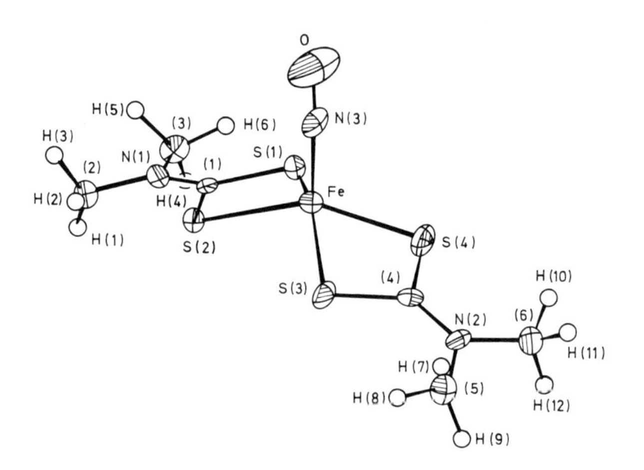

Figure 2.24 $Fe(NO)(S_2CNMe_2)_2$ at −80 °C
(From Davies, G. R. *et al.*[102], by permission of The Chemical Society)

methyl structure was studied at +20 and −80 °C and is shown in Figure 2.24. In $[IrH(NO)(PPh_3)_3]^+$ (Figure 2.22)[130], $RuH(NO)(PPh_3)_3$ [129], and $[Ru(NO)(Ph_2PC_2H_4PPh_2)_2]^+$ [129] the geometry is trigonal-bipyramidal with the NO ligand in an axial position.

Finally, in the series of compounds $Mn(NO)(CO)_{4-x}(PPh_3)_x$, $x = 0,1,2$, the geometry is trigonal-bipyramidal with an equatorial nitrosyl group (Figure 2.25)[143–145]. In this series the Mn remains in the same formal oxidation state while the ligands are systematically varied. A comparison of interatomic

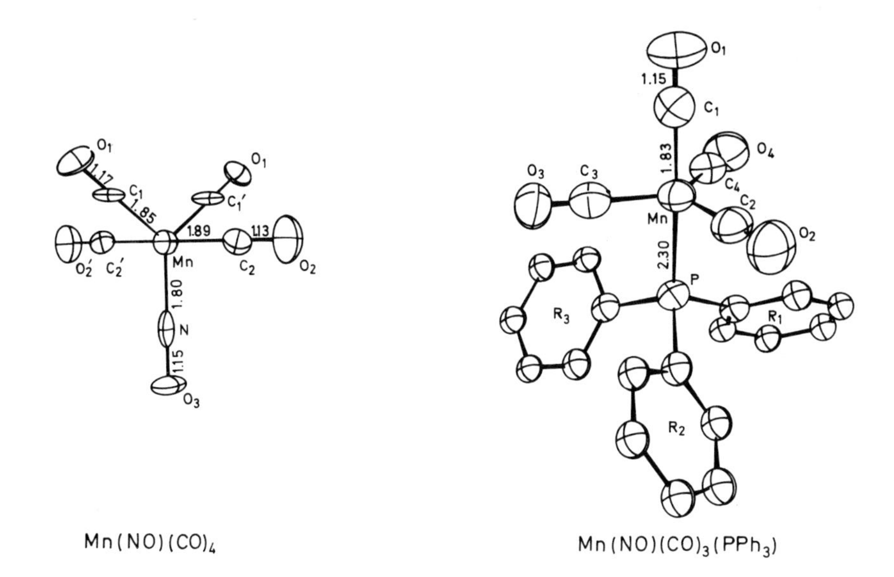

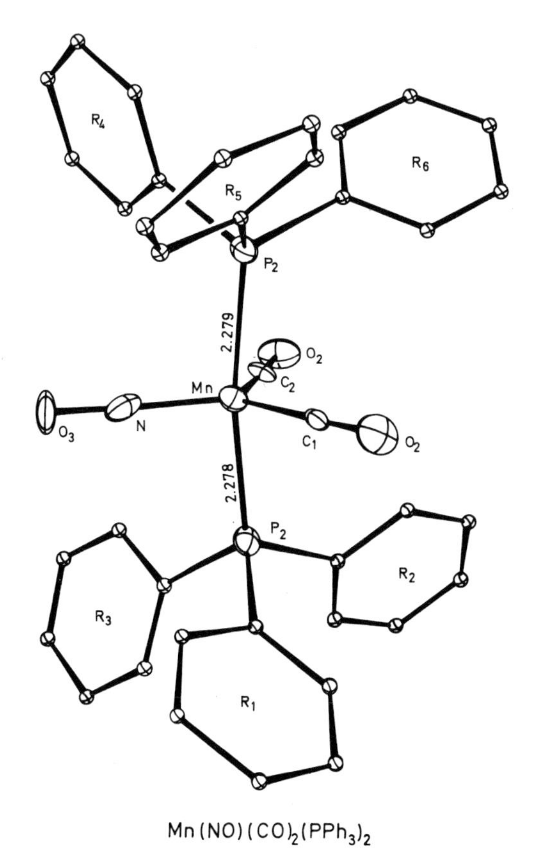

Figure 2.25 $Mn(NO)(CO)_{4-x}(PPh_3)_x$, $x = 0,1,2$
(From Enemark, J. H. *et al.*[143–145], by permission of The American Chemical Society)

Table 2.13 Comparison of interatomic distances (Å) for the series $Mn(NO)(CO)_{4-x}(PPh_3)_x$

(From Enemark, J.H. *et al.*[143–145] by permission of the American Chemical Society)

Atoms	$Mn(NO)(CO)_4$	$Mn(NO)(CO)_3$-(PPh_3)	$Mn(NO)(CO)_2$-($PPh_3)_2$
Mn—C(apical)	1.886(8)	1.833(11)	—
Mn—C(basal)	1.851(8)	1.798(10)*	1.78(2)
		1.782(11)*	1.75(2)
		1.756(10)*	
Mn—N	1.797(13)		1.73(1)
Mn—P	—	2.305(4)	2.278(5)
			2.279(5)

*Refined as $\frac{2}{3}$ C and $\frac{1}{3}$ N.

distances is given in Table 2.13. Substitution of a CO ligand by a phosphine group causes a significant decrease in Mn—N and Mn—C bond distances, indicating that PPh_3 is a poorer π-bonding ligand than NO and CO.

2.3.6 Bridging nitrosyl complexes

Two structures not included in Table 2.11 are ones that contain bridging nitrosyl groups. The structure[209] of $Mn_3(NO)_4(\pi\text{-}C_5H_5)_3$ is a triangle of manganese atoms with three doubly bridging NO groups and a triply bridging nitrosyl group; numerical details on the structure were not provided. Recently the structure of $Ru_3(NO)_2(CO)_{10}$ was determined[208] and is shown

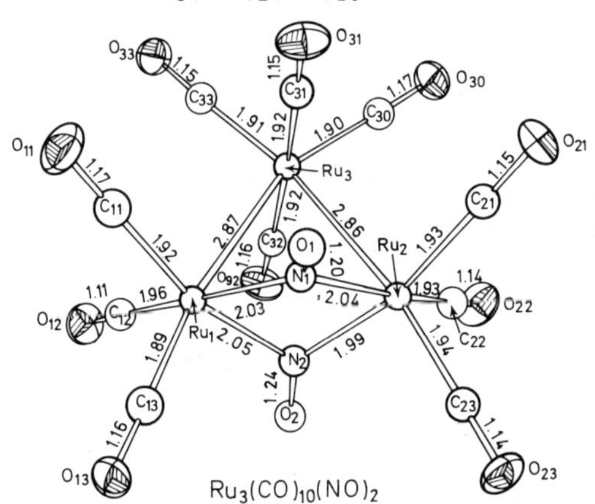

Figure 2.26 $Ru_3(NO)_2(CO)_{10}$
(Reproduced by permission of Dr. W. T. Robinson[208], The University of Canterbury, Christchurch, New Zealand)

In Figure 2.26. The triangle of Ru atoms is isosceles with the nitrosyl bridged Ru—Ru bond 0.28 Å longer than the non-bridged metal–metal bonds. This bond lengthening upon nitrosyl bridging is similar to the lengthening upon hydride bridging but contrasts with the bond shortening observed in carbonyl-bridged bonds[35].

2.4 SUMMARY

Almost 200 crystal structures of monomeric 5-coordinate transition metal complexes and 60 structures of transition metal nitrosyl complexes have been reported to date. Although the approximate structures of many of these could have been predicted from other techniques and from theory, many others of these have depended on diffraction studies to elucidate their detailed structural properties. With the large amount of empirical evidence that now exists on 5-coordinate structures and nitrosyl structures it is possible to distinguish trends and draw analogies to unknown structures. However, the structures of many complexes still remain elusive and future diffraction studies are certain to reveal new and interesting molecular geometries. The existing literature remains as a challenge to the chemist to develop a sound and all-encompassing theory of chemical bonding to account for the diverse structural properties of 5-coordinate transition metal complexes and transition metal nitrosyl complexes.

Acknowledgement

We express our gratitude to the many authors who have submitted short, incomplete communications of such vital interest that they need not be followed by complete papers. This work was supported by the Advanced Research Projects Agency of the Department of Defense through the Northwestern University Materials Research Centre.

Note added in proof

Subsequently it has been reported that the Br and Cl atoms are disordered in $[CuBr_3Cl_2]^{3-}$. See Ref. 260.

References

1. Cross, R. J. and Winfield, J. M. (1970). *Ann. Rep. Progr. Chem.*, **66A,** Chap. 14
2. Gillard, R. D. and Mitchell, P. R. (1970). *Struct. Bonding*, **7,** 46
3. Mason, S. F. (1970). *Pure Appl. Chem.*, **24,** 335
4. Mills, O. S. (1969). ibid., **20,** 117
5. McGinnety, J. A. and Mays, M. J. (1969). *Ann. Rep. Progr. Chem.*, **65A,** Chap. 13; McGinnety, J. A. and Mays, M. J. (1970). ibid., **66A,** Chap. 15
6. Kotz, J. C. and Pedrotty, D. G. (1969). *Organometal. Chem. Rev. Sect. A*, **4,** 479
7. Muetterties, E. L. (1970). *Accounts Chem. Res.*, **3,** 266
8. Pratt, J. M. and Thorp, R. G. (1969). *Advan. Inorg. Chem. Radiochem.*, **12,** 375
9. Ciampolini, M. (1969). *Struct. Bonding*, **6,** 52
10. Orioli, P. L. (1971). *Coord. Chem. Rev.*, **6,** to be published
11. Bruce, M. I. (1969). *Organometal. Chem. Rev. Sect. B*, **5,** 351
12. Spiro, T. G. (1970). *Progr. Inorg. Chem.*, **11,** 1
13. McCleverty, J. (1969). *Organometal. Chem. Rev. Sect. B*, **5,** 419
14. Kang, J. W. and Maitlis, P. M. (1969). ibid., **5,** 475
15. Powell, J. (1969). ibid., **5,** 507
16. Amma, E. L. (1971). *Advan. Chem. Ser.*, **98,** 120
17. Thayer, J. S. (1970). *Organometal. Chem. Rev. Sect. A*, **5,** 53
18. Belluco, U., Crociani, B., Pietropaolo, R. and Uguagliati, P. (1969). *Inorg. Chim. Acta Rev.*, **3,** 19
19. Sykes, A. G. and Weil, J. A. (1970). *Progr. Inorg. Chem.*, **13,** 1
20. Hathaway, B. J. and Billing, D. E. (1970). *Coord. Chem. Rev.*, **5,** 143
21. Hathaway, B. J. and Tomlinson, A. A. G. (1970). ibid., **5,** 1
22. Griffith, W. P. (1970). ibid., **5,** 459
23. Quinn, H. W. and Tsai, J. H. (1969). *Advan. Inorg. Chem. Radiochem.*, **12,** 217

24. Nelson, J. H. and Jonassen, H. B. (1971). *Coord. Chem. Rev.*, **6,** 27
25. Churchill, M. R. (1970). *Progr. Inorg. Chem.*, **11,** 53
26. Johnston, R. D. (1970). *Advan. Inorg. Chem. Radiochem.*, **13,** 471
27. Churchill, M. R. (1970). *Perspect. Struct. Chem.*, **3,** 91
28. Biryukov, B. P. and Struchkov, Yu. T. (1970). *Russ. Chem. Rev.*, **39,** 789
29. Abel, E. W. and Stone, F. G. A. (1969). *Quart. Rev. Chem. Soc.*, **23,** 325
30. Chini, P. (1970). *Pure Appl. Chem.*, **23,** 489
31. Fischer, E. O. (1970). ibid., **24,** 407
32. Bottomley, F. (1970). *Quart. Rev. Chem. Soc.*, **24,** 617
33. Norbury, A. H. and Sinha, A. I. P. (1970). ibid., **24,** 69
34. Sinn, E. and Harris, C. M. (1969). *Coord. Chem. Rev.*, **4,** 391
35. Frenz, B. A. and Ibers, J. A. (1971). *Transition Metal Hydrides*, ed. by E. L. Muetterties, Chap. 3. (New York: Marcel Dekker)
36. James, B. D. (1970). *Record of Chemical Progress*, **31,** 199
37. James, B. D. (1971). *J. Chem. Educ.*, **48,** 176
38. West, B. O. (1969). *Record of Chemical Progress*, **30,** 249
39. Tapscott, R. E., Belford, R. L. and Paul, I. C. (1969). *Coord. Chem. Rev.*, **4,** 323
40. Graddon, D. P. (1969). ibid., **4,** 1
41. Thompson, D. W. (1971). *Struct. Bonding*, **9,** 27
42. Ibers, J. A. (1965). *Ann. Rev. Phys. Chem.*, **16,** 375
43. Muetterties, E. L. and Schunn, R. A. (1966). *Quart. Rev. Chem. Soc.*, **20,** 245
44. Sacconi, L. (1968). *Pure Appl. Chem.*, **17,** 95
45. Hodgson, D. J., Payne, N. C., McGinnety, J. A., Pearson, R. G. and Ibers, J. A. (1968). *J. Amer. Chem. Soc.*, **90,** 4486; Hodgson, D. J. and Ibers, J. A. (1968). *Inorg. Chem.*, **7,** 2345
46. Raymond, K. N., Meek, D. W. and Ibers, J. A. (1968). ibid., **7,** 1111
47. Long, T. V., Herlinger, A. W., Epstein, E. F. and Bernal, I. (1970). *Inorg. Chem.*, **9,** 459
48. Riedel, E. F. and Jacobson, R. A. (1970). *Inorg. Chim. Acta*, **4,** 407
49. Drummond, J. and Wood, J. S. (1969). *Chem. Commun.*, 1373
50. Coyle, B. A. and Ibers, J. A. (1970). *Inorg. Chem.*, **9,** 767
51. Heath, C. and Hursthouse, M. B. (1971). *Chem. Commun.*, 143
52. Raymond, K. N. and Jurnak, F. A. (1970). *American Crystallographic Association Meeting*, New Orleans, 65
53. Terzis, A., Raymond, K. N. and Spiro, T. G. (1970). *Inorg. Chem.*, **9,** 2415
54. Raymond, K. N., Corfield, P. W. R. and Ibers, J. A. (1968). *Inorg. Chem.*, **7,** 1362
55. Bernal, I., Elliott, N. and Lalancette, R. A. (1968). *XIth Intern. Conf. Coord. Chem.*, Haifa, Israel, 518; Bernal, I., Elliott, N. and Lalancette, R. A. (1971). *Chem. Commun.*, 803
56. Hunter, S. H., Emerson, K. and Rodley, G. A. (1969). ibid., 1398
57. Gillespie, R. J. (1963). *J. Chem. Soc.*, 4672
58. Zeeman, J. (1963). *Z. anorg. allg. Chem.*, **324,** 241
59. Gillespie, R. J. (1963). *J. Chem. Soc.*, 4679
60. Hansen, K. W. and Bartell, L. S. (1965). *Inorg. Chem.*, **4,** 1775
61. Cotton, F. A., Dunne, T. G. and Wood, J. S. (1965). ibid., **4,** 318
62. Hanson, A. W. (1962). *Acta Crystallogr.*, **18,** 502
63. Almenningen, A., Haaland, A. and Wahl, K. (1968). Quoted in ref. 64
64. Beagley, B., Cruickshank, D. W. J., Pinder, P. M., Robiette, A. G. and Sheldrick, G. M. (1969). *Acta Crystallogr.*, **B25,** 737
65. Frenz, B. A. and Ibers, J. A. (1971). *American Crystallographic Association Meeting*, Ames, Iowa, 77; Frenz, B. A. and Ibers, J. A. (1972). *Inorg. Chem.*, in press.
66. Drew, M. G. B. and Mandyczewsky, R. (1970). *J. Chem. Soc. A*, 2815
67. Hess, H. and Hartung, H. (1966). *Z. Anorg. Allg. Chem.*, **344,** 157
68. Coetzer, J. (1970). *Acta Crystallogr.*, **B26,** 872
69. Drake, J. E., Vekris, J. and Wood, J. S. (1968). *J. Chem. Soc. A*, 1000
70. Russ, B. J. and Wood, J. S. (1966). *Chem. Commun.*, 745
71. Shiro, M. and Fernando, Q. (1971). ibid., 63
72. Bruins, D. and Weaver, D. L. (1970). *Inorg. Chem.*, **9,** 130
73. Doedens, R. J. and Ibers, J. A. (1967). *Inorg. Chem.*, **6,** 204
74. Greene, P. T. and Orioli, P. L. (1969). *J. Chem. Soc. A*, 1621
75. Corfield, P. W. R., Doedens, R. J. and Ibers, J. A. (1967). *Inorg. Chem.*, **6,** 197

76. Fowles, G. W. A., Greene, P. T. and Wood, J. S. (1967). *Chem. Commun.*, 971
77. Churchill, M. R. and Fennessey, J. P. (1967). *Inorg. Chem.*, **6,** 1213
78. Bennett, M. J. and Mason, R. (1963). *Proc. Chem. Soc., London,* 273; Bennett, M. J. (1965). Ph.D. Thesis, Sheffield University
79. Chaiwasie, S. and Fenn, R. H. (1968). *Acta Crystallogr.*, **B24,** 525
80. Wilford, J. B. and Powell, H. M. (1969). *J. Chem. Soc. A,* 8
81. Churchill, M. R. and Fennessey, J. P. (1968). *Inorg. Chem.*, **7,** 953
82. Bush, M. A., Hardy, A. D. U., Manojlović-Muir, Lj. and Sim, G. A. (1971). *J. Chem. Soc. A,* 1003
83. McPhail, A. T., Knox, G. R., Robertson, C. G. and Sim, G. A. (1971). ibid., 205
84. Manojlović-Muir, Lj., Muir, K. W. and Ibers, J. A. (1970). *Inorg. Chem.*, **9,** 447
85. Epstein, E. F., Bernal, I. and Balch, A. L. (1970). *Chem. Commun.*, 136; *Chem. Eng. News* (Jan. 5, 1970). 30
86. Davies, J. E. and Gatehouse, B. M. (1970). *Chem. Commun.*, 1166
87. Lindley, P. F. and Smith, A. W. (1970). ibid., 1355
88. Hoskins, B. F. and White, A. H. (1970). *J. Chem. Soc. A,* 1668
89. Di Vaira, M. and Orioli, P. L. (1967). *Inorg. Chem.*, **6,** 955; (1968). *Acta Crystallogr.*, **B24,** 595; (1968). ibid., **24,** 1269; Ciampolini, M., Di Vaira, M. and Orioli, P. L. (1968). *XIth Intern. Conf. Coord. Chem.*, Haifa, Israel, 217
90. Hodgson, D. J. and Ibers, J. A. (1969). *Inorg. Chem.*, **8,** 1282
91. Waters, J. M. and Whittle, K. R. (1971). *Chem. Commun.*, 518
92. Pierpont, C. G., Van Derveer, D. G., Durland, W. and Eisenberg, R. (1970). *J. Amer. Chem. Soc.*, **92,** 4760
93. Mingos, D. M. P. and Ibers, J. A. (1971). *Inorg. Chem.*, **10,** 1035
94. Mingos, D. M. P., Robinson, W. T. and Ibers, J. A. (1971). ibid., **10,** 1043
95. Masters, C., McDonald, W. S., Raper, G. and Shaw, B. L. (1971). *Chem. Commun.*, 210
96. Troughton, P. G. H. and Skapski, A. C. (1968). ibid., 575; see also ref. 94
97. La Placa, S. J. and Ibers, J. A. (1965). *Inorg. Chem.*, **4,** 778
98. Skapski, A. C. and Troughton, P. G. H. (1968). *Chem. Commun.*, 1230
99. Muir, K. W. and Ibers, J. A. (1970). *Inorg. Chem.*, **9,** 440
100. Meek, D. W., Alyea, E. C., Stalick, J. K. and Ibers, J. A. (1969). *J. Amer. Chem. Soc.*, **91,** 4920; Stalick, J. K. and Ibers, J. A. (1970). *Inorg. Chem.*, **9,** 453
101. Bertrand, J. A. and Plymale, D. L. (1966). ibid., **5,** 879
102. Davies, G. R., Mais, R. H. B. and Owston, P. G. (1968). *Chem. Commun.*, 81; Davies, G. R., Jarvis, J. A. J., Kilbourn, B. T., Mais, R. H. B. and Owston, P. G. (1970). *J. Chem. Soc. A,* 1275
103. Colapietro, M., Domenicano, A., Scaramuzza, L., Vaciago, A. and Zambonelli, L. (1967). *Chem. Commun.*, 583
104. Dapporto, P., Di Vaira, M. and Sacconi, L. (1969). ibid., 153
105. Orioli, P. L. and Sacconi, L. (1969). ibid., 1012
106. Di Vaira, M. and Orioli, P. L. (1967). *Inorg. Chem.*, **6,** 955
107. Blundell, T. L., Powell, H. M. and Venanzi, L. M. (1967). *Chem. Commun.*, 763
108. Dori, Z., Eisenberg, R. and Gray, H. B. (1967). *Inorg. Chem.*, **6,** 483
109. Di Vaira, M. and Orioli, P. L. (1969). ibid., **8,** 2729
110. Gerloch, M. (1966). *J. Chem. Soc.*, 1317
111. Goldfield, S. and Raymond, K. N. (1971). Private communication
112. Kilbourn, B. T., Raeburn, U. A. and Thompson, D. T. (1969). *J. Chem. Soc. A,* 1906
113. Bright, D. and Mills, O. S. (1966). *Chem. Commun.*, 211
114. Luxmoore, A. R. and Truter, M. R. (1962). *Acta Crystallogr.*, **15,** 1117
115. Corradini, P., Pedone, C. and Sirigu, A. (1966). *Chem. Commun.*, 341; Pedone, C. and Sirigu, A. (1967). *Acta Crystallogr.*, **23,** 759
116. Robiette, A. G., Sheldrick, G. M., Simpson, R. N. F., Aylett, B. J. and Campbell, J. A. (1968). *J. Organometal. Chem.*, **14,** 279
117. Emerson, K., Ireland, P. R. and Robinson, W. T. (1970). *Inorg. Chem.*, **9,** 436
118. Robinson, W. T. and Ibers, J. A. (1967). ibid., **6,** 1208
119. Blundell, T. L. and Powell, H. M. (1971). *J. Chem. Soc. A,* 1685
120. Dahl, L. F. (1957). *J. Chem. Phys.*, **26,** 1750
121. Wilford, J. B. and Powell, H. M. (1967). *J. Chem. Soc. A,* 2092
122. Stalick, J. K. and Ibers, J. A. (1970). *J. Organometal. Chem.*, **22,** 213
123. Titus, D. D., Orio, A. A., Marsh, R. E. and Gray, H. B. (1971). *Chem. Commun.*, 322

124. Frenz, B. A. and Ibers, J. A. (1970). *Inorg. Chem.*, **9,** 2403
125. Baker, R. W. and Pauling, P. (1969). *Chem. Commun.*, 1495
126. Baker, R. W., Ilmaier, B., Pauling, P. J. and Nyholm, R. S. (1970). ibid., 1077
127. Davis, B. R., Payne, N. C. and Ibers, J. A. (1969). *Inorg. Chem.*, **8,** 2719
128. La Placa, S. J. and Ibers, J. A. (1965). *Acta Crystallogr.*, **18,** 511
129. Pierpont, C. G., Pucci, A. and Eisenberg, R. (1971). *J. Amer. Chem. Soc.*, **93,** 3050
130. Mingos, D. M. P. and Ibers, J. A. (1971). *Inorg. Chem.*, **10,** 1479
131. Stalick, J. K. and Ibers, J. A. (1969). ibid., **8,** 1084
132. Stalick, J. K. and Ibers, J. A. (1969). ibid., **8,** 1090
133. Powell, H. M., Watkin, D. J. and Wilford, J. B. (1971). *J. Chem. Soc. A*, 1803
134. Stalick, J. K. and Ibers, J. A. (1969). *Inorg. Chem.*, **8,** 419
135. Payne, N. C. and Ibers, J. A. (1969). ibid., **8,** 2714
136. Ciechanowicz, M., Skapski, A. C. and Troughton, P. G. H. (1969). *VIIIth International Congress of Crystallography*, Stony Brook, New York, s172, or *Acta Crystallogr.*, **22,** s172
137. Muir, K. W. and Ibers, J. A. (1969). *J. Organometal. Chem.*, **18,** 175
138. Manojlović-Muir, Lj., Muir, K. W. and Ibers, J. A. (1969). *Discuss. Faraday Soc.*, 84
139. McGinnety, J. A. and Ibers, J. A. (1968). *Chem. Commun.*, 235
140. Ricci, J. S., Ibers, J. A., Fraser, M. S. and Baddley, W. H. (1970). *J. Amer. Chem. Soc.*, **92,** 3489; Ricci, J. S. and Ibers, J. A. (1971). ibid., **93,** 2391
141. La Placa, S. J. and Ibers, J. A. (1965). ibid., **87,** 2581
142. La Placa, S. J. and Ibers, J. A. (1966). *Inorg. Chem.*, **5,** 405
143. Enemark, J. H. and Ibers, J. A. (1967). ibid., **6,** 1575
144. Enemark, J. H. and Ibers, J. A. (1968). ibid., **7,** 2339
145. Frenz, B. A., Enemark, J. H. and Ibers, J. A. (1969). ibid., **8,** 1288
146. Ross, F. K. and Stucky, G. D. (1969). ibid., **8,** 2734
147. Raymond, K. N. (1969). *Chem. Commun.*, 1294
148. Huq, F. and Skapski, A. C. (1971). *J. Chem. Soc. A*, 1927
149. Goebel, C. V. and Doedens, R. J. (1970). *Chem. Commun.*, 839; (1971). *Inorg. Chem.*, **10,** 2607
150. Blundell, T. L. and Powell, H. M. (1971). *J. Chem. Soc. A*, 1685
151. Dapporto, P. and Di Vaira, M. (1971). ibid., 1891
152. Epstein, E. F., Bernal, I. and Balch, A. L. (1970). *Chem. Commun.*, 136
153. Wilford, J. B. and Powell, H. M. (1969). *J. Chem. Soc. A*, 8
154. McDonald, W. S. and Raper, G., quoted in Deeming, A. J. and Shaw, B. L. (1970). ibid., 2705
155. Bonds, W. D. and Ibers, J. A. (1971). Unpublished results
156. Muir, K. W. and Ibers, J. A. (1969). *Inorg. Chem.*, **8,** 1921
157. Taylor, I. F., Jr., Weininger, M. S. and Amma, E. L. (1971). Quoted in reference 16
158. Spofford, W. A., III and Amma, E. L. (1971). Quoted in reference 16
159. Coucouvanis, D. (1970). *Progr. Inorg. Chem.*, **11,** 233
160. Larsen, E. M. (1970). *Advan. Inorg. Chem. Radiochem.*, **13,** 1
161. Tsibris, J. C. M. and Woody, R. W. (1970). *Coord. Chem. Rev.*, **5,** 417
162. Fleischer, E. B. (1970). *Accounts Chem. Res.*, **3,** 105
163. Alderman, P. R. H. and Owston, P. G. (1956). *Nature (London)*, **178,** 1071; Alderman, P. R. H., Owston, P. G. and Rowe, J. M. (1962). *J. Chem. Soc.*, 668
164. Shetty, P. S., Ballard, R. E. and Fernando, Q. (1969). *Chem. Commun.*, 718; Shetty, P. S. and Fernando, Q. (1970). *J. Amer. Chem. Soc.*, **92,** 3964
165. Mathew, M. and Palenik, G. S. (1969). ibid., **91,** 4923
166. McGinnety, J. A., Payne, N. C. and Ibers, J. A. (1969). ibid., **91,** 6301
167. Immirzi, A. and Allegra, G. (1969). *Acta Crystallogr.*, **B25,** 120
168. Mague, J. T. (1969). *J. Amer. Chem. Soc.*, **91,** 3983; Mague, J. T. (1970). *Inorg. Chem.*, **9,** 1610
169. McGinnety, J. A. and Ibers, J. A. (1968). *Chem. Commun.*, 235
170. Bonds, W. D. and Ibers, J. A. (1971). *162nd American Chemical Society Meeting*, Washington, D. C., INOR 156; *J. Amer. Chem. Soc.*, in press
171. Meek, D. W. and Ibers, J. A. (1969). *Inorg. Chem.*, **8,** 1915
172. Rodgers, J. and Jacobson, R. A. (1970). *J. Chem. Soc. A*, 1826
173. Orioli, P. L. and Di Vaira, M. (1968). ibid., 2078
174. Mair, G. A., Powell, H. M. and Henn, D. E. (1960). *Proc. Chem. Soc. London*, 415
175. Orioli, P. L. and Sacconi, L. (1968). *Chem. Commun.*, 1311; Orioli, P. L. and Ghilardi, C. A. (1970). *J. Chem. Soc. A*, 1511

176. Di Vaira, M. and Sacconi, L. (1969). *Chem. Commun.*, 10; Di Vaira, M. (1971). *J. Chem. Soc. A*, 148
177. Bosnich, B., Nyholm, R. S., Pauling, P. J. and Tobe, M. L. (1968). *J. Amer. Chem. Soc.*, **90,** 4741
178. Orioli, P. L., Di Vaira, M. and Sacconi, L. (1965). ibid., **87,** 2059; Orioli, P. L., Di Vaira, M. and Sacconi, L. (1966). ibid., **88,** 4383
179. Stevenson, D. L. and Dahl, L. F. (1967). ibid., **89,** 3424
180. Bailey, N. A., Gibson, J. G. and McKenzie, E. D. (1969). *Chem. Commun.*, 741
181. Dapporto, P. and Sacconi, L. (1969). ibid., 1091; Dapporto, P. and Sacconi, L. (1970). *J. Chem. Soc. A*, 1804
182. Fleischer, E. B. and Hawkinson, S. W. (1968). *Inorg. Chem.*, **7,** 2312
183. Bianchi, A. and Ghilardi, C. A. (1971). *J. Chem. Soc. A*, 1096
184. Dapporto, P., Morassi, R. and Sacconi, L. (1970). ibid., 1298
185. Haugen, L. P. and Eisenberg, R. (1969). *Inorg. Chem.*, **8,** 1072
186. Legg, J. I., Nielson, D. O., Smith, D. L. and Larson, M. L. (1968). *J. Amer. Chem. Soc.*, **90,** 5030
187. Blundell, T. L. and Powell, H. M. (1967). *J. Chem. Soc. A*, 1650
188. Di Vaira, M., Orioli, P. L. and Sacconi, L. (1966). *Chem. Commun.*, 300; Di Vaira, M., Orioli, P. L. and Sacconi, L. (1971). *Inorg. Chem.*, **10,** 553
189. Seleborg, M., Holf, S. H. and Post, B. (1971). ibid., **10,** 1501
190. Mazurek, W., Phillip, A.T., Hoskins, B. F. and Whillans, F. D. (1970). *Chem. Commun.*, 184
191. Preston, H. S. and Kennard, C. H. L. (1969). *J. Chem. Soc. A*, 2955
192. Weeks, C. M., Cooper, A. and Norton, D. A. (1969). *Acta Crystallogr.*, **B25,** 443
193. van der Helm, D. and Franks, W. A. (1969). ibid., **25,** 451
194. Prout, C. K., Armstrong, R. A., Carruthers, J. R., Forest, J. G., Murray-Rust, P. and Rossotti, F. J. C. (1968). *J. Chem. Soc. A*, 2791
195. Musker, W. K. and Hussain, M. S. (1968). *XIth Intern. Conf. Coord. Chem.*, Haifa, Israel, 262
196. Hall, D., Sheat-Rumball, S. V. and Waters, T. N. (1968). *J. Chem. Soc. A*, 2721
197. Bonamico, M. and Dessy, G. (1970). *Chem. Commun.*, 1218
198. Cannas, M., Carta, G. and Marongiu, G. (1971). ibid., 673
199. Bell, J. D., Freeman, H. C., Wood, A. M., Driver, R. and Walker, W. R. (1969). ibid., 1441
200. Dori, Z. (1968). ibid., 714
201. Hoskins, B. F. and Whillans, F. D. (1970). *J. Chem. Soc. A*, 123
202. Bailey, N. A., McKenzie, E. D. and Mullins, J. R. (1970). *Chem. Commun.*, 1103
203. Marsh, W. C. and Trotter, J. (1971). *J. Chem. Soc. A*, 1482
204. Bentley, G. A., Waters, J. M. and Waters, T. N. (1968). *Chem. Commun.*, 988
205. Johnson, B. F. G. and McCleverty, J. A. (1966). *Prog. Inorg. Chem.*, **7,** 277
206. Bertinotti, F., Corradini, P., Diana, G., Ganis, P. and Pedone, C. (1963). *Ric. Sci. Rend A*, **3,** 210
207. Jagner, S. and Vannerberg, N.-G. (1967). *Acta Chem. Scand.*, **21,** 1183
208. Norton, J., Collman, J. P., Dolcetti, G. and Robinson, W. T. (1972). *Inorg. Chem.*, **11,** 382
209. Elder, R. C., Cotton, F. A. and Schunn, R. A. (1967). *J. Amer. Chem. Soc.*, **89,** 3645
210. Ward, D. L. and Caughlan, C. N. (1971). *American Crystallographic Association Meeting*, Columbia, S. Carolina, 62
211. Albano, V., Bellon, P., Ciani, G. and Manassero, M. (1969). *Chem. Commun.*, 1242
212. Brockway, L. O. and Anderson, J. S. (1937). *Trans. Faraday Soc.*, **33,** 1233
213. Cox, A. P., Thomas, L. F. and Sheridan, J. (1958). *Nature (London)*, **181,** 1157
214. Meek, D. W. and Ibers, J. A. (1970). *Inorg. Chem.*, **9,** 465
215. Cotton, F. A. and Lippard, S. J. (1966). ibid., **5,** 416
216. Pauling, L. (1967). *The Chemical Bond*, 149. (New York: Cornell University Press)
217. Tullberg, A. and Vannerberg, N.-G. (1966). *Acta Chem. Scand.*, **20,** 1180; Tullberg, A. and Vannerberg, N.-G. (1967). ibid., **21,** 1462
218. Manoharan, P. T. and Hamilton, W. C. (1963). *Inorg. Chem.*, **2,** 1043
219. Svedung, D. H. and Vannerberg, N.-G. (1968). *Acta Chem. Scand.*, **22,** 1551
220. Enemark, J. H., Quinby, M. S., Reed, L. L., Steuck, M. J. and Walters, K. K. (1970). *Inorg. Chem.*, **9,** 2397
221. Khodashova, T. S. and Bokii, G. B. (1960). *Zhur, Strukt. Khim.*, **1,** 151, or *J. Struct. Chem. USSR*, **1,** 138; see also ref. 257; Hodgson, D. J. (1971). Private communication

222. Pratt, C. S., Coyle, B. A. and Ibers, J. A. (1971). *J. Chem. Soc. A*, 2146
223. Snyder, D. A. and Weaver, D. L. (1969). *Chem. Commun.*, 1425; Snyder, D. A. and Weaver, D. L. (1970). *Inorg. Chem.*, **9**, 2760
224. McWeeny, R., Mason, R. and Towl, A. D. C. (1969). *Discuss. Faraday Soc.*, **47**, 20
225. Kettle, S. F. A. (1965). *Inorg. Chem.*, **4**, 1662
226. Carter, O. L., McPhail, A. T. and Sim, G. A. (1966). *J. Chem. Soc. A*, 1095
227. Dahl, L. F., Rodulfo de Gil, E. and Feltham, R. D. (1969). *J. Amer. Chem. Soc.*, **91**, 1653
228. McPhail, A. T. and Sim, G. A. (1968). *J. Chem. Soc. A*, 1858
229. Enemark, J. H. (1971). *Inorg. Chem.*, **10**, 1952
230. Mingos, D. M. P. and Ibers, J. A. (1970). *Inorg. Chem.*, **9**, 1105
231. Manoharan, P. T. and Gray, H. B. (1966). ibid., **5**, 823
232. Masek, J. (1969). *Inorg. Chim. Acta Rev.*, **3**, 99
233. Mingos, D. M. P. (1971). *Nature (London)*, **229**, 193
234. Eller, P. G. and Corfield, P. W. R. (1970). *American Crystallographic Association Meeting*, Ottawa, Canada, 85
235. Harrison, W. and Trotter, J. (1971). *J. Chem. Soc. A*, 1542
236. Chan, L. Y. Y. and Einstein, F. W. B. (1970). *Acta Crystallogr.*, **B26**, 1899
237. Johansson, G. and Lipscomb, W. N. (1958). *Acta Crystallogr.*, **11**, 594
238. Jagner, S. and Vannerburg, N.-G. (1968). *Acta Chem. Scand.*, **22**, 3330; Jagner, S. and Vannerburg, N.-G. (1970). ibid., **24**, 1988
239. Albano, V. G., Bellon, P. L. and Sansoni, M. (1970). Personal communication.
240. Thomas, J. T., Robertson, J. H. and Cox, E. G. (1958). *Acta Crystallogr.*, **11**, 599
241. Davis, B. R. and Ibers, J. A. (1971). *Inorg. Chem.*, **10**, 578
242. Domenicano, A., Vaciago, A., Zambonelli, L., Loader, P. L. and Venanzi, L. M. (1966). *Chem. Commun.*, 476
243. Brennan, T. F. and Bernal, I. (1970). ibid., 138
244. Simonsen, S. H. and Mueller, M. H. (1965). *J. Inorg. Nucl. Chem.*, **27**, 309
245. Calderon, J. L., Cotton, F. A. and Legzdins, P. (1969). *J. Amer. Chem. Soc.*, **91**, 2528
246. Carty, P., Walker, A., Mathew, M. and Palenik, G. J. (1969). *Chem. Commun.*, 1374
247. Bentley, G. A., Laing, K. R., Roper, W. R. and Waters, J. M. (1970). ibid., 998
248. Bush, M. A., Sim, G. A., Knox, G. R., Ahmad, M. and Robertson, C. G. (1969). ibid., 74
249. Strouse, C. E. and Swanson, B. I. (1971). ibid., 55
250. Papiev, N. A. and Bokii, G. B. (1957). *Zh. Neorg. Khim.*, **2**, 8, 1972, or *Russ. J. Inorg. Chem.*, **2**, 8, 414; Papiev, N. A. and Porai-Koshits, M. A. (1959). *Kristallografiya*, **4**, 30, or *Sov. Phys. Crystallogr.*, **4**, 26
251. Cotton, F. A. and Edwards, W. T. (1968). *Acta Crystallogr. Sect.*, **B24**, 474
252. Palenik, G. J. (1964). *Acta Crystallogr.*, **17**, 360
253. Rae, A. I. M. (1967). *Chem. Commun.*, 1245
254. Jarvis, J. A. J., Mais, R. H. B., Owston, P. G. and Taylor, K. A. (1966). ibid., 906
255. Bright, D. and Ibers, J. A. (1969). *Inorg. Chem.*, **8**, 703
256. Bright, D. and Ibers, J. A. (1968). ibid., **7**, 1099
257. Khodashova, T. S. (1963). *Zh. Strukt. Khim.*, **4**, 111; or *J. Struct. Chem. USSR*, **4**, 98; Khodashova, T. S. (1965). *Zh. Strukt. Khim.*, **6**, 716; or *J. Struct. Chem. USSR*, **6**, 678
258. Bokii, G. B. and Parpiev, N. A. (1957). *Kristallografiya*, **2**, 691; or *Sov. Phys. Crystallogr.*, **2**, 681; Parpiev, N. A. and Bokii, G. B. (1959). *Zh. Neorg. Khim.*, **4**, 2453; or *Russ. J. Inorg. Chem.*, **4**, 1127
259. Mori, M., Saito, Y. and Watanabe, T. (1961). *Bull. Chem. Soc. Jap.*, **34**, 295
260. Goldfield, S. A. and Raymond, K. N. (1971). *Inorg. Chem.*, **10**, 2604

3
Investigations of Crystal Structures in the U.S.S.R.

B. K. VAINSHTEIN
and
G. N. TISHCHENKO
Institute of Crystallography, Moscow, U.S.S.R.

3.1 INTRODUCTION

Investigation in the U.S.S.R. into the atomic structure of crystals is carried out in many directions, from the determination of the simple structures of metals and alloys to the investigation of proteins and viruses. Along with the basic methods of such investigations – x-ray-, electron- and neutron-diffraction – use is made of some other techniques, as for instance, resonance methods, optical spectroscopy, electron microscopy, etc. It would be certainly impossible to reflect the extensive, many-sided complex of such work in one review article. Therefore, in order to give the reader an idea of the development of structural investigations and crystal chemistry in the U.S.S.R. in recent years, we have concentrated our attention on some of the main original lines of development, and have made only brief mention of the others. For the same reason, we give references not only to the original works, but also to reviews, collected articles and books on related topics.

3.2 INVESTIGATIONS OF STRUCTURE OF SILICATES AND THEIR ANALOGUES

In recent years, a series of new investigations of silicates and their analogues has been carried out which has substantially broadened our knowledge of the structure of these principal minerals of the earth's crust. Most of these works have been performed by Belov and co-workers.

As is known, W. L. Bragg and his school investigated mainly the silicates, principally containing small cations such as Mg, Fe and Al. The main architectural detail of these structures is a silicon–oxygen tetrahedron $[SiO_4]$, whose edge is close in size to the edge of the octahedron occupied by some or another of the cations mentioned below (Figure 3.1).

On the other hand, the coordination polyhedron around large cations, e.g. Na, Ca, K, has the edges to which the diortho group $[Si_2O_7]$ appears to be better attached (Figure 3.1). From recent investigations it has become clear that the determining role in the structure of silicates is played not by the silicon-oxygen radical as such, but by the number and sort of cations in a given mineral. The architectural components $[SiO_4]$ and $[Si_2O_7)$ may be present in the appropriate structures both in the elementary form and as fragments of chains, ribbons, one- and two-storied rings, two-dimensional chainmails and elements of the three-dimensional framework.

The division of silicates into two groups by reference to the dominant cations in a given mineral proved very fruitful in practical mineralogy.

An example has been the analysis of the isomorphous replacement of ordinary cations by the cations of rare and scattered elements in silicates. The concept of the silicon–oxygen radical retains its significance in the

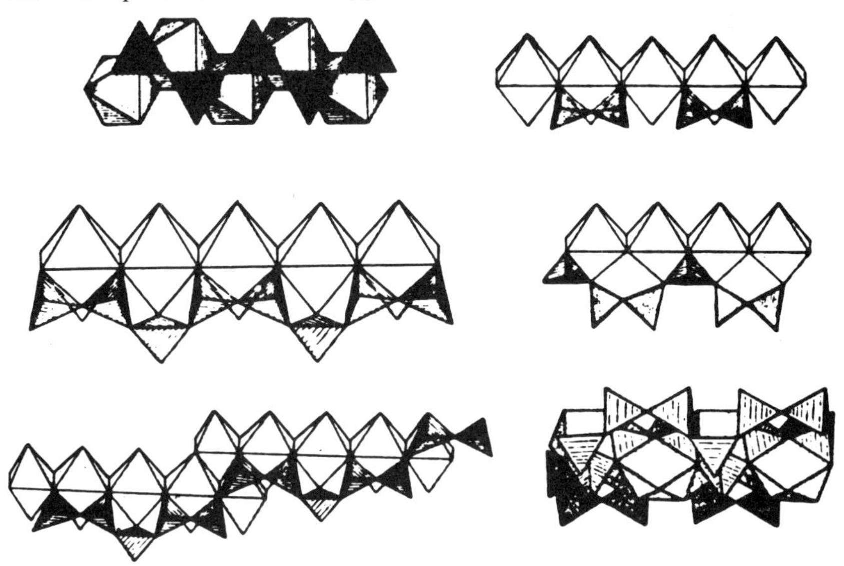

Figure 3.1 Fragments of the structures of silicates in which the SiO_4 tetrahedron is conjugated with octahedra, or the Si_2O_7 diortho-group is conjugated with octahedra containing large cations
(From Belov[1], by courtesy of Izd. Akad. Nauk SSSR)

formal classification of silicates. However, it is as if the passive role of silica in solution continues in a solid body as well, where the strong but flexible radical adjusts itself to the crystal chemical situation being dictated by cations[1–3].

3.2.1 New data on crystal chemistry of silicates

In recent years, a number of structures of silicates and their analogues with large alkali- and rare-earth cations as well as with average-sized cations (with partly-filled d and f shells[4–30]) has been determined. This has enabled us to elucidate some problems of the crystal chemistry of these elements and to establish the new laws governing isomorphous replacement.

The question of the role of titanium in natural minerals has always been the subject of a difference in opinion between mineralogists on the one hand, who saw tetravalent titanium as an isomorphous equivalent of silicon, like germanium and aluminium, and chemical crystallographers on the other, who, starting from a rather abundant experimental material, opted for 6-coordination. The situation was cleared up by determining the structure of bafertisite $BaFe_2TiO[Si_2O_7]$[4], the structure of Na_2TiSiO_5[5] and others. Bafertisite is an analogue of mica-like minerals, but the central octahedral layer is occupied by iron cations. This layer is coated above and below by a net consisting of the $[Si_2O_7]$ diortho-groups and Ti-octahedra. In bafertisite it was demonstrated for the first time that silicon and titanium play the same role in the structure, although titanium retains its usual sixfold coordination. Still more interesting results were obtained for lomonosovite $Na_5Ti_2(PO_4)[Si_2O_7]O_2$ in which half of the Ti atoms (the average Ti—O distance is 1.96 Å) is joined with silicon in a net similar to that of bafertisite, while the remaining Ti atoms (Ti—O = 2.04 Å) together with Na(Na—O = 2.51 Å) appears in the usual role of cations whose octahedra form the central layer of a mica-like packet[6]. The analogous division of titanium takes place in inyelite: Na_2Ba_3 (BaK)(Ca,Na)Ti$(TiO_2)[SO_4]_2[Si_2O_7]$ ($P1$, $a = 14.76$, $b = 7.14$, $c = 5.40$ Å, $\alpha = 90°$, $\gamma = 99°$, $\beta = 95°$, $Z = 1$)[7]. Here there are distinct walls made up of Ba-polyhedra which divide the cell into two unequal piers. Here we see the cation role of titanium: the chainmail (net) composed of $[Si_2O_7]$ diortho-groups and Ti-pyramids armour the walls.

The cationic role of Ti is also clearly seen in tinaksite (also triclinic silicate), $NaK_2Ca_2TiSi_7O_{19}(OH)$[8] ($P\bar{1}$, $a = 10.35$, $b = 12.17$, $c = 7.05$ Å, $\alpha = 91°$, $\beta = 99°$, $\gamma = 92° 30'$; $Z = 2$). In it a silicon–oxygen radical, Si_7O_{18} (OH), of previously unknown type was discovered. Ti plays the analogous role in tundrite—a rare-earth acentric mineral, Na_{3-y}, $TR_3(TR, Ca)Ti_2Si_2O_{14}(OH)_2{\cdot}8H_2O$[9] ($P\bar{1}$, $a = 7.57$, $b = 13.98$, $c = 5.03$ Å; $\alpha = 101° 31$) in vinogradowite[10], neptunite $Na_2FeTi[Si_4O_{12}]$[11].

The investigation of baotite, $Ba_4Ti_4(Ti, Nb)_4[Si_4O_{12}]O_{16}Cl$ ($a = 19.99$; $c = 5.908$ Å; $I4_1/a$; $Z = 4$; $R = 4\%$)[12] serves as an example to illustrate how an isomorphous Nb atom may enter into the mineral. The rutile-like infinite square columns made up of octahedra (Figure 3.2) constitute the

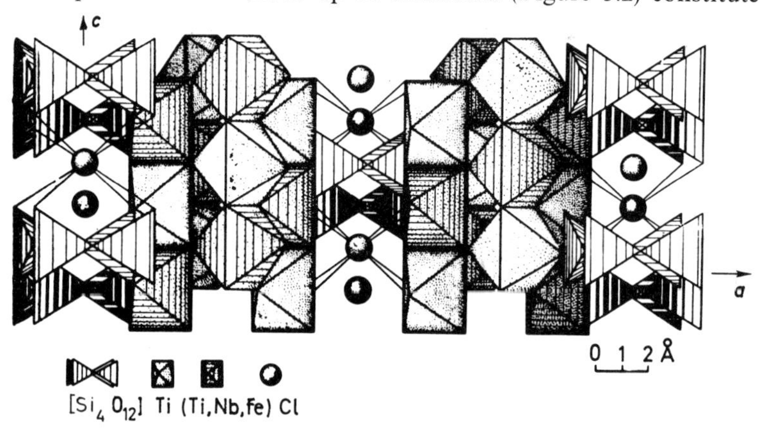

Figure 3.2 Lateral projection of the structure of baotite (From Simonov 1969). *Kristallografiya*, **14**, 602, by courtesy of Izd. Akad. Nauk SSSR)

main feature of the structure. The difference is that, in a rutile structure, all octahedra are equivalent, whereas in baotite the period is doubled and two independent octahedra are displaced relative to it. Due to the difference in the position of Ba atoms with respect to the octahedra the Nb admixture is concentrated in only one of these. The difference occurs from the fact that the nearest Ba atoms are situated from Ti_1 and Ti_2 atoms at distances of

3.51, 3.90, 3.82 and 4.14 Å, respectively, and that the replacement of Ti^{4+} by Nb^{5+} in the Ti_1 position is hindered by Ba cations to a greater extent than in the Ti_2 position in which the Nb is found to be concentrated. The analysis of baotite clearly reveals the high structural sensitivity of the process of isomorphous replacement in crystals.

Of interest are works on cement-forming phases. One of them, namely: phase Y = $Ca_6\ [Si_2O_7]\ [SiO_4](OH)_2$ [13], should be considered as the most traditional one. The phase Y – one of the most frequently occurring cement compounds – arises in a wide range of measurements on p and T, and is always found to be present in the cement stone.

Its triclinic structure can be represented as being composed of separate frameworks of three important cement minerals (Figure 3.3); two blocks are

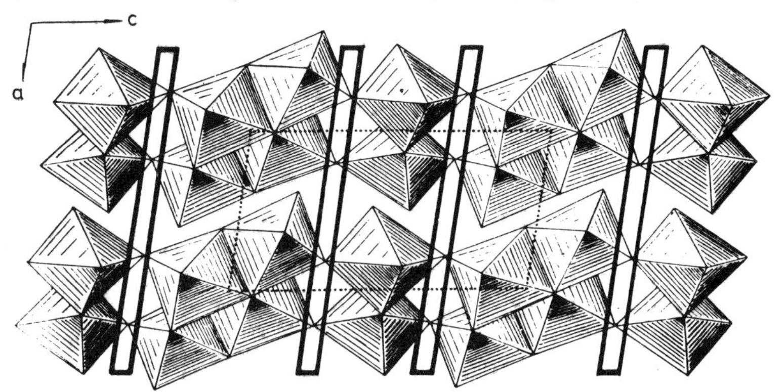

Figure 3.3 *xz*-Projection of the structure of the Y-phase in polyhedra. Portlandite walls and blocks of single and double Ca-ribbons are shown (From Ilyukhin, by courtesy of Izd. Akad. Nauk SSSR)

separated from each other by a portlandite wall (it is a 'trellis', not a solid). One block corresponds to the ribbon from Ca-polyhedra in $\beta(\gamma)-Ca_2SiO_4$, the other to the doubled cuspidine ribbon. Just this combination of the architectural elements from Ca octahedra of three principal cement minerals accounts for the wide range of the γ-phase stability. An addition of it in small amounts (up to 10–15%) to cement substantially improves the durability of concrete. The γ-phase is the first Ca silicate in which $[SiO_4]$ ortho-groups and $[Si_2O_7]$ diortho-groups are found to co-exist. Similarly, in Ca silicates triortho-groups $Si_3O_{10}(Na_2Ca_3Si_3O_{10})$ [14] were observed for the first time. The entry of a large Cl anion into the Ca environment (equally with oxygens) was discovered in $Ca_5[SiO_4]Cl_2$ [15]. Interesting new radicals were found in zeolite-like structures of eudialite[16], lomontite[17] and others. The structures with rare earth elements were studied[18, 19].

3.2.2 Investigations of layer silicates

In the Soviet Union considerable progress has been made into the structural investigation of layer silicates and clay minerals involving both x-ray and electron diffraction and electron microscopy[20, 21]. The use of high accelerating voltages (up to 400 kV)[22] has played an important role in increasing the

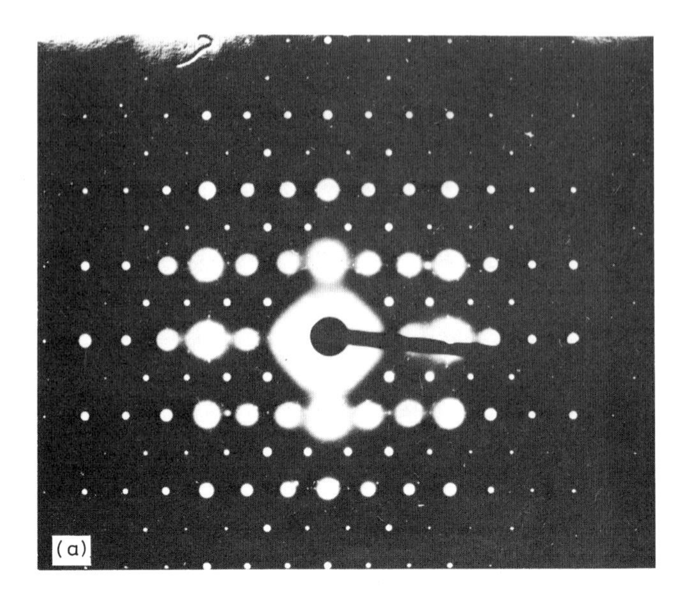

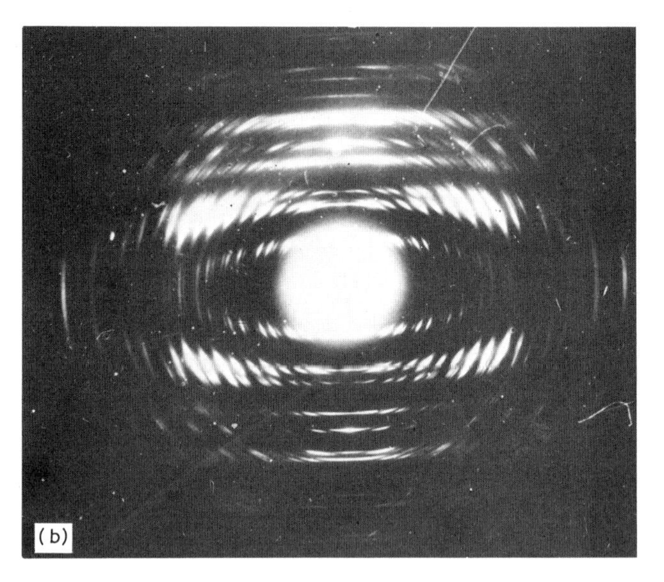

Figure 3.4 Electron diffraction patterns obtained under accelerating voltage of 400 kV: (a) pattern of the serpentine single crystal; (b) texture pattern of nakrite (the angle of inclination 73 degrees)
(From Zvyagin *et al.*[22], by courtesy of Izd. Akad. Nauk SSSR)

diffraction effectiveness. The diffraction patterns rich in reflections (Figure 3.4) illustrate the possibilities of high-voltage electron diffraction. The consideration of the layer structures and the derivation of their polytypes are based on the method in which the individual layers and their sequences are given in an analytical form. A system of conventional symbols is used, characterising the structures, orientation of layers and their mutual arrangement, symmetry, atomic coordinates and structure factors[20]. The polytype

Figure 3.5 Three-dimensional synthesis of the lattice potential of the dioctahedral Al mica
(From Zvyagin *et al.*[26], by courtesy of Izd. Akad. Nauk SSSR)

theory constitutes a basis for rapid identification of various modifications from experimental data.

Studies on polytyping of kaolin minerals resulted in the solution of complicated problems which had been the subject of conflicting opinions and inexact concepts. It thus became possible to throw light on the structural vagueness of nakrite, and to refine the azimuthal orientations and relative displacements of its layers[20]; it was established that the structure of halloysite can be approximated by a two-layer monoclinic structure distorted by some random displacements of the layers by $\pm b/3$ [23].

The existence of previously unknown rare modifications of serpentine-like minerals was established[24]; the structural nature of the main mineral of zinc clays was revealed, zinalcite, which turned out to be a Zn analogue of serpentines, being a mixture of semi-random modifications A, B[25].

The establishment of the polytype theory made it possible to substantially refine and consolidate the structures of pyrophyllites and talcs[26]. A systematic derivation of all the possible chlorite polytypes was carried out[27], and a number of structures of these polytypes were studied[28].

Interesting data were also obtained on polytyping of micas. An example of electron-diffraction analysis of such structures is that of dioctahedral Al mica[29]. Figure 3.5 shows the scheme for its formation constructed as a sequence of sections of the lattice potential. The detailed structural data established for this modification gave a clear insight into the interrelation of different mica polytypes. The possible modifications of pyrosmalite minerals (Mn silicates) with repeats of 1, 2 and 3 'pyrosmalite' layers were derived[30]. Various theoretically derived MoS_2 modifications were investigated experimentally, and some important conditions for their formation were established[31].

Summarising, the data of the precise determination of layer silicate structures permitted some general characteristics of the structure of these minerals[32] to be established. It was shown that the position of a tetrahedral cation depends not only on the degree of replacement of Si by Al tetrahedra, but also on the position and distribution of the compensating positive charge. Here, the displacement of tetrahedral cations from the geometrical centre of tetrahedra is accompanied by a change in the individual cation–anion interatomic distances, whereas the average distance is determined only by the composition of tetrahedra and can be calculated beforehand for tetrahedra of a known composition.

The individual anion–anion interatomic distances in tetrahedra of layer silicates vary within wide limits and depend on the position of a tetrahedral cation—its displacement from the centre. It turns out that the average length of the edges of each individual tetrahedron, l_{av}, is always connected with the average distance d_{av} between the cation and anion by the relationship

$$l_{av} = 2\sqrt{2/3}\, d_{av}$$

An analysis of the octahedral interatomic distances in layer silicate structures shows the shared edges of two octahedra are 0.1–0.3 shorter than the other edges of the octahedron. In all cases the average length of the edges of a given octahedron, l_{av}, satisfies the relationship

$$l_{av} = \sqrt{2}\, r_{av}$$

where r_{av} is the average cation–anion interatomic distance in a given octahedron. This relationship holds true for octahedra in any inorganic compound.

On the basis of the revealed regularities of the layer silicate structure, one can take into account beforehand the nature of the distortions of the octahedra and tetrahedra, and thus develop structural models close to real structures using the data on the content and size of the unit cells[32].

The nature of the disordered and ordered mixed-layer structures of clay minerals has been investigated[33].

The microdiffraction method has broken new ground in the field of structural investigations of minerals[21]. Thus, as a result of the analysis of the intensity of the reflections on microdiffraction patterns from micro-monocrystals, the structure of a new, rather peculiar, layer mineral tochilinite ($a = 5.37$, $b = 10.75$ Å, $\beta = 95°$, CI) was determined[34]. This mineral is built up from layers of iron and brucite sulphides which alternate with each other in a regular way. The structural type of sulphide layers is analogous to the layers of mackinawite with the ordered distribution of vacancies.

The disordered highly dispersive Cu silicate, chrysocolla[35], was investigated by the microdiffraction method, and as a result a structure model from two-dimensional nets of distorted Cu octahedra was proposed for this mineral.

3.2.3 The structures of borates

Interesting chemical crystallographic data have been obtained during the investigation of a new class of borate structures. To these belong inderite[36] and kurnakovite[37]—monoclinic and triclinic modifications of the compounds $Mg[B_3O_2(OH)_5{\cdot}5H_2O]$, inderborite $CaMg[B_3O_3(OH)_5]_2{\cdot}6H_2O$ [38], hydroboracite $CaMg[B_3O_4(OH)_3]_2{\cdot}3H_2O$ [39], probertite $CaNa[B_5O_7(OH)_4]{\cdot}3H_2O$ [40], preobrazhenskite $Mg_3[B_{11}O_{14}(OH)_8HO_2]$ [41], dimorphic *p*-veatchite and veatchite $Sr_2[B_5O_3(OH)]_2{\cdot}B(OH)_3{\cdot}H_2O$ [42, 43]. The characteristic frameworks of these structures at BO_4 tetrahedra, with an average B—O distance of 1.47 Å and O—O distance of 2.41 Å, and BO_3 triangles (△) with the B—O distance of 1.37 Å and O—O distance of 2.38 Å.

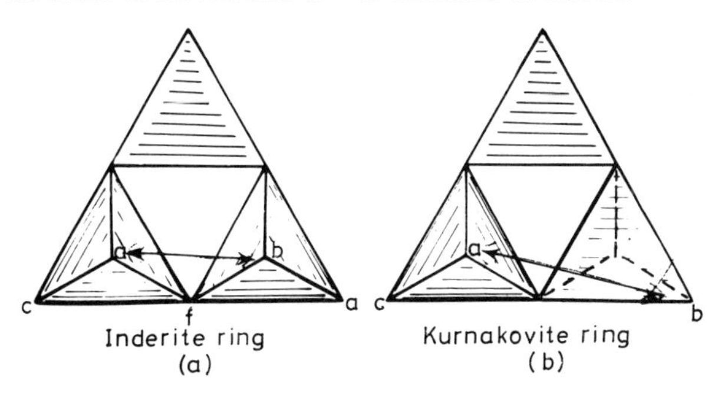

Figure 3.6 Three-membered boron–oxygen ring (2t, △) $B_3O_3(OH)_5$: (a) inderite form; (b) kurnakovite form
(From Rumanova *et al.*[43], by courtesy of Izd. Akad. Nauk SSSR)

In inderite and kurnakovite the isolated rings (△, 2t) were found. However, these rings appear to be different: in inderite the vertices of two tetrahedra, not lying in the plane of the ring, are arranged on one side of this plane, and the distance ab is 2.92 Å (Figure 3.6(a)), whereas in kurnakovite (Figure

3.6(b)) they face in opposite directions, and the distance is now 3.59 Å. This fact is responsible for the difference in structural motifs of both compounds. In inderite the distance of 2.92 Å is commensurable with an edge of the Mg octahedron. Therefore, each boron–oxygen ring in inderite is joined to one Mg octahedron through unshared OH vertices. As a result, the isolated complexes $MgB_3O_3(OH)_5{\cdot}4H_2O$ are formed. In kurnakovite the unshared vertices are drawn apart and, consequently, each ring turns out to be connected to two Mg octahedra forming the infinite chain $MgB\ O_3(OH){\cdot}4H_2O$ made up of alternating Mg octahedra and boron–oxygen rings. The chain structure of kurnakovite is more compact than the insular structure of inderite. A three-membered isolated ring (△, 2t) is present in inderborite also.

In hydroboracite an infinite chain radical $[B_3O_4(OH)_3]_n^{2n-}$ was discovered. Its repeat link is a three-membered ring (△, 2t). Hydroboracite is the second mineral, colemanite being the first[44], in which analogous chains were found. In probertite a new type of chain $[B_5O_7(OH)_4]_n^{3n-}$ was detected, the repeat link of which consists of three BO_4 tetrahedra linked at corners; each neighbouring pair of these tetrahedra is attached to a BO_3 triangle. Preobrazhenskite is a borate of high complexity; it contains a net polyion $[B_7^{t}B_4^{\triangle}\ O_{14}(OH)_3HO_2]_n^{6n-}$, in the elementary link of which five tetrahedra linked at corners are strung in pairs by BO_3 triangles. Here the first and last triangles adjoin the single tetrahedra. Dimorphic p-veatchite and veatchite $Sr_2[B_5O_3(OH)]_2$ are borates of mixed type with new layer radicals $[B_5O_8(OH)]^{2-}$. The repeat link of such a radical consists of two rings (2△, t) and (2t, △), connected by a common tetrahedron, and of independent insular triangular complexes $B(OH)_3$. p-Veatchite is a mixed borate, the first to be known.

The analysis of complicated chain and layer polyions in hydroboracite, probertite, preobrazhenskite, p-veatchite (and veatchite) shows that the three-membered boron–oxygen ring (△, 2t) is the principal building unit in their composition.

3.2.4 Some applications to geology

Studies of the structure of silicates and their analogues, as well as of other minerals[45, 46], permit general rules to be made, important not only for crystal chemistry as such, but also for geology and mineralogy.

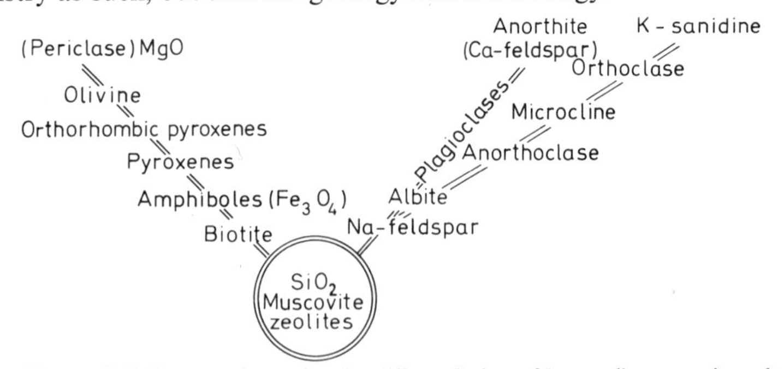

Figure 3.7 Bowen scheme for the differentiation of 'normal' magmatic melt (From Belov *et al.*[3], by courtesy of Moscow University (Geol.))

Using data on the atomic structure of minerals and the accepted Bowen scheme (Figure 3.7), N. V. Belov suggested the general geocrystal chemical picture of the formation of minerals from magnetic melts, and their further transformations[3]. The 'fight' for O atoms leads to the formation of a skeletal framework net in the magmatic glass. The first to abandon this net are the average cations of the Mg and Fe type. It is precisely these cations that at the early stage of magmatic mineral formation are engaged in construction of their own crystalline buildings: the heavy phases with close-packed O atoms and occupancy of their octahedral vacancies (periclase and others). After that, the neutral SiO_2 molecules separate out from the magma, solidify the masses of Mg octahedra with further formation of fersterite—olivine, followed by iron and other cations[3].

3.3 STUDIES ON CRYSTAL CHEMISTRY OF SEMICONDUCTORS

A series of electron- and x-ray-diffraction studies on the structure of multi-component semiconductors has been carried out in the Soviet Union in the last few years. These investigations have led to new conclusions being formed about the crystal chemistry of this class of compounds which has important practical application. The results obtained have been published in a series of monographs[47–51], in the Proceedings of the Conference devoted to the discussion of chemical binding in crystals[52] and in a number of original papers.

It is known that the composition of two- and multi-component semi-conducting compounds often deviates from stoichiometry. This leads to the formation of various kinds of defect structures. Semiletov[53] has recently considered the question of the formation of point defects in semiconducting crystals. On the basis of published data the author has reached the conclusion that the deviation from stoichiometric composition depends not only on the chemical nature of interacting atoms but also on the structure type to which the given compound belongs. The overwhelming majority of semiconducting compounds of variable composition which deviate from stoichiometry have octahedral coordination (i.e. PbTe, SnTe, SnSb, Bi_2Te_3, CeTe etc.). In the case of tetrahedral coordination, the deviations from stoichiometry are very small. It is very difficult to fix such deviations and to determine an excess of one of the components.

However, there is an approach to overcome this difficulty. In compounds with structures of the type Ga_2Te_3 and $ZnGa_2Te_4$ there are stoichiometric vacant positions: $\frac{1}{3}$ and $\frac{1}{4}$ of positions not occupied by Ga and Zn atoms. For these compounds the connection between the lattice constant and concentration of vacancies or excess atoms were established.

Using this dependence it was shown that in semiconductors with the tetrahedral coordination the presence of 1 % of cation vacancies is accompanied by a decrease in the lattice constant, on the average, by 0.0044 Å. This method was used for determining the concentration of point defects and the range of homogeneity in gallium arsenide for which the precision x-ray data on the magnitude of the lattice constant are known. It was established that 1 cm^3 of GaAs contains $\sim 10^{19}$ excess As atoms or cation vacancies, and $\sim 10^{18}$ interstitial Ga atoms. In other words, gallium arsenide is the phase of a variable composition with the homogeneity range between 49.99 and 50.06 atomic %As.

Apart from elementary semiconductors and binary compounds, various three-component semiconducting compounds have been extensively studied[47, 48, 50]. The rules for the formation of such compounds with tetrahedral coordination, and some questions on their crystal chemistry have been elucidated in the monograph by Goryunova[50]. Abrikosov and colleagues[47] have established that a large group of ternary compounds is formed in the pseudo-binary systems of the type Bi_2Te_3–PbTe. These compounds have high electric conductivity and low thermal conductivity, and could have application in thermoelectric generators, refrigerators etc.

The structure of many of these compounds was determined by the electron diffraction method. Crystal chemical analysis of these structures was carried out by Imamov, Semiletov and Pinsker[54]. They established that the peculiarities inherent in binary compounds (atomic coordination, interatomic distances, volume per molecule or density of packing) are retained in ternary compounds. This permits calculation of lattice constants of ternary compounds on the basis of the constants of the initial binary compounds. The type of the structure of the compounds under consideration is defined by the ratio of the number of cations (Ge,Sn,Pb,Bi) to the number of anions (Se,Te).

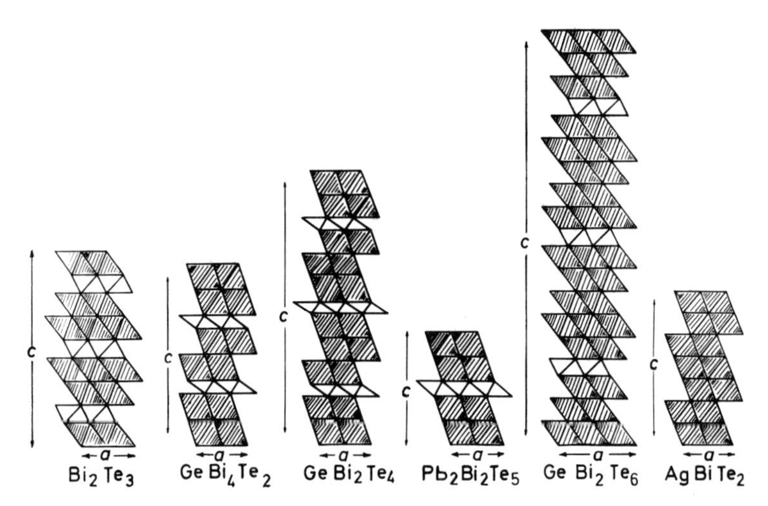

Figure 3.8 Projections of the structures Bi_2Te_3, $GeBi_4Te_2$, $GeBi_2Te_4$, $PbBi_2Te_5$, $GeBi_2Te_6$ and Ag $BiTe_2$ in coordination polyhedra
(From Imanov *et al.*[54], by courtesy of Izd. Akad. Nauk SSSR)

If this ratio is 2:3, the structure of the type Bi_2Te_3 is formed, if it is equal to 1, the structure of the NaCl or $NaInO_2$ type arises. For the compounds AB_4X_7, AB_2X_4, $A_2B_2X_5$ and $A_3B_2X_6$ the ratio lies between 2:3 and 1. Accordingly, these structures are intermediate between the structures of the type Bi_2Te_3 and PbTe(NaCl) (Figure 3.8).

Along with structures in which the cations are located in tetrahedra or octahedra are a number of semiconducting compounds with a mixed atomic coordination. These are, primarily, double sulphides of spinel type with the general formula AIn_2S_4, where A is a Group II or Group VIII element. These compounds are based on the structure In_2Sr, in which part of the In atoms are replaced by Group II (Mg,Cd,Hg) or transition metal (Mn,Fe,Co,Ni) atoms. In the first case, the normal spinel is formed in which the atoms of Group II are located in tetrahedra, whereas in the second case, an inverse spinel makes its appearance where the atoms of a transition metal, together with a part of the In atoms, statistically occupy the 16-fold position in octahedra[54].

Among sulphide spinels of the type $A^{II}B_2^{III}X_4^{VI}$ the structure $ZnIn_2S_4$ remained uncertain for a long time. It has been established[55] that this compound has the rhombohedral structure ($R3m$, $a = 3.85$ $c = 37.06$ Å). More recently the compounds $Zn_3In_2S_6$ and $Zn_2In_2S_5$ have been discovered in the system $ZnS–In_2S_3$, and it has been suggested that other compounds are possible[57]. In investigating this system many polytypic forms were established, and the crystal structure of some of them was determined[58–60]. All these structures are built up of packets in which the number of layers is determined by the chemical formula. The Zn atoms are situated in tetrahedra, the In atoms in tetrahedra and octahedra. In the $ZnIn_2S_4$ structure the In—S distance in octahedra is 2.62 Å, in tetrahedra it equals 2.34; 2.77 Å; the Zn—S distances in tetrahedra are 2.43; 2.36 Å; the S—S distances fluctuate between 3.56 and 3.92 Å. In Zn and In tetrahedra these distances are close to the sums of the tetrahedral (covalent) radii, whereas in the In octahedron they are closer to the sum of the ionic radii. This indicates the different nature of chemical bonding in such polyhedra. The existence of polytypic forms is associated with the different orientation of packets with respect to one another. Thus, for ZnI_2Se_4 16 modifications were found, each modification having $P3m1$ or $R3$, $a = 3.85$ Å. For the simplest four-layer modifications $c = 12.34$ Å; for other modifications the number of layers is a multiple of four, the greatest number of layers is 96, and, correspondingly, $c = 296.16$ Å.

The apparent difference between the spinel-type cubic structure and hexagonal structures of the compounds $ZnIn_2S_4$ is actually quite small. In all these structures the sulphur atoms arrange themselves according to some closest packing type, whereas the cations, as in spinel, occupy the tetrahedral and octahedral holes.

The new regularities in the crystal structure of the Bi–Se, Bi–Te and Sb–Te systems were established in the work of Imamov and Semiletov[61]. For the above-mentioned compounds of the A_2X_3 type (Bi_2Te_3 and others) the composition satisfies the classical valence rules. But apart from them, a large number of other phases were recorded. At first sight, the composition of these phases (for instance, Bi_8Se_9, Bi_2Se_2, Bi_8Se_7, Bi_4Se_3) has nothing in common with the valence of elements[62, 63].

The crystal structure of all these phases is based on the following principle. The five-layer packets TeBTeB '5', characteristic for A_2X_3 structures, are retained in them almost unchanged. The two-layer '2' packets of sulphur or bismuth, the number of which depends on the chemical composition, are situated between the '5' packets. The shortest interatomic distances characteristic of the structures Sb,Bi and A_2X_3 compounds are retained almost unchanged in the structures of the intermediate phases. For instance, the

Bi—Bi distance lies between 3.04 and 3.09 Å, and the Bi—Se between 2.87 and 3.15 Å etc.

Combining the five-layer '5' packets and the two-layer '2' packets the structures established can be described as follows:

5555555...	A_2X_3	(type Bi_2Te_3)
525 525 525...	A_2X_2	
52525252...	A_4X_3	
522522522...	AX_2	
52225225222...	$A_{14}X_6$	
52225222...	A_8X_3	
2222222...	A_2	(type Bi)

It is thus seen that in the given phases, a continuous transition from the structures of the type Bi_2Te_3 to the structures Bi(Sb) and vice versa is possible. It is obvious that both an ordered and disordered alternation of packets may exist, the probability of the ordered phase depending on the ratio between the quantities of the '5' and '2' packets.

Some structural features of copper, silver and gold tellurides are discussed in References 64 and 65. In Reference 65 the structure of Tl_5Te_3 is considered and an analysis of the ordering in these compounds is described. In contrast to the majority of the elements of Groups I and II of the Periodic System, copper, silver and gold are found to form rather complex, often disordered structures with sulphur, selenium and tellurium, which cannot be described in terms of close packing.

The common feature of these compounds is the complexity of the structural motif, differences in the short-range order of atoms of various elements, large variety in the interatomic distances in the similar pairs of atoms and absence of coordination polyhedra built up of atoms of the same sort. Thus, the Cu—Cu distances fluctuate between 1.82 and 3.10 Å and Cu—Te between 2.39 and 3.40 Å. In explaining such a peculiarity the authors put forward a new approach, which considers these structures as the packing of 'clusters' of nearest and most strongly bonded atoms which are in the super-cells. The choice of such cells should be made in such a way that the clusters, characteristic of each structure, arrange themselves at their vertices. Figure 3.9 shows an arrangement of the clusters in structures of the β'-phase of the Cu–Te system.

On referring to the shape of clusters and the structural motif of their arrangement, one can compare the structures of a given class with one another as well as with other structures, and also analyse the structural transformations during phase transitions and obtain information on the interaction forces.

The above-mentioned features of the structure of tellurides could be associated with the different valence of cations (Cu,Ag,Te) entering into them. The cations of different valence, especially Cu^+ and Cu^{2+}, should have different coordination numbers in accordance with the crystal chemical rules revealed by Belov in the analysis of the crystal structures of sulphides and other chalcogenides[66].

Analysis of the interatomic distances in the octahedral structures was carried out by Semiletov[67]. By now about a thousand semiconducting inorganic compounds have been studied. Most of them show the tetrahedral or octahedral coordination of atoms. In the case of tetrahedral coordination the bond is covalent. It has always been considered that the chemical bond in compounds with octahedral coordination is mainly ionic, and hence they were called polar semiconductors. However, analysis of interatomic distances for coordination number 6 has shown that they differ noticeably from the sum

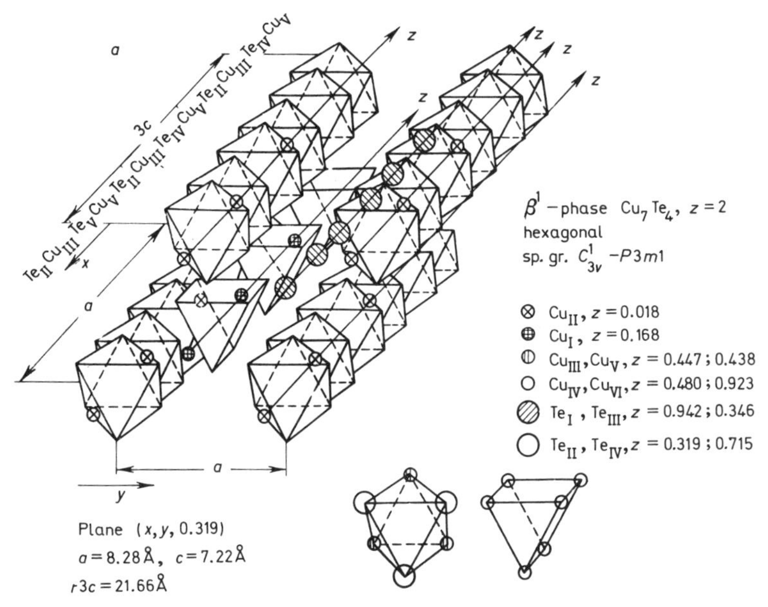

Figure 3.9 Structure of the β'-phase of the Cu–Te system, represented by the groups of atoms (clusters). The vertices of the prisms and octahedra are occupied by the atoms as is shown at the bottom right-hand corner of the diagram
(From Baranova and Pinsker[64], by courtesy of Izd. Akad. Nauk SSSR)

Table 3.1 Octahedral covalent radii (in Å)

Cu 1.25	Mg 1.42	Al 1.41	C 0.97	N 0.95	O 0.90
Ag 1.43	Zn 1.27	Ga 1.35	Si 1.37	P 1.35	S 1.30
Au 1.40	Cd 1.45	In 1.55	Ge 1.43	As 1.43	Se 1.40
	Hg 1.45	Ib 1.73	Sn 1.60	Sb 1.60	Te 1.56
	Mn 1.31		Pb 1.67	Bi 1.65	

of ionic radii. This suggested the predominance of the covalent component of the chemical bond in octahedral semiconductors. This is evidenced by the physical properties of the given compounds, namely metallic lustre, relatively high electric conductivity (for instance, in PbS, PbSe and PbTe), carrier scattering by acoustic vibrations of the lattice etc. Taking the aforesaid into account, Semiletov[67] made up a table of octahedral covalent radii, which describes well enough the interatomic distances in such types of compound. These radii may be useful in the analysis of further unknown structures.

3.4 NEUTRON DIFFRACTION STUDIES OF NUCLEAR AND MAGNETIC STRUCTURE

A large number of crystal studies by neutron diffraction have been carried out in the U.S.S.R. in recent years. In particular, the structures of crystal hydrates and magnetic alloys have been studied, and theoretical problems of neutron diffraction have been considered[68].

It is known that the position of the hydrogen atoms in crystal structures are most easily determined by neutron diffraction (in this sense, the electron diffraction method is also very effective[69, 70]), whereas the low electron

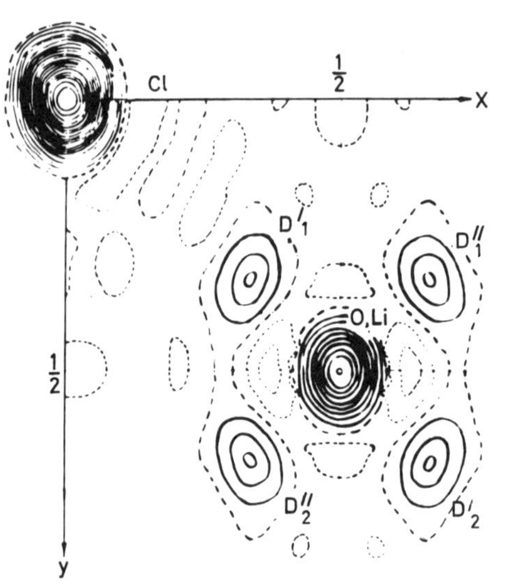

Projection of the nuclear density of deuterated lithium chloride monohydrate along the *z*-axis
(From Ozerov *et al.*[73], by courtesy of Izd. Akad. Nauk SSSR)

density of hydrogen atoms makes the detection of these atoms by x-ray diffraction less accurate.

New data on the structure of some crystal hydrates and the functions of the H_2O molecules in them were obtained with the aid of neutron diffraction by Ozerov, Datt and others. Because of the very large cross-section of incoherent scattering of neutrons by protons, there arises another possibility, that of obtaining data on the vibrations of water molecules.

The following series of crystal hydrates has been studied:

$Li_2SO_4 \cdot H_2O$ [71]; $LiClO_4 \cdot 3H_2O$ [72]

$LiCl \cdot H_2O$ [73]; $Li_2Cr_2O_7 \cdot 2H_2O$ [74]

In Figure 3.10 is shown an example of the projection of the nuclear density of deuterated $LiCl \cdot H_2O$ along the *z* axis.

As for the configuration of water molecules (the H—O—H angle, O—H distances), the results obtained appear to be in good agreement with the known values (see, for instance References 75 and 76): the H—O—H angle is 103–109 degrees, the O—H—O distance 0.94–0.97 Å. In all the compounds, weak hydrogen bonds with O—H distances fluctuating between 2.8–3.0 Å were found. The crystals investigated differ substantially in configuration of the intermolecular hydrogen bonds and their systems. In $Li_2SO_4{\cdot}H_2O$ the water molecules are combined with each other by hydrogen bonds in such a way that an infinite chain is formed which extends all over the crystal along the 2_1 axis (Figure 3.11). At definite levels the SO_4 and LiO_4 tetrahedra join this chain. In $LiClO_4{\cdot}3H_2O$ the hydrogen bonds link the $Li(OH)_6$ and ClO_4 polyhedra together (Figure 3.12). In $LiCl{\cdot}H_2O$ the bonds also connect the OH_2 and Cl ions. In this compound, which has a relatively simple structure, the general situation is seen to be substantially complicated on account of the dynamic or static (domain) reorientation of water molecules[73].

The different hydrogen bonding systems have a considerable effect on the physical properties of crystals. Thus, the presence of infinite chains (Figure 3.11) is associated with a strong pyroelectric effect in $Li_2SO_4{\cdot}H_2O$.

It has been established by the inelastic neutron scattering method that the hydrogen bonding system also produces a strong influence on the dynamics of the crystal as a whole. The cross-section of the incoherent scattering of neutrons is very large; as a result, the spectra of inelastic incoherent scattering of neutrons (INSN) of crystal hydrates mainly contain the information about the vibration of hydrogen atoms. The crystal hydrates just mentioned have been investigated in this way on the basis of the 'time of flight' method on the Pulsen Reactor in the Joint Institute for Nuclear Research at Dubna[77, 80].

In Figure 3.13 the spectrum of $Li_2SO_4{\cdot}H_2O$ is shown and, for comparison, that of ice. Three regions of frequency vibration have been recorded: the well-pronounced region I (300–1000 cm^{-1}) for rotational vibrations of water molecules (twisting, wagging and rocking) denoted in Figure 3.13 by T, W and R respectively; region II (130–300 cm^{-1}) for translational vibrations and low frequency region III (80 cm^{-1}) for acoustic vibrations.

The available structural data has also permitted the interpretation of the low-frequency maximum in the INSN spectra. The fact is that this peak has been found to be present in the INSN spectra of ice and $Li_2SO_4{\cdot}H_2O$ (see Figure 3.13), and to be absent from spectra of a large number of other crystal hydrates[77]. Comparison of the structures of these compounds shows that in ice and $Li_2SO_4{\cdot}H_2O$ the hydrogen bonds form respectively three- and one-dimensional continuous chains, which turn out to be the principal 'armouring' element. In all other crystal hydrates the hydrogen bonds form a closed network.

On this basis, it has been assumed that the degree of participation of protons in the acoustic vibrations of the lattice depends on the particular type of hydrogen bonding system. In the case where this system forms a continuous framework, which determines the structure, the proton participates actively in the unit cell vibrations as a whole (acoustic vibrations). If the hydrogen bonds are closed, the proton participates only in the optical local vibrations.

The hydrogen atoms have also been determined by electron diffraction.

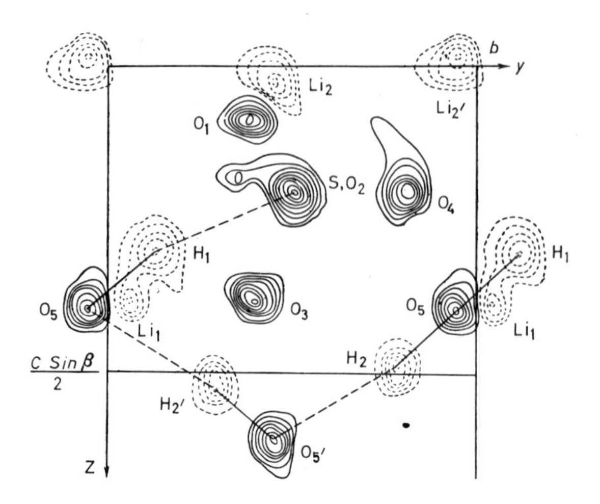

Figure 3.11 Scheme of the hydrogen bonds in the structure of lithium sulphate monohydrate
(From Ozerov (1963). *Dokl. Akad. Nauk SSSR*, **148,** 1069, by courtesy of Izd. Akad. Nauk SSSR)

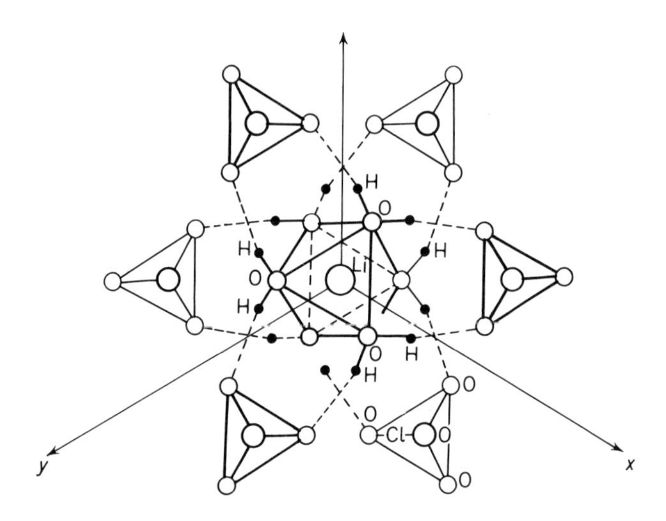

Figure 3.12 Structure of lithium perchlorate trihydrate: the $Li(OH)_6$- and ClO_4-polyhedra are linked together by hydrogen bonds
(From Ozerov *et al.*[72], by courtesy of Izd. Akad. Nauk SSSR)

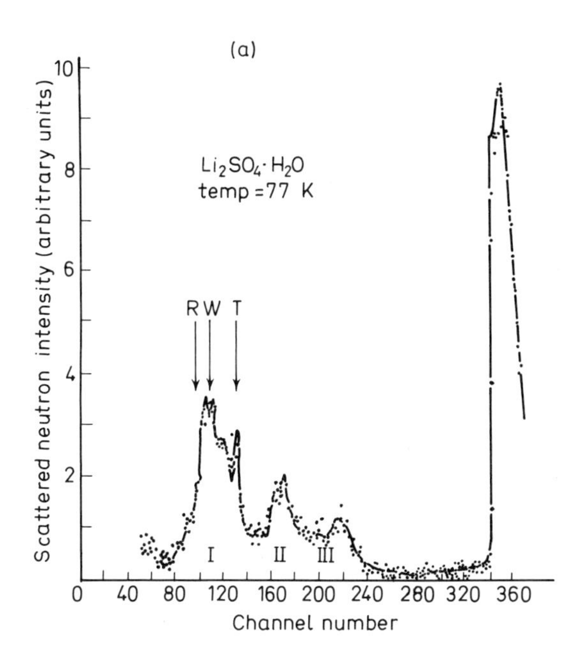

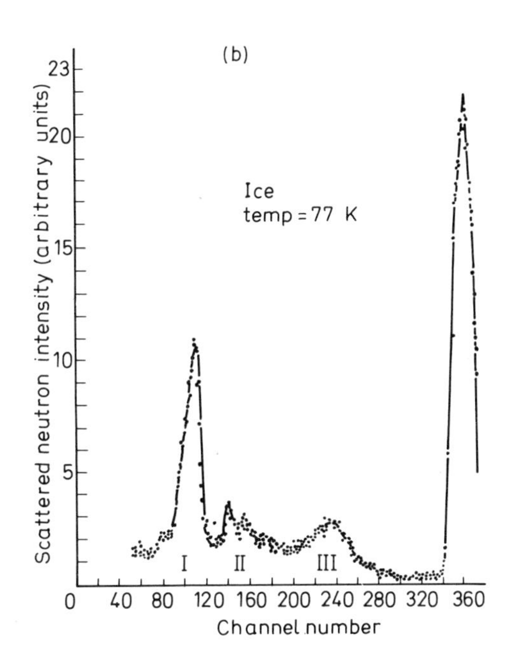

Figure 3.13 Spectra of inelastic incoherent scattering of neutrons obtained by the 'time of flight' method: (a) lithium sulphate monohydrate; (b) hexagonal ice
(From Ozerov *et al.*[78], by courtesy of Izd. Akad. Nauk SSSR)

In the extension of such investigations[69, 70] new theoretical calculations of the integral characteristics of the pótential of the hydrogen atoms were carried out[79]. The crystal structure of NH_4Br was investigated from the point of view of the thermal motion of the atoms and the termination of the Fourier series[80]. In this structure the NH_4 tetrahedron occupies two equally probable positions so that the 'half-atoms' H/2 lie on the cube diagonals; the Br atoms are arranged at the vertices of the cube, whereas the N atoms are at its centre. Figure 3.14 shows the distribution of the potentials of the N and H/2 atoms separately and as a sum of the distributions of the potentials along the N—H line. It can be seen from Figure 3.14 that the termination wave from

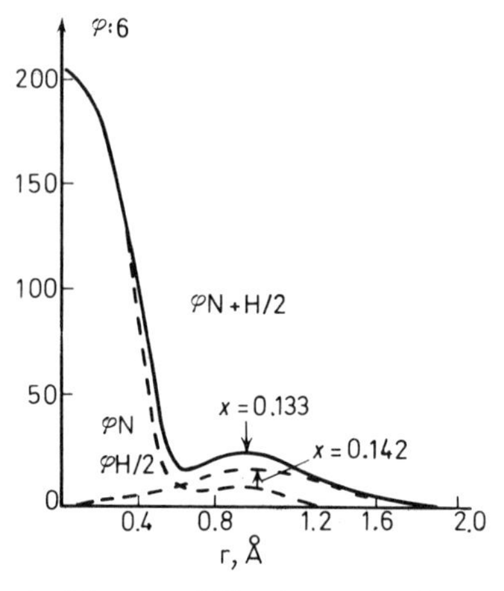

Figure 3.14 Potential distribution in the lattice of NH_4Br along the N—H bond (From Dvoryankin[80], by courtesy of Izd. Akad. Nauk SSSR)

the N atom shifts the peak of H/2 potential and increases it by about 5–6 V. The bond length N—H = 0.99 ± 0.04 Å was obtained from the difference three-dimensional synthesis. The electron diffraction study of the low-temperature (below -38.5 °C) phase of ammonium bromide ($P4/pmm$, $a = 5.809$ Å, $c = 4.130$ Å, $Z = 2$) is described in Reference 81. It has been found from the Fourier synthesis that the parameters of the hydrogen atom are $x = 0.175$, $z = 0.135$ (the eightfold position $X0Z$), and the N—H distance is 1.01 ± 0.03 Å. The height of the hydrogen peak is equal to 34 V, which is close to the theoretical value of 32 V. This indicates the ordering of the hydrogen position in the structure. Thus, the mechanism of the phase transition in NH_4Br from phase II to phase III consists of the displacement of the Br atoms and ordering of the position of the NH_4 groups.

A number of works have been devoted to the structures of metals and alloys. A series of investigations by Goman'kov, Loshmanov and others has been devoted to permendure-type alloys in the systems Fe–Co and Fe–Ni, and their analogues[82–83].

An investigation of the ordering in the Ni_3Fe alloys containing the isotope ^{62}Ni with a negative amplitude of scattering has enabled the intensity of the superstructure reflections to be increased by 14 times. It has been found that the ordering in Ni_3Fe represents a first-order phase transition, but the short-range order is retained up to a temperature of about 1000 °C ($\sim 2\ T_C$).

An investigation into the effect of alloying elements on the magnetic momenta of iron and nickel showed[83] that Mo, W, Gr, V mainly influence the nickel momentum, whereas Si produces the influence on the iron momentum. These results, and the investigations of the influence of the alloying elements on the electronic structure of the Ni atoms, have enabled us to explain the degradation of the superstructure on alloying[83]. A series of investigations has been devoted to the alloys Ni–Cr, Fe–Cr[84, 85], in particular, use has been made of small-angle scattering techniques which enabled the investigators to refine the phase diagram as well as to elucidate the connection of the kinetics of splitting with the initial state of the solid solution.

It follows from the general theory of magnetic structures[86, 87] that the most general type of ordering of magnetic moments in crystals is the helical ordering whereas the parallel and antiparallel ordering corresponding to ferro- and antiferro-magnetism are only its particular cases. Consideration

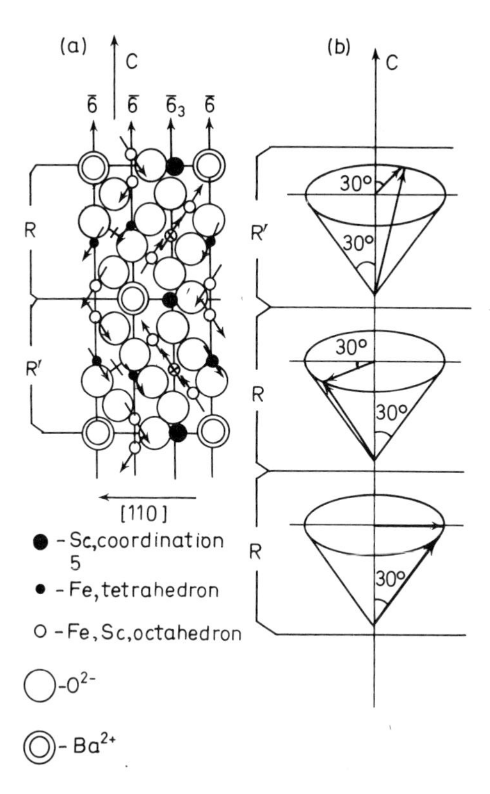

Figure 3.15 Magnetic structure of the hexagonal ferrite $BaSc_{1.8}Fe_{10.2}O_{19}$
(From Jamzin *et al.*, by courtesy of Izd. Akad. Nauk SSSR)

of neutron diffraction results obtained for MnO_2[88] and metallic chromium[89, 90] enabled the existence of helical magnetic ordering to be established for the first time.

An important class of magnetic compounds are the ferrites, by which are usually meant the complex many-component oxides with the atomic structure of the spinel or garnet type, or their hexagonal analogues.

Magnetic investigations[91] of ferrites predicted a structure of the simple helical type, observable in hexagonal ferrites of the $Ba_{2-x}Sr_xZn_2Fe_{12}O_{22}$ system.

The true picture of spin configurations in the system of the indicated compounds was revealed by Jamzin and his colleagues[92]. To explain the diffraction pattern, a model of quasi-helical spin-ordering was proposed. Using the terminology now accepted, it is more conveniently called a flat block helix. In this structure the unit cell breaks up along the *c*-axis into three equivalent magnetic blocks, each of which contains six oxygen layers.

A new type of helical spin-ordering in hexagonal ferrite crystals of the $BaSc_xFe_{12-x}O_{19}$ system was established by Aleshko-Ozhevskii, Jamzin and co-workers[93]. The investigations were carried out in the temperature interval from 4.2 K to the Curie temperature, and in a wide range of magnetic field. It has been established that, beginning with a definite scandium concentration ($x \geqslant 1.2$) in the low-temperature region, a new type of a magnetic structure arises; a block conic ferromagnetic helix. The preferred distribution of non-magnetic scandium ions at definite sites in the crystal structure leads to the formation of non-magnetic layers between the spinel blocks of the unit cell. The considerable decrease and modification of the strength of the exchange interaction leads to the fact that the total magnetic moments of these blocks begin to behave as independent magnetic structure elements, thereby forming the angular antiphase, helical and other types of magnetic structures. In the detected helix (Figure 3.15) the total magnetic moments of the spinel blocks arranged themselves along the rotational line forming a cone whose axis coincided with the *c*-axis. Here, the moments rotated with respect to each block in such a way that their projections on the family of the basal planes formed the simple helix.

The principal magnetic state of mixed oxides having the spinel structure, in which the paramagnetic ions are arranged both in tetrahedral (A-sublattice) and octahedral (B-sublattice) sites, is ferrimagnetic and defined by the predominance of the A–B interaction. However, in the case where in the A-sublattice a small number of paramagnetic ions is present, the weaker B–B magnetic interaction becomes comparable with the A–A interaction, and even stronger than the latter. In these cases there may arise antiferromagnetic and non-collinear structures of different type. Thus, in zinc-manganese ferrite, with the cation distribution $Zn_{0.77}Fe_{0.10}Mn_{0.13}[Fe_{1.90}Zn_{0.10}]O_4$, an antiferromagnetic structure similar to the structure of pure zinc ferrite was found[94–96]. The magnetic cell is doubled along one of the cubic axes, and coincides along two remaining directions with the chemical cell. The Néel temperature is 11 ± 0.5 K.

In order to elucidate the effect of the paramagnetic ions on the magnetic structure of zinc ferrite, the mixed oxides $Zn_{0.88}Fe_{0.12}[Fe_2]O_4$, $Zn_{0.86}Fe_{0.14}[Ni_{0.10}Fe_{1.90}]O_4$, $Zn_{0.90}Fe_{0.10}[Co_{0.10}Fe_{1.90}]O_4$ and $Zn[Mn_{0.12}Fe_{1.88}]O_4$ were investigated. It turned out that the antiferromagnetic structure is retained only for the first compound, and is absent in the other three; but in $Zn[Mn_{0.12}Fe_{1.88}]O_4$ the short-range magnetic order is observed at 4.2 K, whereas

$Zn_{0.86}Fe_{0.14}[Ni_{0.10}Fe_{1.90}]O_4$ and $Zn_{0.90}Fe_{0.10}[Co_{0.10}Fe_{1.90}]O_4$ are paramagnetic at 4.2 K.

3.5 STRUCTURES OF COORDINATION COMPOUNDS

Studies in the field of crystal chemistry of complex compounds began in the Soviet Union in 1940 and early 1950 under the influence of a group of chemical synthesists led by academician Chernyayev, in co-operation with chemical theoreticians directed by Syrkin and Dyatkina, who were developing the methods of quantum chemistry. Investigations in this field, started by Bokii, have been intensively developed by the Porai-Koshits school, who are engaged in the study of complex compounds of transition metals, and by other investigators.

The structural data on 15 ethylenediamine compounds of nickel, of composition $NienXX^I$ where X and X^I are the same or different acidoligands[97], explained the specific features of the structures of these compounds: namely the tendency to form *cis*-octahedral configuration, variable structure types (molecular or cation complexes; monomeric and polymeric structures, etc.). Owing to the great strength of M—en bonds in comparison with M—X in solution the metastable complexes $[Nien_2]^{2+}$ and $[Nien_3]^{2+}$ should be expected to arise. They may serve as the basis for crystallisation of the compounds $[Nien]_2X_2$ and *cis*-$Nien_2XX^I$.

The particular form of the *cis*-octahedral structure is defined by the nature of ligands X and X^I, their volume, shape, tendency to bridge function and strength of bonding with a metal.

Analysis of literature data and that of his own work on the structure of ethylenediamine cycles led Porai-Koshits[98] to the conclusion that conformation in complexes with ethylenediamine cycles having the *gauche*-structure is defined by three factors: non-valent atomic interactions in neighbouring ligands, asymmetry of a ligand field, arising from its non-equivalence, and advantages of the centrosymmetric molecules during crystallisation. The first factor prevails in $[M(en)_3]^{n+}$ type complexes (KKK conformation), the first and second factors in *cis*-$[M(en)_2X_2]$ complexes (KK conformation), the second in *cis*-$[M(en)_2XX^I]$ complexes (KK′ conformation), and the third in *trans*-$[M(en)_2X_2]$ complexes (KK′ conformation). The fourth factor is the tendency to form intermolecular hydrogen bonds.

Work on the studies of nitroso- and oxo-complexes of transition metals[99–114] proceeds on a broad front. Comparison of the data on the structure of the nitroso-complex of ruthenium[99] allowed conclusions to be made about the effective charge distribution in the nitroso-group, and provided evidence that the structure of the Ru—NO grouping is close to a linear one. The structural formulae were established for oxo-ethylenediamine compounds of rhenium ReO_3en and $Re_3O_9(OH)en_2$, and the structure of these complexes was determined[100].

Substantial numbers of works, mainly of Atovmyan and co-workers have been devoted to the determination of the structure of oxo-complexes of transition metals: Mo [101–106], Os [107, 108], V [109, 110], Fe [111] as well as U [112]. The results of the structural investigations served as a basis of the crystal chemical generalisations made by Poray-Koshits and Atovmyan[113, 114]. They have shown that: (a) At the maximal formal valence of metals the dioxo group must have a *cis*-structure, whereas at a valence two units less a *trans*-structure exists. At an intermediate valence the most stable complexes must be the dimeric complexes with a single oxygen bridge and *cis*-arrangement of the bridge and terminal oxygen atoms. (b) Effect of the lengthening of the bonds being in a *trans* position to a multiple bond is a characteristic property for the majority of oxo-complexes. This effect is a consequence of non-valent ligand–ligand interactions, resulting in the stretching of the most labile bond. The bond lability, in its turn, is defined by the specific character of the electronic structure of a complex. (c) There is a tendency to lower the co-ordination number of a metal as its degree of oxidation increases.

The review material on the crystal chemistry of coordination compounds of transition metals is also given in References 113, 114.

A substantial amount of work has been devoted to the investigation of acetate and the formate compounds of Co [115], Sc [116], Th [117] and Ce [118]. This work is closely related to the study of the structure of neodymium nitrileacetate $Nd(NTA)_2$ [119], neodymium triglycinate $Nd(NH_2CH_2COO)_3Cu{\cdot}3H_2O$ [120], and the glycinate[121] and propionate[122] of copper carried out (except Reference 121) under the direction of Malinovskii.

The analysis of the structural data showed that, by reference to the mode of attachment of the carboxyl group RCOO to a metal, they can be divided

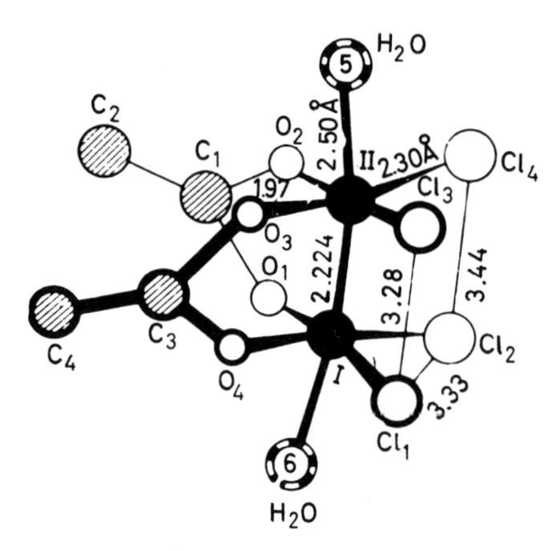

Figure 3.16 Fragment of the structure of rhenium tetrachlorodiacetate dihydrate $Re_2Cl_4[CH_3COO(H)]_2{\cdot}2H_2O$ (From Kuznetsov *et al.*[124], by courtesy of Izd. Akad. Nauk SSSR)

into four types: monodentate, bidentate-bridging, bidentate-cyclic and tridentate bridge-cyclic. In transition metal compounds with a small radius, only the first two types of COO groupings are encountered, but in metal compounds with large atoms the last two exist. It should be noted that in acetates, either the monodentate groups or bidentate-bridging groups with a *syn–syn* structure are encountered, whereas in formates bidentate-bridging groups with the *syn–anti-* and *anti–anti* structure occur. The *syn–syn* structure enables the atoms to draw together, and leads to the M—M bond formation, although the *syn–syn* structure can be realised without the M—M formation as such[116].

The *anti–syn-* and *anti–anti* structure enables the formation of net or framework structures without the M—M bonds.

Increasing the size of a metal atom makes it necessary to increase its coordination number. The coordination requirements of a metal are satisfied partly at the expense of water molecules, and partly at the expense of the increase in dentation of acetic groups. The coordination number of cerium[118] and thorium[117] in their acetate compounds is 9–12.

The structural features of sulphate compounds of zirconium[123] bear strong resemblance to those of cerium acetate. The structure analysis of Zr compounds confirmed the absence in them of zirconyl groupings suggested by chemical synthesists.

In the structure $Re_2Cl_4[CH_3COO(H)]_2{\cdot}2H_2O$ solved by Kuznetsov and co-workers[124], separate complexes are linked in dimers, both by multiple Re—Re bonds and bridging acetate groups (Figure 3.16). The same authors investigated the complexes of the type $(PyH)_2Re_2X_8$ (where X = Cl, Br)[125] in which the fourfold Re—Re bonds were discovered for the first time.

The investigation of acetate–Ce compounds and analysis of literature data on the structure of the rare-earth elements allowed Sadikov, Poray-Koshits, *et al.*[118] to formulate the crystal chemical features of rare-earth elements; namely the tendency to high coordination numbers (9–10), changeability of a coordination number in compounds with the same ligands, difference between coordination polyhedra at one and the same coordination number, complexity and asymmetry of coordination polyhedra, large variety in modes of coordination of ligands having several functional atoms, large scatter in the metal-oxygen distances, tendency to involve the water molecules into the coordination sphere, tendency towards polymerisation and formation of chain and layer structures.

A number of complex compounds of rare-earth elements belonging to the class of β-diketonates, i.e. tris-acetylacetonate hydrates of neodymium, europium, holmium, ytterbium (complexed with acetylacetonimine), tetrakis-benzoylacetonates of europium, gadolinium (compound with piperidinium and diethylamine)[126] have been studied.

The important result of the investigation is the establishment of the nature of isomerism of the compounds of rare-earth tetrakis-β-diketonates with such large and readily deformed cations as piperidinium, diethylammonium etc.; the α-form is a salt containing the anions [Ln(β-diketonate)$_4$] and one of the cations listed above; the β-form is a crystallosolvate containing the neutral complexes, HLn(β-diketonate)$_4$, and the molecules of piperidinium, diethylamine and others located in the channels of the structure. This explains why their partial removal is so easy (Figure 3.17). Here the geometrical parameters of the base molecules determine the form of the channels and, thereby, the mode of 'stamping' of the structure on these molecules.

The second important result is the explanation of insignificance of the quantum yield of luminescence of tris-β-diketonate hydrates as compared

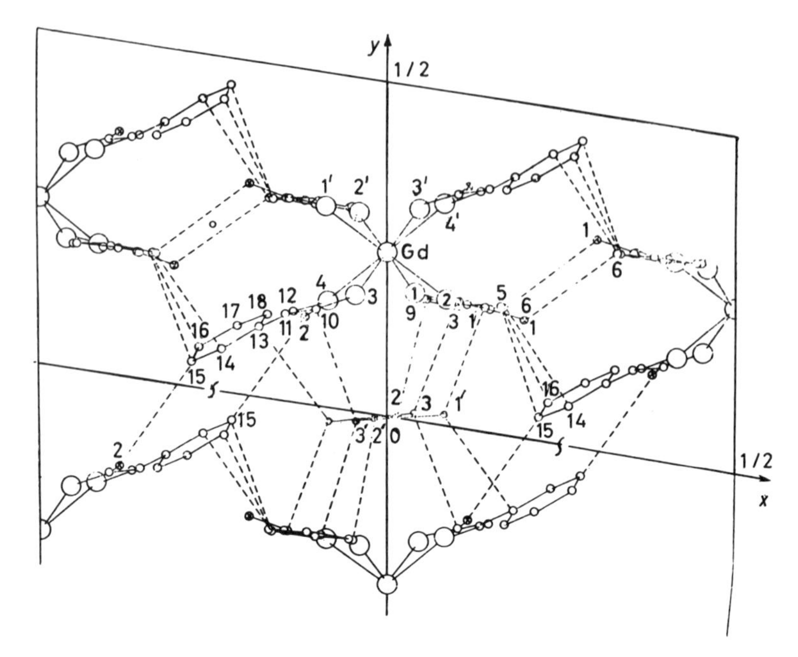

Figure 3.17 *xy*-Projection of the structure of $HGd(BA)_4 \cdot x$ Pip ($x \approx 1$)
The piperidinium is located in the channels of the structure (the filled circles are Pip atoms)
(From Poray-Koshits *et al.*[126], by courtesy of Izd. Akad. Nauk SSSR)

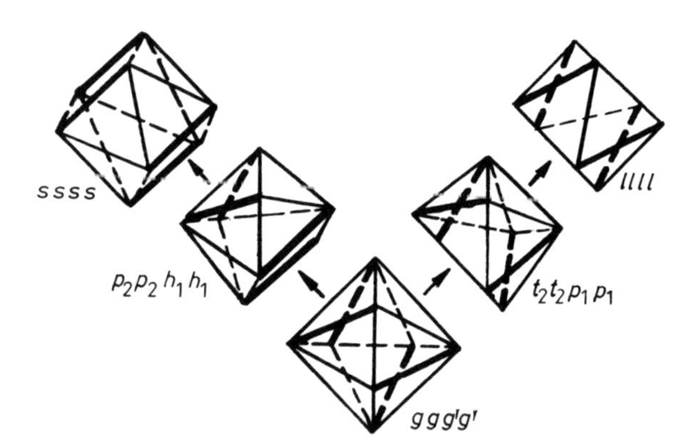

Figure 3.18 The instance of interrelation of all possible stereoisomers of a dodecahedron. The edges embraced by chelate ligands are drawn with thicker lines
(From Poray-Koshits, by courtesy)

with tetrakis-compounds in terms of the presence of H_2O molecules in the inner sphere of these hydrates.

In Reference 127 the classification of the eight-vertex polyhedra occurring in rare-earth chelates, as well as the criteria for the description of the form of the polyhedron belonging to the dodecahedron class are suggested. The problems of stereoisomerism of the eight-vertex complexes with the chelate ligands were considered. In particular, the interrelation of all possible stereoisomers of the dodecahedron, two-cap trigonal-prism and antiprism obtained one from another via deformational transformations was established. (In Figure 3.18 is given an example of such an interrelation.) The chains of stereoisomers, for which the probability of realisation is maximal, were separated out.

The studies of the structure of complex compounds of bi- and tri-valent metals with ethylenediaminetetra-acetic acid (EDTA) are in progress[128]. The results of the investigations in combination with the data of other scientists, mainly of Hoard with co-workers, permit conclusions to be made about the structural role of various components of these compounds, and to reveal the definite regularities in structural rebuilding during transition from one compound to another[129].

It was established that the coordination number of the complex-forming atom is determined, not only by the steric factor, but depends substantially on the valence state of a metal. The degree of the EDTA dissociation is not the decisive factor determining the EDTA dentation and coordination number of the central atom.

The coordination polyhedra in transition metal complexonates are notable for a large variety of forms (octahedron with a different degree of distortion, square, square-pyramid, trigonal face-centred prism, pentagonal-bipyramid, nine- and ten-vertex figures of great complexity). The choice of a polyhedron is determined by the electronic structure, the size of the central atom and the magnitude of stress at the valence angles of the ligand, arising from the process of complex formation. The structural role of water molecules in the complexes with EDTA is discussed in Reference 129.

In Figure 3.19 is shown a fragment of the structure of the binuclear complex $Cu_2A{\cdot}4H_2O$ (A = EDTA) in which EDTA performs the bridge function.

The structure of chelate compounds with the ligands belonging to the different groups of organic compounds – β-diketonates, salicylaliminates, alkyldithiocarbamic acids etc. were investigated by Shugam and Shkol'nikova[130–133].

Starting from the results of structural investigations, Shugam proposed a classification of extra metal–ligand bonds, depending on the interatomic distances, and showed that the extra bonds possess a tendency towards selectivity. Among the donors, the trend to form such bonds is characteristic of O, S and C; the last named entering into the aromatic system. For nitrogen, the tendency to participate in extra bonds is not characteristic. Copper forms the bonds with O, C, S and, sometimes, N; Ni does not form bonds with C, nor Pd with O. The extra bonds possess the directivity and participate in the formation of the coordination polyhedron.

The geometry of metallocycles and the nature of bonds in them were also considered. In particular, it was established that the four-membered metallocycles as well as the five- and six-membered ones can be bent over down to 20 degrees. The monodithiobenzoylmethanate of palladium[132] serves as a rare example of the manifestation of the *trans*-influence in chelate compounds. Here, the effect of *trans*-influence manifests itself not only in the increase of the metal–ligand distance, but also in the change of distances in the metallocycle.

On the basis of structural data Zvonkova and co-workers investigated the intramolecular donor–acceptor interaction in the chelate compounds of

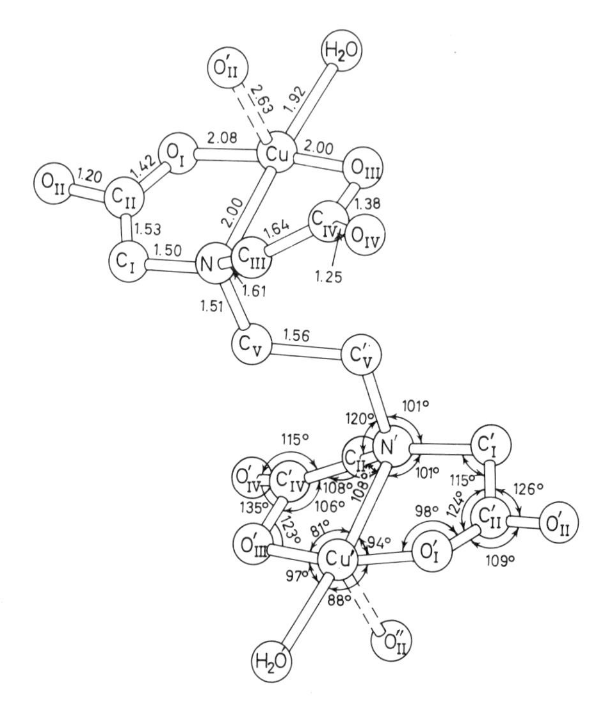

Figure 3.19 Fragment of the structure $Cu_2A{\cdot}4H_2O$ (A = EDTA)
(From Poray-Koshits, *et al.*[128], by courtesy of Izd. Mosk. Univ.)

metals with S-ligands: diethyldithiocarbamates of zinc and cadmium[134] and copper hexamethylenedithiocarbamate[135]. The conclusion was made that the photoelectric properties of these compounds are mainly defined by the short-range order – the nature of the chemical metal–ligand bonds in the chelate node and, to a considerably lesser degree, depend on the long-range order – the molecular packing in a crystal[136, 137].

The calculations of the electronic structures of organic molecules with the N-, S- and O-containing ligands, by means of MO–LCAO methods of the Hartree–Fock–Roothan self-consistent field in the Parr–Pariser–Pople approximation, confirmed the regularities obtained by generalisation of the experimental data on the atomic structures[138].

3.6 ORGANIC CRYSTAL CHEMISTRY AND CONFORMATIONAL ANALYSIS

3.6.1 The 'packing' and interaction energy of molecules in the molecular crystal

The theoretical and experimental work on organic chemistry in the U.S.S.R. has made a considerable contribution to the development of the principal lines of organic crystal chemistry. These are investigations of definite classes of compounds and studies of the 'architecture' of the molecular crystal, i.e. the mutual arrangement of the internal molecules. These regularities define many physical and physico-chemical properties of organic substances.

Kitaigorodskii[139] was the first to work out a consistent theory of the architecture of molecular crystals. He formulated the principle of close packing of the molecules and derived the most probable space groups of symmetry of molecular crystals.

The geometrical aspects of the close-packing theory has been extended by Zorkii and Poray-Koshits[140]. The development of the close-packing theory took the course of transition to the physical and thermodynamical representations, i.e. to the search for minimum free energy for the equilibrium structures.

Kitaigorodskii proposed the method of atom–atom potentials[141,142], as a means of calculating interaction energies of molecules in a crystal. The essence of the method is that the interaction energy of a pair of molecules is reckoned as additive with respect to the interaction energies of atoms; the increment in interaction was taken as an atom–atom potential curve, the shape of which does not depend on the choice of the molecule into which these atoms may enter.

The potential curves of the '6-exp' type

$$f(r) = -Mr^{-6} + N\exp(-gr)$$

are the most convenient for calculations. The constants of the curves are selected empirically, by analysis of the known structures and experimental data on kinetics, adsorption, heat of sublimation and other physico-chemical data which are determined by the interaction of 'neutral' atoms or atoms entering into various molecules.

A number of papers[143–145] present tests of the atom–atom approximation for calculating the heat of sublimation and structure of organic crystals. In References 146 and 147 this method, which only takes into account dispersion interaction of molecules and repulsion from overlap of electron shells, was supplemented by a consideration of electrostatic interaction for molecules with static multipole moments. It was established[146] that the contribution of electrostatic interactions to the lattice energy of many molecular crystals is not the determining factor; it is negligibly small for hydrocarbons, while the establishment of equilibrium molecular orientations is in many cases only slightly dependent on these interactions. This contribution is not proportional to the dipole moment of the molecules, and in some cases can be sufficiently large ($\sim 62\%$ for succinohydrate) to allow the structure to be assigned to a definite subclass (see below)[147].

In principle, the atom–atom potentials method allows one to deduce the structure of a crystal from the shape and size of the molecule. However, the determination of the cell parameters and space group is carried out by experiment, using a computer to find the molecular positions. Other useful applications of the atom–atom approximation are: to 'explain' known structures and to resolve various other problems in crystal physics and crystal chemistry. For example lattice disorder and movement of molecules; calculation of barriers to rotation, surface energies and heats of phase transitions; the study of mutual solubility in the solid state; effects of the crystal field on molecular conformation, etc.[145,148].

An extension of the theory of the structure of molecular crystals is the method of the symmetry of potential functions, proposed by Zorkii[149]. Here, the intermolecular interaction potential is considered in general: only symmetrical properties of the function are analysed without regard for their concrete mathematical expression. This method permits one to derive the optimal and allowable space groups, to separate out the chains and layers of molecules most strongly linked together or to establish the absence of such chains or layers in the structure. This provides a good basis for classification of the molecular crystal structures by introducing the notion of a structural class or subclass[150]. The method of the symmetry of potential functions has led to the derivation of the structural classes and subclasses possible for crystals made up of centrosymmetrical[151] and asymmetrical molecules of one sort equal in the congruent and mirror fashion, as well as of the molecules of two sorts when the symmetry of the molecules of one of these sorts is $1, \bar{1}, 2$ or $2\,2\,2$[152].

The structural classes and subclasses were classified by reference to the degree of probability of their realisation; this classification was confirmed by a statistical analysis of literature data on the structure of molecular crystals (3259 references)[153]. The calculation of the intermolecular interaction energy for a large number of molecular crystals allows one to reckon the structure as belonging to a definite subclass, thereby enabling comparison of the structure of a crystal with the anisotropy of its properties.

3.6.2 Conformational analysis

The simplified mechanical model[154] allows the molecular conformation to be calculated, i.e. the geometry corresponding to the minimum potential energy. Here the bonds between the atoms are assumed to be rigid, and the valence angles elastic. The actual molecular conformation is established as a result of a compromise between the tendency of the atoms not bound by valencies to draw apart at equilibrium distances and the tendency of the valence angles to retain the elastic values.

The molecular conformations are calculated using the method of atom–atom potentials[141,142]. The interaction of atoms not bound by valencies is described by the potential functions of the '6-exp' type. It should be noted that these latter are also used for calculating the conformations of polypeptide chains[155] along with a set of 6–12 functions[156],

$$f(r) = -Mr^{-6} + Nr^{-12}$$

Ramachandran, Venkatachalam *et al.*[157] tested all kinds of potential functions and concluded that further studies in this direction were required.

At the fixed bond lengths the molecular structure is determined by the sum of energies, namely the energy connected with the distortion of the valence angles, plus the energy of non-valent interaction. The appropriate total energy surface is calculated by varying the mutual arrangement of atoms. The presence on this surface of one global minimum which substantially exceeds all other minima in depth, will be indicative of the existence of one stable molecular conformation. Several approximately equal minima will give evidence for the presence of conformational isomers of a given molecule.

Conformations were calculated for very varied hydrocarbons[154]. In nearly every case the agreement with experiment was so good that the predictive value of the method was indisputable.

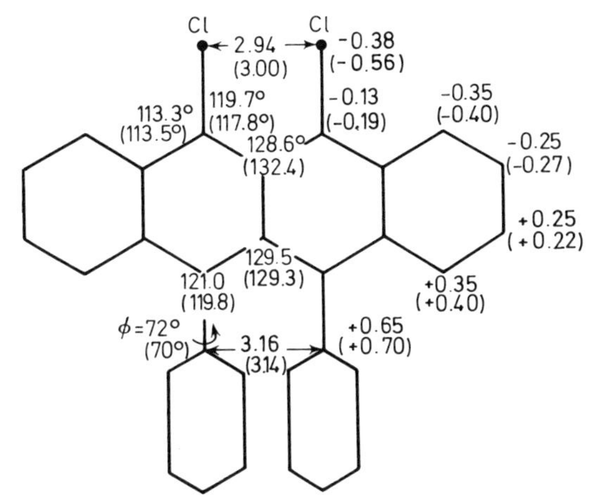

Figure 3.20 Conformation of the molecule of 5,6-dichloro-11,12-diphenylnaphthalene. The experimental data are given in brackets
(From Kitaigorodskii and Dashevskii[160], (1968). *Tetrahedron*, **24**, 5917, by courtesy of Pergamon Press)

The conformational analysis technique has been worked out in detail for many classes of organic molecules including the very interesting class of overcrowded aromatic molecules: derivatives of benzene, naphthalene, acenaphthalene etc.[158–160] (Figure 3.20). It has been shown[159], that, when the ideal models experience a large strain, the conformational analysis results are in good agreement with the experimental data obtained from molecular geometry. In the case of too small a strain, the effects of molecular packing in crystals and possibility of hydrogen-bond formation should be taken into account. (Conformations were also calculated for some cycloalkanes, hydrocarbons with multiple bonds, halogen substituents of methane[161]).

Work on the refinement of the parameters needed for conformation calculations – potential curves for non-valent interaction and elastic constants using the experimental data on the structures of the compounds – is in progress[159, 160].

The mechanical model was applied to the derivation of the primary and secondary structures of polypeptides and stereoregular polymer chains[162].

3.7 STRUCTURE OF ORGANIC AND ORGANOMETALLIC COMPOUNDS

3.7.1 Structures of organic compounds with intramolecular strains

Since 1960 there have been x-ray investigations of the true conformation of overcrowded molecules with steric hindrances, mainly the molecules of aromatic compounds with spherical (halogen atoms) and non-spherical (nitro groups) substituents.

In early 1966, Avoyan, Struchkov and Dashevskii[163] published a review which classified the classical cases of spacial strain in aromatic systems. The analysis of abundant experimental material allowed the authors to reveal some general empirical regularities and to propose a model for conformational distortion. According to the model, the tendency to diminish steric hindrances in overcrowded aromatic molecules with polyatomic substituents should lead, first of all, to the rotation of the substituent with respect to the plane of the aromatic ring, then to the deviation of C—X bonds from their ideal directions (X is the 'key' atom in the substituent). It should further lead to the distortion of the valence angles within the ring itself and disruption of its coplanarity.

During the period 1965–1970 a great deal of experimental material has accumulated on the structure of the substituents of benzene[164, 165], naphthalene[166, 167], anthraquinone[168, 169], diphenyl[170], stilbene[171]. In 1,8-dinitronaphthalene steric hindrance leads to the rotation of the plane of the nitro group through 43 degrees relative to the plane of the ring, withdrawal of N atoms from it to a distance of 0.4 Å and substantial distortion of the valence angles.

In a conformational distortion of the *p,p'*-dimethyl-α,α'-difluorostilbene molecule[171] the valence angle C—C=C increases up to 131–132 degrees and the molecule ceases to be coplanar on account of the rotation of benzene rings through 25 and 27 degrees with respect to the plane of the ethylene chain.

The results of the investigations carried out in the period 1965–1970 confirmed the correctness of the conformational distortion model proposed by Struchkov and Avoyan, and provided additional information on the distances between non-bonded atoms in the molecule needed for conformation calculations.

The molecules of the compounds mentioned in this section are the classical models of overcrowded systems, although many other molecules of organic and organometallic compounds which will be considered in a later section of this review can also be reckoned among overcrowded molecules.

3.7.2 Structure of organic compounds containing nitrogen, oxygen and sulphur

Kondrashov[172] has studied the structures of diazoaminobenzenes, whose molecules contain the triazene chain —NH—N=N—. The precise determination of the N—N bond lengths and localisation of hydrogen atoms revealed the nature of tautomeric transformations in this series.

Compounds containing hemin systems with the >N atom, N—C—X (X = OH, OR, OCOR, >N, CN, Hal, SR), have been investigated[173, 174]. The geometrical parameters of the hemin group which were obtained, have revealed the nature of intramolecular interaction of atomic groupings which defines the specific chemical properties of these compounds.

A series of works[175] has been devoted to the investigation of the conformation of dinitromethyl anions. It has been established that the conformation of this anion is greatly influenced by the substituent as well as by the cation and, in some cases, by the molecular packing. The basic regularities of such an influence were revealed.

Of interest is the work on the determination of the structure of nitro-isoxazolisidine[176] – a new binuclear heterocyclic compound

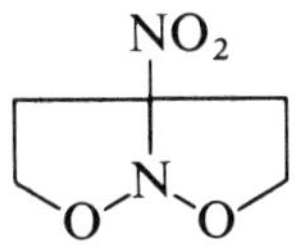

NO_2

N

O O

which contains a previously unknown fragment

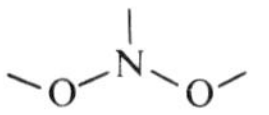

N

O O

Structural studies confirmed the stereochemistry of the molecule proposed on the basis of chemical properties and spectral evidence. Both five-membered cycles have the conformation of an envelope having the corners turned down on the side opposite to the N atom of the nitro-group ('dog-eared' conformation). The NO_2 group is turned out of the plane of the cycles through 12 degrees.

New data were obtained on the double salts of diazonium[177, 178]. It was established that the real charge distribution in a diazonium cation is described only approximately by two terminal forms

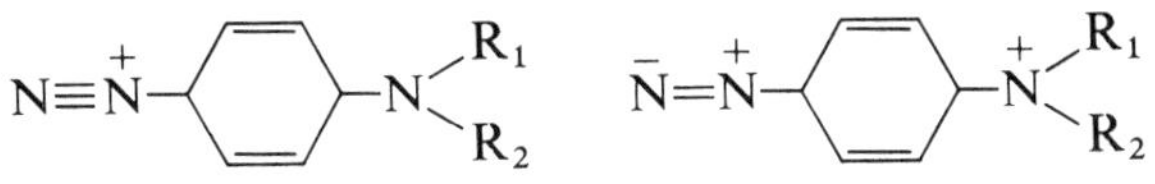

$N{\equiv}\overset{+}{N}$—⟨ ⟩—$N(R_1)(R_2)$ $\overset{-}{N}{=}\overset{+}{N}$—⟨ ⟩—$\overset{+}{N}(R_1)(R_2)$

and is determined by the character of the anion and the radicals R_1 and R_2.

Atovmyan and his co-workers investigated the organic semiconductors: cyclic β-aminovinylketones[179, 180] and salts of the stable anion-radical 7,7,8,8-tetracyanxinodimethane[181]

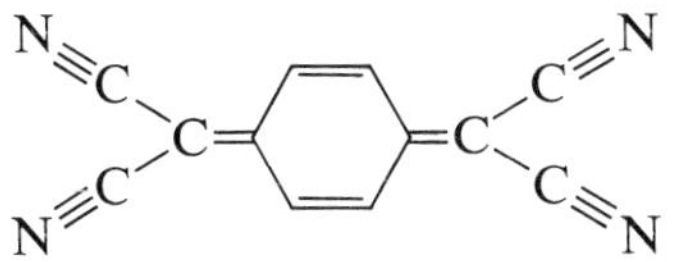

$(N{\equiv}C)_2C{=}$⟨ ⟩$={C}(C{\equiv}N)_2$

which possess a wide range of conductivity. The 'favourable' structural criteria for the conductivity of these salts were revealed.

Studies[182, 183] of the relation of structure to piezoelectric properties are exemplified by two compounds with centrosymmetrical molecules – benzophenone and paraoxyacetophenone.

A large number of works by Zvonkova and co-workers[184, 185] is mainly concerned with the investigation of sulphur-containing organic compounds. In these works the specific features of the structure of sulphur-containing groups, such as the thiocyanide group and thiodiazol ring, have been elucidated, and the regularities in deformation of the benzol ring under the action of various substituents revealed.

Studies of the aptness of the sulphonyl group $>SO_2$ for conjugation are exemplified by several structures[186].

3.7.3 Structure of organometallic compounds

In the early 1960s Vol'pin, Koreshkov, Kursanov and Dulova[187] synthesised the series of organic compounds of germanium and silicon by way of attachment of dihalide derivatives of germanium and silicon to the triple C≡C bond. It was assumed that their molecules include a three-membered heterocycle. These compounds were investigated by various physico-chemical and structural methods (x-ray structure analysis, gas-electron diffraction). The crystal structures of several unsaturated heterocycles were determined[188]

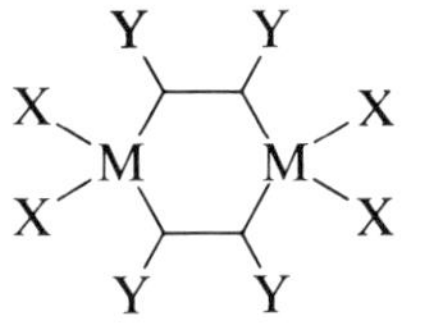

I $M = Si, X = CH_3, Y = C_6H_5$
II $M = Ge, X = CH_3, Y = H$*
III $M = Ge, X = Cl, Y = H$
IV $M = Ge, X = I, Y = H$
V $M = Ge, X = C_6H_5, Y = H$

In all the structures molecules with a six-membered heterocycle were found. This is in agreement with the data of i.r. and u.v. spectra, n.m.r. spectra, saturated steam pressure and mass-spectrometric measurements in the gas phase. At the same time the gas-phase electron diffraction[189] and cryoscopic determination of the molecular weight of germirenes at low concentrations showed that, apparently, one has to admit that the structures with a three-membered cycle do exist, if only as intermediate states.

The lengths of the bonds Ge—C (1.96–2.01 Å) and Si—C (1.88 Å) discovered in the course of the x-ray diffraction investigation correspond with the single bond, the C—C bond length in the heterocycle (1.28–1.35 Å) with the double bond. From the electron diffraction data the length of the Ge—C bond was found to be 1.93–1.98 Å. Here, the lengths of the exo- and endocyclic metal–carbon bonds are equal to one another. It is as if such a bond length distribution indicated the absence of any marked interaction of d orbitals of the heteroatom with the carbon–carbon π-bond of the heterocycle.

However, the coplanarity of heterocyclic systems observed in all germa-

*Melting point of the substance 10 °C; investigation was carried out at −50 °C.

nium compounds is, probably, a manifestation of conjugation, although the effect is too weak to give rise to a noticeable change in bond lengths.

At the end of 1968 a detailed review[190] on the structural chemistry of organic compounds of the Group IV non-transition elements (Si,Ge,Sn,Pb) was published.

This review contains the structural data on the compounds of the Group IVb elements with 'tetrahedral' molecules; compounds with the coordination number above 4; the analogues of linear and cyclic hydrocarbons and, finally, on the compounds of silicon with oxygen, nitrogen and sulphur – siloxans, silazans and silthians.

The authors carried out a very careful analysis of the lengths of the metal–carbon, metal–halogen etc. bonds and came to the conclusion that at present the *a priori* prediction of these values are impossible, and in order to resolve this problem one must take into account all possible valencies inherent in the atom partners.

A brief consideration was also given to π-cyclopentadienyl compounds and metal polynuclear carbonyls with the Group IVb element – transition metal bond.

The structure of a number of polynuclear carbonyl derivatives containing tin[191, 192] and lead[193] were determined. However, it is more expedient to discuss the results of these studies along with the investigations of other π-complexes.

Two structures of tin with 'tetrahedral' molecules have been deciphered: the structure of triphenyl tin chloride[194] and that of tin tetraphenyl[195]. In the first structure the phenyl rings are mutually orientated in a propeller-like manner, in the second compound the molecules have 4 symmetry, the planes of the phenyl rings are rotated about the Sn—C bond through 57 degrees.

In the structure of germanium chloride complex with 1,4-dioxan[196] the Ge atom coordinates two Cl atoms and two O atoms of two dioxan molecules.

Study has been made of interesting structures of icosahedral non-benzenoid aromatic systems[197, 198], the object being to elucidate the laws of substitution in borene derivatives (i.e. boron–hydrogen systems with partial replacements of B by C).

3.7.4 Structure of π-complexes

Although the first π-complex, the Zeise's salt $K[Pt(C_2H_4)Cl_2]$ was discovered as far back as 1828, the rapid development of the chemistry of this class of compounds, representing the area between organometallic chemistry and the chemistry of complex compounds, began in 1951 after the discovery of ferrocene and the establishment of its structure[199]. The Zeise's salt itself was thought to be a compound with the usual σ-bond up to 1960s when its structure was solved by Soviet scientists[200].

π-Complexes have been intensively studied under the direction of Struchkov[201–210].

x-Ray structural analysis plays the leading role in the establishment of the structure of this class of compounds because the great variety of π-complex structural types imposes restrictions on the predictive possibilities of quantum chemistry and hinders the interpretation of the results obtained by other physical methods.

The classical representatives of π-complexes are compounds of the ferrocene type, where the metal atom seems to be 'squeezed' between two flat rings ('sandwich' structure). To this class of compounds belong diferrocenyl and its derivatives. A series of works by Kaluski and Struchkov[201–206] has been devoted to a structural investigation of these compounds, the main purpose being to determine the actual conformation of molecules in a crystal.

It was established[202, 207] that the conformation of 'sandwiches' in the diferrocenyl molecule is intermediate between the prismatic ($\phi = 0°$) conformation and the antiprismatic ($\phi = 36°$) one found in the ferrocene structure: the angle of mutual rotation of five-membered rings of the sandwich about the normal to their plane ϕ is 16 degrees.

In diethyldifferrocenyl[203] and dichlorodiferrocenyl[204] crystals, the conformation of the sandwiches is almost prismatic (the angles are 4 and 5 degrees, respectively). Preliminary data indicates that diacetyldiferrocenyl has the same conformation[205]. In the terferrocenyl molecule, or to be more precise, of 1,1′-diferrocenylferrocene (Figure 3.21) the central sandwich has the

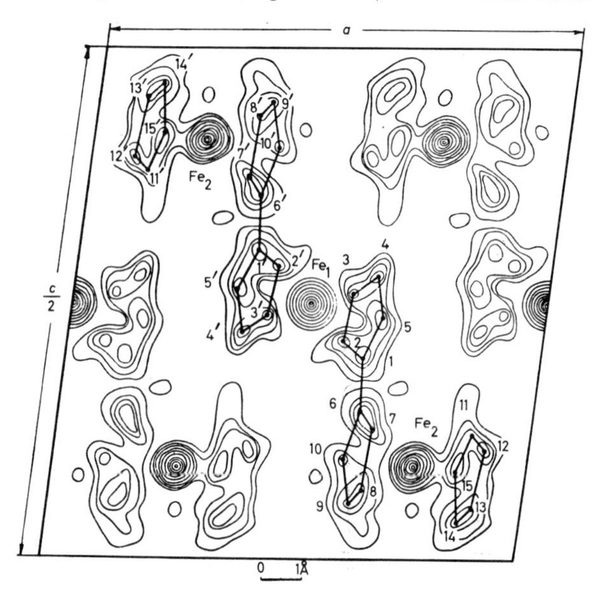

Figure 3.21 Molecular structure of terferrocenyl (From Struchkov[206], by courtesy of Izd. Akad. Nauk SSSR)

antiprismatic conformation, whereas the terminal sandwiches show approximately the prismatic conformation[206]. Earlier, the existence of sandwiches of two types in one molecule of the π-complex was established only in one case: for the compound of composition $Co_2C_{24}H_{25}$ [208]. The structure of this complex compound was established only from structural analysis data, according to which the structural formula can be written as follows: π-$C_5H_5CoC_5H_5$—C_5H_4—C_5H_5—CoC_5H_5-π(2,4-bis(π-cyclopentadienylcobalt-cyclopentadienyl)-cyclopentadiene-2,4). The molecule has the unusual spacial configuration shown in Figure 3.22. It is built up of two sandwich fragments linked together via the planar cyclopentadiene bridge cycle. One of the sandwiches has the prismatic conformation, the other the antiprismatic. The sandwich rings not linked to the bridge are planar. In the sandwich rings linked to the bridge, the bonds with cobalt atoms form with only four carbon atoms in each ring. The fifth atom of these rings, forming the usual

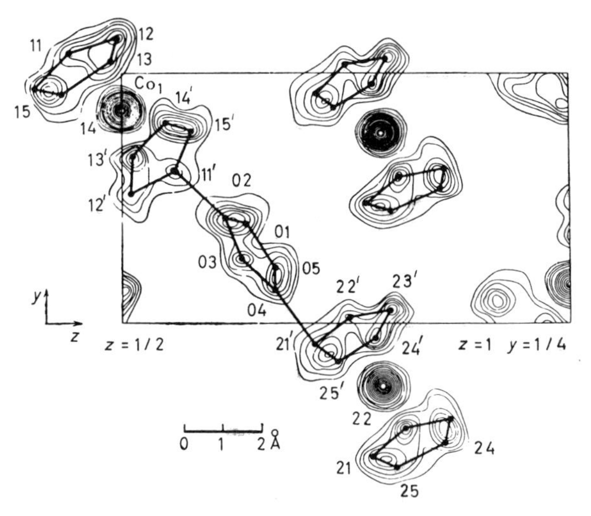

Figure 3.22 *yz*-Projection of $Co_2C_{25}H_{24}$
(From Struchkov[208], by courtesy of Izd. Akad. Nauk SSSR)

bonds with the bridge cycle, do not interact with the cobalt atoms, and deviate noticeably from them, thereby disrupting coplanarity of bonded rings.

In the structure *trans-β*-ferrocenyl-acrylonitrile (*trans*-π-C_5H_5)Fe(π-C_5H_4 CH=CHCN)[209], the acrylonitrile grouping is planar and almost coplanar with the substituted cyclopentadienyl ring (the dihedral angle is 3.5°). Both cyclopentadienyl rings are planar, are parallel to each other and have the conformation intermediate between the antiprismatic and prismatic conformation with the ϕ-angle equal to 20 degrees. It thus follows that the maximal conjugation of π-bonds is reached in the molecule.

Studies of the structure of 'sandwich' compounds confirmed the known conception that the ferrocene rotamers are almost equal energetically, and that their conformation in a crystal is defined by intermolecular interactions. Close to the sandwich structure of diferrocenyl derivatives is the structure of π-cyclopentadienyl-π-tetraphenylcyclobutadiene vanadium dicarbonyl, (π-C_5H_5)V$(CO)_2$(π-Ph_4C_4)[210]; its molecule has the conformation of the wedge-shaped sandwich, the angle between the planes of cyclopentadienyl and cyclobutadiene rings being equal to 43 degrees. The average distances from the V atom to the C atoms of these rings are 2.24 and 2.26 Å respectively. The work is part of the investigation of π-complexes which are formed as a result of the reaction of photochemical substitution of carbonyl groups in the compound of the type (π-C_5H_5)M$(CO)_4$ (M = V,Nb,Ta).

The structure of tetracyclopentadienylzirconium $(C_5H_5)_4$Zr[211] – a representative of the class of polycyclopentadienyl compounds, is of interest. According to the data obtained by various physico-chemical methods (with the exception of x-ray structure analysis) these compounds can have a very varied structure: (1) the molecule adopts the piano-stool conformation – one ring is attached in the sandwich-type mode, and three rings are linked by metal–carbon σ-bonds, (2) two rings are attached via π-bonds and two

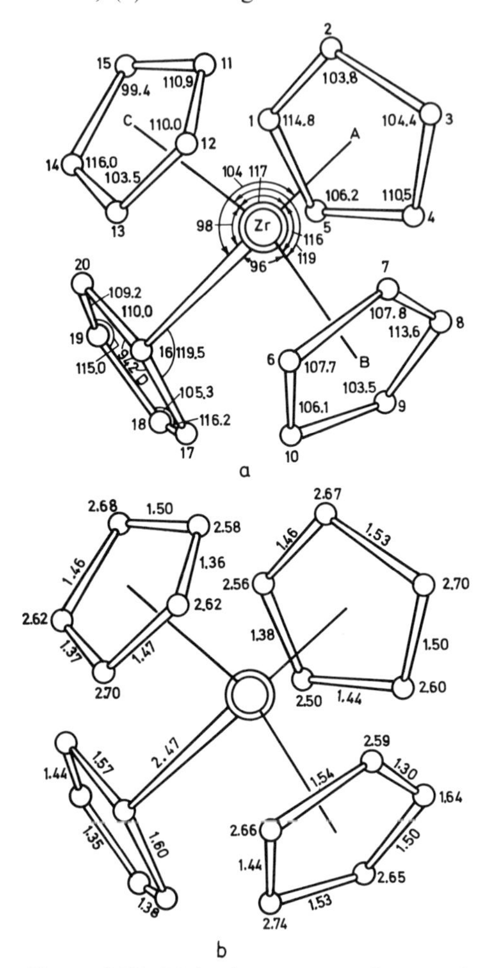

Figure 3.23 Molecular structure of tetracyclopentadienyl zirconium
(From Struchkov *et al.*[211], by courtesy of Izd. Akad. Nauk SSSR)

rings via σ-bonds, (3) all the four rings are equivalent and are linked by σ-bonds. For Zr and Hf compounds, the evidence from paramagnetic resonance and that from i.r. spectroscopy are in conflict with each other (^{1}H n.m.r. – all the four rings are equivalent; i.r. spectroscopy – different modes of attachment are possible).

x-Ray structural analysis showed that the Zr atom is bonded with three cyclopentadienyl rings in the sandwich-type mode (the average Zr—C distance is 2.64 Å), and with the fourth ring it forms the localised Zr—C σ-bond, the bond length being 2.47 Å; the σ-bond is non-coplanar with the plane of its ring and forms an angle of 52 degrees. This indicates sp^3 hybridisation of the σ-bonded C atom. Formally, the Zr atom has tetrahedral coordination. The structure of the molecule $(C_5H_5)_4Zr$ is shown in Figure 3.23.

3.7.5 Semi-sandwich complexes of transition metals

A recent review[212] includes a considerable amount of work on the determination of structures of the simplest representatives of π-complexes of transition metals – olefin and acetylene complexes.

The characteristic feature of the compounds of platinum and palladium – *cis*-$Pt(C_2H_2)(NH_3)Br_2$ (Zeise's salt)[213], $(C_6H_{10})PdCl_2$ (C_6H_{10}-hexadiene-1,5)[214] and (π-C_3H_5)Pd (π-C_5H_5)(C_3H_5-allyl)[215] – is the square-planar coordination of a metal atom with an approximately perpendicular arrangement of coordinated double bonds with respect to the plane of the square. In the Pd^{II} complex, the bidentate ligand 1,5-hexadiene assumes a peculiar conformation, thus obeying the laws of square-planar coordination of the Pd atom (Figure 3.24).

Several π-allyl complexes of transition metals Pd and Fe were investigated. They are interesting in that they combine the features of bonds of both

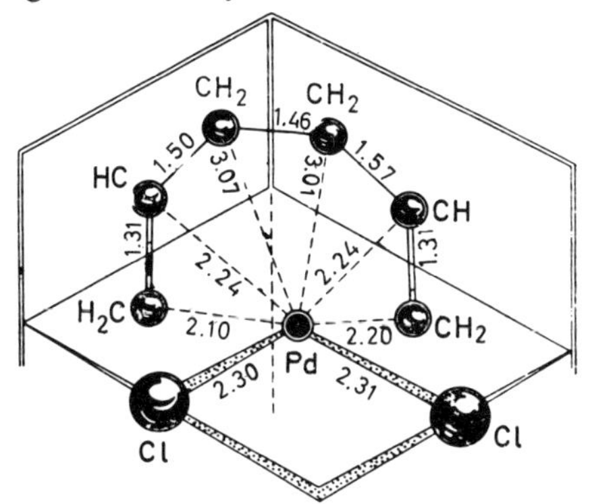

Figure 3.24 Structure of the Pd complex with 1,5-hexadiene
(From Poray-Koshits *et al.*[214], by courtesy of Izd. Akad. Nauk SSSR)

olefin and arene complexes. It was established that all these compounds are true π-allyl complexes, i.e. the allyl group in them is symmetrical, and double bonds are delocalised.

Complexes of rhodium with substituted quinones can be reckoned among the π-olefin complexes[218]. The quinone ligand in these compounds stands out as a diolefin ligand: only double bonds coordinate to the Rh atom. As a result, the six-membered ring, planar in free quinones[219], adopts the distinctly expressed bath conformation. Here, the conformation of quinone ligands does not depend on the properties of the second ligand, but does depend on the electronic structure of the complex-forming metal; replacement of Rh by Ni^{IV} leads to the fact that the duroquinone ligand appears to depart less from coplanarity, and the M—C bonds become considerably longer[220].

Of interest is the structure of another π-complex: the product of the reaction of the equimolar mixture of α- and β-bromomethylnaphthalene with iron nonacarbonyl[221] (Figure 3.25). The structural formula of the

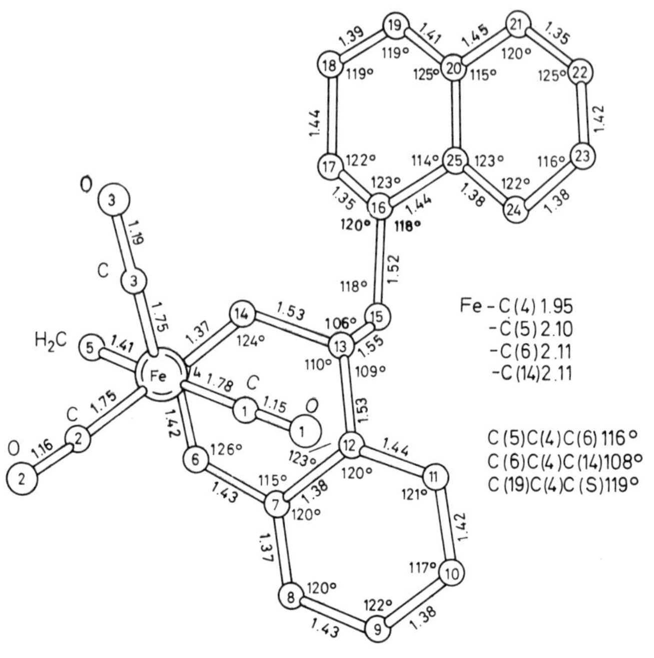

Figure 3.25 Molecular structure of a product of the reaction of the equimolar mixture of the α- and β-bromomethylnaphthalenes with iron nonacarbonyl
(From Struchkov[221], by courtesy of Izd. Akad. Nauk SSSR)

compound was established only from x-ray structural analysis. Apart from three CO-groups, the β-naphthamethyl residue, also coordinates to the Fe atom or, to be more exact, only its trimethylmethane fragment with the β-methylene group. The α-naphthomethyl residue is not capable of such a coordination and is found as a substituent at position 4 of the β-naphthomethyl residue:

H CH_2 4 3 2 1 H_2C $Fe(CO)_3$

Structural studies were made of four π-acetylene complexes of niobium[222, 223], obtained as a result of photochemical reactions of replacement of carbonyl groups in $(\pi\text{-}C_5H_5)Nb(CO)_4$ by the molecule of diphenyl acetylene (Ph_2C_2), tetraphenylcyclobutadiene (Ph_4C_4) and carbomethoxy-acetylene (CH_3OCOC) respectively:

I $(\pi\text{-}C_5H_5)Nb(CO)(Ph_2C_2)(\pi\text{-}Ph_4C_4)$
II $(\pi\text{-}C_5H_5)Nb(CO)(Ph_2C_2)$
III $[(\pi\text{-}C_5H_5)Nb(CO)(Ph_2C_2)]_2$
IV $[(\pi\text{-}C_5H_5)Nb(CO)(CH_3OCOC)_2]_2$

In complexes I and II the non-bridge coordination of the triple bond is realised; in bimolecular complexes III and IV, bridge coordination takes place. The structure of complexes III and IV is shown in Figure 3.26.

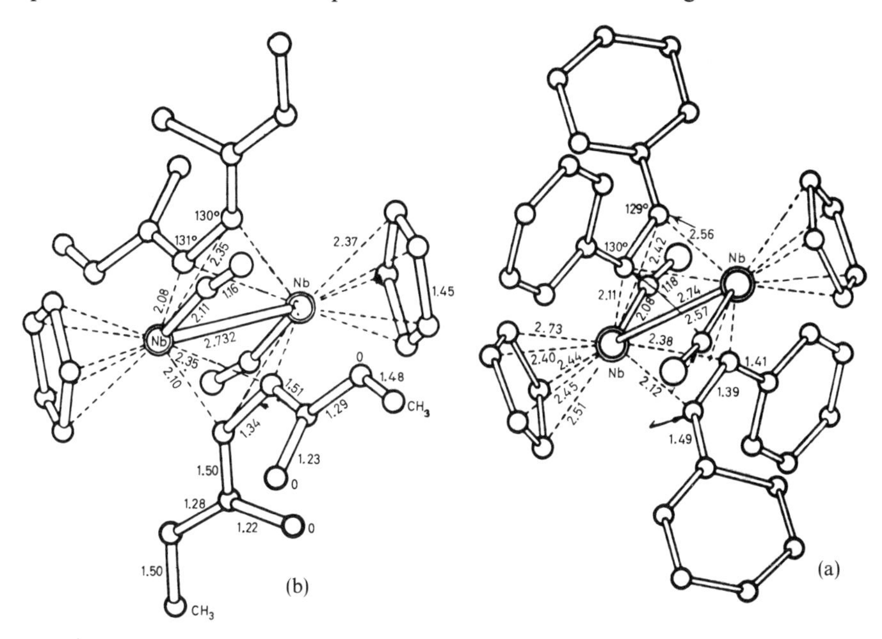

Figure 3.26 Structure of two binuclear acetylene complexes of Nb: (a) $[(\pi\text{-}C_5H_5)Nb(CO)(Ph_2C_2)]_2$; (b) $[(\pi\text{-}C_5H_5)Nb(CO)(CH_3OCOC)_2]_2$
(From Struchkov[212], by courtesy of Izd. Akad. Nauk SSSR)

For vanadium, whose radius is less than that of Nb by 0.15–0.20 Å, of compounds of type I and II, only type II exists; its structure is described in Reference 224.

In recent years the structure of semi-sandwich π-complexes of transition metals has been investigated, mainly of iron with the metal carbon σ-bond and the overall formula $(\pi\text{-}C_5H_5)Fe(CO)(L)(\sigma\text{-}R)$ where $L = CO$ or $P(C_6H_5)_3$ (replacement of the carbonyl ligand by the triphenylphosphine molecule stabilises the molecule); σ-R = aryl, acyl etc.[225–227]. The main object of these investigations is to elucidate the nature of the metal–carbon σ-bonds depending on the electronic structure of the metal and the character of the σ-ligand.

The formulae of the compounds investigated are given below.

I $(\pi\text{-}C_5H_5)Fe(CO)[P(C_6H_5)_3](\sigma\text{-}C_6H_5)$
II $(\pi\text{-}C_5H_5)Fe(CO)[P(C_6H_5)_3](\sigma\text{-}COC_6H_5)$
III $(\pi\text{-}C_5H_5)Fe(CO)[P(C_6H_5)_3](\sigma\text{-}\alpha\text{-}C_4H_3S)$
IV $(\pi\text{-}C_5H_5)W(CO)_3(\sigma\text{-}C_6H_5)$

In all the complexes the Fe atom has the piano-stool coordination characteristic of semi-sandwich π-complexes, the P atom the distorted tetrahedral coordination and the W atom the 7-coordination analogous to that found for semi-sandwich π-cyclopentadiene complexes of molybdenum.

The authors have analysed in detail the nature of bonds in these compounds.

For compound IV the covalent radius of tungsten (1.57–1.58 Å) is established from the values of W—C bond lengths.

A new π-complex of iron with σ-bonds has a peculiar structure[229]. In this complex, which is a product of the interaction of σ-(β-acetylvinyl)-π-cyclopentadienyl iron dicarbonyl with iron nonacarbonyl, the organic radical is linked simultaneously by π- and σ-bonds. The complex is binuclear with the bridge CO-group.

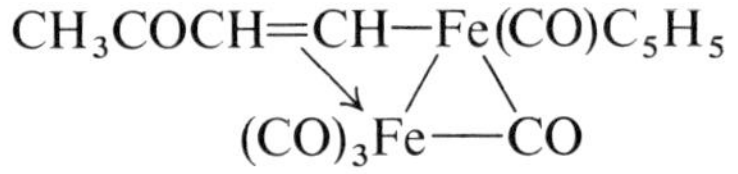

$CH_3COCH{=}CH{-}Fe(CO)C_5H_5$
$(CO)_3Fe{-}CO$

One of the Fe atoms is linked with the acetylvinyl ligand by a π-bond, the other by a σ-bond. The acetylvinyl radical is planar and adopts *cis*-conformation, whereas *trans*-conformation is more favourable energetically.

Several semi-sandwich π-complexes of transition metals without metal–carbon σ-bonds were investigated[230–232].

The polynuclear polymetallic carbonyls which combine in their molecules the fragments of π-complexes of transition metals and σ-organometallic compounds can be reckoned among the semi-sandwich compounds. These fragments can be linked together by means of direct bonding between metal atoms. Special interest has also been attached to studies of compounds with bonding between the transition metal and Group IV elements (Si,Ge, Sn,Pb), because in these compounds, the nature of the metal–metal bond and stereochemistry of the complex are extremely sensitive to the character of ligands attached to the atom of the Group IV element.

Four compounds of the type $[\pi\text{-}C_5H_5Fe(CO)_2]_2MX$ were investigated; here M = Sn or Pb, X = the inorganic (NO_2) or organic (CH_3,C_2H_5) ligand[191–193].

I M = Sn, X = NO_2
II M = Sn, X = CH_3
III M = Sn, X = C_2H_5
IV M = Pb, X = CH_3

The molecules of all these compounds comprise a chain of covalently bonded atoms, e.g. Fe—M—Fe.

The shortening of the Sn—Fe (but not Pb—Fe) bonds was discovered as well as a certain degree of rehybridisation ($sp^3 \rightarrow sp^2$) of tetrahedral hybrid orbitals of the Sn atom in accordance with the Bent law[234].

The cyclopentadienyl rings bonded to the Fe atom are planar in all the compounds. In compound III, two rings out of four are linked with the Sn atom by a σ-bond (2.18 Å) and have the conformation of an envelope with a corner turned down through 27 degrees.

The coordination of the iron atom is the piano-stool coordination characteristic for semi-sandwiches, while the coordination of the Sn and Pb atoms

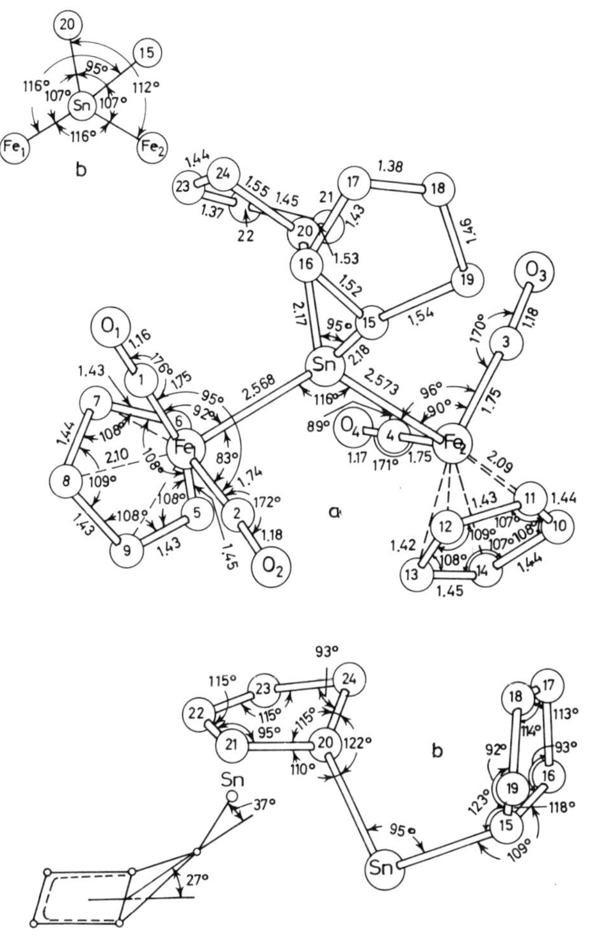

Figure 3.27 Molecule of $[(\pi\text{-}C_5H_5)Fe(CO_2)]_2Sn(C_5H_5)_2$ (From Struchkov[191], by courtesy of Izd. Akad. Nauk SSSR)

is the strongly distorted tetrahedron. The NO_2 ligand in I coordinates to the Sn atom with its oxygen atom as the nitrite group O—N=O. The structure of molecule III is shown in Figure 3.27.

Also belonging to the polymetallic organometallic class of compounds is $(CO)_4CoSn(C_6H_5)_2Mn(CO)_5$ [191, 192] in which there is a chain of atoms

Co—Sn—Mn and where the Sn—Co bond (2.66 Å) was found for the first time (the Sn—Mn bond length is 2.73 Å). Also the dimeric molecule of $(\pi\text{-}C_5H_5)Mo(CO)_3Mn(Co)_5$ [235, 236] in which the monomeric fragments are linked together by Mo—Mn bonds (3.08 Å).

Systematic studies on the structure of semi-sandwich π-complexes showed that the metal–carbon distances in cyclopentadienyl are sufficiently constant in value and can be used for estimating the single-bonded covalent radii of the transition metals in their π-complexes[237].

The present review on the investigations in the field of organic crystal chemistry gives evidence that these are concentrated on very important lines of research, which enables one to obtain results valuable both for theoretical and experimental chemistry.

3.8 THE STRUCTURE OF BIOLOGICAL OBJECTS

The investigation of the structure of biologically important compounds from relatively low-molecular compounds, such as amino acids, nucleotides, antibiotics etc., to the most complex structures of proteins and viruses represents one of the most important and rapidly developing branches of the analysis of the structure of matter. Modern molecular biology owes, to a large extent, its success to these investigations.

Here we shall briefly consider some investigations of the structure of nucleic acids, proteins and related compounds which have been recently carried out in the U.S.S.R. These investigations are concerned with all levels of the organisation of biological molecules: from the primary structure to the quaternary one.

3.8.1 Nucleic acids

The question concerning the possibility of the hydrogen bond formation between complementary nucleotides of various sorts is of interest. In addition to numerous pairs of such a type[238], Baklagina and co-workers[239] have investigated the purine–pyrimidine complexes 1-methyl-5-bromouracil–9-methyladenine. Besides the complexes having hydrogen bonds indicated in Figure 3.28, a small quantity (about 4%) of complexes in which the second compound contributes to the hydrogen bonds by the O_2 and N_3 atoms is formed. The molecules in these latter are turned over, thus causing the CH_3 and Br to change places.

The primary (chemical) structures of individual tRNAs have been investigated in recent years. In particular, Bayev and co-workers have established the structure of valine tRNA and a number of its fragments[240–242]. The tRNA molecule (molecular weight 6000) consists of about 80 nucleotides. The generally accepted model of the secondary structure is the so-called clover leaf[243]. On the basis of conformational calculations a number of models of the tertiary structure has been proposed (see, for example, Reference 244). Attempts to carry out direct x-ray diffraction determination of the RNA structure have so far been unsuccessful. However, the small-angle x-ray

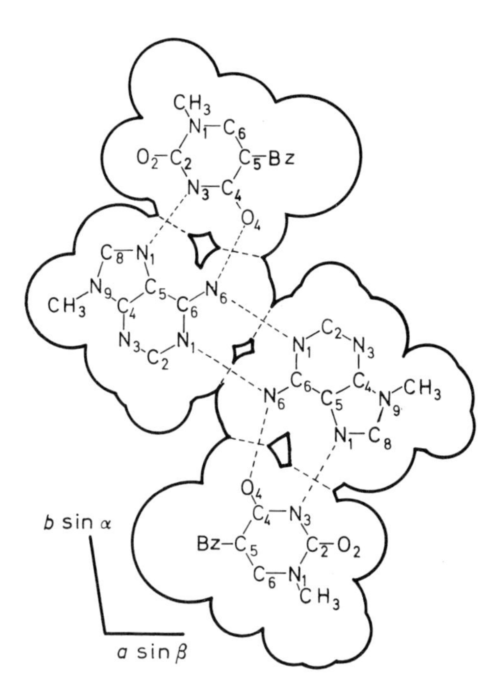

Figure 3.28 Formation of the purine–pyrimidine complex of the four molecules through H-interaction
(From Baklagina, by courtesy of Institut Vysokomol. Soedin)

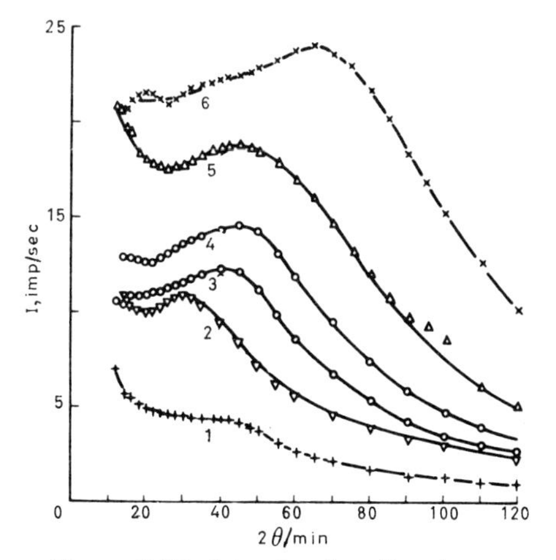

Figure 3.29 Intensity of small-angle x-ray scattering of t-RNA solutions of different concentrations: Curve 1, 0.5%; Curve 2, 1.3%; Curve 3, 1.6%; Curve 4, 2.1%; Curve 5, 3.5%; Curve 6, 6%
(From Vainshtein[245], by courtesy of Izd. Akad. Nauk SSSR)

scattering technique gives information about the structure of tRNA molecules and their behaviour in solution.

By this method Feigin and colleagues[245, 246] have investigated the tRNA isolated from brewer's yeast. A series of experiments has been performed over a wide range of tRNA concentrations in water. For concentrations higher than 1%, a maximum on the scattering curve was observed (Figure 3.29) which becomes more distinct with the increase in concentration, and is displaced to the high-angle region.

The appearance of the maximum indicates the presence of intermolecular interference arising as a result of some ordering of tRNA molecules in

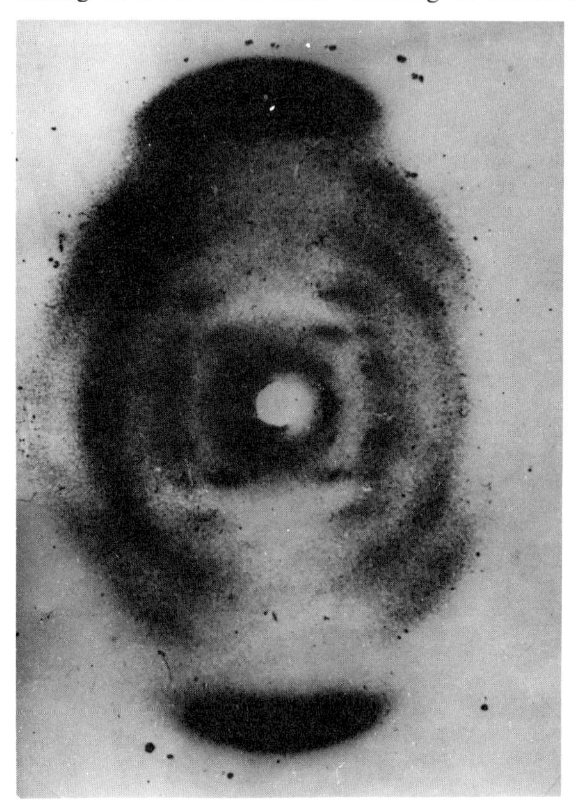

Figure 3.30 x-Ray diffraction pattern of DNA phage T_2, T conformation (humidity 66%)
(From Mokul'sky, by courtesy of Institut Atomnoi Energii)

solution. The average distance, $\bar{a}$, between the molecules, caluclated from small-angle scattering, for example, for concentration 1.6% $\bar{a} = 135$ Å turned out to be close to the average distance calculated from concentration.

Such an ordering was explained as being due to the negative electric charge carried by tRNA molecules, so that their mutual repulsion leads to a certain statistical equilibrium arrangement of molecules. In fact, on adding cations to solution the ordering was observed to disappear.

Repulsion of tRNA molecules at quite small concentrations also explains the fact that crystals of this substance always contain a very large amount (up to 80 %) of water of crystallisation. On the other hand, the repulsion may shed light on the biological function of tRNA whose molecules do not have to be necessarily in contact with each other, but only with other molecules.

The addition of electrolyte to the solution made it possible to eliminate the

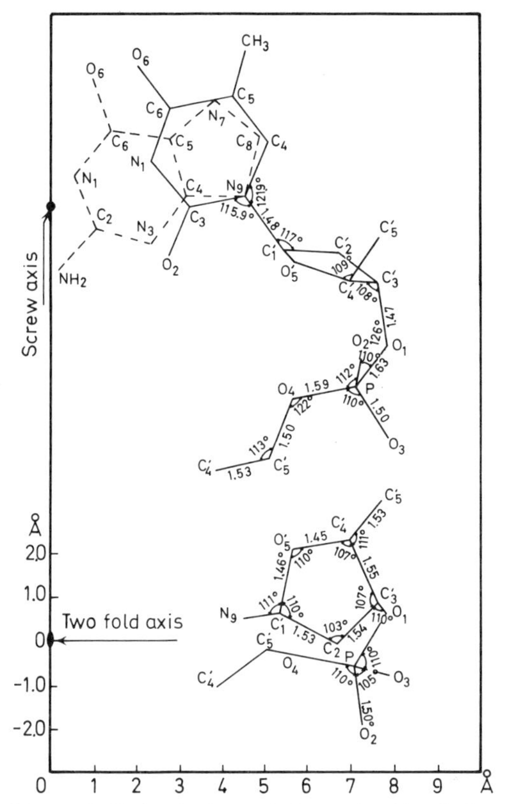

Figure 3.31 Projections of the nucleotide of the DNA molecule with T-conformation; top-projection along the screw axis; bottom-projection along the twofold axis
(From Mokul'sky *et al.* (1972). Molecular Biology, by courtesy of Plenum Publishing Corp.)

intermolecular interference effects, and to determine from the shape of the zero maximum the geometrical parameters of the molecule. Its radius of inertia is 21 Å, the radius of inertia of the cross-section is 9.5 Å and the cross-section area is 480 Å^2. The molecule can be approximated by an ellipsoid with axis dimensions 18.33 and 85 Å.

The structure of bacteriophage DNA was investigated. As is known, the DNA molecule has a double helix structure which assumes, depending on the degree of humidity, two main conformations: A and B.

In the course of the investigation of the DNA of phage T_2, Mokul'skii, Mokul'skaya *et al.*[248, 249] found the new so-called T-conformation (Figure 3.30). This DNA differs from the majority of other DNA molecules in the anomalous primary structure: the base complementary to guanine in it is glycosidic oxymethylcytosine.

On the usual conformational transition B–A with the decrease in humidity the number of base pairs N per turn changes from 10 to 11. Instead, in the DNA phage T_2 the decrease in humidity involves conformational transitions which lead, in the end, to the appearance of the previously unknown T-conformation. During the conformational transition the number N decreases from 10 to $8\frac{1}{3}$–8; also, intermediate conformations have been recorded where N is close to $9\frac{1}{3}$ and 9. The presence of glucosidic residues, which lie in broad grooves of the DNA phage T_2, is accounted for by the absence of the usual conformational transition. To construct stereochemically acceptable atomic models of DNA with a given value of N, use was made of the atomic models and computer calculations. Among conformations constructed, only those were selected which gave the best agreement with diffraction data. In this way the coordinates of atoms in the DNA molecule in conformations with $N = 8$, $8\frac{1}{3}$, 9 and $9\frac{1}{3}$ were determined and the picture of the conformational transition B–T was also elucidated.

The unit cells and the molecular positions in them were determined for the structures observed (Figure 3.31). Analysis of the x-ray patterns and calculations of the intensities enabied one to establish that for the cases of B-conformation ($N = 10$) as well as for conformations with $N = 9\frac{1}{3}$ and 9 the helical disorder in the molecular arrangement is possible. For the cases with $N = 8\frac{1}{3}$ and 8 (T-conformation) the helical disorder is absent and the values for relative helical displacements can be determined with good accuracy.

The x-ray diffraction study has been made of the structure of bacterial viruses[250, 251]. The scattering curves for phage CD clearly show a maximum in the 24 Å region, which tells of the ordered arrangement of DNA in the heads of the phage particles. The size of the ordered regions is about 100 Å. Investigations of diluted solutions of phage CD allowed one to establish the radius of inertia (300 Å) for this phage and determine the phage particle dimensions.

3.8.2 Investigations of proteins and related compounds

Proteins are chains of amino acid residues. The structures of all basic amino acids and many related molecules are now known. The results of these investigations are summarised in the monograph of Gurskaya[252]. The structures of proline[253, 254] and phenylalanine[255] have been determined by x-ray diffraction, and those of glycinate and alaninate of copper by electron diffraction[256].

The investigations of cyclic peptides have been carried out both by x-ray diffraction and other physical methods. The investigation of gramicidin S, a cyclic decapeptide, is in progress; a number of its crystalline derivatives containing heavy atoms have been obtained. The F^2 series give evidence that this peptide is built up on the basis of β-conformation[257, 258]. Ovchinnikov and co-workers[259] carried out mass spectrometric and spectroscopic investigations supplemented with conformational analysis[260]. This has

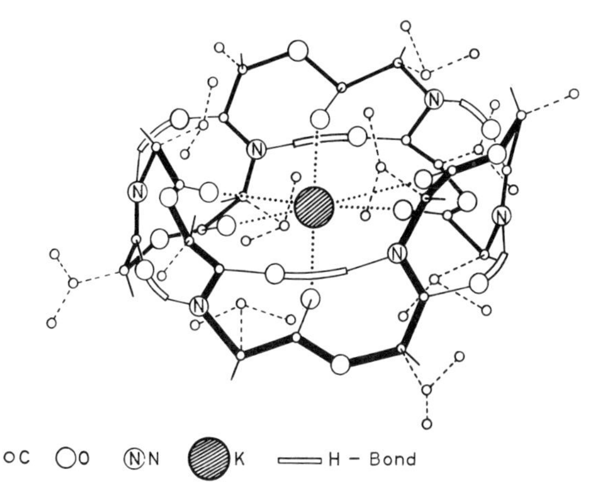

Figure 3.32 Three-dimensional molecular structure of the valinomycine K† complex (side view)
(From Ivanov *et al.* (1971). *Khim. Prirod. Soedin*, 221, by courtesy of Pran U.S.S.R.)

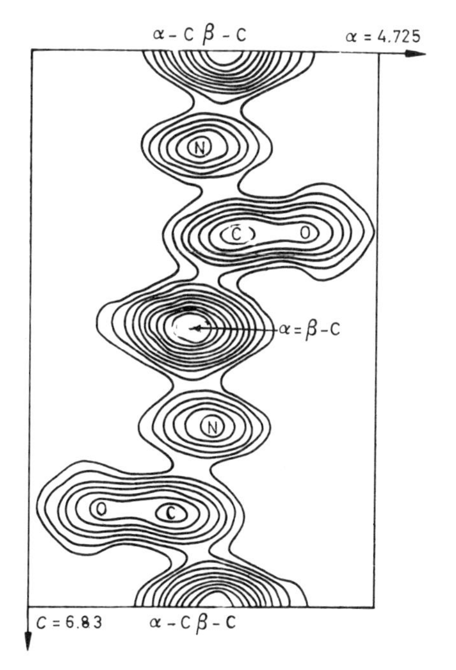

Figure 3.33 Fourier projection of potential of an extended polypeptide chain of the poly-γ-methyl-L-glutamate in the β-form
(From Vainshtein[261], by courtesy of Academic Press)

enabled a model for a number of structures of cyclic peptides and related structures to be prepared which play an important role in the active ion transfer in biological membranes. As an example, Figure 3.32 shows a model of valinomycine. The substances simulating the main types of the secondary structure of proteins are synthetic polypeptides in which the R radicals attached to the basic backbone of the polypeptide chain are only

$$-\overset{O}{\overset{\|}{C}}-\underset{H}{N}-\overset{R_1H}{C}-\underset{\underset{O}{\|}}{C}-\overset{H}{N}-\underset{R_2}{C}(H)-$$

of one, two or three sorts (in contrast to natural proteins in which the number of main R radicals is 22). Electron diffraction studies of poly-γ-methyl-L-glutamate (PMG) allowed[261] two main conformations to be recorded: the α-helix and the stretched β-structure (Figure 3.33). In the β-structure, the molecules are packed together in a parallel array, the H-bonds being formed between neighbouring chains. Pauling and Corey[262] predicted the two main types of the β-chain packing – parallel and antiparallel. In PMG the chains are laid out in parallel; however, the framework of statistical disorder areas in which the chains are running antiparallel may be encountered. The β-layers are superimposed with an approximate preservation of parallelism, but with random shifts in the plane of the layers themselves.

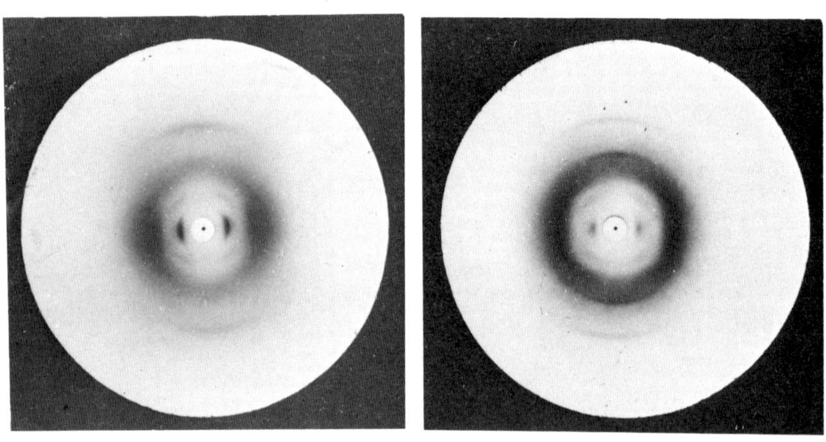

Figure 3.34 x-Ray diffraction patterns of: (a) collagen; (b) oriented fibres of the synthetic polymer (-Gly-Pro-Oxypro)
(From Andreeva[265], by courtesy of Izd. Akad. Nauk SSSR)

A large series of works by Andreeva, Shibnev *et al.*[263–265] has been devoted to the investigation of synthetic polypeptides simulating the collagen structure. This fibrous protein is made up of three polypeptide chains forming a helix which contains 10 structural units per three turns[266, 267]. x-Ray diffraction patterns of natural collagen are rather poor in reflections and therefore the simulation of this structure by synthetic analogues should give the answer to the question about the reasons of the collagen helix formation.

Characteristic of the primary structure of collagen is the presence of triplets containing the residues of glycine and the imino acids prolyne and oxyproline. The polymers of the type $(\text{-Gly-Pro-Oxypro})_n$ and $(\text{-Gly-Oxypro-Oxypro})_n$ were synthesised as well as a number of other polymers close in structure. Some of them (Figure 3.34) gave x-ray diffraction patterns typical of collagen, and, undoubtedly formed the same secondary structure. A change in humidity may cause some variation in the 10/3 helix parameters. It thus follows that the collagen helix is formed if the first residue in the triplet is glycine and two others, i.e. imino acids; the number of triplets should exceed five.

A series of investigations, in particular with the aid of the small-angle x-ray scattering technique has been carried out in order to study the conformational transitions of polypeptides in solution. Studies have been made of the conditions of the transition, the extended conformation-helix-coil. The role of hydrophobic interactions and other factors in these transitions were studied[162, 268–270]

As is known, the secondary and tertiary structure of globular proteins is defined by the primary structure – the sequence of amino acid residues. Stereochemical analysis[272] and investigations of model synthetic peptides indicate that some amino acid residues permit α-helix formation quite readily while others, on the contrary, are antihelical. Ptitsyn and co-workers[271] carried out the statistical analysis of the distribution of amino acid residues in globular proteins for which the tertiary structure was established by x-ray diffraction[273, 274]. From this analysis a number of regularities were revealed. It was possible in particular, to establish that close correlation in the arrangement of groups of residues along the chain is practically absent. This means that a given section of the chain assumes a helical configuration depending only on the presence in it of definite sorts of amino acid residues, i.e. the helix formation is determined by interaction of their side radicals R with the main chain, but not by their interaction with each other: for example, by the presence of definite pairs or triplets of residues.

The α-helix is formed with assistance from ala, leu, met, while asn, cys, glu, ser, thr, trp offer opposition to helix formation, asp, glu located within the N-terminal sections of the chain aid in helix formation, but they are found to counteract this formation in the O-terminal section, etc. Each residue can be ascribed a certain 'helical potential', and summation of such potentials shows suggestion of α-helix formation. The method predicts the helical (or non-helical) state of the polypeptide chain with the probability of about 80%.

The x-ray diffraction study of liquid–crystal gels formed by contracting proteins isolated from muscles has been carried out[275]. In actine liquid crystals, a structural transition initiated by definite specific ions was found. During the reconstruction of an element of the muscle functional apparatus – thick fibres of myosin – the effect of ordering of these fibres in solution was revealed. Work with native protein, F-actine, confirmed the double helical model of its structure. The x-ray patterns of such gels resemble the x-ray patterns of the native muscle. The effective diameter of myosin is 22 Å, that of F-actine 85 Å.

A series of works has been devoted to the x-ray diffraction study of globular proteins. In the investigation of such proteins, especially at high molecular weights, electron microscopy can also give structural information, especially if it is used in combination with new methods of mathematical treatment of electron micrographs, and optical diffraction. In this respect, the method of small-angle x-ray scattering is informative, too.

Data have been obtained on the unit cells and symmetry of glycerophosphate dehydrogenase[277], and leghaemoglobin[278]. The inhibited trypsin has been investigated by the introduction of heavy atoms[279]. Using the method of small-angle scattering, data on the structure of amylase has been obtained[280].

The investigation of solutions of pepsinogen and pepsin was carried out[276]. The radius of inertia is 23 Å for pepsinogen, 20.5 Å for pepsin; the

Figure 3.35 Molecular model of pepsin at 5 Å resolution (From Andreeva[281], by courtesy of Plenum Publishing Corp.)

axis dimensions of ellipsoids of revolution simulating the molecules are 42 and 84 Å for pepsinogen, 37 and 74 Å for pepsin; the ratio of the surface of the molecule to its volume is 27 $Å^{-1}$ for pepsinogen, and 0.26 $Å^{-1}$ for pepsin.

For single crystals of pepsin, Andreeva, Borisov, Melik-Adamyan and Shutskever[281, 282] obtained the isomorphous derivatives and determined the structure at 5.5 Å resolution. Pepsin is an acid proteolytic enzyme (molecular weight 35 000) which crystallises at pH = 2 from 20% ethyl alcohol ($P2_1$, $a = 54.7$ Å, $b = 36.3$ Å, $c = 73.5$ Å, $\beta = 104°$). The derivatives of K_2HgI_4, $K_2Pt(NO_2)^{2-}$, $KPt(C_2O_4)_2^{2-}$ were obtained. The heavy atom positions were found and then refined from Patterson functions. Then, the phases of reflections of protein were calculated. On the electron density distribution at the 5.5 Å resolution, the individual molecules cannot be separated out quite distinctly due to their being in very close contact with each other; however, the most probable separation appears to be that shown in Figure 3.35 in which the regions with the electron density values exceeding 0.52 Å are represented. On this distribution one can separate out the compact rectilinear section, apparently corresponding to the α-helix. All three sites of heavy-atom attachment are concentrated around this section. Also, there is a large number of other extended dense segments which, in all probability, correspond to the polypeptide chain sections. But at the given resolution it seems impossible to join them into a continuous sequence. It can be seen from available biochemical data that the HgI_3 ion inhibits the pepsin activity and its position may serve as an indication of the active centre position in this protein.

On the basis of the combined use of electron microscopy, optical diffraction and x-ray analysis, study has been made of catalase which is an enzyme catalysing the decomposition of hydrogen peroxide. Its molecular weight is 250 000. The molecule contains four hemes, and can be dissociated into two, and then a further four, subunits. Under different crystallisation conditions this protein gives different modifications and forms plane monomolecular layers as well as tubes with monomolecular walls[283–288]. The space group of hexagonal catalase is $P3_121$, the number of molecules in the unit cell $z = 6$. The data on the structure are summarised in the table below. Here ω is the unit cell volume, X the coefficient of filling the unit cell with protein.

Table 3.2

Method	*x-Ray diffraction*		*Electron microscopy*
Crystals	*Wet I*	*Intermediate II*	*Dry III*
a (Å)	173	135–155	130
c	237	235	200–180
ω (10^6 Å^3)	6.14	5.5–4.7	2.9–2.6
X	0.30	0.36–0.42	0.67–0.75

A series of electron microscope photographs of the catalase structure was obtained. One of these is shown in Figure 3.36. The position of the molecules in the unit cell and their shape were obtained from these micrographs by the method of three-dimensional reconstruction[286, 287, 290]. These data were used for calculating the phases of x-ray reflections[288]; the results obtained were followed by the construction of Fourier syntheses of electron density at the ~30 Å resolution. A model of the catalase structure is shown in Figure 3.37 and the structure of the molecule in Figure 3.38. The molecule possesses tetrahedral symmetry 222, its volume is approximately 300 000 Å^3, half-axis dimensions are 90, 80 and 95 Å. Four subunits of the molecule stand out clearly.

Electron microscopy of preparations of muscle phosphorylase B revealed three types of crystalline formations: plane layers of particles, tubes and three-dimensional crystals. In contrast to catalase tubes[284, 285], whose walls

Figure 3.36 Electron micrograph of a crystal of hexagonal catalase along [0001] (From Vainshtein[287], by courtesy of Izd. Akad. Nauk SSSR)

Figure 3.37 Model of the molecular packing in a crystal of hexagonal catalase obtained from electron microscopic and x-ray data (From Vainshtein[288], by courtesy of Izd. Akad. Nauk SSSR)

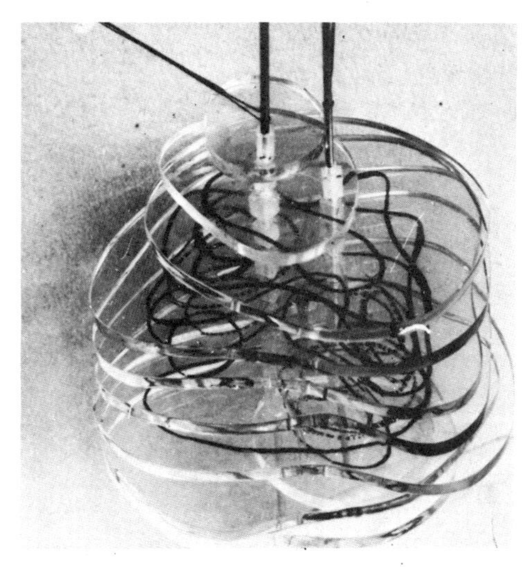

Figure 3.38 Structure of the catalase molecule from the Fourier synthesis of electron density at ~30 Å resolution
(From Vainshtein[288], by courtesy of Izd. Akad. Nauk SSSR)

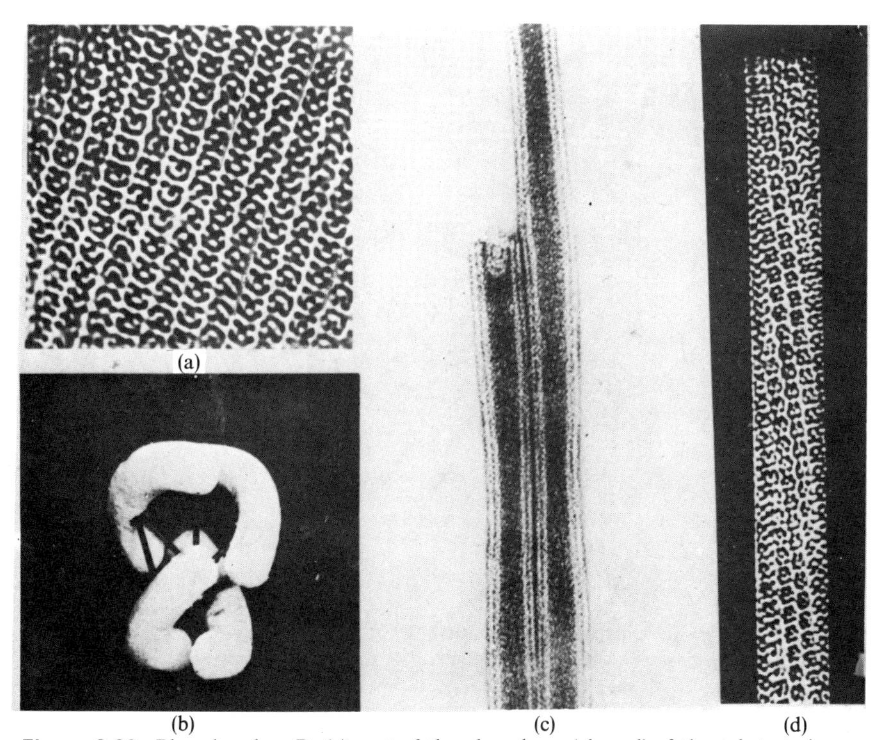

(a) (b) (c) (d)

Figure 3.39 Phosphorylase B: (a) part of the plane layer 'cleared' of the substrate image (protein is black, ×460 000); (c) tubes with two walls, ×200 000; (d) images of a separate wall of the phosphorylase B tube (protein is black, ×300 000); (b) model of the molecule
(From Kiselev[292], by courtesy of Plenum Publishing Corp.)

represent a monomolecular layer, phosphorylase can form tubes with one, two, three and more walls. By means of the methods of optical diffraction and optical filtering, which are used for investigating electron microscope images, an image was obtained of layers 'cleared' of the substrate image (Figure 3.39a). After that, the images of separate walls (Figure 3.39d) were extracted from the composite image of tubes (Figure 3.39c). It was established that the phosphorylase B molecule consists of four bent subunits arranged with the point-group symmetry 222 at the vertices of a tetrahedron (Figure 3.39b)[292].

As already mentioned, the electron microscope image of a negatively stained object represents its projection. If the object is symmetrical, one

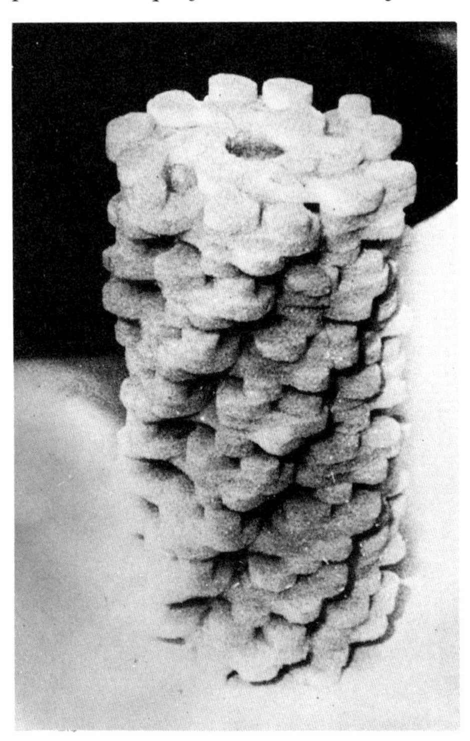

Figure 3.40 Three-dimensional model of the structure of the phage T_6 tail
(From Mikhailov and Vainshtein[297], by courtesy of Izd. Akad. Nauk SSSR)

can then carry out the three-dimensional reconstruction of its structure on the basis of one projection[289]. The method of reconstruction, the so-called 'synthesis of projecting functions' has been developed[290, 291].

Reconstruction of the three-dimensional structure of the object by this method is based on algebraic ideas and is carried out directly in the real space of an object without using the Fourier transforms. This method was applied to the study of helical tails of bacteriophages T_2, T_6 and DD_6 [293, 294]

The tail consists of a hollow cylindrical rod of molecular weight 2.3×10^6–3.2×10^6, and of a sheath of molecular weight 7.8×10^6–8.1×10^6. The tails can be considered as a pile of discs with their own symmetry 6, turned through $4\pi/7 \approx 103$ degrees with respect to each other. All three phages possess symmetry group $S_{\frac{7}{2}}.6$.

The tails are cylindrical-shaped bodies on the outer surface of which one can observe two families of helical grooves forming 'parallelograms'—'units cells of radial projections'. Along the equator the unit cells are arranged in groups of six corresponding to the number of asymmetrical units in the discs. Apart from the central channel, there are six more helical channels located at some distance from the axis of the tails. The main features of the structure have very much in common with the features revealed for bacteriophage T_4 of the same group[289]. In Figure 3.40 a model of the three-dimensional reconstruction of the phage T_6 tail is shown (resolution about 30 Å). Its parameters are as follows: length of the tail 1150 Å, maximal external diameter 175 ± 10 Å, internal diameter 80 Å, number of rows 24, number of sub-units in the sheath 144. The shape of protein molecules of the sheath suggests that they consist of two sub-units.

3.9 CONCLUSION

The limitations of space allowed us to include only some of the important results of the study on crystal structure and crystal chemistry obtained in the U.S.S.R. in the period from 1966 to 1970, and to give some examples by way of illustration. The same reason prevented us from giving the proper attention to many works and even some fields of research that received development in the period under consideration.

In addition to the works under review, we should mention systematic x-ray diffraction studies of the phase equilibria and the structure of many phases in intermetallic systems carried out by Kripyakevich and Gladyshevskii with co-workers[295,296]. Equally we must mention the electron diffraction studies of Pinsker, Khitrova and others on the structure and crystal chemistry of transition metal oxides, and on the processes of phase formation in metal–oxygen systems[297,298]; the works of Tatarinova, Nabitovich and others on the study of the amorphous state by x-ray and electron diffraction methods[299,300], the investigations of polymer structure carried out by Kargin, Markova and Tsvankin[301,302]. Closely related to the investigations of crystalline matter are electron diffraction studies on the determination of the structure of inorganic and organic substances in vapour carried out by Spiridonov, Vilkov and Rambidi[304–307].

In this review the literature citations are in no way meant to be exhaustive; but the references given here readily give access to the other papers of Soviet workers.

Acknowledgement

We wish to record our thanks to V. I. Simonov, V. V. Ilyukhin, I. M. Rumanova, B. B. Zvyagin, V. A. Drits, S. A. Semiletov, R. P. Ozerov, I. I. Yamzin, V. F. Dvoryankin, V. G. Dashevskii, M. A. Mokul'skii, N. A. Kiselev, M. A.

Poray-Koshits, L. O. Atovmyan, T. N. Polynova, L. A. Aslanov who have made available their articles, pictures and other materials. We wish to use this opportunity for thanking E. M. Voronkova for great help in preparing the manuscript; we also thank L. A. Demidova for technical assistance.

References

1. Belov, N. V. (1961). *Kristallokhimiya Silikatov s Krupnymi Kationami.* (Moscow: Izd. Akad. Nauk SSSR)
2. Belov, N. V. (1967). *Problemy Kristallokhimii Mineralov Eandogennogo Mineraloobrazovaniya,* 15. (Moscow: Nauka)
3. Belov, N. V. *et al.* (1970). *Vestn. Mosk. Univ., Geol.,* **8,** 4
4. Guan' Ya-Syan', Simonov, V. I. and Belov, N. V. (1963). *Dokl. Akad. Nauk SSSR,* **149,** 446
5. Nikitin, A. V. *et al.* (1964). *Dokl. Akad. Nauk SSSR,* **154,** 1355
6. Rastsvetayeva, R. K., Simonov, V. I. and Belov, N. V. (1971). *Dokl. Akad. Nauk SSSR,* **197,** 81
7. Chernov, A. I., Ilyukhin, V. V., Maksimov, B. A. and Belov, N. V. (1971). *Kristallografiya,* **16,** 86
8. Petrunina, A. A., Ilyukhin, V. V. and Belov, N. V. (1971). *Dokl. Akad. Nauk SSSR,* **198,** 575
9. Shumyatskaya, N. G. *et al.* (1969). *Dokl. Akad. Nauk SSSR,* **185,** 1289
10. Rastsvetayeva, R. K., Simonov, V. I. and Belov, N. V. (1967). *Dokl. Akad. Nauk SSSR,* **177,** 832
11. Borisov, S. V. *et al.* (1965). *Kristallografiya,* **10,** 816
12. Nekrasov, Ju. V. *et al.* (1969). *Kristallografiya,* **14,** 602
13. Ganiyev, R. M., Ilyukhin, V. V. and Belov, N. V. (1970). *Dokl. Akad. Nauk SSSR,* **190,** 831
14. Treushnikov, E. I., Ilyukhin, V. V. and Belov, N. V. (1971). *Kristallografiya,* **16,** 76
15. Treushnikov, E. I., Ilyukhin, V. V. and Belov, N. V. (1970). *Dokl. Akad. Nauk SSSR,* **193,** 1048
16. Golyshev, V. M., Siminov, V. I. and Belov, N. V. (1971). *Kristallografiya,* **16,** 93
17. Amirov, S. G. *et al.* (1971). *Zap. Vses. Mineral. Obshchest.,* Vol. 1, 19
18. Maksimov, B. A. *et al.* (1968). *Dokl. Akad. Nauk SSSR,* **181,** 591
19. Solov'eva, L. P., Borisov, S. V. and Belov, N. V. (1968). *Dokl. Akad. Nauk SSSR,* **183,** 432
20. Zvyagin, B. B. (1967). *Electron-diffraction Analysis of Clay mineral Structures.* (New York: Plenum Press)
21. Gritsayenko, G. S., Zvyagin, B. B., Boyarskaya, R. K., Gorshkov, A. I., Samotoin, N. D. and Frolova, K. E. (1969). *Metody Elektronnoi Mikroskopii Mineralov.* (Moscow: Nauka)
22. Fedotov, A. F. and Zvyagin, B. B. (1971). *Rentgen Mineral Syr'ya,* No. 10 (in press)
23. Zvyagin, B. B., Berkhin, S. I. and Gorshkov, A. I. (1966). *Rentgen, Mineral Syr'ya,* No. 5, 69
24. Zvyagin, B. B., Mishchenko, K. S. and Shitov, V. A. (1966). *Fiz. Metody Issled. Miner. Osad. Porod,* 130. (Moscow: Nauka)
25. Chukhrov, F. V., Zvyagin, B. B. *et al.* (1971). *Izuchenie Odnorodnosti i Neodnirodnosti Mineralov* (in press). (Moscow: Nauka)
26. Zvyagin, B. B., Mishchenko, K. S. and Soboleva, S. V. (1968). *Kristallografiya,* **13,** 599
27. Drits, V. A. and Karavan, Ju. V. (1969). *Acta. Crystallogr.,* **B25,** 632
28. Drits, V. A., Aleksandrova, V. A. and Sokolova, G. V. *Kristallografiya,* (in press)
29. Soboleva, S. V. and Zvyagin, B. B. (1968). *Kristallografiya,* **13,** 605
30. Kashayev, A. A. and Drits, V. A. (1970). *Kristallografiya,* **15,** 52
31. Chukhrov, F. V. *et al.* (1968). *Geol. Rud. Mestorozhd.,* **X,** 12
32. Drits, V. A. (1970). *Kristallografiya,* **15,** 913
33. Frank-Kamenetskii, V. A., Logvinenko, N. V. and Drits, V. A. (1963). *Zap. Vses. Mineral. Obshchest,* **XCII,** 560
34. Organova, N. I. *et al. Zap. Vses. Mineral. Obshchest.* (in press)
35. Chukhrov, F. V., Zvyagin, B. B. *et al.* (1968). *Izv. Akad. Nauk SSSR, Ser. Geol.,* No. 6, 29

36. Rumanova, I. M. and Ashirov, A. (1963). *Kristallografiya,* **8,** 517
37. Razmanova, Z. P., Rumanova, I. M. and Belov, N. V. (1969). *Dokl. Akad. Nauk SSSR,* **189,** 1003
38. Kurkutova, E. N., Rumanova, I. M. and Belov, N. V. (1965). *Dokl. Akad. Nauk SSSR,* **164,** 90
39. Rumanova, I. M. and Ashirov, A. (1963). *Kristallografiya,* **8,** 828
40. Rumanova, I. M., Kurbanov, H. M. and Belov, N. V. (1965). *Kristallografiya,* **10,** 601
41. Rumanova, I. M., Razmanova, Z. P. and Belov, N. V. (1971). *Dokl. Akad. Nauk SSSR,* **199,** 3
42. Rumanova, I. M. and Gandymov, O. (1971). *Kristallografiya,* **16,** 99
43. Rumanova, I. M., Gandymov, O. and Belov, N. V. (1971). *Kristallografiya,* **16,** 286
44. Christ, C. L., Clark, J. R. and Evans, H. T. (1958). *Acta Crystallogr.,* **11,** 761
45. Belov, N. V. (1950–1970). *Mineral. Sb. L'vovsk. Univ. Ocherki Strukt. Mineral.,* **4-22**
46. Smirnova, N. L., Akimova, N. V. and Belov, N. V. (1967). *Zh. Strukt. Khim.,* **8,** 80
47. Abrikosov, M. H., Bankina, V. E., Poretskaya, L. V., Skudnova, E. V. and Shelimova, L. E. (1967). *Poluprovodnikovye Soedineniya, ikh Poluchenie i Svoistva.* (Moscow: Nauka)
48. Berger, L. I. and Prokuchan, V. D. (1968). *Troinye Almazopodobnye Poluprovodniki.* (Moscow: Metallurgiya)
49. *Kristallicheskie Struktury Arsenidov i Arsenosul'fidov i ikh Analogov,* (1964), ed. Bokii, G. B. (Novosibirsk: Izd. Sib. Otd. Akad. Nauk SSSR)
50. Goryunova, N. A. (1968). *Slozhnye Almazopodobnye Poluprovodniki.* (Moscow: Sov. radio)
51. Medvedeva, Z. S. (1968). *Khal'kogenidy Elementov IIIB Podgruppy Period. Sistemy.* (Moscow: Nauka)
52. *Khimicheskaya Svyaz' v Kristallakh,* (1969), red. Sirota, N. N. (Minsk: Nauka i Tekhnika)
53. Semiletov, S. A. (1967). *Kristallografiya,* **12,** 333
54. Imamov, R. M., Semiletov, S. A. and Pinsker, Z. G. (1970). *Kristallografiya,* **15,** 287
55. Lappe, F., Niggli, A., Nitsche, R. and White, Z. C. (1962). *Z. Krystallogr.,* **117,** 146
56. Zhitar', V. F., Goryunova, N. A. and Radautsan, S. I. (1965). *Izv. Akad. Nauk SSSR,* **2,** 9
57. Radautsan, S. I. *et al.* (1969). *V Nauchnoteknicheskaya Konferentsiya Kishinevskogo Politekhnicheskogo Instituta,* 146. (Kishinev)
58. Donika, F. G. *et al.* (1970). *Kristallografiya,* **15,** 813
59. Donika, F. G., Kiosse, G. A. *et al.* (1967). *Kristallografiya,* **12,** 854
60. Donika, F. G. *et al.* (1970). *Kristallografiya,* **15,** 816
61. Imamov, R. M. and Semiletov, S. A. (1970). *Kristallografiya,* **15,** 972
62. Eckerlin, P. and Stegherr, A. (1966). *XII International Congress and Symposium on Crystal Growth,* A 82. (Copenhagen: Munksgaard)
63. Stasova, M. M. (1967). *Zh. Strukt. Khim.,* **8,** 655
64. Baranova, R. V. and Pinsker, Z. G. (1970). *Zh. Strukt. Khim.,* **11,** 690; (1971). *Kristallografiya,* **16,** 127
65. Man, L. I., Imamov, R. M. and Pinsker, Z. G. (1971). *Kristallografiya,* **16,** 122
66. Belov, N. V. and Pobedimskaya, E. A. (1968). *Kristallografiya,* **13,** 969
67. Semiletov, S. A. (1969). *Dissertatsiya Institut Kristallografii,* Moscow
68. Izyumov, J. and Ozerov, R. (1969). *Magnetic Neutron Diffraction.* (New York: Plenum Press)
69. Vainshtein, B. K. (1964). *Structure Analysis by Electron Diffraction.* (Oxford: Pergamon)
70. Vainshtein, B. K. (1963). *Advances in Structure Research by Diffraction Methods,* Vol. 1. (New York-London: Interscience)
71. Rannev, N. V., Datt, I. D., Tovbis, A. B. and Ozerov, R. P. (1968). *Kristallografiya,* **10,** 914
72. Datt, I. D., Rannev, N. V. and Ozerov, R. P. (1968). *Kristallografiya,* **13,** 261
73. Datt, I. D., Rannev, N. V., Ozerov, R. P. and Kuznets, V. M. (1971). *Kristallografiya,* **16,** 631
74. Datt, I. D., Rannev, N. V., Balicheva, T. G. and Ozerov, R. P. (1970). *Kristallografiya,* **15,** 949
75. Hamilton, W. and Ibers, J. (1968). *Hydrogen Bond in Solids. (New York: Benjamin)*
76. Datt, I. D. and Ozerov, R. P. *Kristallografiya* (in press)
77. Bajorek, A., Janik, J. A. *et al.* (1968). *Neutron Inelastic Scatter,* **11,** 143. (Vienna: IAEA)
78. Bajorek, A. *et al.* (1970). *Kristallografiya,* **15,** 1156
79. Dvoryankin, V. F. and Dvoryankina, G. G. (1968). *Kristallografiya,* **13,** 508

80. Kolomiichuk, V. N. and Dvoryankin, V. F. (1964). *Kristallografiya*, **9**, 50
81. Kolomiichuk, V. N. (1965). *Kristallografiya*, **10**, 565
82. Goman'kov, V. I., Loshmanov, A. A. and Puzei, I. M. (1965). *Kristallografiya*, **10**, 416
83. Goman'kov, V. I., Loshmanov, A. A. and Puzei, I. M. (1966). *Fiz. Metal. Metalloved.*, **22**, 128; (1967). *Fiz. Metall. Metalloved.*, **23**, 1124; (1968). *Fiz. Metall. Metalloved.*, **25**, 36; (1969). *Fiz. Metall. Metalloved.*, **28**, 262
84. Vintaikin, E. Z. and Loshmanov, A. A. (1966). *Fiz. Metal. Metalloved.*, **22**, 473; (1967). *Fiz. Metal. Metalloved.*, **24**, 754
85. Vintaikin, E. Z. and Loshmanov, A. A. (1968). *Uporyadochenie Atomov i ego Vliyanie na Svoistva Splavov.* (Kiev: Naukova Dumka)
86. Borovik-Romanov, A. S. and Orlova, M. P. (1956). *Zh. Eksp. Teor. Fiz.*, **31**, 579
87. Dzyaloshinsii, I. E. (1957). *Zh. Eksp. Teor. Fiz*, **32**, 1547; **33**, 1454; (1964). *Zh. Eksp. Teor. Fiz.*, **46**, 1420
88. Yoshimori, A. (1959). *J. Phys. Soc. Japan*, **14**, 807
89. Bykov, V. N. *et al.* (1959). *Dokl. Akad. Nauk SSSR*, **128**, 1153
90. Corliss, R. M. and Elliot, N. (1959). *J. Hastings Phys. Rev. Lett.*, **3**, 211
91. Enz, V. (1961). *J. Appl. Phys.*, **32S**, 3, 22
92. Sizov, V. A., Sizov, R. A. and Jamzin, I. I. (1967). *Zh. Eksp. Teor. Fiz.*, **53**, 1256
93. Aleshko-Ozhevskii, O. P., Sizov, R. A., Jamzin, I. I. and Lyubimtsev, V. A. (1968). *Zh. Eksp. Teor. Fiz.*, **55**, 820; (1969). *Zh. Eksp. Teor. Fiz.*, **56**, 4
94. Fayek, M. K., Leciejewicz, J., Murasik, A. and Jamzin, I. I. (1969). *Phys. Stat. Sol.*, **34**, K 29; (1970). *Phys. Stat. Sol.*, **37**, 843
95. König, V., Bertant, E. F., Grosg, Y., Mitrikov, M. and Chol, G. (1970). *Solid State Commun.*, **8**, 759
96. Kocharov, A. G., Leciejewicz, J., Fazek, K. and Murasik, A. (1971). *Phys. Stat. Sol.*, **4**, 53
97. Poray-Koshits, M. A., Antsishkina, A. S. *et al.* (1965). *Zh. Strukt. Khim.*, **6**, 168, 170; (1967). *Zh. Strukt. Khim.*, **8**, 296, 365; (1968). *Zh. Strukt. Khim.*, **9**, 646; (1968). *Zh. Neorg. Khim.*, **13**, 1245; (1969). *Zh. Strukt. Khim.*, **10**, 79, 650; (1970). *Zh. Strukt. Khim.*, **11**, 936
98. Poray-Koshits, M. A. (1968). *Zh. Neorg. Khim.*, **13**, 1233
99. Khodashova, T. S. (1965). *Zh. Strukt. Khim.*, **6**, 716
100. Khodashova, T. S., Poray-Koshits, M. A. *et al.* (1970). *Zh. Strukt. Khim.*, **11**, 783
101. Atovmyan, L. O. *et al.* (1965). *Izv. Akad. Nauk SSSR, Ser. Khim.*, **2**, 257; (1968). *Zh. Strukt. Khim.*, **9**, 1097; (1969). *Zh. Strukt. Khim.*, **10**, 505; (1969). *Chem. Commun.*, 649; (1970). *Zh. Strukt. Khim.*, **11**, 469, 782; (1970). *Dokl. Akad. Nauk. SSSR*, **195**, 107; (1971). *Dokl. Akad. Nauk SSSR*, **196**, 91
102. Sokolova, Ju. A. (1966). *Acta Crystallogr.*, **21**, A185, 11
103. Plyasova, L. M., Klevtsova, R. F., Borisov, S. V. and Kefeli, L. M. (1966). *Dokl. Akad. Nauk SSSR*, **167**, 87; (1967). *Kristallografiya*, **12**, 939
104. Egorov-Tismenko, Ju. K., Simonov, M. A. and Belov, N. V. (1967). *Kristallografiya*, **12**, 511
105. Pinsker, G. Z. and Kuznetsov, V. G. (1968). *Zh. Strukt. Khim.*, **13**, 74
106. Poray-Koshits, M. A., Aslanov, L. A., Ivanova, G. V. and Polynova, T. N. (1966). *Zh. Strukt. Khim.*, **7**, 812; (1968). *Zh. Strukt. Khim.*, **9**, 475
107. Poray-Koshits, M. A., Atovmyan, L. O. and Andrianov, V. G. (1961). *Zh. Strukt. Khim.*, **2**, 743; (1962). *Zh. Strukt. Khim.*, **3**, 685
108. Atovmyan, L. O. *et al.* (1967). *Zh. Strukt. Khim.*, **8**, 169; (1968). *Zh. Strukt. Khim.*, **9**, 29
109. Markin, V. N. (1968). *Vestn. Leningr. Univ. Fiz. Khim.*, **4**, 151
110. Atovmyan, L. O. *et al.* (1970). *Zh. Strukt. Khim.*, **11**, 782; *Chem. Commun.* (in press)
111. Atovmyan, L. O. *et al.* (1970). *Zh. Strukt. Khim.*, **11**, 557
112. Mikhailov, Ju. N., Kuznetsov, V. G. and Kovaleva, E. S. (1965). *Zh. Strukt. Khim.*, **6**, 787; (1968). *Zh. Neorg. Khim.*, **13**, 710; (1968). *Zh. Strukt. Khim.*, **9**, 710
113. Poray-Koshits, M. A. and Atovmyan, L. O. (1966). *Kristallokhimiya, Itogi Nauki.* (Moscow: VINITI); (1968). *Kristallokhimiya, Itogi Nauki.* (Moscow: VINITI)
114. Poray-Koshits, M. A. and Gelinskaya, E. L. (1966). *Kristallokhimiya, Itogi Nauki.* Vol. 1, Moscow: VINITI)
115. Antsishkina, A. S., Gusseinova, M. K. and Poray-Koshits, M. A. (1967). *Zh. Strukt. Khim.*, **8**, 365
116. Gusseinova, M. K., Antsishkina, A. S. and Poray-Koshits, M. A. (1968). *Zh. Strukt. Khim.*, **9**, 1040

117. Arutyunyan, E. G. *et al.* (1966). *Zh. Strukt. Khim.*, **7,** 471, 813
118. Sadikov, G. G., Kukina, G. A. and Poray-Koshits, M. A. (1967). *Zh. Strukt. Khim.*, **8,** 551; (1968). *Zh. Strukt. Khim.*, **9,** 145; *Zh. Strukt. Khim.* (in press)
119. Belyayeva, K. F., Poray-Koshits, M. A. *et al.* (1966). *Zh. Strukt. Khim.*, **7,** 130
120. Belyayeva, K. F., Poray-Koshits, M. A., Malinovskii, T. I. *et al.* (1969). *Zh. Strukt. Khim.*, **10,** 557
121. Vainshtein, B. K., D'yakon, I. A. and Ablov, A. V. (1967). *Kristallografiya,* **12,** 354
122. Simonov, Yu. A. and Malinovskii, T. I. (1970). *Kristallografiya,* **15,** 370
123. Rogachev, D. L., Antsishkina, A. S. and Poray-Koshits, M. A. (1965). *Zh. Strukt. Khim.*, **6,** 791; (1969). *Zh. Strukt. Khim.*, **10,** 280, 645
124. Koz'min, P. A., Surazhskaya, M. D. and Kuznetsov, V. G. (1970). *Zh. Strukt. Khim.*, **11,** 313
125. Kuznetsov, V. G., Koz'min, P. A. *et al.* (1963). *Acta Crystallogr.*, **16,** 41; (1963). *Zh. Strukt. Khim.*, **4,** 55; (1965). *Zh. Strukt. Khim.*, **6,** 651
126. Aslanov, L. A., Poray-Koshits, M. A. *et al.* (1967). *Zh. Strukt. Khim.*, **8,** 705, 1106; (1968). *Zh. Strukt. Khim.*, **9,** 331, 540; (1969). *Zh. Strukt. Khim.*, **10,** 285, 345; (1970). *Zh. Strukt. Khim.*, **11,** 46, 311
127. Poray-Koshits, M. A. and Aslanov, L. A. *Zh. Strukt. Khim.* (in press)
128. Polynova, T. N., Poray-Koshits, M. A., Martynenko, L. I. *et al.* (1967). *Zh. Strukt. Khim.*, **8,** 553; (1970). *Zh. Strukt. Khim.*, **11,** 164, 555, 558; (1971). *Zh. Strukt. Khim.*, **12,** 335
129. Polynova, T. N., Poray-Koshits, M. A. *et al.* (1969). *X Vsesoyuznoe Soveshchanie po Khimii Komplexnykh Soedinenii,* 104. (Kiev)
130. Shugam, E. A. *et al.* (1966) *Zh. Strukt. Khim.*, **7,** 897; (1967). *Zh. Strukt. Khim.*, **8,** 171; (1971). *Zh. Strukt. Khim.*, **12,** 102
131. Shugam, E. A., Shkol'nikova, L. M. *et al.* (1968). *Zh. Strukt. Khim.*, **9,** 222; (1969). *Zh. Strukt. Khim.*, **10,** 83; (1970). *Zh. Strukt. Khim.*, **11,** 54, 886, 938
132. Shugam, E. A., Shkol'nikova, L. M. and Livingston, S. E. (1967). *Zh. Strukt. Khim.*, **8,** 550; *Zh. Strukt. Khim.* (in press)
133. Shkol'nikova, L. M. *et al.* (1967). *Zh. Strukt. Khim.*, **8,** 89, 194; (1968). *Zh. Strukt. Khim.*, **9,** 543
134. Zvonkova, Z. V. *et al.* (1967). *Kristallografiya,* **12,** 1065
135. Zvonkova, Z. V. and Jakovenko, V. I. (1968). *Kristallografiya,* **13,** 169
136. Terent'yev, A. P., Bozzhennikov, V. M., Kolninov, O. V., Zvonkova, Z. V., Rukhadze, E. G., Glushkova, V. I. and Berezkin, V. V. (1965). *Dokl. Akad. Nauk SSSR,* **160,** 405
137. Kolninov, O. V., Terent'yev, A. P., Zvonkova, Z. V. and Rukhadze, E. G. (1966). *Dokl. Akad. Nauk. SSSR,* **168,** 1327
138. Ainbinder, B. Ju., Zvonkova, Z. V. and Zhdanov, G. S. (1967). *Dokl. Akad. Nauk. SSSR,* **174,** 115
139. Kitaigorodskii, A. I. (1955). *Organicheskaya Kristallokhimiya.* (Moscow: Izd. Akad. Nauk SSSR)
140. Zorkii, P. M. and Poray-Koshits, M. A. (1961). *Kristallografiya,* **6,** 655; (1967). *Kristallografiya,* **12,** 989
141. Kitaigorodskii, A. I. (1961). *Dokl. Akad. Nauk. SSSR,* **137,** 116
142. Kitaigorodskii, A. I. and Mirskaya, K. V. (1961). *Kristallografiya,* **6,** 507
143. Kitaigorodskii, A. I. *et al.* (1964). *Kristallografiya,* **9,** 174; (1965). *Fiz. Tverd. Tela,* **7,** 643; (1970). *Kristallografiya,* **15,** 405
144. Kitaigorodskii, A. I. (1965). *Acta Crystallogr.*, **18,** 585; (1966). *J. Chim. Phys.*, **63,** 9
145. Mirskaya, K. V. and Kozlova, I. E. (1969). *Kristallografiya,* **14,** 412
146. Kitaigorodskii, A. I. and Mirskaya, K. V. (1964). *Kristallografiya,* **9,** 634; (1965). *Kristallografiya,* **10,** 162
147. Zorkii, P. M., Lazareva, S. G. and Zefirov, Ju. F. (1970). *Kristallografiya,* **15,** 698
148. Kitaigorodskii, A. I. *et al.* (1967). *Kristallografiya,* **12,** 349; (1968). *Kristallografiya,* **13,** 889
149. Zorkii, P. M. (1968). *Kristallografiya,* **13,** 26
150. Zorkii, P. M., Bel'skii, V. K., Lazareva, S. G. and Poray-Koshits, M. A. (1967). *Zh. Strukt. Khim.*, **8,** 312
151. Zorkii, P. M. and Bel'skii, V. K. (1968). *Zh. Strukt. Khim.*, **9,** 1102; (1969). *Zh. Strukt. Khim.*, **10,** 73; (1970). *Zh. Strukt. Khim.*, **11,** 564
152. Zorkii, P. M. *et al.* (1967). *Zh. Strukt. Khim.*, **8,** 670; (1968). *Zh. Strukt. Khim.*, **9,** 95; (1968). *Kristallografiya,* **13,** 26; (1969). *Zh. Strukt. Khim.*, **10,** 882, 1085; (1970). *Zh. Strukt. Khim.*, **11,** 903

153. Bel'skii, V. K. and Zorkii, P. M. (1970). *Kristallografiya,* **15,** 704
154. Kitaigorodskii, A. I. (1951). *Izv. Akad. Nauk. SSSR, Ser. Fiz.,* **15,** 157; (1959). *Dokl. Akad. Nauk SSSR,* **124,** 1267; (1960). *Tetrahedron,* **9,** 183
155. Brant, D. A. and Flory, P. J. (1965). *J. Amer. Chem. Soc.,* **87,** 2791
156. Scott, R. A. and Scheraga, H. A. (1966). *J. Chem. Phys.,* **45,** 2091
157. Venkatachalam, C. M. and Ramachandran, S. N. (1967). *Conformation of Biopolymers.,* **1,** 83. (London–New York: Academic Press)
158. Dashevskii, V. G. (1963). *Zh. Strukt. Khim.,* **4,** 637
159. Dashevskii, V. G. *et al.* (1965). *Zh. Strukt. Khim.,* **6,** 888; (1966). *Zh. Strukt. Khim.,* **7,** 93, 594
160. Kitaigorodskii, A. I. and Dashevskii, V. G. (1967). *Teor. Eksp. Khim.,* **3,** 35
161. Dashevskii, V. G. (1970). *Zh. Strukt. Khim.,* **11,** 489, 746, 912
162. Birshtein, T. M. and Ptitsyn, O. B. (1965). *Konformatsiya Makromolekul.* (Moscow: Nauka)
163. Avoyan, R. L., Struchkov, Ju. T. and Dashevskii, V. G. (1966). *Zh. Strukt. Khim.,* **7,** 289
164. Hotsyanova, T. L. *et al.* (1965). *Zh. Strukt. Khim.,* **6,** 307; (1966). *Zh. Strukt. Khim.,* **7,** 634; (1968). *Zh. Strukt. Khim.,* **9,** 148; (1968). *Kristallografiya,* **13,** 787
165. Akopyan, Z. A., Struchkov, Ju. T. and Dashevskii, V. G. (1966). *Zh. Strukt. Khim.,* **7,** 408
166. Davydova, M. A. and Struchkov, Ju. T. (1965). *Zh. Strukt. Khim.,* **6,** 113, 922; (1968). *Zh. Strukt. Khim.,* **9,** 258, 547
167. Akopyan, Z. A., Kitaigorodskii, A. I. and Struchkov, Ju. T. (1965). *Zh. Strukt. Khim.,* **6,** 729
168. Chetkina, L. A. and Gol'der, G. A. (1967). *Kristallografiya,* **12,** 42, 404
169. Klimasenko, N. L., Gol'der, G. A. and Zhdanov, G. S. (1969). *Kristallografiya,* **14,** 266
170. Neronova, N. N. (1968). *Zh. Strukt. Khim.,* **9,** 147
171. Chetkina, L. A. and Gol'der, G. A. (1968). *Zh. Strukt. Khim.,* **9,** 250
172. Kondrashov, Ju. D. *et al.* (1965). *Kristallografiya,* **10,** 822; (1967). *Kristallografiya,* **12,** 416; (1968). *Kristallografiya,* **13,** 622, 1076
173. Andrianov, V. I., Kostyanovskii, R. G., Shibaeva, R. P. and Atovmyan, L. Ø. (1967). *Zh. Strukt. Khim.,* **8,** 10
174. Vorontsova, L. G., Andrianov, V. I. and Tarnopol'skii, B. L. (1969). *Zh. Strukt. Khim.,* **10,** 872
175. Grigor'eva, N. V., Margolis, N. V. *et al.* (1966). *Zh. Strukt. Khim.,* **7,** 278; (1967). *Zh. Strukt. Khim.,* **8,** 175; (1968). *Zh. Strukt. Khim.,* **9,** 547, 550; (1969). *Zh. Strukt. Khim.,* **10,** 559, 943; (1970). *Zh. Strukt. Khim.,* **11,** 165, 562
176. Ginzburg, S. L., Neigauz, M. G. *et al.* (1969). *Zh. Strukt. Khim.,* **10,** 887
177. Polynova, T. N., Bokii, N. G. and Poray-Koshits, M. A. (1965). *Zh. Strukt. Khim.,* **6,** 878
178. Nesterova, Ja. M. and Poray-Koshits, M. A. (1971). *Zh. Strukt. Khim.,* **12,** 108
179. Andrianov, V. I. *et al.* (1969). *Zh. Strukt. Khim.,* **10,** 859
180. Shvets, A. E. *et al. Zh. Strukt. Khim.,* (in press)
181. Shibaeva, R. P., Atovmyan, L. O. *et al.* (1969). *Chem. Commun.,* 1494
182. Lobanova, G. M. (1968). *Kristallografiya,* **13,** 984
183. Lobanova, G. M., Gel'fand, I. M. *et al.* (1970). *Dokl. Akad. Nauk SSSR,* **195,** 341
184. Zvonkova, Z. V. *et al.* (1965). *Kristallografiya,* **10,** 194, 734; (1966). *Kristallografiya,* **11,** 385; (1969). *Kristallografiya,* **14,** 696
185. Vorontsova, L. G., Zvonkova, Z. V. and Zhdanov, G. S. (1967). *Dokl. Akad. Nauk SSSR,* **172,** 57
186. Vorontsova, L. G. (1966). *Zh. Strukt. Khim.,* **7,** 240, 280
187. Vol'pin, M. E., Koreshkov, Ju. D., Dulova, V. G. and Kursanov, D. N. (1962). *Tetrahedron,* **18,** 107
188. Bokii, N. G. and Struchkov, Ju. T. (1965). *Zh. Strukt. Khim.,* **6,** 571; (1967). *Zh. Strukt. Khim.,* **8,** 122, 501; (1968). *Zh. Strukt. Khim.,* **9,** 838
189. Vilkov, L. V., Mastryukov, V. S. *et al.* (1965). *Zh. Strukt. Khim.,* **6,** 811; (1970). *Zh. Strukt. Khim.,* **11,** 3
190. Bokii, N. G. and Struchkov, Ju. T. (1968). *Zh. Strukt. Khim.,* **9,** 722
191. Biryukov, B. P. and Struchkov, Ju. T. (1968). *Zh. Strukt. Khim.,* **9,** 228, 488; (1969). *Zh. Strukt. Khim.,* **10,** 95
192. Biryukov, B. P., Struchkov, Ju. T. *et al.* (1967). *Chem. Commun.,* 749, 750; (1968). *Chem. Commun.,* 159; (1969). *Chem. Commun.,* 667

193. Biryukov, B. P., Struchkov, Ju. T. *et al.* (1968). *Zh. Strukt. Khim.*, **9**, 922
194. Bokii, N. G., Zakharova, G. N. and Struchkov, Ju. T. (1970). *Zh. Strukt. Khim.*, **11**, 895
195. Akhmed, M. A. and Aleksandrov, G. G. (1970). *Zh. Strukt. Khim.*, **11**, 891
196. Kulishov, V. I. *et al.* (1970). *Zh. Strukt. Khim.*, **11**, 71
197. Stanko, V. I. and Struchkov, Ju. T. (1965). *Zh. Obshch. Khim.*, **35**, 930; (1967). *Zh. Strukt. Khim.*, **8**, 558, 707; (1969). *Zh. Strukt. Khim.*, **10**, 1063
198. Stanko, V. I., Struchkov, Ju. T. and Klimova, A. I. (1966). *Zh. Strukt. Khim.*, **7**, 629; (1966). *Zh. Obshch. Khim.*, **36**, 1707
199. Dunitz, J. D., Orgel, L. E. and Rich, A. (1956). *Acta Crystallogr.*, **9**, 373
200. Bokii, G. B. and Kukina, G. A. (1957). *Kristallografiya*, **2**, 400; (1965). *Zh. Strukt. Khim.*, **6**, 706; Kukina, G. A. (1962). *Zh. Strukt. Khim.*, **3**, 474
201. Kaluski, Z. L., Avoyan, R. L. and Struchkov, Ju. T. (1962). *Zh. Strukt. Khim.*, **3**, 599
202. Kaluski, Z. L., Struchkov, Ju. T. and Avoyan, R. L. (1964). *Zh. Strukt. Khim.*, **5**, 743
203. Kaluski, Z. L. and Struchkov, Ju. T. (1965). *Zh. Strukt. Khim.*, **6**, 104; (1966). *Zh. Strukt. Khim.*, **7**, 283
204. Kaluski, Z. L. and Struchkov, Ju. T. (1965). *Zh. Strukt. Khim.*, **6**, 475, 745
205. Kaluski, Z. L. and Struchkov, Ju. T. (1965). *Zh. Strukt. Khim.*, **6**, 6, 921
206. Kaluski, Z. L. and Struchkov, Ju. T. (1965). *Zh. Strukt. Khim.*, **6**, 316
207. Macdonald, A. C. and Trotler, J. (1964). *Acta Crystallogr.*, **17**, 872
208. Starovskii, O. V. and Struchkov, Ju. T. (1965). *Zh. Strukt. Khim.*, **6**, 248
209. Borovyak, T. E. *et al.* (1970). *Zh. Strukt. Khim.*, **11**, 1087
210. Gusev, A. I., Aleksandrov, G. G. and Struchkov, Ju. T. (1969). *Zh. Strukt. Khim.*, **10**, 655
211. Kulishov, V. I., Bokii, N. G. and Struchkov, Ju. T. (1970). *Zh. Strukt. Khim.*, **11**, 700
212. Gusev, A. I. and Struchkov, Ju. T. (1970). *Zh. Strukt. Khim.*, **11**, 368
213. Kukina, G. A., Bokii, G. B. and Brusentsov, F. A. (1964). *Zh. Strukt. Khim.*, **5**, 730
214. Zakharova, I. A., Kukina, G. A. *et al.* (1966). *Zh Neorg. Khim.*, **11**, 2543
215. Minasyants, M. H. and Struchkov, Ju. T. (1968). *Zh. Strukt. Khim.*, **9**, 481
216. Minasyants, M. H. and Struchkov, Ju. T. (1968). *Zh. Strukt. Khim.*, **9**, 665
217. Minasyants, M. H., Andrianov, V. G. and Struchkov, Ju. T. (1968). *Zh. Strukt. Khim.*, **9**, 1055
218. Aleksandrov, G. G. and Struchkov, Ju. T. (1969). *Zh. Strukt. Khim.*, **10**, 672; (1970). *Zh. Strukt. Khim.*, **11**, 708, 1094; (1971). *Zh. Strukt. Khim.*, **12**, 120
219. Trotler, J. (1960). *Acta Crystallogr.*, **13**, 86
220. Glick, M. D. and Dahl, L. F. (1965). *J. Organometall. Chem.*, **3**, 200
221. Astakhova, I. S. and Struchkov, Ju. T. (1970). *Zh. Strukt. Khim.*, **11**, 472
222. Gusev, A. I. and Struchkov, Ju. T. (1969). *Zh. Strukt. Khim.*, **10**, 107, 294, 515
223. Nesmeyanov, A. N., Gusev, A. I. *et al.* (1969). *Chem. Commun.*, 277, 739
224. Gusev, A. I., Aleksandrov, G. G. and Struchkov, Ju. T. (1969). *Zh. Strukt. Khim.*, **10**, 655
225. Avoyan, R. L., Chapovskii, Ju. A. and Struchkov, Ju. T. (1966). *Zh. Strukt. Khim.*, **7**, 900
226. Semion, V. A. and Struchkov, Ju. T. (1968). *Zh. Strukt. Khim.*, **9**, 1046; (1969). *Zh. Strukt. Khim.*, **10**, 88, 664
227. Chapovskii, Ju. A. *et al.* (1968). *Zh. Strukt. Khim.*, **9**, 1100
228. Andrianov, V. G. *et al.* (1970). *Zh. Strukt. Khim.*, **11**, 168
229. Andrianov, V. G. and Struchkov, Ju. T. (1968). *Zh. Strukt. Khim.*, **9**, 845
230. Chapovskii, Ju. A. *et al.* (1967). *Zh. Strukt. Khim.*, **8**, 559
231. Andrianov, V. G. and Struchkov, Ju. T. (1968). *Zh. Strukt. Khim.*, **9**, 240, 503
232. Aleksandrov, G. G. and Struchkov, Ju. T. (1970). *Zh. Strukt. Khim.*, **11**, 479, 1094
233. Sudarikov, V. S. *et al.* (1969). *Zh. Strukt. Khim.*, **10**, 941
234. Bent, H. A. (1961). *Chem. Rev.*, **61**, 271
235. Biryukov, B. P. and Struchkov, Ju. T. (1968). *Zh. Strukt. Khim.*, **9**, 655
236. Biryukov, B. P., Struchkov, Ju. T. *et al.* (1968). *Chem. Commun.*, 667
237. Andrianov, V. G., Biryukov, B. P. and Struchkov, Ju. T. (1969). *Zh. Strukt. Khim.*, **10**, 1129
238. Rich, A. (1959). *Reviews of Modern Physics. Biophysical Science*, **31**, 23. (Lancaster, New York: American Physical Society)
239. Baklagina, Ju. G., Vol'kenshtein, M. V. and Kondrashev, Ju. D. (1966). *Zh. Strukt. Khim.*, **7**, 399; (1966). *Dokl. Akad. Nauk SSSR*, **169**, 229
240. Aksel'rod, V. D., Fodor, I. I. and Bayev, A. A. (1967). *Dokl. Akad. Nauk SSSR*, **174**, 707

241. Bayev, A. A. *et al.* (1967). *Molec. Biol.*, **1**, 754; (1967). *Molec. Biol.*, **1**, 859
242. Mirzabekov, A. D., Levina, E. S. and Bayev, A. A. (1969). *FEBS Letters*, **5**, 218
243. Holley, R. M. *et al.* (1965). *Science*, **147**, 1462
244. Arnott, S. *et al.* (1968). *Nature (London)*, **220**, 561
245. Vainshtein, B. K. and Feigin, L. A. (1965). *Dokl. Akad. Nauk SSSR*, **161**, 1444
246. Dembo, A. T., Sosfenov, N. I. and Feigin, L. A. (1966). *Kristallografiya*, **11**, 581
247. Wilkins, M. H. F. *et al.* (1955). *Nature (London)*, **175**, 834; (1956). *Cold Spring Harbor Symposium*, **21**, 75
248. Mokul'skaya, T. D. (1966). *Biofizika*, **11**, 528
249. Mokul'skaya, T. D., Gorlenko, Zh. M., Zamchuk, L. A., Bogdanova, E. S., Mokul'skii, M. A., Gol'dfarb, D. M. and Hesin, R. B. (1966). *Biokhimiya*, **31**, 749
250. Dembo, A. T. *et al.* (1965). *Biofizika*, **10**, 404
251. Vainshtein, B. K., Feigin, L. A. *et al.* (1970). *Vopr. virusolog.*, 739
252. Gurskaya, G. V. (1968). *The Molecular Structure of Amino Acids.* (New York: Consultants Bureau)
253. Kayushina, R. L. and Vainshtein, B. K. (1965). *Kristallografiya*, **10**, 833
254. Vainshtein, B. K., Gel'fand, I. M., Kayushina, R. L. and Fedorov, Ju. G. (1963). *Dokl. Akad. Nauk. SSSR*, **153**, 11
255. Vainshtein, B. K. and Gurskaya, G. V. (1964). *Dokl. Akad. Nauk SSSR*, **156**, 312
256. Vainshtein, B. K., D'yakon, I. A. and Ablov, A. V. (1967). *Kristallografiya*, **12**, 354; (1970). *Dokl. Akad. Nauk. SSSR*, **193**, 330
257. Tishchenko, G. N. *et al.* (1967). *Kristallografiya*, **12**, 868; (1969). *Kristallografiya*, **14**, 270
258. Zykalova, K. A., Tishchenko, G. N. *et al.* (1970). *Izv. Akad. Nauk SSSR, Ser. Khim.*, **7**, 1547
259. Shemyakin, M. M., Ovchinnikov, Ju. A. *et al.* (1969). *J. Membr. Biol.*, **1**, 402
260. Ovchinnikov, Ju. A. *et al.* (1970). *Biochem. Biophys. Research Comm.*, **39**, 217
261. Vainshtein, B. K. and Tatarinova, L. I. (1967). *Conformation of Biopolymers*, **1**, 569. (London–New York: Academic Press)
262. Pauling, L., Corey, R. B. and Branson, H. R. (1951). *Proc. Natl. Acad. Sci. (USA)*, **37**, 729
263. Andreeva, N. S., Esipova, N. G., Millionova, M. I., Rogulenkova, V. N. and Shibnev, V. A. (1967). *Molec. Biol.*, **1**, 657
264. Andreeva, N. S., Esipova, N. G., Millionova, M. I., Rogulenkova, V. N. and Shibnev, V. A. (1967). *Conformation of Biopolymers*, **2**, 469. (London–New York: Academic Press)
265. Andreeva, N. S., Esipova, N. G., Millionova, M. I., Rogulenkova, V. N., Tumanyan, V. G. and Shibnev, V. A. (1970). *Biofizika*, **XV**, 198
266. Rich, A. and Crick, F. H. C. (1955). *Nature (London)*, **176**, 915
267. Ramachandran, G. N. and Kartha, G. (1955). *Proc. Indian Acad. Sci.*, **A42**, 215
268. Ptitsyn, O. B. (1967). *Usp. Sovrem. Biol.*, **63**, 3; (1970). *Usp. Sovrem. Biol.*, **69**, 26
269. Ptitsyn, O. B. (1969). *Molec. Biol.*, **3**, 627
270. Ptitsyn, O. B. (1969). *Zh. Molec. Biol.*, **42**, 501
271. Ptitsyn, O. B. and Finkel'shtein, A. V. (1970). *Biofizika*, **XV**, 757; (1970). *Dokl. Akad. Nauk SSSR*, **195**, 1
272. Ramachandran, G. N. and Sasosekharan, V. (1968). *Advan. Prot. Chem.*, **23**, 283
273. Vainshtein, B. K. (1966). *Usp. Fiz. Nauk*, **88**, 527
274. Dickerson, R. E. and Geis, I. (1969). *The Structure and Action of Proteins.* (New York, London: Evanston)
275. Vazina, A. A., Lednev, V. V. and Lemazhikhin, B. K. (1966). *Bioklimiya*, **31**, 720
276. Vazina, A. A., Frank, G. and Lemazhikhin, B. K. (1965). *J. Molec. Biol.*, **14**, 373
277. Goryunov, A. I. and Andreeva, N. S. (1967). *Molec. Biol.*, **1**, 302
278. Arutyunyan, E. G., Zaitsev, V. N. *et al.* (1971). *Kristallografiya*, **16**, 237
279. Vainshtein, B. K., Arutyunyan, E. G. *et al.* (1970). *Kristallografiya*, **15**, 167
280. Karpukhina, S. Ja. and Sosfenov, N. I. (1967). *Dokl. Akad. Nauk. SSSR*, **175**, 723
281. Andreeva, N. S. *et al.* (1971). *Molec. Biol.*, **5**, 905
282. Borisov, V. V. *et al.* (1968). *Kristallografiya*, **13**, 403; (1966). *Acta Crystallogr.*, **A21**, 157
283. Gurskaya, G. V., Karpukhina, S. Ja. and Lobanova, G. M. (1971). *Biofizika*, **XVI**, 553
284. Vainshtein, B. K., Kiselev, N. A. and Shpitsberg, V. L. (1966). *Dokl. Akad. Nauk SSSR*, **167**, 212; (1971). *J. Molec. Biol.*, **25**, 433
285. Kiselev, N. A., De Rosier, D. J. and Klug, A. (1968). *J. Molec. Biol.*, **35**, 561
286. Vainshtein, B. K., Barynin, V. V. and Gurskaya, G. V. (1968). *Dokl. Akad. Nauk SSSR*, **182**, 569

287. Barynin, V. V. and Vainshtein, B. K. (1971). *Kristallografiya,* **16,** 751
288. Gurskaya, G. V., Lobanova, G. M. and Vainshtein, B. K. (1971). *Kristallografiya,* **16,** 764
289. De Rosier, D. I. and Klug, A. (1968). *Nature (London),* **217,** 130
290. Vainshtein, B. K. (1970). *Kristallografiya,* **15,** 894
291. Vainshtein, B. K. (1971). *Dokl. Akad. Nauk SSSR,* **196,** 1072
292. Kiselev, N. A., Lerner, F. Ja. and Livanova, N. B. (1971). *Molec. Biol.,* **5,** 672
293. Mikhailov, A. M. (1970). *Kristallografiya,* **15,** 818
294. Mikhailov, A. M. and Vainshtein, B. K. (1971). *Kristallografiya,* **16,** 505
295. Kripyakevich, P. I. (1970). *Kristallografiya,* **15,** 690
296. Kripyakevich, P. I., Gladyshevskii, E. I. *et al.* (1967). *Kristallografiya,* **12,** 600; (1968). *Kristallografiya,* **13,** 781; (1970). *Kristallografiya,* **15,** 268
297. Khitrova, V. I., Klechkovskaya, V. V. and Pinsker, Z. G. (1967). *Kristallografiya,* **12,** 1044
298. Khitrova, V. I. and Pinsker, Z. G. (1970). *Kristallografiya,* **15,** 540
299. Tatarinova, L. I. (1970). *Kristallografiya,* **15,** 853
300. Nabitovich, N. D. *et al.* (1968). *Ukr. Fiz. Zh.,* **13,** 1596; (1970). *Fizichna Electronika,* **2,** 60
301. Ovchinnikov, Ju. K., Markova, G. S. and Kargin, V. A. (1969). *Vysokomol. Soedin.,* **11,** 329
302. Tsvankin, D. Ja. (1969). *Kristallografiya,* **14,** 431
303. Tsvankin, D. Ja. *et al.* (1970). *Vysokomol. Soedin.,* **12,** 2136, 2588
304. Vilkov, L. V., Rambidi, N. G. and Spiridonov, V. P. (1967). *Zh. Strukt. Khim.,* **8,** 786
305. Spiridonov, V. P. and Lutoshkin, V. I. (1970). *Vestn. Mosk. Univ. Ser. Khim.,* **5,** 509
306. Vilkov, L. V. and Tarasenko, N. A. (1969). *Chem. Commun.,* 1176
307. Hargittai, I. and Vilkov, L. V. (1970). *Acta Chem. Acad. Sci. Hung.,* **63,** 143

4
Co-operative Phenomena in Inorganic Materials

S. C. ABRAHAMS
Bell Telephone Laboratories Inc., Murray Hill, New Jersey

4.1 INTRODUCTION

Solid-state phase transformations are generally complete within small intervals of temperature or pressure, due to the characteristically long-range nature of the interatomic forces that lead to the transition. The phase below the transition necessarily differs from that above in structure, degree of order or composition. In this chapter, phase transformations accompanied by the onset of such co-operative phenomena as ferro- or antiferromagnetism, ferroelectricity, ferroelasticity and superconductivity are considered. Particular emphasis is placed on a structural basis for these properties. x-Ray and neutron diffraction studies have been made on numerous magnetic and ferroelectric materials. Recognition of ferroelasticity is quite recent and, apart from a few direct experiments, this property is largely known only by prediction from earlier structural investigations. Crystallographic study of materials in their superconducting phase is at present confined to a small number of structures. Space limitation allows no more than a highly selective sampling from the large literature on co-operative phenomena in inorganic materials.

4.2 MAGNETIC MATERIALS

4.2.1 General

Magnetic ordering in inorganic solids has, in the past 20 years, been extensively investigated both directly by neutron diffraction and indirectly by x-ray diffraction methods. The interaction of atomic magnetic moments with the magnetic moment of the neutron gives rise to magnetic scattering. The magnitude of this scattering is proportional to that of the atomic moment, and also depends on the relationship between the moment direction, the scattering vector and the polarisation vector of the neutron. Neutron diffraction hence allows both the magnitude and the direction of the atomic moments, in the ordered state, to be determined[1]. Although the possibility of using the magnetic scattering of x-rays by the electrons in magnetically ordered solids has recently been discussed[2], x-rays have generally been confined to a study of the atomic arrangement in the higher symmetry paramagnetic phase and to the lower symmetry alone of the ordered magnetic phase.

Three basic types of magnetic ordering may be distinguished. In an *antiferromagnet*, pairs of atomic moments are antiparallel by symmetry and hence cancel, giving zero net moment at temperatures below the ordering or Néel temperature (T_N). A simple example of antiferromagnetic order is

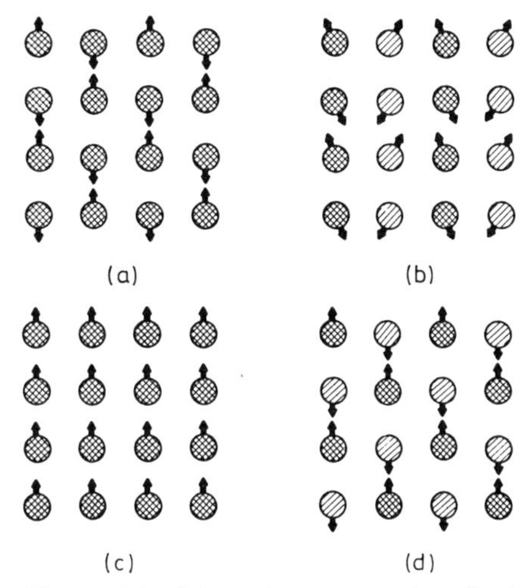

Figure 4.1 Schematic representation of: (a) Simple antiferromagnetically ordered array of identical atoms; spins up represent one magnetic sublattice, spins down, a second. (b) Four-sublattice antiferromagnetic array, containing two atomic species. The spins on each specie are canted with respect to those on the other. (c) Simple ferromagnetic array containing single magnetic sublattice of identical atoms. (d) Ferrimagnetic two-sublattice array, with two atomic species; the spins on one are unequal and antiparallel to the other

illustrated in Figure 4.1a for the case of two sublattices. A real crystal may contain additional pairs of antiparallel magnetic sublattices with moments canted to those of the initial sublattice (Figure 4.1b). A *ferrimagnet* has a spontaneous magnetic moment and contains, in the simplest cases, an array of parallel moments based on a single sublattice below the ordering or Curie temperature (T_c), as illustrated in Figure 4.1c. A *ferromagnet* also has a spontaneous magnetic moment, but contains more than one magnetic sublattice; in the original use of the term, the magnetic sublattices were parallel, or antiparallel but of unequal magnitude, giving a net resultant moment (illustrated in Figure 4.1d). In current usage, there is no angle restriction between moment directions on the different sublattices, provided there is a net spontaneous moment.

The magnetic ordering illustrated in Figure 4.1, on extension to three dimensions, produces sharp maxima in reciprocal space of elastic neutron scattering comparable to that caused by three-dimensional atomic ordering in scattering x-rays. Magnetic order in less than three dimensions is stimulating considerable current interest. A crystal containing isolated lines of antiferromagnetically ordered atoms, as represented in Figure 4.2a, produces planes or sheets of neutron scattering in reciprocal space; in this case, the planes are normal to the ordering direction and occur, for example, at $h = 2n+1$ as in Figure 4.2b. A crystal containing isolated planes of atoms ordered

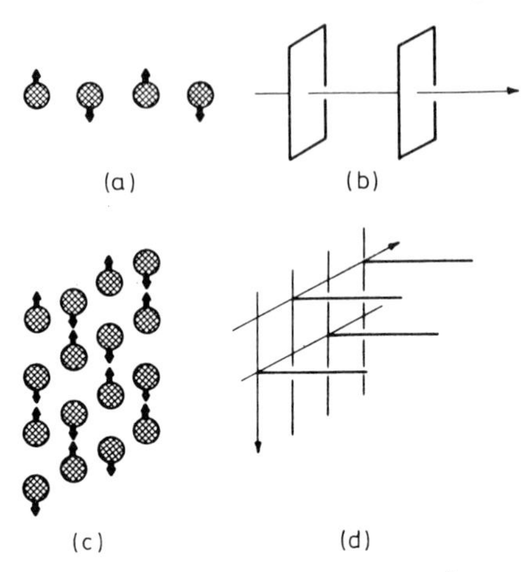

Figure 4.2 Representation of: (a) Linear antiferromagnetically ordered array of identical atoms. (b) Sheets of magnetic scattering in reciprocal space corresponding to the one-dimensional order of (a). (c) Planar antiferromagnetically ordered array. (d) Lines of magnetic scattering in reciprocal space corresponding to the two-dimensional order of (c)

antiferromagnetically, as illustrated in Figure 4.2c, produces lines or rods of neutron scattering in reciprocal space normal to the plane. Scattering corresponding to one- and two-dimensional magnetic ordering has only recently been observed, and is discussed in the following sections together with some recent three-dimensional examples.

The theory of two-dimensional Heisenberg ferromagnets and antiferromagnets has made considerable advances in the last year or two with the application of Green's function approximations[3]. The earlier theoretical and experimental knowledge of magnetism in two dimensions has been conveniently reviewed[4].

4.2.2 One-dimensional magnetic arrays

Crystalline magnetic ordering in one dimension is characterised by the presence of broad maxima both in magnetic susceptibility and in heat capacity, at temperatures well above the long-range three-dimensional ordering-temperature corresponding to T_N. There are still only a few well established one-dimensional magnetic systems, two of which are discussed below.

4.2.2.1 $CsMnCl_3 \cdot 2H_2O$

$CsMnCl_3 \cdot 2H_2O$ is orthorhombic, space group *Pcca* with $a = 9.060$, $b = 7.285$, $c = 11.455$ Å at room temperature, and four formula units in the unit cell[5]. The measured magnetic susceptibility[6] between 0.35 and 77 K is anisotropic below 9 K (suggestive of long-range antiferromagnetic order), with a broad maximum near 30 K. The Néel temperature was accurately determined by neutron scattering[7] to be 4.893 K. Below T_N, neutron diffraction confirms the anti-ferromagnetic ordering arrangement deduced by Spence *et al.* from n.m.r. data[8], illustrated in Figure 4.3. The chain of Mn^{2+} ions forms a gentle zigzag,

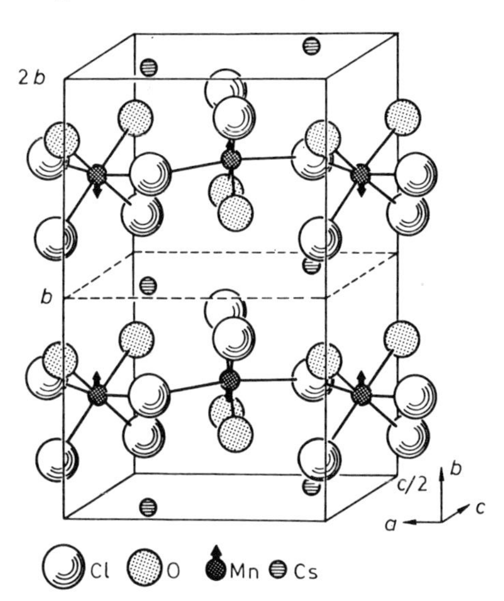

Figure 4.3 Part of $CsMnCl_3 \cdot 2H_2O$ structure viewed along the *c*-axis. The magnetic cell is double the chemical cell along the *b*-axis. The remaining part of the cell is related by an *a′* glide perpendicular to *c*. The Shubnikov group, below $T_N = 4.893$ K, is $P_{2b}c'ca'$

with ions at $0,0.467,\frac{1}{4}$ and $\frac{1}{2},0.533,\frac{1}{4}$ and spins alternately parallel and anti-parallel. Each metal ion is surrounded by a distorted octahedron of four Cl^- anions and two H_2O molecules. The distance between nearest Mn^{2+} ions within a chain is 4.56 Å; the distance between nearest pairs of Mn^{2+} in adjacent chains is 7.29 Å. Neutron measurement between 2.5 and 60 K reveals the disappearance of long-range magnetic Bragg reflections at *hkl* with $k = (2n+1)/2$ as the temperature reaches T_N, and the appearance of *sheets* in reciprocal space above T_N. The sheets of diffuse scattering above T_N are observed at $h = 2n+1$, and a scan at (100) established the planarity of the scattering. The spin system is hence ordered in three dimensions below T_N and, as the temperature reaches *c.* $2T_N$, the interaction between chains becomes almost negligible. Above $3T_N$, Skalyo *et al.*[7] show that $CsMnCl_3 \cdot 2H_2O$ possesses spin correlations in one dimension only; within a chain of Mn^{2+} ions at $3T_N$, there is an average of *c.* 25Mn^{2+} spin correlations. Since the only transition in $CsMnCl_3 \cdot 2H_2O$ is magnetic in origin, the high temperature phase is the prototype and the crystal is represented[9] by PPP(*mmm*)PPA(*mmm*), as the paraelectric paraelastic paramagnetic prototype of point group *mmm* transforms to a paraelectric paraelastic antiferromagnet with the same point group.

4.2.2.2 $RbNiCl_3$

$RbNiCl_3$ is hexagonal, space group $P6_3/mmc$, with $a = 6.95$, $c = 5.90$ Å at room temperature and two formula units per unit cell[10]. The magnetic susceptibility of $RbNiCl_3$, and of isomorphous $CsNiCl_3$, $CsCoCl_3$ and $RbFeCl_3$, has been measured between 1.4 K and room temperature; the Néel temperature for $RbNiCl_3$ is 11 K and for $CsNiCl_3$ is 4.5 K[10]. Both crystals exhibit broad susceptibility maxima, at *c.* 42 and 32 K respectively. The Ni^{2+} ions, separated by only 2.95 Å along the *c*-axis, form infinite linear chains. Each Ni^{2+} ion is located within a face-sharing octahedron of Cl^- ions, as illustrated

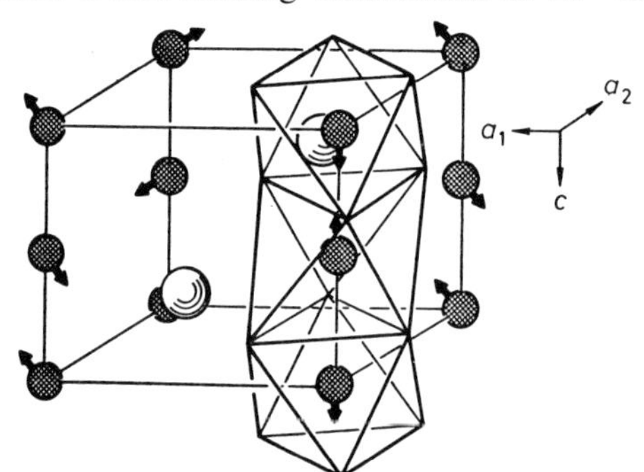

Figure 4.4 View of $RbNiCl_3$ structure, showing linear chains of Ni^{2+} ions with face-sharing chlorine octahedra outlined for one chain. The Rb^+ ions are indicated by larger circles. The spin directions on the Ni^{2+} are suggested by the arrows on the cross-hatched circles

by Figure 4.4. $RbNiCl_3$ forms a layer structure with both Cl^- and Rb^+ at $\pm c/4$ and Ni^{2+} at 000 and $00\frac{1}{2}$. Powder neutron measurements below T_N reveal[11] an antiferromagnetic sequence of spins along the *c*-axis, forming a spiral array with the spins rotating in a plane normal to (001). The spiral array, the validity of which has recently been challenged[11a], has wavelength $3a/2$; the magnetic cell has dimensions $a_M = \sqrt{3}\,a$, $c_M = c$. Above T_N, sheets of magnetic scattering density form perpendicular to [001], located at $l = 2n+1$. Three-dimensional peaks superposed on the sheets persists a little above T_N, but are no longer detectable in the neighbourhood of $4T_N$. Long-range antiferromagnetic spin correlations thus exist above T_N but, as the temperature is increased, these progressively weaken and the characteristics of a linear antiferromagnetic chain are ultimately reached.

4.2.3 Two-dimensional magnetic arrays

Magnetic ordering in two dimensions is possible if the magnetic array forms a plane sufficiently separated from similar adjacent planes that interlayer interactions are reduced to a small fraction of the intralayer interactions. Such a crystal will be characterised by a broad maximum in magnetic susceptibility as a function of temperature. Several two-dimensional magnetic systems are known, of which two are discussed.

4.2.3.1 K_2NiF_4 *family*

Included in this tetragonal family are K_2MnF_4, Rb_2MnF_4, K_2CoF_4, Rb_2FeF_4 and Ca_2MnO_4. K_2NiF_4 has lattice constants at 80 K of $a = 3.994$, $c = 13.04$ Å, space group *I4/mmm*, with two formula units per unit cell[12]. The magnetic susceptibility of K_2NiF_4 has a broad maximum at *c*. 250 K and is suggestive of antiferromagnetic ordering at low temperatures. Neutron scattering shows that three-dimensional antiferromagnetic ordering sets in at $T_N = 97.1$ K. Above T_N, *rods* of neutron scattering are observed in reciprocal space, in which reciprocal lattice points such as $h00$, $h02$, etc. with $h = 2n+1$ become drawn out into continuous $h0r$ rods, with r extending at least over the range $-\frac{1}{2} \leqslant r \leqslant +\frac{1}{2}$. Thus at 99 K for example, K_2NiF_4 scatters neutrons as would a purely two-dimensional antiferromagnet, the correlations within planes of Ni^{2+} ions persisting over at least 1000 Å, but with unobservably small correlations between planes.

On cooling K_2NiF_4 through T_N the rods persist and sharp peaks at 100 etc. appear *discontinuously*. The scattering intensity of the rods rapidly declines as the temperature is lowered further and the intensity of the sharp peaks increases. It appears that the integrated intensity lost by the rods is equal to that gained by the peaks. The phase change in K_2NiF_4 is hence one in which the purely two-dimensional magnetic order present over a wide temperature range above T_N reaches a magnitude large enough to cause magnetic interactions between planes below T_N. At this critical temperature three-dimensional order sets in, i.e. this appears to be a genuine two-dimensional phase transition.

The K_2NiF_4 structure, illustrated in Figure 4.5, consists of corner-sharing sheets of NiF_6 octahedra with $Ni^{2+}-Ni^{2+}$ separations of 4 Å. The sheets are $c/2$ apart, resulting in intersheet closest $Ni^{2+}-Ni^{2+}$ separations of 6.8 Å. The antiferromagnetic arrangement of Ni^{2+} spins is indicated by the arrows in Figure 4.5. The magnetic unit cell clearly does not correspond to the chemical cell, but is *C*-face centred, with $\boldsymbol{a}_M \approx \boldsymbol{a}_1 + \boldsymbol{a}_2$, $\boldsymbol{b}_M \simeq \boldsymbol{a}_1 - \boldsymbol{a}_2$. The

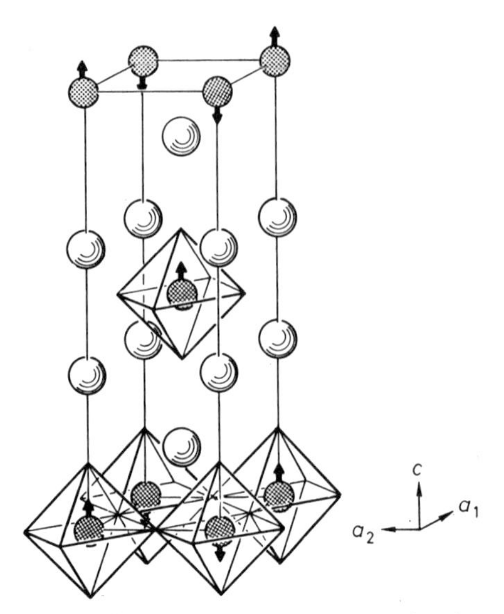

Figure 4.5 View of K_2NiF_4 structure, showing one sheet of corner-sharing NiF_6 octahedra in the $z = 0$ level and a single octahedron at the $z = \frac{1}{2}$ level. The fluorine anions are represented by larger shaded circles. The wide separation between Ni^{2+} sheets along the c-axis is clearly seen. The spin directions on the Ni^{2+} are indicated by arrows

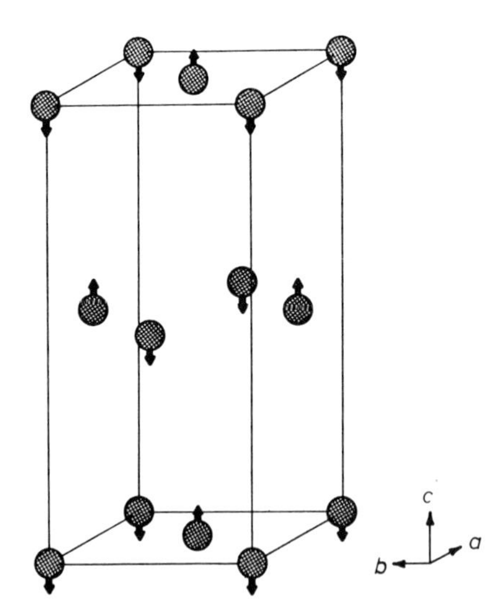

Figure 4.6 A_C magnetic lattice of K_2NiF_4 below T_N. Spins on Ni^{2+} ions are indicated, the K and F atoms are omitted for clarity. In this magnetic cell $\boldsymbol{a} \approx \boldsymbol{a}_1 + \boldsymbol{a}_2$, $\boldsymbol{b} \approx \boldsymbol{a}_1 - \boldsymbol{a}_2$, cf. Figure 4.5

magnetic Bravais lattice is hence orthorhombic, (A_C), with $a_M \neq b_M$. It is most probable that at T_N the crystal becomes ferroelastic as well as antiferromagnetic, i.e. that K_2NiF_4 is PPP(4/*mmm*)PF_CA_C(*mmm*). In the absence of an externally applied stress, the crystal is likely to be twinned below T_N. The potential coupling between ferroelasticity and antiferromagnetism (denoted by the subscript 'c') is of particular interest, since planes in the A_C lattice normal to *a* contain ferromagnetically coupled Ni^{2+} ions whereas those normal to *b* are antiferromagnetically coupled, as illustrated in Figure 4.6. Stress applied along *a* or *b* is hence predicted to reverse the sense of all spins in sheets at $z = \frac{1}{2}$ relative to those at $z = 0$.

4.2.3.2 $MnTiO_3$

$MnTiO_3$ has the ilmenite structure with $a = 5.137$, $c = 14.283$ Å, six formula units in the unit cell and space group $R\bar{3}$ [13]. $MnTiO_3$ has been reported to develop a two-dimensional antiferromagnetic character above the Néel temperature of 64 K [14]. The magnetic susceptibility of $MnTiO_3$ powder shows a typical broad maximum, at *c.* 100 K: an early neutron diffraction study[13] indicated the structure to be antiferromagnetic with antiparallel spins in each layer, at $z = 0$ and $\frac{1}{2}$, directed along the *c*-axis. A single-crystal neutron diffraction experiment has now shown[14] that a magnetic reciprocal-lattice rod is formed at the 100 position; the rods are parallel to the *c*-axis, lying along [10*l*], and extend over an angular range of >16 degrees (2θ). The rods, which have maximum intensity at T_N, persist to at least $2T_N$; they are also observable at temperatures as low as $0.3T_N$, although with intensity that decreases sharply above and below T_N. The two-dimensional order below T_N decreases as the three-dimensional antiferromagnetic long-range order increases.

4.2.4 Three-dimensional magnetic arrays

The great majority of magnetic materials in which magnetic-moment arrays have been investigated crystallographically are characterised by the presence

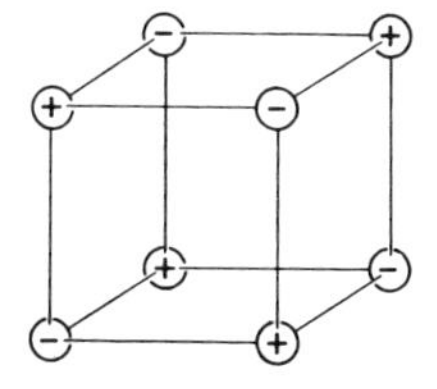

Figure 4.7 Collinear antiferromagnetic ordering in some perovskite-structure ABF_3 compounds, the + and − indicating parallel or antiparallel spins. Other types of collinear antiferromagnetic ordering known in ABF_3 and ABO_3 perovskites are illustrated in Reference 13

of three-dimensional long-range order. Several recent publications provide a valuable guide to the extensive literature on such materials, including Landolt–Börnstein[13], Izyumov and Ozerov[1], Oleś *et al.*[15] and the comprehensive *Index to the Literature of Magnetism*[16]. In this section, a few examples are discussed to illustrate the principles of antiferromagnetic, ferromagnetic and ferrimagnetic ordering. Many cubic perovskite-structure ABF_3 compounds, with A an alkali and B a 3d-metal, order antiferromagnetically at low temperatures with a collinear spin-structure as indicated in Figure 4.7. In spin ordering, the cubic point-symmetry is lowered and the cell becomes slightly distorted, in close analogy to the ferroelectric phases of the cubic ABO_3 perovskites (see Section 4.3.3). In many cases, the distortion is so small that the resulting phase is potentially ferroelastic with a resulting ferroelastic–antiferromagnetic coupling, since both properties simultaneously originate in the same ordering process. Three typical ABF_3 compounds are briefly considered, followed by three additional examples.

4.2.4.1 $KCoF_3$

This is cubic above the Néel temperature of 109.5 K. Below T_N, $KCoF_3$ is antiferromagnetic with the spin arrangement of Figure 4.7 in a tetragonal cell with $a = 4.060$, $c = 4.050$ Å at 78 K [17] and Shubnikov space group $P_I4/mm'm'$. The approach of c/a to unity, together with the small departure in atomic positions from the cubic sites, suggests that reorientation of the unique axis may be expected by application of stress along an a-axis (see Section 4.4), giving a transformation as represented[9] by $PPP(m3m)PF_CA_C$ $(4/mmm)$.

4.2.4.2 $KFeF_3$

$KFeF_3$ is also cubic in the paramagnetic phase above $T_N = 115$ K but transforms to a rhombohedral cell in the antiferromagnetic phase, with spin structure as indicated in Figure 4.7 and $a = 4.108$ Å, $\alpha = 89$ degrees 51 minutes at 78 K and space group $R_R\bar{3}m'$ [15]. The crystal is probably ferroelastic, and the unique axis is expected to be stress reorientable along previously cubic [111] directions; the transition may be represented by $PPP(m3m)PF_CA_C(\bar{3}m)$.

4.2.4.3 $KMnF_3$

$KMnF_3$ is paramagnetic above $T_N = 88$ K and antiferromagnetic below, with a monoclinic unit cell and $a = 4.159$, $b = 4.174$, $c = 4.189$ Å and $\beta = 89$ degrees 54 minutes at 78 K [15]. The spin arrangement is given in Figure 4.7, with a potentially ferroelastic and reorientable unit cell. In addition to the first-order magnetic phase transition at 88 K, a second-order transition from the room-temperature cubic phase to tetragonal symmetry occurs at 184 K. Diffuse x-ray scattering, located along [100] rods in reciprocal space, has been reported close to the upper transition point. The diffuse scattering has been interpreted as due to rotation of the MnF_6 octahedra about their fourfold axes; rotations in a given (100) layer would have low correlation with other layers[18].

4.2.4.4 *Vivianite,* $Fe_3(PO_4)_2 \cdot 8H_2O$

Vivianite has a *non-collinear antiferromagnetic* structure. The crystal is monoclinic with $a = 10.08$, $b = 13.43$, $c = 4.70$ Å and $\beta = 104.5$ degrees at room temperature, space group $C2/m$ and two formula units per unit cell. Vivianite is paramagnetic above $T_N = 8.8$ K and antiferromagnetic below, although earlier studies based on nuclear magnetic resonance, specific heat and Mössbauer spectral data gave conflicting results. A single-crystal neutron diffraction investigation[19] at 4.2 K shows the magnetic c-axis to be double that of the chemical cell, resulting in a C_c lattice. There are two independent sets of Fe atoms, with Fe_I at 0,0,0 and Fe_{II} at 0,0.3898,0; each has a spin that, in principle, is also independent. Interpretation of a spin-density

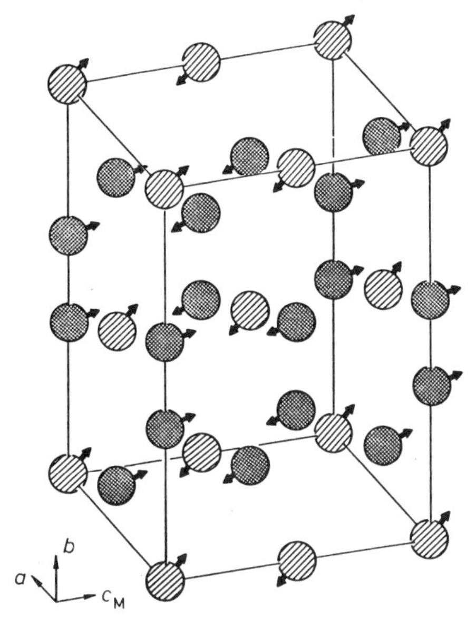

Figure 4.8 Spin arrangement in the vivianite, $Fe_3(PO_4)_2{\cdot}8H_2O$, unit cell at 4.2 K. The hatched circles represent Fe_I, the cross-hatched circles Fe_{II}. The spins both on Fe_I and Fe_{II} are normal to b: those on Fe_I make an angle of 77 degrees to c, and on Fe_{II} an angle of 35 degrees

Patterson function, using Wilkinson's approach[20], enabled an approximate magnetic structure to be derived based on the assumption that both spins are equal. Subsequent least-squares refinement resulted in a model with a moment of 3.46 μ_B on Fe_I and 4.74 μ_B on Fe_{II}, making angles of 77 and 35 degrees respectively to [001], for β obtuse. The symmetry, in magnetic space group C_c2/c, requires the spins to be normal to the monoclinic b-axis. Figure 4.8 illustrates the magnetic structure; only the iron atoms are shown. Pairs of Fe_{II}, separated by only 2.96 Å along b (at $0,y,0$ and $0,1-y,0$), are ferromagnetically aligned. The flattened peak in the specific heat curve at c. 10 K may be due[19] to short-range ordering of the spins on these pairs of Fe_{II} atoms above T_N.

4.2.4.5 Ludlamite $Fe_3(PO_4)_2{\cdot}4H_2O$

Ludlamite, the related iron phosphate hydrate $Fe_3(PO_4)_2{\cdot}4H_2O$, forms a canted spin-array at low temperatures, with a resulting *ferrimagnetic* moment. At 4.2 K, ludlamite is monoclinic with $a = 10.541$, $b = 4.638$, $c = 9.285$ Å

$\beta = 100.73$ degrees, space group $P2_1/a$ and two formula units per cell. The Curie temperature is 15.1 K and the effective moment per iron atom is 0.81 μ_B. A single-crystal neutron study[21] showed the magnetic space group to be identical to the chemical-cell space group. Discrete linear Fe—Fe—Fe groups, with intragroup Fe—Fe separations of 3.267 Å, are ferromagnetically coupled. Alternate triads along the *a*-axis are antiferromagnetically coupled. All spins are slightly canted out of the *ac*-plane in the same sense, giving rise to the net moment along [010]. A polarised neutron-beam experiment[22] has confirmed the magnetic structure and has shown that the central Fe atom in the triad produces the larger contribution to the spontaneous moment, the spins of the other Fe atoms lying closer to the *ac*-plane.

Antiferromagnets consist of either collinear or non-collinear arrays of magnetic spins; in the latter group, equivalent spins are related either by the symmetry elements of the magnetic space group, as illustrated by the case of vivianite, or they form more generalised arrays that are not simply described by two-colour symmetry. The rare-earth elements, such as terbium[23], provide many excellent examples.

4.2.4.6 Au_2Mn

Au_2Mn is a recent example of a *helical* magnetic structure. This compound is tetragonal, space group $I4/mmm$, with $a = 3.369$, $c = 8.756$ Å and two formula units per cell. Below 363 K, Au_2Mn is paramagnetic in weak fields, but in magnetic fields greater than 10 kOe the magnetisation increases rapidly, reaching saturation above 20 kOe. An early neutron diffraction study[24] showed Au_2Mn to form a helical magnetic array, with the moments of the Mn atoms ferromagnetically aligned in the planes normal to *c*, but with rotations of *c*. 50 degrees between successive planes. Two models are consistent with this basic arrangement; firstly a rigid-spin model, with an abrupt change in magnetisation direction from one atom to the next and, secondly, a flexible-spin model, with a continuous change in magnetic-moment direction within the extended cloud of magnetic electrons. The two models are distinguishable by neutron diffraction; a rigid-spin model would result in satellite peaks at $\boldsymbol{B}_{hkl} \pm \tau$, related to each reciprocal lattice vector $\boldsymbol{B}_{hkl}$ by the helical propagation vector τ. The flexible-spin model would not result in satellite peaks. A diffraction study on polycrystalline Au_2Mn (single crystals could not be prepared), by use of the polarisation analysis technique to distinguish coherent magnetic scattering from possible impurity nuclear scattering[25], unambiguously favoured the rigid-spin model.

4.3 DIELECTRIC MATERIALS

4.3.1 General

The dielectric properties of piezoelectricity, pyroelectricity and ferroelectricity are each structure dependent and all have been the subject of numerous crystallographic studies. In a *piezoelectric* crystal the application of stress causes an electrical polarisation that is proportional to the strain and, conversely, the application of an electric field causes the crystal to become strained by an amount proportional to the polarising field. With the exception of point group 432, any crystal that belongs to one of the twenty remaining non-centrosymmetric point groups exhibits piezoelectricity. Detailed properties, including elastic, piezoelectric and electro-optic constants, of over 500 inorganic piezoelectric crystals are given in Landolt–Börnstein[26]. The property of *pyroelectricity* is restricted to materials that crystallise in one of the ten non-centrosymmetric point groups that also contain a unique polar axis[27]. Pyroelectric materials are characterised by the presence of a spontaneous polarisation $\boldsymbol{P}_s$; detection is usually accomplished by means of the change in $|P_s|$ that results from a change in temperature, since the magnitude of $\boldsymbol{P}_s$ is temperature dependent.

Any crystal that belongs to one of the ten pyroelectric point groups[27] is potentially *ferroelectric*; the crystal is ferroelectric only if the spontaneous polarisation is capable of reversal or reorientation either experimentally, by application of an electric field, or conceptually, in terms of the atomic rearrangement required by the structure in reversing or reorienting the polar direction. Ferroelectric crystals have been divided into three classes, based only on the predominant nature of the atomic displacements involved in polarisation reversal[28]. Atomic displacements that are essentially parallel to the polar axis, in reversing the polarity, correspond to the one-dimensional class; displacements that are parallel to a plane containing the polar axis form the two-dimensional class; those in which the displacements are generally oriented comprise the three-dimensional class.

The magnitude of the spontaneous polarisation is found to have a characteristic range for each class. One-dimensional ferroelectrics have $P_s > 25 \times 10^{-2}$ C m^{-2}; for two-dimensional ferroelectrics, $10 \times 10^{-2} > P_s > 3 \times 10^{-2}$ C m^{-2}; and for three-dimensional ferroelectrics, 5×10^{-2} C m$^{-2} > P_s$. These ranges are not rigid, and some overlap has been found between the two- and three-dimensional classes. An accurately determined crystal structure enables P_s to be calculated. The resulting values agree best with experiment for two-dimensional ferroelectrics, assuming either ionic point charges or molecular dipoles. Both the point charge model and the linear relation with atomic displacement[29] give P_s values in somewhat poorer agreement with experiment for one-dimensional ferroelectrics.

Several earlier classifications of ferroelectric materials, including a useful one based on the ferroelectric transition being either of a displacive or an order–disorder type, have been proposed. Together with an excellent treatment of the status of ferroelectric crystals as of a decade ago, these have been reviewed by Jona and Shirane[30]. A more recent review has been presented by Zheludev[31]. Cochran's lattice dynamical theory[32] and Lines' statistical theory[33] of ferroelectricity have provided major advances in theoretical understanding. An interesting model in which inversion centres in centrosymmetric crystals may be lost due to paramagnetic–antiferromagnetic transitions has recently been proposed[34]. In this model, an antiferromagnetic transition is considered from Shubnikov space group *Pmma* to $P_{2a}m'm'2$; a spontaneous polarisation on the order of 0.01×10^{-2} C m^{-2} may thereby be produced. Magnetic ordering of all types (Section 4.2) could result in similar polarisations.

4.3.2 Piezoelectric materials

An understanding of the origin of the piezoelectric effect in a given crystal requires not only the crystal structure to be well determined but also the absolute orientation of the atomic arrangement to be known with respect to the charges produced on compression along a specified direction. In addition, both the ionic charge distribution and the electronic polarisation contribution must be assessed. Although any acentric atomic arrangement may

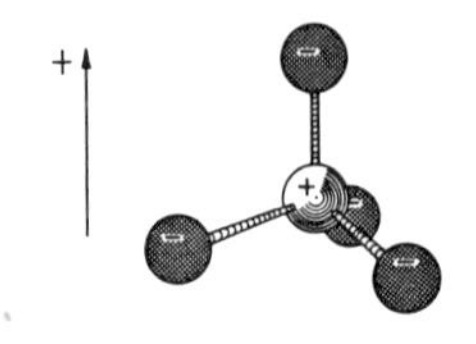

Figure 4.9 Tetrahedral arrangement of negatively charged atoms about a central atom of positive charge. Uniaxial extension along the vertical direction produces a positive polarity as shown

lead to a measurable piezoelectric effect, a motif common to many piezoelectric crystals is the tetrahedrally coordinated atom, Figure 4.9. Of the examples discussed below, the sense of the piezoelectric constant is simply derived in all cases but one from a consideration only of the ionic charge distribution. The exception is taken first.

4.3.2.1 InSb *and related Group II–VI and I–VII compounds*

InSb crystallises in the zinc blende structure with space group $F43m$; the body diagonals of the unit cell are coincident both with the tetrahedral bonds and the polar directions. If the tetrahedral coordination angles In—Sb—In and Sb—In—Sb are more easily deformed than the nearest-neighbour interatomic contact distance In—Sb, then compressive stress applied along [111] effectively displaces the charge associated with Sb towards the (111) face; conversely, expansion along [111] would displace the In charge toward (111). Experimentally, the (111) face develops a negative polarity on expansion[35]; the purely ionic polarisation in the Group III–V compounds is hence more than compensated by an opposite polarisation, probably largely electronic in origin.

The Group II–VI compounds with zinc blende structure such as ZnS, as well as the comparable Group I–VII compounds such as CuCl [36], all produce a positive polarity on the (111) face on expansion along [111], as predicted by the simple ionic-charge model, neglecting electronic polarisation. The polarity produced in the würtzite form of ZnO on expansion is similarly that predicted on the basis of Zn^{2+} and O^{2-} ions.

4.3.2.2 $Bi_{12}GeO_{20}$

$Bi_{12}GeO_{20}$ is available as large single crystals and is an example of a structure with greater complexity in which the piezoelectric effects can still be directly related to a tetrahedrally coordinated atom (see Figure 4.10). $Bi_{12}GeO_{20}$ crystallises in the cubic system, space group $I23$ and $a = 10.1455$ Å, with Ge atoms at the origin and body centre[37]; each Ge is surrounded by a tetrahedron of oxygen atoms with identical orientation. In point group 23, each of the four triad axes is polar. Uniaxial tension applied to [111] provides a

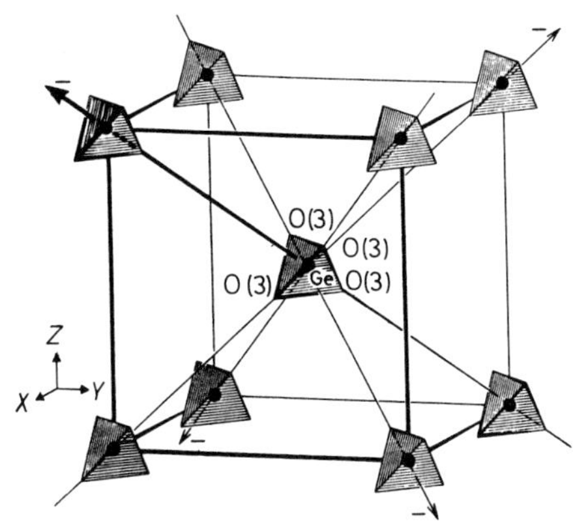

Figure 4.10 GeO_4 tetrahedra in $Bi_{12}GeO_{20}$, showing the polarisation resulting from uniaxial stress applied along [111]. The 24 bismuth and 32 other oxygen atoms in the unit cell are omitted for clarity (Adapted from Fig. 1 of Ref. 37)

stress component along the three remaining triads, effectively displacing the positive charge associated with Ge away from the tetrahedral GeO_4 face normal to [111]. A negative polarity as shown in Figure 4.10 is thus expected, in agreement with experiment[37].

4.3.2.3 Tourmaline

One of the oldest known piezoelectrics, tourmaline, is considered in this section although it is also pyroelectric. Several varieties of tourmaline have been studied crystallographically, including dravite[38], elbaite and an iron

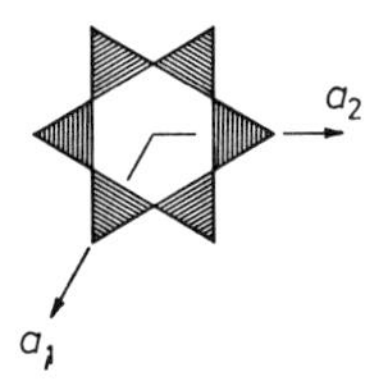

Figure 4.11 Six-membered rings of corner-sharing SiO_4 tetrahedra in tourmaline, the tetrahedra apexes pointing below the plane of the figure, along $-c$. Extension along c causes a negative polarity to develop on the (0001) face, as expected for positively charged Si

tourmaline, buergerite[39]. Buergerite, ideally $NaFe_3B_3Al_6Si_6O_{30}$, is rhombohedral with space group $R3m$ and $a = 15.869$, $c = 7.188$ Å. Buergerite contains six-membered rings of silicon–oxygen tetrahedra, with all six tetrahedra pointing in the same c direction. A face of each tetrahedron is nearly normal to c, and each apex points along $-c$ as illustrated in Figure 4.11. Extension along the polar axis would result in a negative polarity on the (0001) face if the silicon atoms within the tetrahedra depicted in Figure 4.11 are, as expected, positively charged. Experiment shows this is indeed the polarity produced[39].

4.3.3 One-dimensional ferroelectrics

In the one-dimensional class of ferroelectric crystals, all vectors that connect atomic positions in a structure of given polarity to the equivalent positions in the corresponding structure with opposite polarity are characteristically parallel to the polar axis[28]. Typically, an atom at xyz, in a one-dimensional ferroelectric structure with polar c-axis, is displaced to the position $x,y,1-z$ on reversing polarity across a pseudomirror plane at $z = \frac{1}{2}$. The displacement $\Delta = 1 - 2z$ is usually on the order of 0.1 Å but, as seen from examples given below, may be as large as 1.2 Å. The mirror plane is generally formed of closely packed anions, particularly oxygen. Dipoles originating on cations displaced from the pseudomirror are generally of the same sense in one-dimensional ferroelectric crystals.

The three most important families of one-dimensional ferroelectrics are discussed.

4.3.3.1 ABO_3 *perovskite family*

This family (A includes Na, K, Ca, Sr, Ba and Pb and B includes many of the transition metals) is typified by $BaTiO_3$, a material with properties sufficiently interesting that over 600 papers on it have now appeared in the literature. The highest symmetry form of $BaTiO_3$ is cubic, space group $Pm3m$ with $a = 4.012$ Å at 435 K, and structure shown schematically in Figure 4.12c. Below 408 K [40], the structure transforms to tetragonal, space group $P4mm$ with $a = 3.9920$, $c = 4.0361$ Å at 293 K. The Ba and Ti atoms are each displaced from their mean oxygen planes by 0.074 and 0.128 Å respectively, in the same sense[41], as indicated in Figures 4.12a, and 4.12b. Figure 4.12d represents the corresponding displacements of Ba and Ti in a single domain with polarity opposite to that of Figures 4.12a or 4.12b. The observed spontaneous polarisation of 26×10^{-2} C m^{-2} originates in the dipoles associated with these displacements along the polar axis. The calculated value of P_s is 33×10^{-2} C m^{-2}, based on the relation[29]:

$$P_s = (258 \pm 9)\mathbf{\Delta}/2 \times 10^{-2}\ \mathrm{C\,m^{-2}}$$

where $\mathbf{\Delta}/2$ is the out-of-plane displacement of the Ti atom: that based upon point charges of 2+ on Ba, 4+ on Ti and 2− on O is 15×10^{-2} C m^{-2}, by use of the relation:

$$P_s = (e/2V)\Sigma\, Z_i \mathbf{\Delta}_i$$

where the summation is taken over the volume V containing i ions of charge Z_i, with Δ_i the atomic displacement of the ith ion. It is of particular interest that, despite a 1.1% difference in length between the a and c axes, the c axis may be transformed ferroelastically into an a axis at room temperature by the application of uniaxial pressure[42]. Compressibility by as much as 10^{-2} is unlikely, suggesting that the ferroelastic transformation of axes in tetragonal $BaTiO_3$ is triggered by a much smaller critical compression.

At 278 K, $BaTiO_3$ undergoes a transition to orthorhombic symmetry, space group $Amm2$ with $a = 3.990$, $b = 5.669$, $c = 5.682$ Å at 263 K. The structure of the ferroelectric orthorhombic phase, determined at 263 K [26],

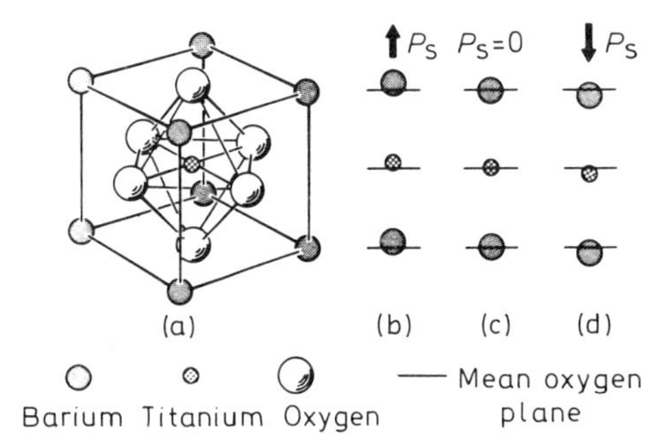

Figure 4.12 (a) $BaTiO_3$ perovskite structure. (b) Schematic of Ba and Ti ionic displacement from mean oxygen planes, with resultant P_s. (c) Schematic of paraelectric cubic $BaTiO_3$. (d) As in b with displacements and polarity reversed

differs from that of the tetragonal phase only by the displacement of one independent oxygen atom through less than 0.02 Å normal to the polar axis; the displacement parallel to the polar axis of the Ba and Ti atoms from the mean oxygen planes is nearly identical with the tetragonal displacement. The small spontaneous strain (Section 4.4), given by $c-b/c+b = 1.15 \times 10^{-3}$, together with atomic displacements of <0.17 Å required to interchange the b and c axes, suggests that this phase is also ferroelastic. It is noteworthy that in both tetragonal and orthorhombic phases, the interchange of axes results in reorientation of the polar axis by 90 degrees. Full coupling is expected between ferroelasticity and ferroelectricity in orthorhombic $BaTiO_3$. Below 183 K $BaTiO_3$ becomes rhombohedral, space group $R3m$ with $a = 4.001$ Å and $\alpha = 89.85$ degrees at 105 K [26]. The spontaneous polarisation in the rhombohedral phase lies along a body diagonal; the small distortion of α from 90 degrees suggests that this phase is also ferroelastic, with the possibility of reorienting the polar axis by 109.47 degrees on application of uniaxial compression along the polar direction. The sequence of phases and properties in $BaTiO_3$ is represented by

$$\mathrm{PPP}(m3m)\mathrm{F_C F_C P}(4mm)\mathrm{F_C F_C P}(mm2)\mathrm{F_C F_C P}(3m)$$

Similar phase sequences and relationships are found in other ABO_3 perovskites[26]. The recent observation[43] of x-ray diffuse scattering and also inelastic neutron-scattering measurements[44, 45] on several ferroelectric perovskites have been interpreted both in terms of static and of dynamic disorder. The cubic-to-tetragonal phase transformation in $BaTiO_3$ has been shown by the neutron experiment[45] to be completely associated with an instability in the excitation spectrum, i.e. to be dynamical in nature. It is expected that further diffuse-scattering experiments will provide additional insight into the ferroelectric ordering process.

4.3.3.2 *Ferroelectric tungsten-bronze type family*

The ferroelectric tungsten-bronze type family (a widely used misnomer since neither tungsten nor bronze is associated with these compounds) has the general formula $(Al)_2(A2)_4C_4(B1)_2(B_2)_8O_{30}$. The five different kinds of site in this structure, illustrated in Figure 4.13, may be occupied by alkali or alkali earth atoms in the A1 and A2 positions, by Li or perhaps Be in the C position and by Ti, Nb or Ta in the B1 and B2 positions. The structures of

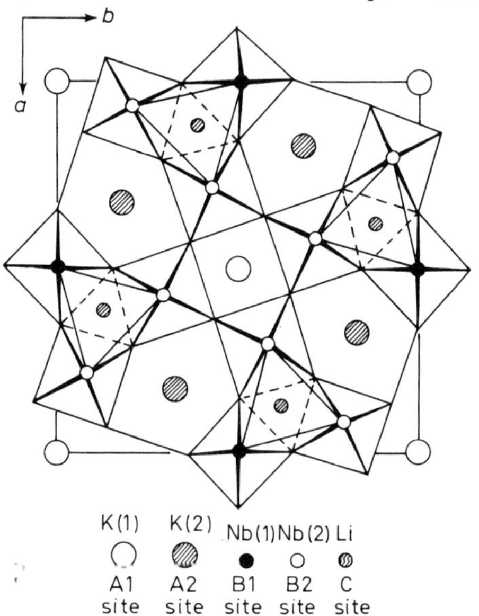

Figure 4.13 Schematic diagram of potassium lithium niobate structure along the polar axis, with the five kinds of metal atom site and oxygen atom positions indicated
(From Abrahams, S. C. *et al.*[47] by permission of the American Institute of Physics)

several ferroelectric tungsten bronzes have now been elucidated; that of $K_{(6-x-y)}Li_{(4+x)}Nb_{(10+y)}O_{30}$ (KLN) is typical, and is taken to exemplify the family.

KLN is non-stoichiometric, with a composition range extending from 51 to 67 mole % Nb_2O_5 and from 26 to 36 mole % K_2O [46]. KLN, with $x \approx 0.07$, $y \approx 0.23$ and $T_C = 613$ K is tetragonal, $a = 12.5764$, $c = 4.0149$ Å at 298 K, and space group *P4bm* [47]. The 12-coordinated A1 site is partially occupied by Li as well as K, whereas the 9-coordinated A2 site is almost exclusively populated by K alone. Both B1 and B2 sites contain Nb atoms only, forming characteristic NbO_6 octahedra. The C site is primarily occupied by Li, although some Nb is also present. The 'composite' atom at the C site is slightly displaced from the plane formed by the three nearest oxygen atoms, along the polar axis. Indeed, each metal atom is displaced along the polar

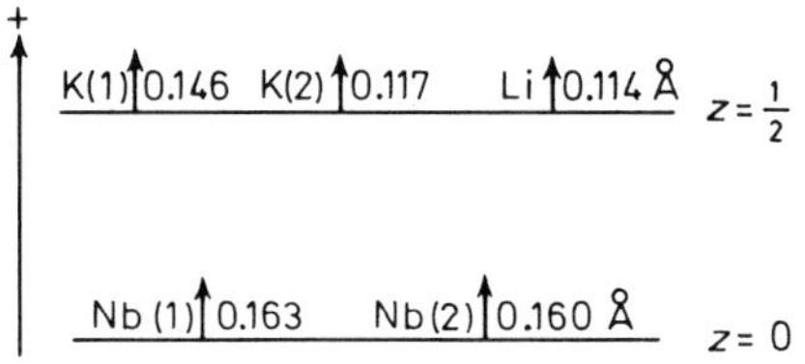

+
K(1) ↑ 0.146 K(2) ↑ 0.117 Li ↑ 0.114 Å $z = \frac{1}{2}$

Nb (1) ↑ 0.163 Nb (2) ↑ 0.160 Å $z = 0$

Figure 4.14 Absolute sense and magnitudes of metal atom displacements in $K_{(6-x-y)}Li_{(4+y)}Nb_{(10+y)}O_{30}$ from mean oxygen atom planes, and macroscopic polarisation sense (From Abrahams, S. C. *et al.*[47] by permission of the American Institute of Physics)

axis from the nearest mean oxygen–atom plane in the same sense. The displacement of the two independent Nb atoms by 0.163 and 0.160 Å, corresponds to $P_s = 42 \times 10^{-2}$ C m^{-2} [29]; the positive polarisation direction is that of the dipoles originating in the metal atom displacements, as indicated by Figure 4.14. The transition in KLN at T_C is represented by PPP(4/*mmm*)FPP(4*mm*).

4.3.3.3 $LiNbO_3$

The lithium niobate family is represented here by $LiTaO_3$ and at present is known to include only $LiNbO_3$, $LiTaO_3$ and $BiFeO_3$. The family is related to the ABO_3 perovskites and the ferroelectric tungsten bronzes in that the

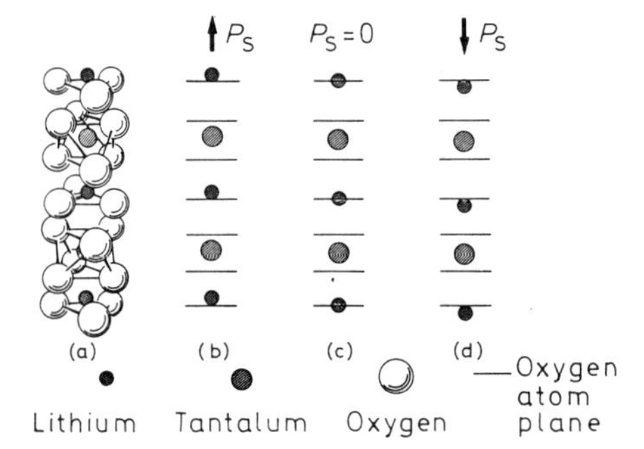

Figure 4.15 (a) Octahedra in ferroelectric $R3c$ phase of $LiTaO_3$ stacked along the trigonal axis, containing in sequence Li, Ta, empty. (b) Schematic representation of Li and Ta ionic displacement from mean oxygen planes, with resultant P_s. (c) Schematic representation of high temperature paraelectric $R\bar{3}c$ phase. (d) As in (b), with displacements and polarity reversed (Adapted from Fig. 3 of Ref. 28)

basic building block in all three families is the BO_6 octahedron with the oxygen atoms closely packed. $LiTaO_3$ is rhombohedral, $a = 5.1543$, $c = 13.7835$ Å at 298 K with space group $R3c$ [48]. The TaO_6 octahedra share faces both with empty and with LiO_6 octahedra, forming stacks as illustrated in Figure 4.15a.

Adjacent stacks are connected by sharing edges. In the paraelectric $R\bar{3}c$ phase, the Ta atom occupies the centre of its octahedron and the Li atom is assumed[49] to lie in the nearest oxygen plane, Figure 4.15c. At 298 K, within a single domain, the Ta atom is displaced parallel to the polar axis by 0.201 Å and the Li atom is displaced in the same sense by 0.601 Å. The calculated spontaneous polarisation, based only on the Ta atom displacement[29], is 51×10^{-2} C m^{-2}; the point charge model gives 49×10^{-2} C m^{-2}. The observed value is 50×10^{-2} C m^{-2} [26].

Neutron diffraction measurements at several temperatures below and above $T_C = 908$ K have shown the oxygen atom to move, with respect to Ta at 000, to 0.055, $\frac{1}{3}, \frac{1}{12}$ from the room-temperature position, a distance of *c*. 0.2 Å [50]. This motion is equivalent to Ta moving to the inversion centre as depicted in Figure 4.15c. The Li atom position, however, does not move appreciably as a function of temperature until T_C is reached, whereupon it apparently becomes disordered and equally distributed about ± 0.37 Å on either side of the inversion centre at $00\frac{1}{4}$. The large difference in T_C between $LiTaO_3$ (908 K) and $LiNbO_3$ (1483 K) may be related to a difference in Li behaviour at T_C, with $LiNbO_3$ more closely represented by Figure 4.15c. However, a recent n.m.r. study[51] of $LiNbO_3$ suggests that Li may have a temperature dependent to-and-fro motion through the nearest triangular arrangement of oxygen atoms; a time-average of such motion would be equivalent to space disorder. The statistical theory of ferroelectricity[33] places the major differences in ferroelectric properties between $LiNbO_3$ and $LiTaO_3$ primarily on the greater covalency and strength of the Ta—O bond, as compared with the Nb—O bond.

4.3.4 Two-dimensional ferroelectrics

The two-dimensional class of ferroelectrics is characterised by vectors connecting positions of identical atoms, in a structure of given and of reversed polarity, that lie in parallel planes containing the polar direction[28]. All two-dimensional ferroelectrics known at present crystallise in point group *mm*2, although there is no restriction to this particular point group. Two representative materials belonging to the two-dimensional ferroelectric class are discussed.

4.3.4.1 $BaCoF_4$ *family*

This is typified by $BaCoF_4$; Co may be replaced by Mn, Fe, Ni, Zn or Mg. $BaCoF_4$ is orthorhombic, $a = 5.852$, $b = 14.628$, $c = 4.210$ Å and space group $A2_1am$, with $P_s = 8.0 \times 10^{-2}$ C m^{-2} [52]. The spontaneous polarisation has been reversed at room temperature, but T_C lies above the melting point[53]. The structure contains infinite sheets of CoF_6 octahedra in (010), consisting of corner-sharing chains parallel to [001] and adjacent corner-sharing chains running along [100], as indicated in Figure 4.16. Polarity reversal is suggested in Figure 4.16 by the arrows connecting the structure given in bold outline to that in dashed outline. All atoms lie in mirror planes at $z = 0$ or $\frac{1}{2}$, and hence structures of both polarity and the connecting atomic displacement vectors lie in the (001) planes, where a is the polar direction.

The calculated value of P_s, based on the point charge model (the linear relation model[29] is applicable only to the one-dimensional ferroelectric class) is 8.6×10^{-2}C m^{-2}, in excellent agreement with measurement. The

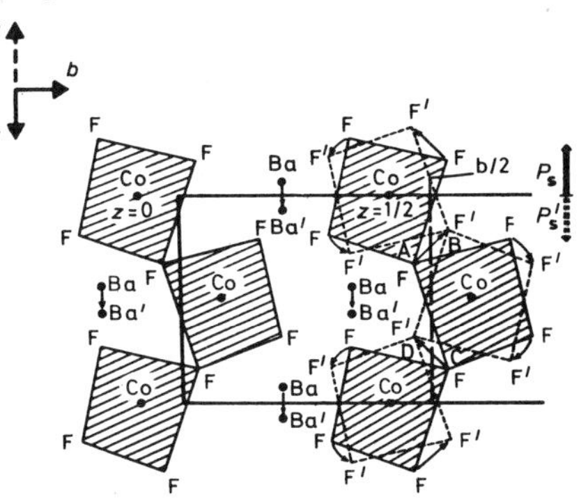

Figure 4.16 Schematic representation of CoF_6 octahedra in ferroelectric $BaCoF_4$. Right half of figure represents both polarities, by full and broken lines or by unprimed and primed symbols. The corresponding macroscopic polarisation sense is similarly represented (From Fig. 6 of Ref. 28)

sense of P_s is calculated from the atomic coordinates[52] to be positive in the $-a$ direction, a result in agreement with absolute sense measurements, and is indicated by the matching solid and dashed P_s arrows of Figure 4.16.

4.3.4.2 $NaNO_2$

$NaNO_2$ crystallises in space group *Immm* above $T_C \approx 437$ K and in *Im2m* below, with $a = 3.560$, $b = 5.563$, $c = 5.384$ Å at 298 K; the simple polar structure is shown in Figure 4.17 [26]. Above T_C the structure is disordered with an additional mirror plane generated normal to the *b*-axis. Polarity reversal can take place either by a rotation of the NO_2^- ion about [001], violating a

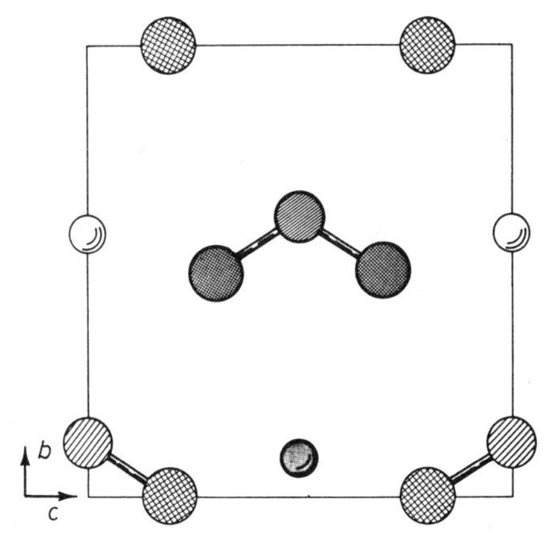

Figure 4.17 Schematic view of ferroelectric $NaNO_2$. Polarisation along *b* is reversible by one of the mechanisms given in text. Bold outline circles represent atoms at $x = \frac{1}{2}$, light outline atoms at $x = 0$

mirror plane; by rotation about [100], violating both a mirror plane and a diad axis, or by the ion passing through a linear configuration and hence maintaining all symmetry elements. Each of these ferroelectric mechanisms is two-dimensional; the first appears to be favoured by recent infrared[54] and neutron diffraction[55] measurements, although definitive evidence is still needed. The experimental $P_s = 8.5 \times 10^{-2}$C m^{-2}, that calculated on the basis of Na^+ and $(NO_2)^-$ is 7.7×10^{-2}C m^{-2}.

4.3.5 Three-dimensional ferroelectrics

In this case, vectors connecting corresponding atom positions in structures of reversed polarity have unrestricted orientations[28]. The dipoles in three-dimensional ferroelectrics are necessarily at least partially compensating, leading to characteristically small P_s values. Ferroelasticity (see also Section 4.4) is not uncommon, and the two three-dimensional ferroelectrics discussed below are both ferroelastic.

4.3.5.1 β-$Gd_2(MoO_4)_3$

β-$Gd_2(MoO_4)_3$, GMO, together with several other isomorphous rare-earth molybdates, is ferroelectric with $P_s \approx 0.20 \times 10^{-2}$C m^{-2} [56]. β-GMO is orthorhombic with $a = 10.3858$, $b = 10.4186$, $c = 10.7004$ Å and space group $Pba2$ [57, 58]. The structure contains three crystallographically independent types of MoO_4 tetrahedra, with one set oriented such that an apex points along $+c$, one with apex along $-c$, and the third with pseudo-$\bar{4}$-axis parallel to c, as illustrated in Figure 4.18. Each Gd atom occupies a 7-coordinated

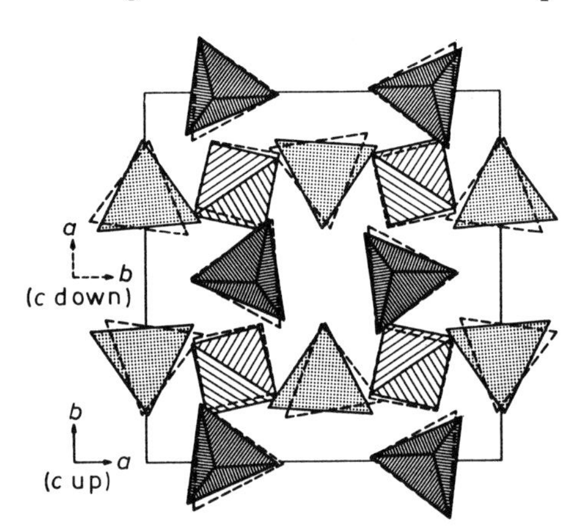

Figure 4.18 Arrangement of MoO_4 tetrahedra in β-$Gd_2(MoO_4)_3$, the bold outlines representing one orientation of single domain, the broken outline the corresponding reversed polarity orientation. The mismatch between corresponding atoms indicates the atomic displacement path taken in reversal

oxygen atom polyhedron. The spontaneous polarisation, calculated on the basis of point charges at the centres of gravity of the Gd^{3+} and $(MoO_4)^{2-}$ ions, is 0.26×10^{-2}C m^{-2} and originates in the mutual displacements of the ionic sublattices; the polarisations due to the Gd^{3+} ions nearly cancel.

As the temperature increases and approaches $T_C = 432$ K, the independent sets of MoO_4 tetrahedra with apices along the polar direction approach positions related by a $\bar{4}$ axis as do the independent pairs of Gd atoms;

simultaneously, the third set of tetrahedra approaches true $\bar{4}$ symmetry. Above T_C, the space group becomes $P\bar{4}2_1m$ as the approach to full symmetry is realised. In the ferroelectric phase, for each atom at $x_1y_1z_1$ there is an equivalent but independent atom at $x_2y_2z_2$ related by $x_1y_1z_1 = (y_2, \frac{1}{2}-x_2, 1-z_2)+\Delta$, where $0.46 \geqslant \Delta \geqslant 0.01$ Å. The application of uniaxial stress along [010] interchanges the a and b axes ferroelastically, and simultaneously reverses the sense of the polar axis, as required by the above relationship. Conversely, the application of an electric field along [001] not only reverses the polarity but also interchanges the two other axes. The axial interchange is readily observable experimentally by the rotation of the biaxial optic figure either on application of field or stress. Figure 4.18 shows the atomic arrangement for both orientations, with the small displacements Δ connecting equivalent atoms easily discernible.

At very low temperatures, GMO becomes antiferromagnetic[59]; reversal of the magnetic field with respect to the ferroelectric polarity has no effect on the magnetic moment. There is no report of a crystallographic investigation on the antiferromagnetic phase; assuming no change from room temperature, GMO may be represented by: $PPP(\bar{4}2m)F_CF_CP(mm2)F_CF_CA(mm2)$.

4.3.5.2 *Boracite family*

$M_3B_7O_{13}X$, where M is a divalent metal including many $3d$ ions and X is Cl, Br or I, has attracted considerable interest since ferroelectricity and ferromagnetism were reported to be simultaneously present in $Ni_3B_7O_{13}I$ [60]. Apart from solid solutions between appropriate perovskite-type structures, no other compounds with both properties are known. Further, reversal of the spontaneous polarisation rotates the magnetisation through 90 degrees, indicating major coupling between the two properties. As in the case of $Fe_3B_7O_{13}I$, the orthorhombic distortion of the $Ni_3B_7O_{13}I$ structure from the paraelectric cubic phase is small. The iron compound is cubic above $T_C = 345$ K with space group $F\bar{4}3c$, $a = 12.2310$ Å, and orthorhombic below T_C, with $a = 8.6595$, $b = 8.6515$, $c = 12.2406$ Å at 343 K and space group $Pca2_1$ [61]. The magnitude of the spontaneous polarisation has not been reported, but is less than 0.2×10^{-4}C m^{-2} for $Mg_3B_7O_{13}Cl$ [26]. The only structural report at present on the ferroelectric boracite phase is on $Mg_3B_7O_{13}Cl$ (1951) in which atoms at $x_1y_1z_1$ in the ferroelectric phase appear related to equivalent but independent atoms at $x_2y_2z_2$ by $x_1y_1z_1 = (y_2, \bar{x}_2, \frac{1}{2}-z_2)+\Delta$, with $\Delta \leqslant 0.5$ Å. An x-ray and neutron powder-diffraction study[62] of the paraelectric phase of $Ni_3B_7O_{13}I$ has confirmed the earlier structure but, with the improved accuracy, both independent boron atoms now appear to occupy only slightly distorted BO_4 tetrahedra. The above relation among atomic coordinates in ferroelectric boracites, together with the observed coupling between ferroelectricity and ferromagnetism, suggests that on reversing polarity not only is the magnetisation reoriented but the a and b axes are also interchanged. These compounds hence have all three properties coupled and may be represented by $PPP(\bar{4}3m)F_CF_CF_C(mm2)$.

Additional three-dimensional ferroelectrics that are also ferroelastic include KH_2PO_4 and NH_4HSO_4; others that are not ferroelastic include

$(NH_4)_2SO_4$, guanidinium aluminium sulphate hexahydrate, $CaB_3O_4(OH)_3 \cdot H_2O$ and $LiH_3(SeO_3)_2$ [28].

4.4 FERROELASTIC MATERIALS

The property of ferroelasticity was first recognised by Aizu in 1969 [63]. He defined a ferroelastic crystal as one that has two or more stable orientational states in the absence of mechanical stress or electric field, and that can be reproducibly transformed or switched from one to another of these states by the application of mechanical stress. The analogy to ferroelectricity is close, with mechanical stress replacing electric field in causing reorientation of the mechanical strain tensor ($\boldsymbol{E}_s$) rather than the electric polarisation vector ($\boldsymbol{P}_s$). The corresponding properties of antiferroelasticity and para-elasticity have been similarly defined by analogy with antiferroelectricity and paraelectricity[64].

All ferroelastic crystals contain pairs of pseudosymmetrically related atoms, i.e. for every atom in the unit cell at $x_1y_1z_1$ there is another at $x_2y_2z_2$

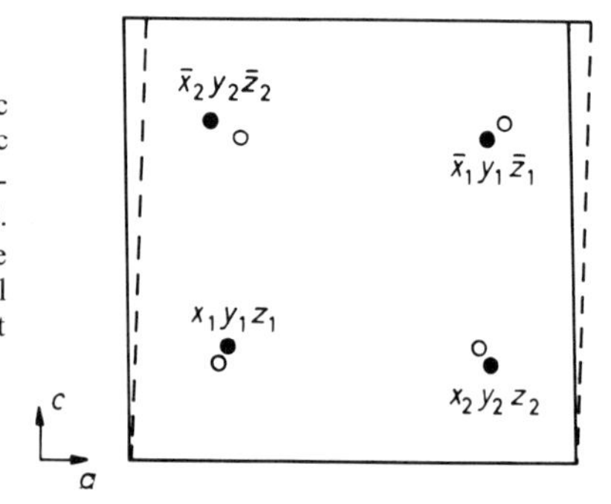

Figure 4.19 Representation of monoclinic unit cell with $\beta = 90$ degrees $+\delta(\delta \approx 10$ arc seconds) containing two symmetry-independent atoms (filled circles) in space group $P2$. For $x_1y_1z_1 = \bar{x}_2y_2z_2 + \Delta$ with $\Delta \approx 0.1$ Å, the crystal is potentially ferroelastic; mechanical stress may switch the orientation to that given by dashed outline and open circles.

such that $x_1y_1z_1 = f(x_2y_2z_2) + \Delta$, where Δ is a small atomic displacement generally on the order of 0.1 Å, and $f(x_2y_2z_2)$ is a transformation that results in a reorientation of the $\boldsymbol{a}$, $\boldsymbol{b}$ and $\boldsymbol{c}$ lattice vectors. These conditions are found in crystals with point group symmetry only slightly lower than that of a parent supergroup, and with a symmetry operator of the supergroup replaced by a pseudosymmetry operator in the subgroup. In consequence, the distortion of the unit cell from dimensional restrictions appropriate to the higher symmetry form must, in the ferroelastic phase, be small.

An illustration of a hypothetical ferroelastic crystal is presented in Figure 4.19, which shows a monoclinic unit cell with space group $P2$ containing two crystallographically independent atoms at $x_1y_1z_1$ and $x_2y_2z_2$, together with two others related by the diad axis. For a small distortion of this cell from orthorhombic, and with pseudomirrors relating $x_1y_1z_1$ to $x_2y_2z_2$ and $\bar{x}_1y_1\bar{z}_1$ to $\bar{x}_2y_2\bar{z}_2$, i.e., for $\beta \rightarrow 90$ degrees and $x_1y_1z_1 = \bar{x}_2y_2z_2 + \Delta$ with $\Delta \rightarrow 0$, mechanical stress could switch the orientation denoted by the bold outline and filled circles in Figure 4.19 to that denoted by the dashed outline and open circles. The parent or prototype supergroup in this example is $mm2$.

Crystallographic methods are particularly appropriate for investigating ferroelastic materials, since any crystal with dimensional distortions from a higher symmetry unit cell on the order of 10^{-3} are candidates for ferroelasticity, if there are also pseudosymmetry relations among the atomic coordinates. Ferroelastic materials have been classified[65] on the basis of additional properties the crystal may possess that couple to the ferroelasticity. A *simple ferroelastic* has zero coupling between the ferroelasticity and other co-operative properties. The absence of other such phenomena necessarily results in zero coupling in the ferroelastic phase; their presence does not necessarily result in coupling unless both phenomena are related by common structural features. Most of the examples in this Section are simple ferroelastics: some *ferroelastic-ordered magnetics* are found in Section 4.2; some *ferroelastic-ferroelectrics* in Section 4.3; and some *ferroelastic-superconductors* in Section 4.5.

4.4.1 $SrNb_2O_6$

$SrNb_2O_6$ is an example close to the hypothetical case of Figure 4.19. Of two known polymorphs, one is pseudo-orthorhombic monoclinic, space group $P2_1/a$ with $a = 11.02$, $b = 7.73$, $c = 5.60$ Å, $\beta = 90.2$ degrees[66]. The crystal is said to be isostructural with orthorhombic $CaTa_2O_6$, space group *Pnma* [67], although atomic coordinates are not given. Pseudomirror operators derived from *Pnma* are thus present in $SrNb_2O_6$ causing the point symmetry $2/m$ to approach *mmm*. This pseudosymmetry, together with the small departure of β from 90 degrees, indicates that monoclinic $SrNb_2O_6$ may be a potential ferroelastic; experimental verification has not yet been reported. The inversion centre present in the structure precludes piezo- or ferroelectricity. At 1723 K monoclinic $SrNb_2O_6$ transforms into a tetragonal polymorph[66]; of the possible choices of space group $P4/mmm$ is most likely, hence the $SrNb_2O_6$ phases may be represented by PPP($4/mmm$)PFP($2/m$).

4.4.2 $LaFeO_3$

$LaFeO_3$ is a ferroelastic that is also a weak ferromagnet at room temperature[13]. $LaFeO_3$ has an orthorhombic perovskite-like structure, space group *Pbnm* with $a = 5.553$, $b = 5.563$, and $c = 7.867$ Å at room temperature[68]. As-grown crystals are highly twinned, with [100] and [010] directions frequently parallel in any given crystal. The spontaneous strain $E_s = (b-a)/(b+a) = 9.0 \times 10^{-4}$ is small, and for each atom in the structure[68] at $x_1y_1z_1$ there is a like atom at $x_2y_2z_2$ such that $x_1y_1z_1 = (y_2\bar{x}_2z_1)+\Delta$, with $0.79 \geqslant \Delta \geqslant 0.23$ Å except for Fe which remains invariant under this transformation. It is hence expected that compressive stress applied along either a [100] or [010] direction will de-twin $LaFeO_3$; similarly, the *a* and *b* axes in a single domain may be interchanged by application of uniaxial pressure along [010].

4.4.3 $LaAlO_3$

$LaAlO_3$ is a simple ferroelastic that twins during crystal growth. $LaAlO_3$ is rhombohedral, space group $R\bar{3}m$ with $a = 5.357$ Å and $\alpha = 60.1$ degrees at room temperature, with a perovskite-type structure. The application of compressive stress along a <111> direction is found[69] to de-twin $LaAlO_3$;

further application of stress to various $\{111\}$ faces causes the optic axis to reorient through 109.47 degrees as the direction under compression becomes $[111]$.

4.4.4 $SrTiO_3$

$SrTiO_3$ is a cubic perovskite at room temperature with $a = 3.905$ Å that transforms to tetragonal at 107 K with $a \approx 3.897$, $c \approx 3.899$ Å [26]. Multiple domains are formed in $SrTiO_3$ below the transition, due to the slightly elongated tetragonal c-axis being randomly oriented along each of the former cube axes[70]. Compressive stress applied along a cubic $[110]$ direction below 107 K produces compressive components along two of the domain orientations containing the tetragonal c-axis. Optical examination along a cube axis direction before stress is applied reveals a domain structure having (110) and $(1\bar{1}0)$ walls, with all domains disappearing above stresses greater than 1.2 kg mm^{-2}. Ferroelectricity has recently been observed in a $SrTiO_3$ crystal that was subjected to a much higher compressive stress along $[110]$ [71]; this appears to be the first example of a stress-induced ferroelectric phase in a crystal that otherwise does not exhibit ferroelectricity at any temperature in the absence of applied stress.

4.5 SUPERCONDUCTIVE MATERIALS

The role of crystallography in investigating the properties of superconductive materials has been primarily that of determining the lattice constants and crystal structure type of the material in its normal conducting state. A comprehensive compilation by Roberts[72] lists the superconducting transition temperatures T_C of a large number of materials. At T_C there is nearly a total loss of resistance, the resistivity in the superconducting state being at most 10^{-17} that of copper at room temperature. Superconductivity has now been observed in more than 570 different materials including elements, compounds and representative solid solutions. The transition temperatures range from a high of 20.05 K for $Nb_3Al_{0.8}Ge_{0.2}$ to very low temperatures. The room temperature symmetry of most of these materials is high, with the cubic beta-tungsten, the Laves and the tetragonal sigma phases occurring most frequently[73]. The possibility that low temperature structural transformations may be associated with the superconducting transition has been investigated only for a very small number of compounds. The results obtained are considered below; clear relationships have not yet emerged, and further crystallographic study is undoubtedly required.

4.5.1 V_3Si

V_3Si was the first superconductor in which a structural transformation was found near T_C(17.0 K) [74]. A single crystal was examined in a cryostat, within the temperature range 30–4.2 K, both by Berg–Barrett topography and by

Bond-diffractometer absolute lattic constant determination. The two methods shows that the cubic beta-tungsten phase, with $a = 4.718$ Å, transforms to a tetragonal phase at about 21.0 K with c/a increasing continuously from unity as the temperature is lowered and reaching a maximum value of 1.0024 within 0.1 K of T_C. There are no additional changes in c or a as the temperature is lowered further. The volume of the unit cell remains constant to within parts per 10^4 throughout the temperature range studied.

Berg–Barrett topographs[74] show that V_3Si is twinned below T_C. The ferroelastic behaviour of tetragonal V_3Si expected on the basis of the small distortion from cubic symmetry has been verified by the observation of c-axis reorientation on application of uniaxial compressive stress along [100] [75]. This experiment is likely to result in ac-domains normal to the stress direction. However, mechanical stress applied along a [110] direction is expected to result in a single domain untwinned crystal, since compressive components are thereby produced along both [100] and [010]. T_C for a single domain V_3Si crystal has not been reported: an early[76] application of uniaxial stress along [100] or [111] was found to shift T_C by as much as 0.4 K.

4.5.2 Nb_3Sn

Nb_3Sn with the cubic β-tungsten type structure, has a superconducting transition temperature of 18.1 K. An x-ray diffraction study on a single crystal of Nb_3Sn by the techniques used on V_3Si shows this material also transforms to a tetragonal phase, at *c.* 43 K [77]. Above this structural transition, $a = 5.281$ Å: the ratio c/a rapidly decreases below 43 K, with a and c reaching a constant value by *c.* T_C. At 4.2 K, $c/a = 0.9938$, i.e., c is less than a in Nb_3Sn, in contrast to V_3Si. The departure of c/a from unity is 2.6 times greater for Nb_3Sn than for V_3Si, but is probably well within the limit required for Nb_3Sn also to be ferroelastic. Application of compressive stress along a cubic [100] direction, either on cooling through the 43 K transition or at liquid helium temperature, is expected to eliminate all domains, the stressed direction becoming the tetragonal c-axis. The effects of such stress on T_C for Nb_3Sn has not been reported.

4.5.3 V_3Ir

V_3Ir, another member of the β-tungsten structure-type family, has also been examined for a possible structural transformation by Batterman and Barrett[74]. No indications of transformation were detected down to 4.6 K. It is interesting to note that V_3Ir does not become superconducting at the lowest temperature examined, 0.35 K.

A dynamic electron–phonon interaction theory of the cubic to tetragonal phase transformation in Nb_3Sn and V_3Si has recently been presented[78]. The difference in behaviour between the two compounds is ascribed to different relative positions of the band edges and Fermi level. There is now evidence that the cubic and tetragonal phases of Nb_3Sn and V_3Si may each have characteristic T_C values[79]. Powder samples subjected to about 8.5 kbar of nearly uniaxial pressure at room temperature gave T_C curves with a sharp upper break in relative susceptibility and an additional well formed inflection in the curve as much as 6 K lower. The phenomena of superconductivity and ferroelectricity are considered by Matthias[80] to be mutually exclusive.

References

1. For a general review see Izyumov, Yu. A. and Ozerov, R. P. (1970). *Magnetic Neutron Diffraction.* (New York: Plenum Press)
2. Platzman, P. M. and Tzoar, N. (1970). *Phys. Rev.*, **B2,** 3556
3. Lines, M. E. (1971). *Phys. Rev.*, **B3,** 1749; (1970) *J. Phys. Chem. Solids*, **31,** 101
4. Lines, M. E. (1969). *J. Appl. Phys.*, **40,** 1352
5. Jensen, S. J., Andersen, P. and Rasmussen, S. E. (1962). *Acta Chem. Scand.*, **16,** 1890
6. Smith, T. and Friedberg, S. A. (1968). *Phys. Rev.*, **176,** 660
7. Skalyo, J., Shirane, G., Friedberg, S. A. and Kobayashi, H. (1970). *Phys. Rev.*, **B2,** 1310 and 4632
8. Spence, R. D., de Jonge, W. J. M. and Rama Rao, K. V. S. (1969). *J. Chem. Phys.*, **51,** 4694
9. Keve, E. T. and Abrahams, S. C. (1970). *Ferroelectrics*, **1,** 243
10. Achiwa, N. (1969). *J. Phys. Soc. Jap.*, **27,** 561
11. Minkiewicz, V. J., Cox, D. E. and Shirane, G. (1970). *Solid State Commun.*, **8,** 1001

11a. Epstein, A., Makovsky, J. and Shaked, H. (1971). *Solid State Commun.*, **9,** 249

12. Birgeneau, R. J., Guggenheim, H. J. and Shirane, G. (1970). *Phys. Rev.*, **B1,** 2211; (1969). *Phys. Rev. Lett.*, **22,** 720
13. Landolt–Börnstein, New Series, Group III, Vol. **4a** and **4b.** (1970). (Berlin: Springer–Verlag)
14. Akimitsu, J., Ishikawa, Y. and Endoh, Y. (1970). *Solid State Commun.*, **8,** 87
15. Olés, A., Bombik, A., Kajzar, F., Kucab, M. and Sikora, W. (1970–1971). *Tables of Magnetic Structures Determined by Neutron Diffraction.* (Krakow: Institute of Nuclear Techniques)
16. *Index to the literature of Magnetism*, published semiannually. (New York, N.Y.: American Institute of Physics)
17. Vinnik, M. A. and Selezneva, L. N. (1970). *Sov. Phys. Cryst.*, **14,** 928
18. Comes, R., Denoyer, F., Deschamps, L. and Lambert, M. (1971). *Phys. Lett.*, **34A,** 65
19. Forsyth, J. B., Johnson, C. E. and Wilkinson, C. (1970). *J. Physics C: Solid State Phys.*, **3,** 1127
20. Wilkinson, C. (1968). *Phil. Mag.*, **17,** 609
21. Abrahams, S. C. (1966). *J. Chem. Phys.* **44,** 2230
22. Forsyth, J. B. and Wilkinson, C. (1970). Abstract MeG5, International Conference on Magnetism, Grenoble, France, September 1970
23. Koehler, W. C. (1967). *Trans. Amer. Cryst. Assoc.*, **3,** 53
24. Herpin, A. and Meriel, P. (1961). *J. Phys. Rad.*, **22,** 337
25. Felcher, G. P., Aldred, A. T., Moon, R. M. and Koehler, W. C. (1970). *Int. J. Mag.*, **1,** 85
26. Landolt-Börnstein, New Series, Group III, Vol **1,** 1966, and Vol. **2,** 1969. (Berlin: Springer-Verlag)
27. These point groups are: 1, 2, *m*, 2*mm*, 4, 4*mm*, 3, 3*m*, 6 and 6*mm*. In 1 and *m*, the polar direction is not necessarily a crystallographic axis.
28. Abrahams, S. C. and Keve, E. T. (1971). *Ferroelectrics*, **2,** 129
29. Abrahams, S. C., Kurtz, S. K. and Jamieson, P. B. (1968). *Phys. Rev.*, **172,** 551
30. Jona, F. and Shirane, G. (1962). *Ferroelectric Crystals.* (New York: Macmillan Company)
31. Zheludev, I. S. (1971). *Physics of Crystalline Dielectrics.* (New York: Plenum Press)
32. Cochran, W. (1969). *Advan. Phys.*, **18,** 157; (1960). **9,** 387
33. Lines, M. E. (1969). *Phys. Rev.*, **177;** 797; 812; 819; (1970). **B2,** 698
34. Goshen, S., Mukamel, D., Shaked, H. and Shtrikman, S. (1970). *Phys. Rev.*, **B2,** 4679
35. Arlt, G. and Quadflieg, P. (1968). *Phys. Stat. Sol.*, **25,** 323
36. Miller, R. C., Abrahams, S. C., Barns, R. L., Bernstein, J. L. and Turner, E. H. (1971). *Solid State Comm.*, **9,** 1463

37. Abrahams, S. C., Jamieson, P. B. and Bernstein, J. L. (1967). *J. Chem. Phys.*, **47,** 4034
38. Buerger, M. J., Burnham, C. W. and Peacor, D. R. (1962). *Acta Crystallogr.*, **15,** 583
39. Barton, R. (1969). *Acta Crystallogr.*, **B25,** 1524
40. The 1945 value of $T_C = 393$ K for $BaTiO_3$ is often quoted[26], but Camlibel, I., DiDomenico, M. and Wemple, S. H. (1970). *J. Phys. Chem. Sol.*, **31,** 1417 give a value of 408 K which refers to colourless melt-grown single-domain crystals.
41. Harada, J., Pedersen, T. and Barnea, Z. (1970). *Acta Crystallogr.*, **A26,** 336
42. Remeika, J. P., unpublished results
43. Comes, R., Lambert, M. and Guinier, A. (1970). *Acta Crystallogr.*, **A26,** 244
44. Axe, J. D. (1971). *Trans. Amer. Cryst. Assoc.*, **7,** 89
45. Shirane, G., Axe, J. D. and Harada, J. (1970). *Phys. Rev.*, **B2,** 3651
46. Scott, B. A., Giess, E. A., Olson, B. L., Burns, G., Smith, A. W. and O'Kane, D. F. (1970). *Mat. Res. Bull.*, **5,** 47
47. Abrahams, S. C., Jamieson, P. B. and Bernstein, J. L. (1971). *J. Chem. Phys.*, **54,** 2355
48. Abrahams, S. C. and Bernstein, J. L. (1967). *J. Phys. Chem. Sol.*, **28,** 1685
49. Abrahams, S. C., Hamilton, W. C. and Sequiera, A. (1967). *J. Phys. Chem. Sol.*, **28,** 1693
50. Abrahams, S. C., Buehler, E., Hamilton, W. C. and LaPlaca, S. J. (1970). Abstract 16, ACA Summer Meeting, Ottawa, Canada, Program and Abstract Book
51. Halstead, T. K. (1970). *J. Chem. Phys.*, **53,** 3427
52. Keve, E. T., Abrahams, S. C. and Bernstein, J. L. (1970). *J. Chem. Phys.*, **53,** 3279
53. DiDomenico, M., Eibschütz, M., Guggenheim, H. J. and Camlibel, I. (1969). *Solid State Commun.*, **7,** 1119
54. Barnoski, M. K. and Ballantyne, J. M. (1968). *Phys. Rev.*, **174,** 946
55. Kay, M. I. and Gonzalo, J. A., Abstract Cl, ACA Winter Meeting, Columbia, S. C., Program and Abstract Book
56. Keve, E. T., Abrahams, S. C., Nassau, K. and Glass, A. M. (1970). *Solid State Commun.*, **8,** 1517
57. Jeitschko, W. (1970). *Naturwissenschaften*, **57,** 544
58. Keve, E. T., Abrahams, S. C. and Bernstein, J. L. (1971). *J. Chem. Phys.*, **54,** 3185
59. Fisher, R. A., Hornung, E. W., Brodale, G. E. and Giauque, W. F. (1972). *J. Chem. Phys.*, **56,** 193
60. Ascher, E., Rieder, H., Schmid, H. and Stössel, H. (1966). *J. Appl. Phys.*, **37,** 1404
61. Kobayashi, J., Mizutani, I., Schmid, H. and Schachner, H. (1970). *Phys. Rev.*, **B1,** 3801
62. Becker, W. J. and Will. G. (1970). *Z. Krist.*, **131,** 139
63. Aizu, K. (1969). *J. Phys. Soc. Jap.*, **27,** 387
64. Aizu, K. (1969). *J. Phys. Soc. Jap.*, **27,** 1171
65. Abrahams, S. C. (1971). *Mat. Res. Bull.*, **6,** 881
66. Brusset, H., Pandraud, G. and Voliotis, S. D. (1971). *Mat. Res. Bull.*, **6,** 5
67. Jahnberg, L. (1963). *Acta Chem. Scand.*, **71,** 2548
68. Marezio, M. and Dernier, P. D. (1971). *Mat. Res. Bull.*, **6,** 23
69. Fay, H. and Brandle, C. D. (1967). *J. Phys. Chem. Sol.*, **28,** Supp. 1, 51
70. Chang, T. S., Holzrichter, J. F., Imbusch, G. F. and Schawlow, A. L. (1970). *Appl. Phys. Lett.*, **17,** 254
71. Burke, W. J. and Pressley, R. J. (1971). *Solid State Commun.*, **9,** 191
72. Roberts, B. W. (1969). *Superconductive Materials and Some of their Properties*, National Bureau of Standards Technical Note 482
73. Matthias, B. T., Geballe, T. H. and Compton, V. B. (1963). *Rev. Mod. Phys.*, **35,** 1
74. Batterman, B. W. and Barrett, C. S. (1964). *Phys. Rev. Lett.*, **13,** 390; (1966). *Phys. Rev.*, **145,** 296
75. Patel, J. R. and Batterman, B. W. (1966). *J. Appl. Phys.*, **37,** 3447
76. Weger, M., Silbernagel, B. G. and Greiner, E. S. (1964). *Phys. Rev. Lett.*, **13,** 521
77. Mailfert, R., Batterman, B. W. and Hanak, J. J. (1967). *Phys. Lett.*, **24A,** 315
78. Pytte, E. (1970). *Phys. Rev. Lett.*, **25,** 1176
79. Matthias, B. T., Corenzwit, E., Cooper, A. S. and Longinotti, L. D. (1971). *Proc. Nat. Acad. Sci.*, **68,** 56
80. Matthias, B. T. (1970). *Mat. Res. Bull.*, **5,** 665

5
Structural Studies of Large Molecules of Biological Importance

MARJORIE M. HARDING
University of Edinburgh

5.1 INTRODUCTION

This chapter selects for discussion some groups of large molecules whose crystal structures have been studied on account of their biological significance.

Protein structures are described in another chapter and studies of oriented paracrystalline samples of polysaccharides and nucleic acids could not be included here. The largest molecules whose structures have been successfully determined by conventional crystallographic methods are still around one hundred atoms in size, for example thiostrepton[1] with 110 atoms other than hydrogen and solvent molecules. There is an intriguing gap between this and insulin[2], the smallest structure determined by the methods of protein crystallography, and it will be interesting to see how, and how soon, the phase problem is solved for structures of intermediate size. The molecules included in this chapter have 15–20 carbon atoms or more. Generally, work published between 1965 and the end of 1970 and listed by Kennard and Watson for their Bibliography[5] has been included, together with a number of structures determined earlier and belonging to chemical groups under discussion.

In many cases, e.g. vitamin B_{12} and many of the antibiotics, the primary reason for the crystal structure analysis has been to establish the chemical structure; in other cases studies of several closely related compounds have been undertaken with the intention of making detailed comparisons of stereochemistry and in the hope of being able to relate this to biological activity; the porphyrin studies by Hoard and his co-workers[3] or the study of cephalosporins by Sweet and Dahl[4] are perhaps the best examples of this. Some of the molecules are synthetic ones, designed as models for parts of biologically important structures, e.g. some of the peptides.

Following Kennard and Watson[5] the molecules are grouped according to their chemical types but the antibiotics section includes a variety of chemical types. Chemical formulae, crystallographic details and reference numbers are collected in tables, one for each group of compounds. Each compound has been given a number, such as M3, and these will be found to run serially through the tables.

5.1.1 Methods used in data measurement and structure analysis

The crystallographic details given in the tables include the number of reflections observed and used in the analysis, and the method of recording the data, by automatic diffractometer (D) or by Weissenberg photographs (W). The accuracy of the structure determined at the end of the structure analysis is traditionally indicated by the R factor, but much more satisfactorily by the estimated standard deviations (σ) of the atomic positions.

Often there have been difficulties in obtaining really good crystals with the result that the number of reflections observable is rather small, in relation to molecular size, or the accuracy of intensity measurements is limited, or both. Sometimes this can be attributed to disorder of a particular part of the molecule or of solvent molecules in large solvent spaces; in other cases the packing of large, awkwardly shaped molecules in the crystal may just be a little irregular, leading to a rapid fall off of the average intensity of x-ray reflections with scattering angle and so, at the end of the structure analysis, to apparent large vibration parameters for the atoms. In spite of such problems chemical structure and conformation have often been established, but the accuracy of the atom positions is seriously affected; estimated standard deviations for bond lengths may be 0.05 Å or more, and so detailed comparison of the structure with others has little significance.

The solution of the phase problem in crystal structures of this size is the biggest challenge. The heavy atom method has been much used. It used to be said that for success the scattering power of the 'heavy' atom, approximately proportional to Z_H^2, should approach half the total scattering estimated by summing the squares of the atomic numbers, $\sum Z^2$. But many structures have been solved with a much smaller heavy atom contribution than this; in vitamin B_{12} coenzyme, M18, $Z_{Co}^2/\sum Z^2$ is 0.13; there are five sulphur atoms in thiostrepton, M43, but for each $Z_S^2/\sum Z^2$ is only 0.043. In a few cases anomalous dispersion effects have been recorded and have assisted in the solution, for example in cobyric acid, M14.

Recently, the symbolic addition method has been impressively successful in solving structures of medium size such as raffinose, a trisaccharide, with 34 atoms apart from hydrogen[6]. Of course, as with the heavy atom method applied to structures of this size, the first electron density series (or *E* map) calculated with trial phases has usually revealed only a part of the structure; further cycles of phase calculation and electron density maps have then gradually revealed the complete structure. The use of the symbolic addition method has the advantage that no 'heavy' atom (e.g. bromine) need be introduced into the crystal; problems of chemical modification of the natural product are avoided and atom positions are determined with greater accuracy in a crystal containing light atoms only.

5.2 PORPHYRINS AND CORRINS

5.2.1 Porphyrins

The porphyrins and metallo-porphyrins listed in Table 5.1 have been studied either on account of their biological origin and importance or as simpler model compounds whose structures are likely to be determined with greater accuracy then those of the large molecules whose stereochemistry we wish to understand. That slight non-planarity is normal for the porphyrin system has been recognised in the last 8 years or so, and it is also common for the metal atom to be displaced from the plane of the four nitrogens.

The dimensions of the metal-free porphine skeleton have been established most satisfactorily in tetraphenyl porphine, TPP, M11. Unfortunately the x-ray analysis of porphine itself, M10 (Figure 5.1), has shown the ring orientation to be disordered: the final difference electron density series clearly contained peaks corresponding to four partial hydrogen atoms, one attached to each of the four nitrogens and so the dimensions determined must be averages over chemically non-equivalent bonds. For TPP the two central hydrogen atoms have been unambiguously located. In the dimensions of TPP (Figure 5.2) the only significant difference in bond length between the pyrroles I and II is in one C—C bond which is 1.455 Å in ring I and 1.428 Å in ring II; a simple valence-bond approach indicates that this bond should be longer in ring I than in ring II but cannot go much further to account for

Table 5.1 Porphyrins

M—N is the distance from the metal atom to the porphine nitrogens. Δ is the displacement of the metal atom from the mean plane of the four porphine nitrogens

	Short name	*Formula (excluding solvent)*	*Data*	*R*	*σ for C—C bond/ Å*	*Coordinates published*	*M—N/Å*	*Δ/Å*	*Reference*
M1	Nickel(II) etioporphyrin-I*	$C_{32}H_{36}N_4Ni$	200D	—	—	Yes	1.975	0	7
M2	Nickel deuteroporphyrin	$C_{36}H_{36}N_4NiO_6$	3122D	0.10	0.01 Å	Yes	1.960	0	8
M3	Chlorohemin (α-form)	$C_{34}H_{32}ClFeN_4O_4$	1566D	0.095	0.02–0.03	Yes	2.062	0.48	9
M4	Methoxy iron(III) mesoporphyrin*	$C_{37}H_{43}FeN_4O_5$	1702D	0.12	—	Yes	2.073	0.46	10
M5	Aquomagnesium TPP	$C_{44}H_{30}MgN_4O$	431D	0.085	0.01–0.02	Yes	2.072	0.27	11
M6	Pyridine zinc TPyP	$C_{45}H_{29}N_9Zn$	6434D	0.09	0.004	Yes	2.073	0.33	12
M7	Chloroiron(III) TPP	$C_{44}H_{31}FeN_4O_2$	399	0.073	—	Yes	2.049	0.38	13
M8	Bis(imidazole)iron(III) TPP	$[C_{50}H_{36}FeN_8]^+, Cl^-$	5551D	0.077	0.01	No	1.989	<0.01	14
M9	Vanadyl DPEP*	$C_{32}H_{34}N_4OV$	2000D	0.055	0.02	Yes	2.06	0.48	15
M10	Porphine	$C_{20}H_{14}N_4$	2421D	0.049	0.004	Yes			16
M11	Tetraphenyl porphine	$C_{44}H_{30}N_4$	1726D	0.056	0.005–0.01	Yes			17
M12	Phyllochlorin ester	$C_{33}H_{38}N_4O_2$	2000W			Yes			18

*TPP is α,β,γ,δ-tetraphenyl porphine (see Figure 5.1 for numbering of positions).
TPyP is α,β,γ,δ-tetra(4-pyridyl)porphine.
In M1 etioporphyrin-I is 1,3,5,7-tetramethyl-2,4,6,8-tetraethyl porphine.
In M2 deuteroporphyrin is 2,4-diacetyl deuteroporphyrin-IX-dimethyl ester.
In M4 mesoporphyrin is mesoporphyrin-IX-dimethyl ester.
In M9 DPEP is deoxyphylloerythro-etioporphyrin (see Figure 5.3b).

M3, M4 and M7 contain high spin Fe^{III} but M8 contains low spin Fe^{III}
M3, M4, M5, M7 and M10 all suffer from some disorder in the crystals.

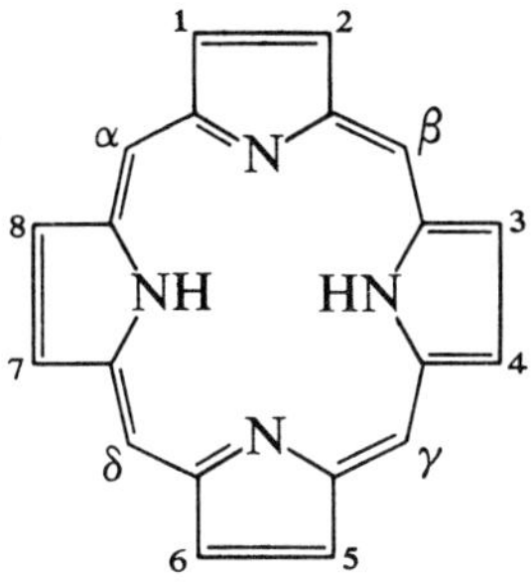

Figure 5.1 Porphine, M10. In metalloporphyrins the metal ion displaces the two central hydrogens and is coordinated by four nitrogens

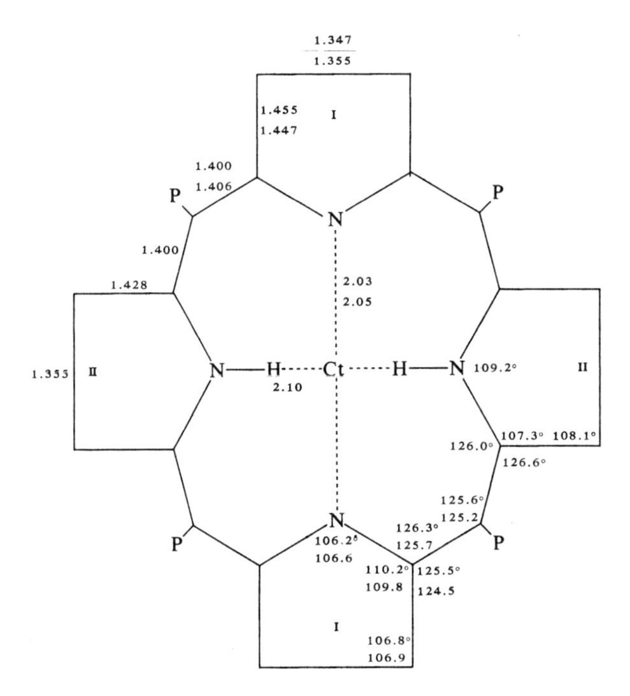

Figure 5.2 Dimensions of tetraphenyl porphine, TPP(M11), and a metalloporphyrin (PyrZnTPyP, M6). P in the diagram represents phenyl or pyridyl; Ct represents the ring centre, not necessarily an atom position. In M6 the zinc atom is 0.33 Å above the plane of the nitrogens and there are, of course, no central hydrogen atoms. In both compounds chemically equivalent bond lengths and angles have been averaged (bonds have $\sigma \approx .005$ Å for both). The metalloporphyrin dimensions are the lower numbers where there are pairs, and have been inserted in pyrrole rings I; they agree noticeably better with the TPP dimensions here

the variations in length. There are also differences in bond angles in the pyrrole rings; the angle at N in ring I is much smaller than the angle at NH in ring II. The individual pyrrole rings of TPP are planar, and the central 16-membered ring is nearly so but the molecule as a whole is not; the two pyrroles II are tilted 6.6 degrees out of the mean plane (and the phenyl groups are rotated by 60 degrees). This deviation from planarity could be attributed to repulsion by the central hydrogens, but it also relieves 'angle-strain', allowing most of the bond angles to take values closer to the ideal ones for sp^2 hydridised carbon.

In metallo-porphyrins too, the porphine system is normally not quite planar[3, 8]; often the metal atom is displaced by up to $\frac{1}{2}$ Å from the mean plane of the nitrogens (see Table 5.1) and the metal–porphine then has a slightly convex shape; alternatively the porphine may be 'ruffled' in a less regular way. The metallo-porphyrins studied include examples where the metal atom is 4-, 5- and 6-coordinate. In the two nickel compounds, M1, M2, the metal is coordinated to the four porphine nitrogens only; in M8 there are two axial ligands (imidazole), making the iron 6-coordinate; in all the other metallo-porphyrins of Table 5.1 the metal is 5-coordinate, having one axial ligand such as Cl-, MeO- or H_2O. The dimensions for one of the most accurately determined metallo-porphyrins, M6, are given in Figure 5.2, where it may be seen that they are very close to those of the metal-free porphyrin. There are no significant variations in C—C and C—N bond lengths from one metallo-porphyrin to another except in vanadyl DPEP (see below), and slight non-planarity (either a slightly convex shape or less regular 'ruffling') is normal. In the 5-coordinate complexes the metal atom is displaced from the plane of the porphine nitrogens by $\frac{1}{4}$–$\frac{1}{2}$ Å towards the axial ligand; this and the metal–ligand distances, are quite normal for such coordination[19]. No significant displacement of the metal atoms from the plane has been observed in the 4- or 6-coordinate metallo-porphyrins.

Hoard has discussed the dimensions and planarity in these compounds at length in relation to the stereochemistry of the heme group[3, 12]. He uses Ct to represent the centre of the porphine system and estimates that the strain in the system will be a minimum when Ct–N is *c.* 2.01 Å; this is very close to the distance (2.03 Å) found in TPP itself for the nitrogens of a pyrrole ring not carrying hydro (Figure 5.2). If the metal-ion radius requires a larger M—N distance, then either the porphine ring should be buckled or the metal ion displaced from the plane of the nitrogens, or both. This is observed; in all the compounds in Table 5.1 with M—N greater than 2.03 Å there is a substantial displacement of the metal from the plane which allows Ct—N to remain close to 2.03 Å. On these grounds it seems likely that high-spin Fe^{III} in a metallo-porphyrin or a heme group will normally be displaced from the porphine nitrogen plane by up to 0.5 Å. While low-spin Fe^{III}, with a smaller radius, will be in or nearly in, the plane; the spin state will, of course, be determined by the ligand(s) attached to iron in addition to the porphine nitrogens. In the heme group of myoglobin where Fe^{III} has two additional ligands, an imidazole nitrogen and a water molecule, it is 0.25 Å from the mean plane (calculated from the coordinates published in reference 20).

No one has succeeded in crystallising chlorophyll itself (Figure 5.3a). The closest analogues whose structures have been determined are phyllochlorin ester, M12, which contains no metal atom, and vanadyl DPEP, M9 (Figure 5.3b); neither contains the long phytyl chain. In vanadyl DPEP the vanadium atom is very close to the nitrogen of ring C (1.96 Å, $\sigma = 0.01$) while it is at a much more normal distance from each of the other nitrogens (2.06, 2.11,

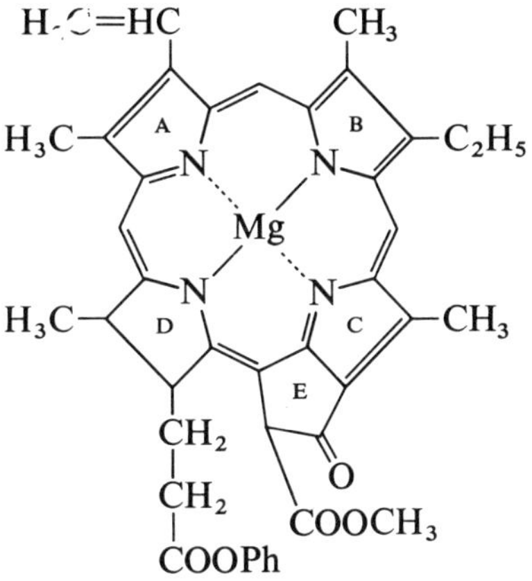

Figure 5.3(a) Chlorophyll *a*, Ph = phytyl, $C_{20}H_{39}$

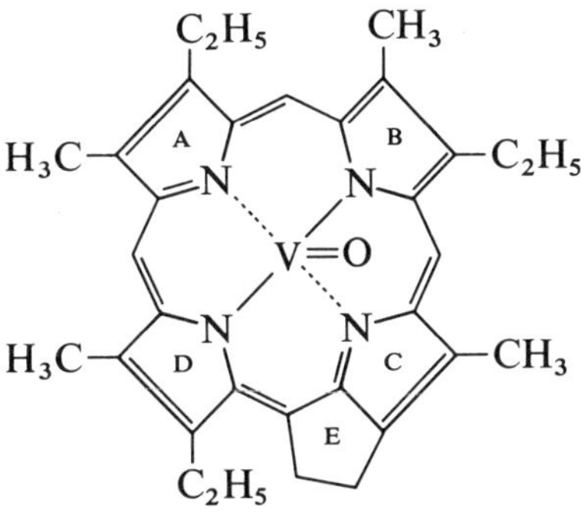

Figure 5.3(b) Vanadyl DPEP, M9

2.13 Å). This distortion is attributed to the exocyclic ring E, and it might very well occur in chlorophyll itself and be important in its mechanism of action. Tantalisingly, one can say no more than this from the study of model compounds.

5.2.2 Vitamin B_{12} and related corrins

The structure of vitamin B_{12} (cyanocobalamin, M17 (Figure 5.4)) was established in the early 1950s by Dorothy Hodgkin and her co-workers. It was larger, by an order of magnitude, than any other structure determined at the time, and it was solved as a result of parallel studies, each by the heavy atom method, of vitamin B_{12} itself in its wet and dry states, and the hexacarboxylic acid derivative, M15. With the primitive computing facilities of the time it was a most daring undertaking and it was impressively successful.

The chemical structure was thus established, and in particular, the

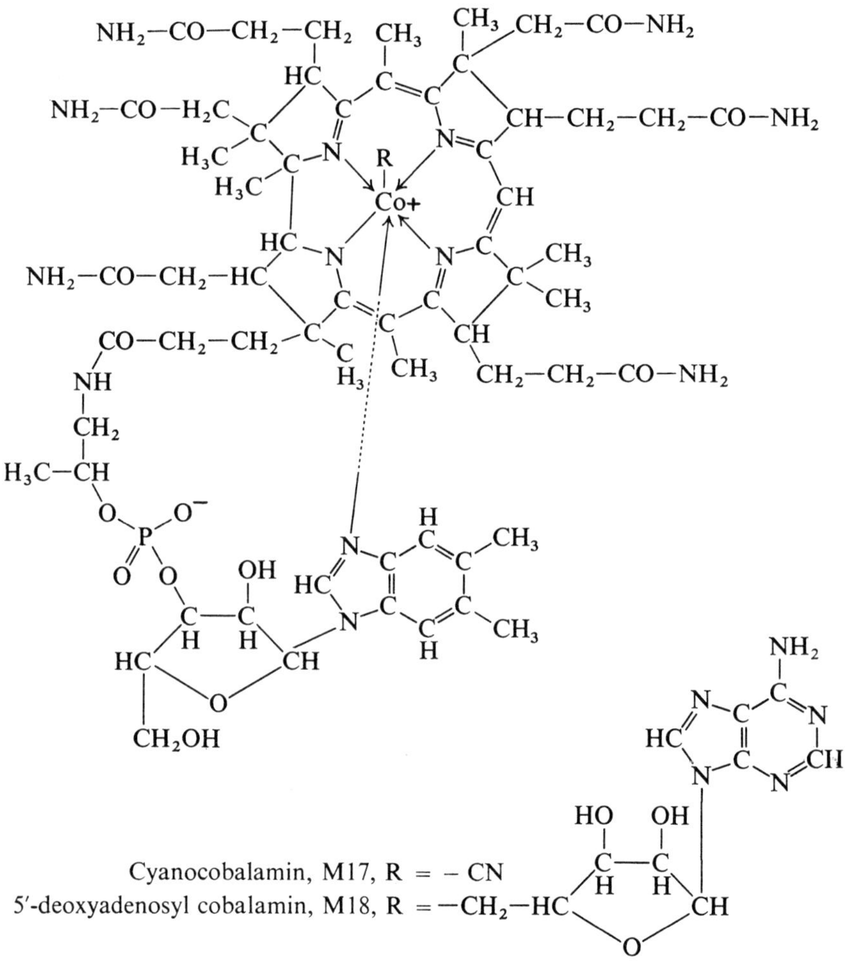

Cyanocobalamin, M17, R = – CN
5′-deoxyadenosyl cobalamin, M18, R = $-CH_2-HC$

Figure 5.4 Cyanocobalamin, M17, and 5′-deoxyadenosyl cobalamin, M18

unexpected structure of the corrin nucleus (Figure 5.5) with its direct link between atoms C(1) and C(19). There remained some uncertainties; there appeared to be a system of alternating single and double bonds around the corrin nucleus, but it was not clear whether there were five or six double bonds; in some of the amide or carboxyl groups oxygen and nitrogen could

Table 5.2 Corrins

	Short name, alternatives	*Formula (excluding solvent of crystallisation)*	*Data*	*R*	*σ for C—C bond/Å*	*Coordinates published*	*Reference*
M13	Nickel corrin	$[C_{25}H_{30}N_5Ni]^+Cl^-$	2700 D	0.16	0.02–0.04 Å	no	21
M14	Cobyric acid, factor V1a	$C_{46}H_{67}CoN_{11}O_9$	4000 D	0.16	—	no†	22, 31
M15	Vitamin B_{12} hexacarboxylic acid degradation product	$C_{46}H_{58}ClCoN_6O_{13}$	3351 W	0.20	0.05	yes	23
M16	Vitamin B_{12} monocarboxylic acid E2	$C_{63}H_{87}CoN_{13}O_{15}P$	6323 W	0.14	0.04	no	24
			1531 neutron D	0.23			25
M17	Cyanocobalamin, vitamin B_{12}	$C_{63}H_{88}CoN_{14}O_{14}P$	2900 W*	0.22	0.05	yes	26
			2086 W	0.32	—	yes	27–29
M18	Vitamin B_{12} coenzyme, 5′-deoxyadenosyl cobalamin	$C_{72}H_{100}CoN_{18}O_{17}P$	3068 W	0.13	0.04	yes	30

*These data refer to the wet crystals, the subsequent data to air-dried crystals.

†Now published[101].

not be distinguished. Later, the coenzyme which is the biochemically active form involved in transmethylation was isolated, crystallised, and shown by x-ray analysis to be 5′-deoxyadenosyl cobalamin, M18, (Figure 5.4). The nucleoside is joined to the rest of the molecule by a Co—C bond, and this bond is in a very non polar environment; all the neighbours are either $—CH_2—$ or $—CH_3$ groups.

An x-ray and neutron study of a monocarboxylic acid derivative of vitamin B_{12} (M16) not only showed that neutron diffraction is feasible for a structure

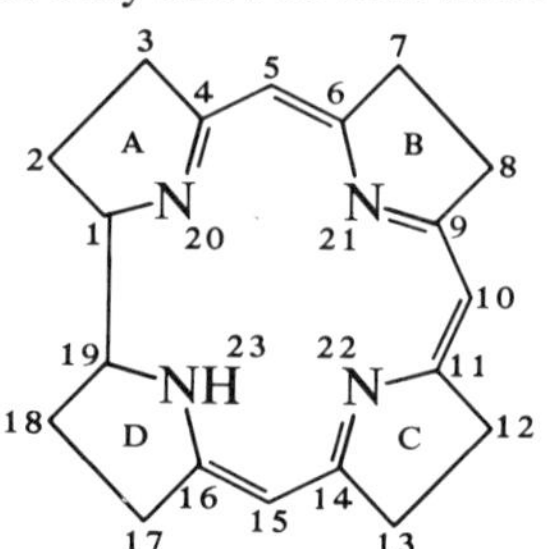

Figure 5.5 The corrin nucleus

of this size but was able to establish hydrogen atom positions with certainty; this confirmed that the system of six alternating single and double bonds runs from N(20), around the corrin ring, to N(23). Nitrogen and oxygen atoms of amides and carboxylic acid groups could also be distinguished easily. The structure of this derivative is shown in Figure 5.4 with R = CN and the amide group on C(13) replaced by —COOH.

x-Ray analysis of cobyric acid (factor V 1a), M14 (Figure 5.6), is a further source of fairly accurate stereochemical information on the corrin nucleus,

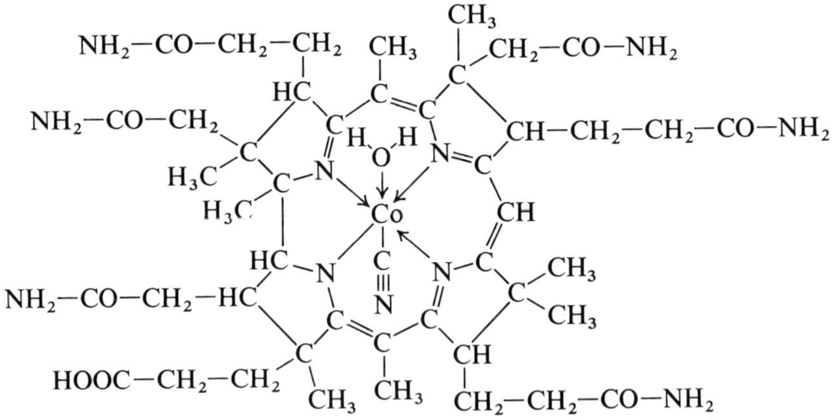

$NH_2—CO—CH_2—CH_2$ CH_3 CH_3 $CH_2—CO—NH_2$
HC C C C
$NH_2—CO—CH_2$ C H O H $CH—CH_2—CH_2—CO—NH_2$
H_3C C C N N=C
H_3C Co CH
HC N C N C CH_3
$NH_2—CO—CH_2—HC$ N C C CH_3
C C CH
$HOOC—CH_2—CH_2$ CH_3 CH_3 $CH_2—CH_2—CO—NH_2$

Figure 5.6 Cobyric acid, M14

especially as it is a smaller molecule; it was possible to identify more than 50 hydrogen atom positions in the last difference electron density series.

In all these molecules the corrin ring is nearly but not quite planar. It is likely that the most stable structure for the corrin ring, like the porphyrin system (see earlier), is a slightly 'ruffled' one, which allows the angles in the ring to have values nearer the ideal ones. In cobyric acid, M14, the corrin ring is slightly convex; the root-mean-square deviation of atoms from the mean plane is 0.15 Å and the maximum is 0.28 Å. In the coenzyme, M18, the pattern of deviations is probably dictated by the more bulky substituents; the mean and maximum deviations are 0.23 and 0.62 Å. In both, bond lengths in the ring approximate to those expected for six single and six double bonds. The conformation of the nucleotide–propanolamine chain is almost the same, despite different crystal environments, in the coenzyme, in cyanocobalamin, and in the hexacarboxylic acid, but it is said to be somewhat different in the monocarboxylic acid derivative, M16. The $Co—N_{corrin}$ distance is 1.94 Å in the coenzyme, comparable with the $Co^{III}—N$ distances in amines and not with the corresponding $Co^{II}—N$ distances. The $Co—N_{benzimidazole}$ distance is 0.2 Å longer in the coenzyme than in vitamin B_{12} itself, a strong *trans* effect due to the $—CH_2—$ ligand.

The corrin whose nickel derivative, M13, has been studied, was synthesised by Eschenmoser. The structure determination has confirmed the chemical structure and conformation and it is hoped that it will also provide accurate values for the bond lengths.

5.3 PENICILLINS AND CEPHALOSPORINS

The penicillins and cephalosporins are a group of antibiotics which interfere with the formation of bacterial cell walls. Details of structural studies of seven of these are given in Table 5.3. They have a range of biological activity, and the different compounds have different solubilities, sensitivities to acid, etc., and thus different relative merits for clinical use as antibiotics. Table 5.3 also gives three compounds with closely related structures which have no biological activity. The earlier studies in the group were undertaken to establish the chemical structures, the others in the search for a structure–function relationship; with the latter aim Sweet and Dahl[4] have studied a biologically inactive Δ^2-cephalosporin, M29, as well as the two biologically active Δ^3-cephalosporins, M25, M26, and have made some very interesting proposals (see later).

Phenoxymethyl penicillin (penicillin V), M21, was the first structure determined after the classic work on the sodium and potassium salts of benzyl penicillin, M19, M20. Phenoxymethyl penicillin is much more stable to acid than benzyl penicillin and so is suitable for oral administration; in the crystals there is an intramolecular hydrogen bond, N—H····O from the amide nitrogen to the phenoxy oxygen to which the acid stability may be attributed. Ampicillin, M23, also widely used orally, has no internal hydrogen bond, but it is a zwitterion which may give it equivalent protection. Hetacillin, M24, was prepared from ampicillin by treatment with acetic anhydride, and its chemical structure was established by x-ray analysis; it has greater acid stability than ampicillin. Detailed comparison of bond lengths and angles in these compounds has not led to any significant conclusions; the errors in their determination are mostly rather large except for those of ampicillin. The stereochemistry around the β-lactam and thiazolidine rings is similar throughout the series: the β-lactam ring is planar; four of the five thiazolidine ring atoms are coplanar, the carbon carrying the carboxyl group is *c.* 0.5 Å out of this plane; the bridgehead nitrogen is pyramidal (the angles around it in M21 add to 337 degrees).

Table 5.3 Penicillins, cephalosporins, and related compounds

Compounds M27–M29 are inactive as antibiotics. The chemical structures are explained in terms of:

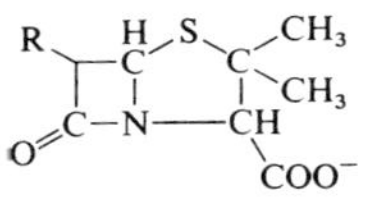

P, in penicillins

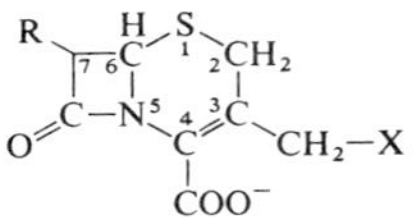

C, in cephalosporins

	Name	Structure	Data	R	σ for C—C bond/Å	Coordinates published	Reference
M19	Benzyl penicillin, sodium salt	P, R = $PhCH_2 \cdot CO \cdot NH$-	434 W	0.22	0.1	yes	32, 33
M20	Benzyl penicillin, potassium salt	P, R = $PhCH_2 \cdot CO \cdot NH$-	1680 W	0.20		yes	34
M21	Phenoxymethyl penicillin (penicillin V)	PH, R = $PhO \cdot CH_2 \cdot CO \cdot NH$-	1670 W	0.126	0.02	yes	35
M22	Hetacillin	PH, R 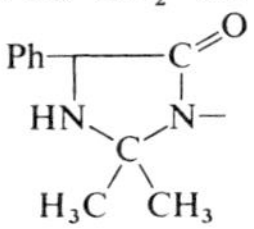 = Ph–(ring: C=O, N–, C(H_3C)(CH_3), HN)	2006 W	0.14		no	36
M23	Ampicillin	P, R = $PhCH(NH_3^+) \cdot CO \cdot NH$-	1470 D	0.046		no	37
M24	Cephalosporin C, sodium salt	C, R = $OOC \cdot CH(NH_3^+) \cdot (CH_2)_3 \cdot CO \cdot NH$- X = $MeCO \cdot O$—	1428 W	0.26	0.07	yes	38
M25	Cephaloridine hydrochloride	CH, R 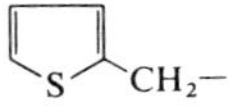 = (2-thienyl)–CH_2— X = (pyridinium) N^+— 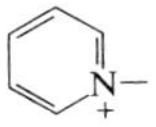	1547 D	0.057	0.01	yes	4
M26	Cephaloglycine	CH, R = $PhCH(NH_3^+) \cdot CO \cdot NH$— X = $MeCO \cdot O$—	1000 D	0.056	0.05	yes	4
M27	6-(*N*-Benzylformamido) penicillanic acid	PH, R = $PhCH_2 \cdot N(CHO)$—	1580 W	0.095		no	79
M28	Phenoxymethyl penicillin sulphoxide methyl ester	formula as M21 except that P contains the group—SO—	2000	0.20		no	80
M29	Phenoxymethyl-Δ^2-desacetoxyl cephalosporin	CH, but with the double bond between carbon atoms 2 and 3. X = H, R = $PhO \cdot CH_2 \cdot CO \cdot NH$—	1269 D	0.18	0.01	yes	4

All the molecules in the cephalosporin series contain an identical β-lactam ring, but it is fused to a six-membered, rather than a five-membered sulphur-containing ring. The greatest problems in the crystal structure analyses have been due to poor crystals. Also, in cephalosporin C, M24, there are two molecules in the asymmetric unit, separated by approximately $c/2$ but with rather different y-coordinates; nevertheless the chemical structure of cephalosporin was first established from these crystals. In cephaloridine hydrochloride, M25, the thiophene groups are disordered and in cephaloglycine,

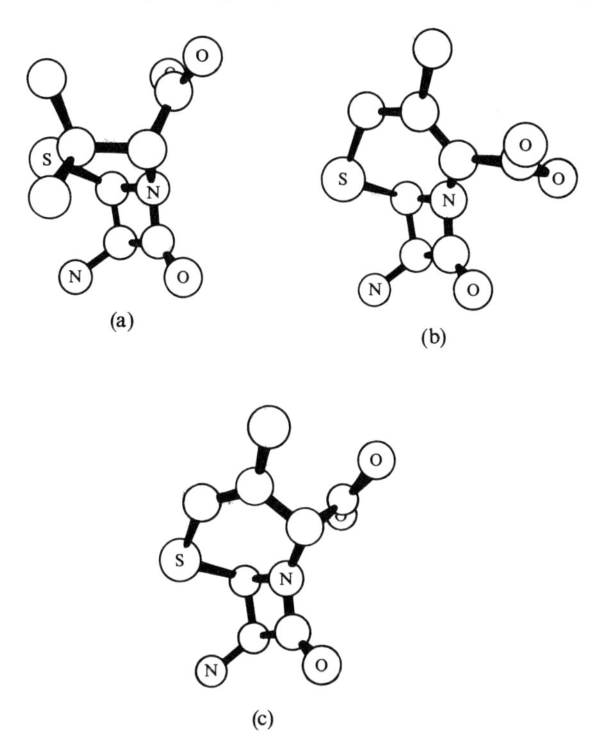

Figure 5.7 The conformations found for the central part of molecules of (a) penicillins, (b) biologically active Δ^3-cephalosporins, and (c) biologically inactive Δ^2-cephalosporins. (From Sweet and Dahl[4], by permission of the American Chemical Society)

M26, the solvent molecules (acetic acid and perhaps water) have indefinite positions.

The conformation of the central part of cephaloridine hydrochloride is illustrated in Figure 5.7b. While the β-lactam ring remains the same as in the penicillins, the adjacent sulphur atom and carboxyl groups take rather different orientations. The bridgehead nitrogen is pyramidal, though now the angles around it add to *c.* 350 degrees. The stereochemistry of the inactive Δ^2-cephalosporin, M29, is shown in Figure 5.7c; the sulphur atom and carboxyl group are again placed differently, though not very much so, but the three bonds to nitrogen are very nearly coplanar.

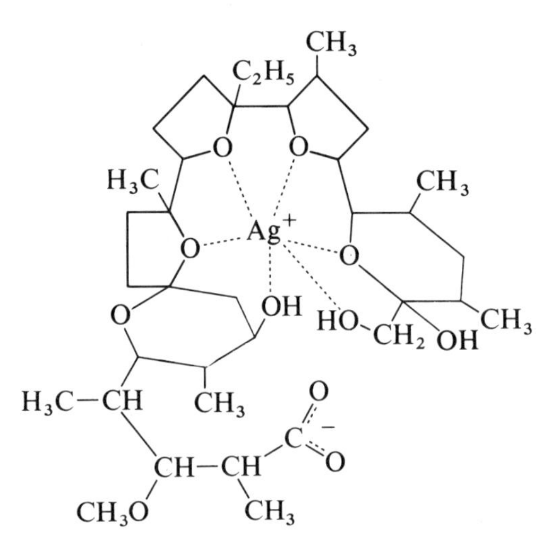

Figure 5.8 The silver salt of monensin, M30

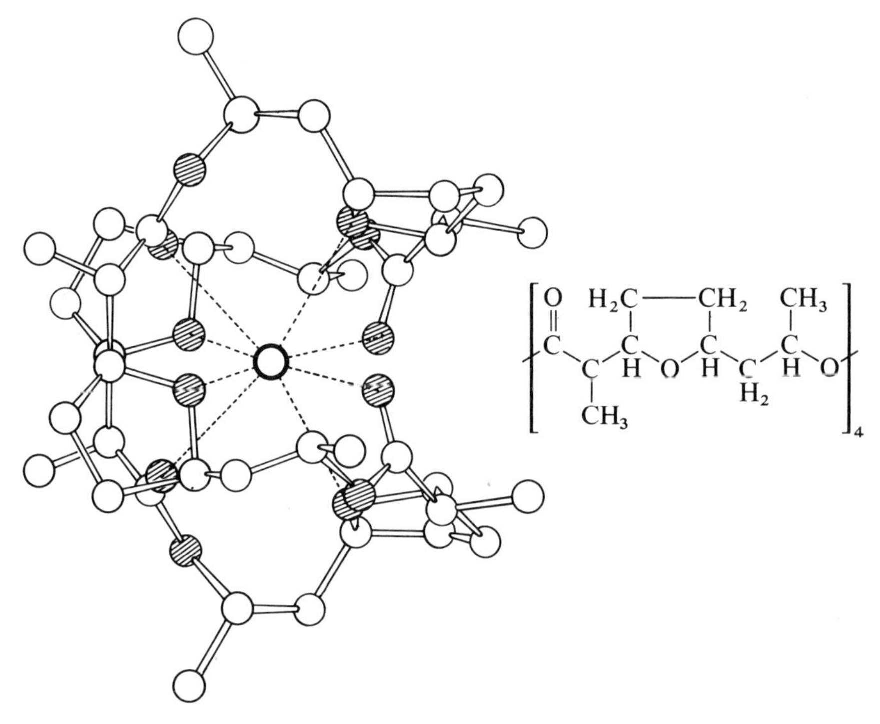

Figure 5.9 Nonactin–K^+ complexion, M33, viewed perpendicular to its approximate $S_4(\bar{4})$ symmetry axis; oxygen atoms are shaded.
(From Kilbourn, B. T., Dunitz, J. D., Pioda, L. A. R., and Simon, W. (1967). *J. Mol. Biol.*, **30,** 561 by permission of Academic Press)

Available evidence suggests that when these compounds interfere with cell-wall synthesis they do so by inhibiting the transfer of the peptide D-alanyl-D-alanine in the three dimensional cross-linking of peptidoglycan strands[39] (and other references cited by Sweet and Dahl[4]). The penicillin or cephalosporin molecule becomes attached to the transpeptidase enzyme in place of D-alanyl-D-alanine and eventually acylates it irreversibly when the amide bond of the β-lactam is cleaved. Sweet and Dahl have most elegantly collated the stereochemical data, the carbonyl bond-stretching frequencies, the biological activity data, and the ease of base hydrolysis of the β-lactam amide bond, to suggest that the pyramidal nitrogen atom of the penicillins and Δ^3-cephalosporins is crucial. The pyramidal nitrogen atom should be associated with less double-bond character in the amide bond than in a simple β-lactam.

5.4 OTHER ANTIBIOTICS

5.4.1 Molecules which form complexes with cations

The first six structures listed in Table 5.4 are of compounds, mostly isolated from various *streptomyces* or similar sources, which affect the transport of alkali-metal cations across biological membranes by forming lipid-soluble complexes with them. The first three are chemically similar polyethers, each with a single carboxyl group, and they all form salts with alkali-metal cations. The crystal structures of the silver salts of monensin and nigericin have been determined, that of nigericin was shown to be isomorphous with the sodium salt[42]. In each case the polyether molecule is wrapped round the cation, presenting a hydrophobic exterior and allowing the cation to be coordinated to five or six oxygens at distances less than 2.7 Å (Figure 5.8). The molecular conformation is held by the metal coordination and by one or two O—H·····O hydrogen bonds. In the barium salt of antibiotic X-537A, two polyether anions are similarly wrapped around one barium ion so that the barium ion is coordinated by six oxygens from one anion and two from the other (and a water molecule); the anions are held together by a hydrogen bond as well.

The compounds M33, M34, M35 complex selectively with potassium ions. They are neutral cyclic molecules. In the nonactin–KCNS complex (Figure 5.9) the potassium ion is 8-coordinated by four keto oxygens and four furane oxygens at 2.75–2.83 Å in a nearly cubic arrangement; the macrotetrolide ring is wrapped round the potassium ion 'like the seam of a tennis ball' with symmetry approximating to $\bar{4}$, and presenting a hydrophobic exterior. The thiocyanate ions do not interact in any specific way with the cations. Crystals of nonactin itself have been prepared, and are quite unlike the KCNS complex; the shortest axis of the unit cell is 5.70 Å so the molecules must have a different conformation from that in the complex. In the valinomycin–$KAuCl_4$ complex, M34, the cyclododecadepsipeptide ring encloses the potassium ion; six valine carbonyl oxygen atoms coordinate to K^+ at 2.7–2.8 Å and the symmetry of the complex is close to threefold; intramolecular N—H·····O bonds help to hold this conformation. A structure based on similar principles has been suggested for the hexadepsipeptide enniatin-B, M35, but the x-ray

Table 5.4 Other antibiotics and closely related compounds

	Name	*Formula of compound studied, excluding solvent*	*Data*	*R*	*σ for C⁻C bond/Å*	*Coordinates published*	*Reference*	
M30	Monensin	$[C_{36}H_{35}O_{11}]^-$, Ag^+	2900	0.08		no	40	molecules which form complexes with alkali metal cations
M31	Nigericin or polyetherin-A	$[C_{40}H_{67}O_{11}]^-$, Ag^+	2066 D	0.065	0.025	yes	41, 42	
M32	Antibiotic X-537A	$[C_{34}H_{53}O_8]_2^- Ba^{2+}$	4753 W	0.11		no	43	
M33	Nonactin	$[C_{40}H_{64}O_{12}K]^+$, CNS^-	1200 D	0.125		yes	44	
M34	Valinomycin	$[C_{54}H_{90}N_6O_{18}K]^+$, $AuCl_4^-$	2100 D	0.19		no	45	
M35	Enniatin-B	$[C_{33}H_{57}N_3O_9K]^+$, I^-	very limited			no	46	
M36	Rifamycin B	$C_{45}H_{53}IN_2O_{13}$	2175 W	0.20		yes	47	large ring systems
M37	Rifamycin Y	$C_{45}H_{51}IN_2O_{14}$	1426 W	0.14	0.07	yes	48	
M38	Tolypomycinone	$C_{53}H_{52}Br_3NO_{16}$	3632 W	0.17		yes	49	
M39	Aureomycin hydrochloride	$[C_{22}H_{24}ClN_2O_8]^+$, Cl^-	2057 W		0.015–0.02	yes	50	tetracyclines
M40	Terramycin hydrochloride (oxytetracycline hydrochloride)	$[C_{22}H_{25}N_2O_9]^+$, Cl^-	2100W	0.13	0.01	yes	51	
M41	Anhydrotetracycline hydrobromide	$[C_{22}H_{23}N_2O_7]^+$, Br^-	2165 D	0.07		no	52	
M42	7-chloro-4-hydroxytetracycloxide (and 7-bromo-)	$C_{19}H_{14}ClNO_9$	1390 D	0.11		no	53	
M43	Thiostrepton	$C_{69}H_{84}N_{18}O_{18}S_5$	5600 D	0.15		no	54	Include several thiazole rings
M44	Micrococcinic acid	$C_{34}H_{21}Br_2N_7O_3S_4$	2488 W	0.12	0.04–0.06	no	55	

data obtainable were rather limited and there was evidence of disorder in the crystals. These structures and some synthetic analogues are discussed in a recent review[56].

5.4.2 Large-ring molecules

Rifamycin B, M36, is an antibiotic from *streptomyces mediterranei*, containing a naphthoquinone and a long aliphatic chain, making together a twenty four-membered ring. The structure determination was undertaken primarily to determine the chemical structure (Figure 5.10). Rifamycin Y, M37, which differs only by oxidation at two carbon atoms is biologically inactive, but although the full stereochemistry of these two large molecules has been established and is very nearly the same except at the two modified carbon atoms, there is not yet any explanation for the difference in activity. Tolypomycinone, M38, biologically active and chemically very similar to rifamycin, shows considerable differences in conformation, for example a swing of

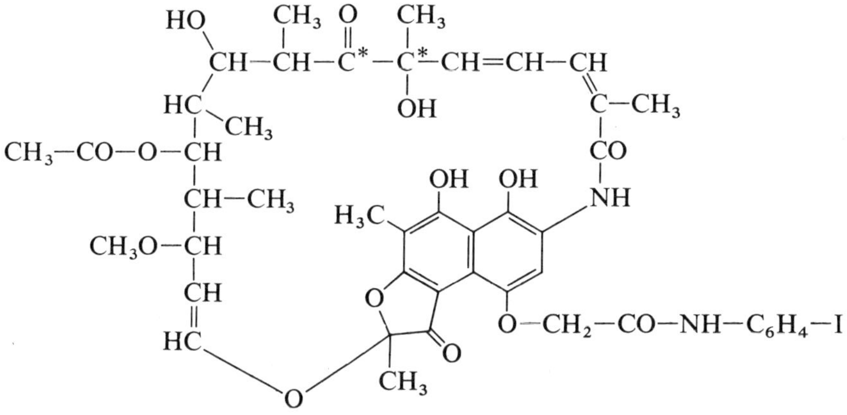

Figure 5.10 Rifamycin Y *p*-iodoanilide. Rifamycin B differs only at the two starred carbon atoms, which are —CHOH—CHMe—

the naphthoquinone plane through 90 degrees. These differences may be dictated by the chemical structure (in which case they would be relevant to biological activity) but they might merely represent the adaptation of a large and rather flexible molecule to different crystal-packing situations which are brought about by the different heavy-atom substituents; both the rifamycins were studied as the *p*-iodoanilides while a tri-*m*-bromobenzoate of tolypomycinone was used. x-Ray analysis has been successfully used to establish the chemical structures of a number of other large ring antibiotic molecules, for example, pyridomycin[57] with a 12-membered ring and kromycin[58] with a 14-membered ring.

5.4.3 Tetracyclines

The chemical formulae of the compounds M39, M40, M41 are explained in Figure 5.11. The crystal structure analyses of aureomycin and terramycin served to establish the stereochemistry and confirm some details of the chemical structure, for example, that there is a localised double bond in ring B. The bond lengths in anhydrotetracycline, M41, are very close to those in

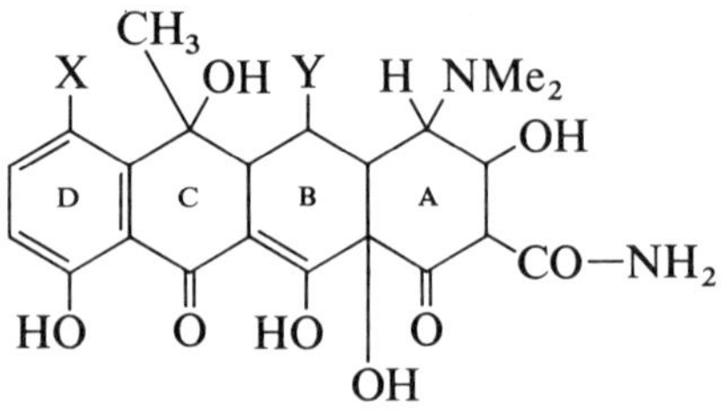

Figure 5.11 Tetracyclines; in aureomycin, M39, X = Cl and Y = H; in terramycin, M40, X = H and Y = OH; in anhydrotetracycline, ring C is aromatic as a result of the removal of H_2O from it

aureomycin and terramycin except in ring C which has become aromatic; the conformation is markedly different, particularly around the junction of rings A and B; its specificity as an antibiotic is also quite different, but there is not yet any explanation in terms of stereochemistry. All the molecules have internal hydrogen bonds, both N—H·····O and O—H·····O.

5.4.4 Thiostrepton, micrococcinic acid

Thiostrepton, M43, with 110 atoms other than hydrogen, represents another feat in structure solving. Its antibiotic properties are similar to penicillin but its solubility in water is low and no medical use has yet been developed. The chemical structure was only partly known from degradation. Four sulphur atom positions were found from the Patterson series, and in the first sulphur-phased electron density series twenty light-atom positions could be found. Repeated structure-factor and Fourier calculations eventually revealed the structure of the whole molecule; the interpretation of many regions of these partial electron density series was made possible by the knowledge of the products of chemical degradation. In the least-squares refinement all atoms were initially treated as carbon and the changes in thermal vibration parameters (B) were used to indicate or confirm the chemical nature of the atom; the refinement procedure should give lower values of B for atoms which are really nitrogen and lower values still for oxygen, since it cannot alter the scattering factor and this is the only way it can make the model approximate more closely to the true structure. The chemical structure thus established is shown in Figure 5.12 and is described as having two 'wings' held in position by hydrogen bonds from three points on them to a water molecule. Tests, such as the calculation of a difference Fourier series using phases calculated for the whole molecule excluding one 16-atom portion of it, and the appearance in the map of the expected atoms and no others, suggest that there can be no doubt about the correctness of this structure for the main part of the molecule. The long side chain caused difficulties though, for it extends into a pool of disordered solvent, and is not rigidly held in place by any inter- or intra-molecular contacts. The five terminal atoms could not be identified

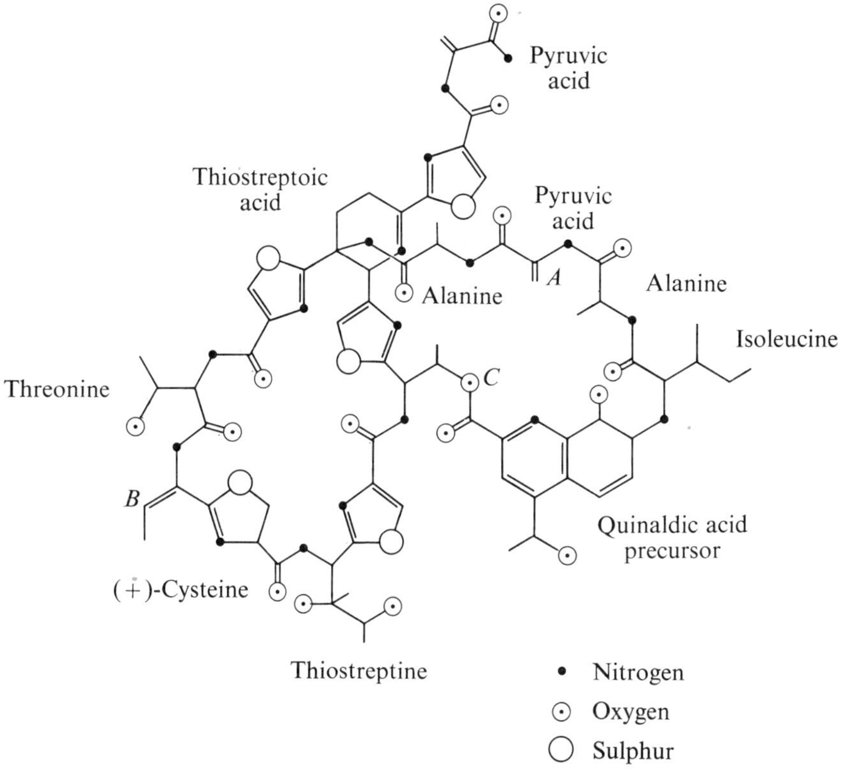

Figure 5.12 The chemical structure of thiostrepton
(From Anderson, Hodgkin and Viswamitra[1], by permission of Macmillan and Company Ltd.)

with certainty from the electron-density series and so chemical information had to be used here.

The structure of the bis-4-bromoanilide of micrococcinic acid, M44, derived from the antibiotic micrococcin P, was established by x-ray analysis and contains three thiazole rings linked to pyridine with a fourth thiazole linked to one of these; the chemical arrangement is like that around the reduced pyridine in thiostrepton.

5.5 PEPTIDES

A selection of peptides studied recently is given in Table 5.5. This includes firstly some linear tri- and tetra-peptides, secondly some cyclic peptides and a

Table 5.5 Peptides

	Name	Formula (excluding solvent)	Data	R	σ for C—C bond/Å	Coordinates published	Reference
M45	*N*(Bromoacetyl)–phe–phe–OEt	$C_{22}H_{25}BrN_2O_5$				no	59
M46	*Z*–Gly–pro–leu–gly	$C_{23}H_{31}BrN_4O_7$	973 D	0.05	0.015–0.03	yes	60
M47	*S*–Benzyl–cys–pro–leu–gly–NH_2	$C_{23}H_{35}N_5O_4S$				no	61
M48	*Se*–Benzyl–selenocys–pro–leu–gly–NH_2	$C_{23}H_{35}N_5O_4Se$				no	61
M49	Cyclohexaglycyl	$C_{12}H_{18}N_6O_6$				yes	62
M50	Cyclotetrasarcosyl	$C_{12}H_{20}N_4O_4$	1204 W	0.078	0.003	yes	63
M51	Cyclotetraglycyl–di–D–alanyl	$C_{14}H_{22}N_6O_6$	1975 D	0.065	0.01	yes	64
M52	Cyclo(D–HyIv–L–MeIleu)$_2$	$C_{24}H_{42}N_2O_6$	1753 D	0.123	0.03	yes	65
M53	Ferrichrome–A	$C_{41}H_{58}N_9O_{20}Fe$	3115 D	0.09	0.02	yes	66

M45 *N*-(bromoacetyl)-L-phenylalanyl-L-phenyl alanine-ethyl ester.
M46 *p*-bromocarbobenzoxy-glycyl-L-prolyl-L-leucyl-glycine.
M47 *S*-benzyl-L-cysteinyl-L-prolyl-L-leucyl-glycinamide.
M48 *Se*-benzyl-L-selenocysteinyl-L-prolyl-L-leucyl-glycineamide.
M52 cyclo-D-α-hydroxy-isovaleryl-L-methyl isoleucyl-D-α-hydroxy-isovaleryl-L-methyl-leucyl)
M53 See Figure 5.15.

cyclic depsipeptide and thirdly ferrichrome-A. The structural chemistry of amino acids and simpler peptides is already well established. In 1967, Marsh and Donohue[67] reviewed the crystal structures of amino acids and peptides and gave the dimensions of the peptide group, derived from a weighted average of the results of three-dimensional crystal structure analyses; these differed just a little from those of Pauling and Corey[68] (Figure 5.13). The largest peptide structures then established were of tripeptides (apart from cyclohexaglycyl, M49 and several proteins!) Ramachandran and his colleagues[69, 70] as well as Scheraga[71] have investigated allowable conformations of peptide groups and hence of polypeptide chains.

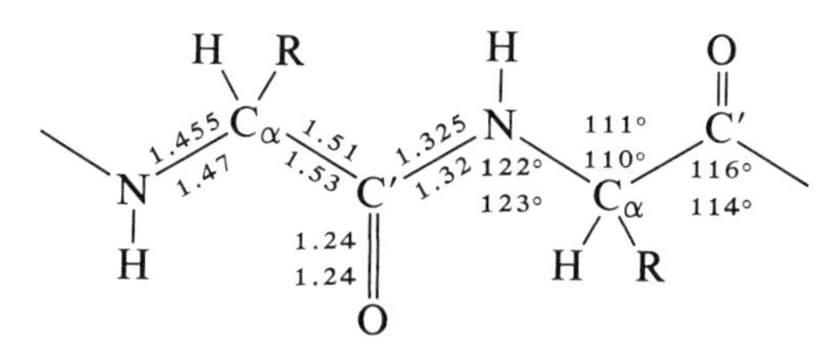

Figure 5.13 Bond lengths and angles within a polypeptide chain. The upper figures are those given by Marsh and Donohue[67], the lower those of Pauling and Corey[68]. R represents different side chains in different amino acids

By use of accepted values for the bond lengths and angles the stereochemistry can be described by the torsion angles ϕ, ψ and ω[72], representing rotations about the bonds N—C_α, C_α—C′, and C′—N. The value $\omega = 0$ is expected; this means that the group C_α—C′—N—C_α is planar and *trans*. Allowable values for the pair of angles ϕ and ψ are those for which no atoms of the peptide chain are closer than acceptable van der Waals contact distances; the atoms are thus treated as hard spheres. An alternative, more realistic, but much more difficult approach is to calculate the potential energy as a function of ϕ and ψ[73]. The results of both types of calculation are mapped and discussed by Ramachandran[71] and values of the torsion angles observed by that time are tabulated.

5.5.1 Tri- and tetra-peptides (M45–48)

The enzyme collagenase can hydrolyse the pentapeptide Z–gly–pro–leu–gly–pro* which is an analogue of a part of the structure of collagen, but it cannot hydrolyse the tetrapeptide Z–gly–pro–leu–gly or the related tri- and di-peptides. x-Ray studies of the series of peptides are being carried out in the hope of understanding better the mechanism of action of the enzyme; the results for the tetrapeptide, M46, are published. The peptide backbone is folded, in a manner that bears some resemblance to parts of the cyclic peptides cyclohexaglycyl and ferrichrome-A. The torsion angles ϕ and ψ at

*Z = carbobenzoxy-; standard abbreviations for amino acids are used, and they are L unless it is otherwise specified.

the leucine residue are said by the authors to fall just outside the region allowed by Ramachandran's calculations; this is *not* because there are contact distances closer than his extreme limits, but because of minor adjustments in other dimensions, the most important being the angles at C′ and C_α of leucine.

The tetrapeptide sequence in M47 and M48 is the same as that at the *C*-terminus of the peptide hormone oxytocin. The conformation is described as like that of M46. M45, analogous to a tripeptide, is also a substrate of the enzyme pepsin; the conformation is said to be like that of gly–phe–gly[74].

5.5.2 Cyclic peptides

M51, a close analogue of the earlier studied cyclohexaglycyl, M49, was a more straightforward structure to solve (it was solved by symbolic addition). The molecule has one conformation in the crystal, very like the commonest of the four conformations found for cyclohexaglycyl. All the peptide groups are *trans*; ω varies from 2 to 8 degrees indicating the slight strain imposed on the formally planar peptide group by the cyclisation. Also the N—C_α—C′ angles have been enlarged slightly; their mean is 2 degrees larger than

(a) $\mathrm{cyclo}\left[\mathrm{N(CH_3)-CH_2-CO}\right]_4$

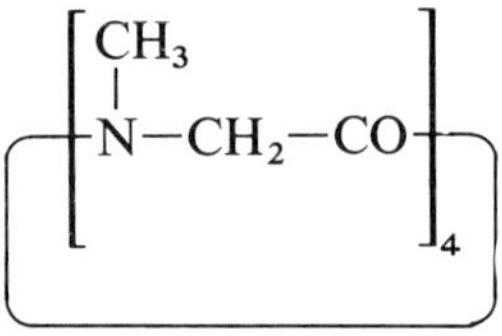

(b) $\mathrm{cyclo}\left[\mathrm{O-CH(CHMe_2)-CO-N(CH_3)-CH(CHMeEt)-CO}\right]_2$ (CH: D; CH: L)

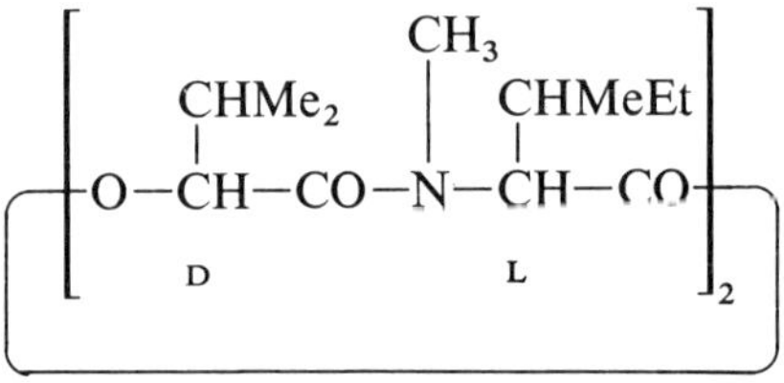

Figure 5.14 (a) M50, cyclotetrasarcosyl; (b) M52, a cyclic depsipeptide

Donohue and Marsh's value for linear peptides. The conformation is stabilised by two rather weak internal hydrogen bonds, NH·····O, and also by two further links using water molecules, NH·····H_2O·····O.

M50 and M52 both contain *N*-methyl peptide groups (Figure 5.14). In M50, cyclotetrasarcosyl, alternate peptide groups are *cis* and *trans* and deviate by 5 and 10 degrees from exact planarity. M52 is a cyclic tetradepsipeptide. Bond lengths are, within experimental error, like those found in peptides, but there is a large opening out of the angles at the valeryl carbonyl groups, from 116 degrees in a normal peptide to 123 degrees here, apparently to relieve overcrowding at the centre of the molecule. Such *cis*-peptide groups have not been found in cyclic hexapeptides, nor in the cyclic tridepsipeptide pyridomycin[75] which is not *N*-methylated; otherwise these features could be expected in larger cyclic peptides and depsipeptides. The structure of the valinomycin–$KAuCl_4$ (M34, see Table 5.4) complex has been determined by x-ray analysis, while there is considerable evidence on the conformations of other cyclic peptides and depsipeptides from n.m.r. and spectroscopic work[76].

5.5.3 Ferrichrome-A

Ferrichrome-A, M53 (Figure 5.15), is a fungal metabolic product and is related to certain growth factors. The x-ray analysis confirmed the chemically determined structure; this includes a cyclic hexapeptide system together with sixfold coordination of the iron by three hydroxamate groups. There are two

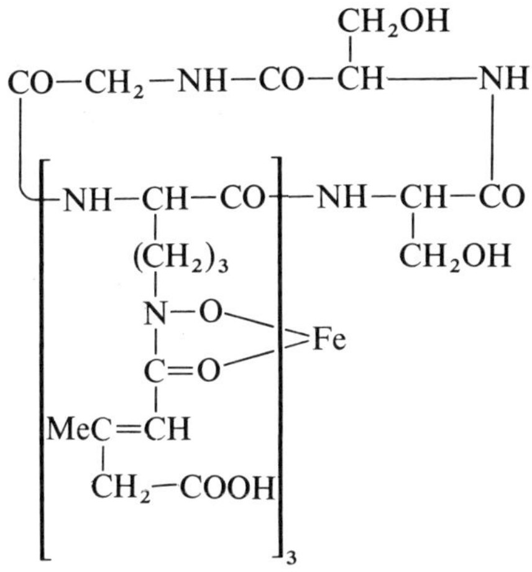

```
                          CH2OH
                            |
CO—CH2—NH—CO—CH————NH
|                                            |
|  ┌                  ┐                      |
└—NH—CH—CO—NH—CH—CO
   |        |          |          |
   |     (CH2)3        |       CH2OH
   |        |          |
   |      N—O——        |
   |        |          >Fe
   |      C=O——        |
   |        |          |
   |MeC=CH             |
   |    |              |
   |   CH2—COOH        |
   └                   ┘3
```

Figure 5.15 Ferrichrome-A, M53

internal hydrogen bonds. Again some of the torsion angles fall up to 20 degrees outside the Ramachandran limits but normal van der Waals contact distances are not violated.

5.5.4 Comments

Thus the peptide chain is fairly versatile from a stereochemical point of view. Bond lengths are nearly unalterable; bond angles may be modified by a few degrees, the 'penalty' in energy terms is *c.* 0.25 kJ mol^{-1} deg^{-1}; the peptide bond is planar, at least within 5–10 degrees, and Ramachandran[70] estimates that 10 degree twist out of plane costs *c.* 2.7 kJ mol^{-1}. (Compare the energy gained on formation of one hydrogen bond, *c.* 20 kJ mol^{-1}.) A range of conformation angles ϕ and ψ is allowable. The calculated 'limits' are rather sensitive to the dimensions assumed for the backbone, especially the bond angles, as Ramachandran has pointed out. He has used Pauling and Corey's dimensions in his calculations; Donohue and Marsh have given slightly different mean values based on more extensive three-dimensional structure analyses (see Figure 5.13). From the point of view of the real structure a small change (2 degrees) from the 'standard' bond angle may extend the range of allowed conformations considerably and so enormously complicate attempts at predicting the structure.

There could well be small changes in conformation in different environments, for example in the crystal, in solution, or when associated with an enzyme. Differences in conformation between chemically identical molecules in different crystal environments have been observed in simpler systems[77, 78].

5.6 SOME OTHER LARGE MOLECULES INCLUDING SUGARS AND NUCLEOSIDES

The selection of compounds for this section is very arbitrary (Table 5.6). Mono- and di-saccharides, nucleosides and nucleotides have been extensively studied in the last few years and deserve a full chapter in a book such as this. The largest single crystal structure determinations in these categories have been included here; for the rest, the bibliography[5] and some earlier reviews must suffice[91–95]. These medium-sized molecules are involved in many parts of the metabolism of living cells, and they are the building bricks for the polymers, nucleic acids and polysaccharides. x-Ray studies of paracrystalline fibres of these polymers are described[96, 97]. The last four compounds in

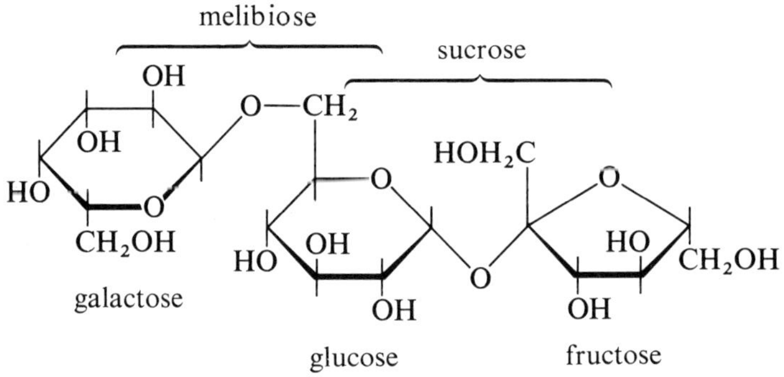

Figure 5.16 Raffinose, M58

Table 5.6, are all antibiotics: erythromycin and amphotericin, M62 and M63 contain large lactone rings, 14-membered and 38-membered respectively, together with an amino-sugar substituent. Kanamycin and streptomycin, M60 and M61, are equivalent to trisaccharides but contain glucosamine and other amino-sugars.

In cyclohexa-amylose, M59, there are six α-D-glucose residues in a ring. The crystal symmetry imposes twofold symmetry on the cyclic molecule but

Table 5.6 Some other large molecules including sugars and nucleosides

	Name	*Formula of compound studied, excluding solvent*	*Data*	*R*	*σ for C—C bond/Å*	*Coordinates published*	*Reference*
M54	Adenosine triphosphate	$[C_{10}H_{14}N_5O_{13}P_3]^{=}, 2Na^+$	1119 D	0.16	0.07	no	81
M55	β-Adenosine-2′-β-uridine-5′-phosphoric acid	$C_{19}H_{24}N_7O_{12}P$	1786 W	0.088	0.01	yes	82
M56	Deoxyuridine disulphide*	$C_{18}H_{22}N_4O_{10}S_2$	2328 D	0.054	0.008	yes	83
M57	4-Thiouridine disulphide	$C_{18}H_{22}N_4O_{10}S_2$	? D	0.079	0.02	no	84
M58	Raffinose	$C_{18}H_{32}O_{16}$	2097 D	0.060	0.008	yes	85
M59	Cyclohexaamylose	$C_{72}H_{120}O_{60}$, 3.8 KOAc	2559 D	0.10	0.013	yes	86
M60	Kanamycin	$C_{18}H_{36}N_4O_{11}, H_2SeO_4$	1154 W	0.106		no	87
M61	Streptomycin oxime	$C_{21}H_{40}N_8O_{12}, 1.5\ H_2SeO_4$	3236 W	0.12		no	88
M62	Erythromycin-A	$C_{37}H_{68}O_{13}NI$	1587 D	0.15		no	89
M63	Amphotericin	$C_{49}H_{73}O_{18}NI$	3702D	0.14		no	90

*5-(1-(2′-deoxy-α-D-ribofuranosyl)uracilyl)disulphide

in fact its symmetry is close to sixfold. The potassium ions were of no use as 'heavy' atoms and the solution was reached by systematic trials of different orientations and positions of the monosaccharide unit. This work allowed the setting up of a detailed structural model for V-amylose, a helical polymer with close to six α-D-glucose units per turn of helix.

Raffinose is a trisaccharide whose chemical structure is indicated in Figure 5.16 and whose components are found in naturally occurring oligosaccharides. Its structure determination, M58, is noteworthy both because it is one of the largest structures in this chapter to be solved by direct methods and because it is the first full publication of the stereochemistry of a trisaccharide. The dimensions and conformations of the component monosaccharides are consistent with those found in smaller molecules. The conformation of each disaccharide portion may be described in terms of the torsion angles around the two C—O bonds to the bridging oxygen (see Figure 5.16, for example). These two conformation angles in the sucrose portion of raffinose are substantially different from the angles in sucrose itself[98], and there is no internal hydrogen bonding in raffinose whereas there is in sucrose. The angles at the bridge oxygens in raffinose are 111 and 122 degrees, the latter being much larger than the equivalent angle of 114 degrees found in sucrose.

Alongside experimental studies such as these there have been calculations of allowed conformations in a manner similar to the calculations described in the previous section for peptides and there have been model building studies of higher polysaccharides[99, 100]. The comparison of raffinose and sucrose is a reminder that more than one conformation may be possible and that the relative stability will depend on other factors in the environment.

M55, is a dinucleoside phosphate, similar, but not identical to, one of the components of natural polynucleic acids. Nevertheless as a model compound it provides much useful conformational information. The structure was solved from the Patterson series where the phosphorus atom position, then the orientation of the PO_4 group was found; although the phosphorus atom represents only 0.11 of the total scattering on a Z^2 basis; the PO_4-phased electron-density map allowed further atom positions to be found. The dimensions and torsion angles found in this and similar nucleoside phosphates are tabulated and discussed by these authors[82] and by Sundaralingam[92].

Two nucleoside disulphides have been studied, M56 and M57; there is evidence that oxidation of such disulphides or the reduction of the corresponding—SH compounds is involved in the determination or regulation of the secondary structure of transfer RNA. The torsion angle around the S—S bond is 87 degrees in M58, similar to values found in simpler disulphides, whereas in M57 it is much smaller, 49 degrees; in the latter, the pyrimidine planes are nearly parallel though they do not overlap.

Crystals of adenosine triphosphate, ATP (Figure 5.17), or one of its salts have been sought for many years on account of the great importance of ATP in biochemical reactions and energy pathways. Kennard and her co-workers have succeeded in preparing crystals of the disodium salt and in determining their structure. The crystals were far from ideal, cell dimensions and reflection intensities were not as reproducible as crystallographers usually expect and only one third of the reflections to a resolution of 0.9 Å were strong enough to be measured. Also there are two molecules in the asymmetric unit of the crystals, so the solution of the phase problem by the Karle procedure, without any atoms heavier than phosphorus and in the face of these difficulties, is a triumph. The two molecules in the asymmetric unit are identical in conformation except for one ribose hydroxyl group. The triphosphate chain is curled, in contrast to the extended form found in inorganic phosphates, but even so the molecule is not particularly compact and has no internal hydrogen

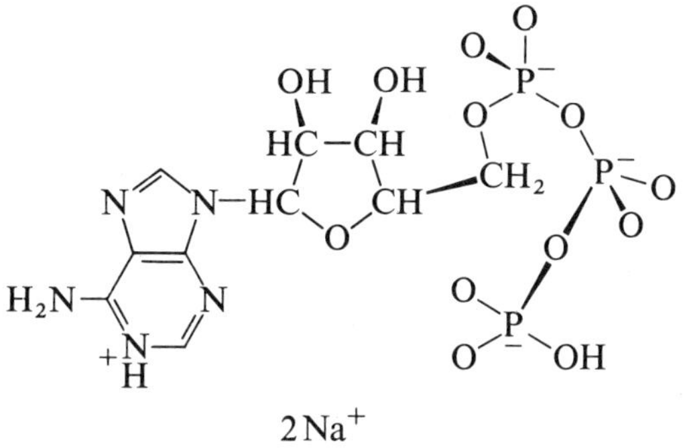

$2Na^+$

Figure 5.17 Adenosine triphosphate, disodium salt, M54. The stereochemistry of the sugar ring and the triphosphate chain are illustrated approximately as they are found in the crystal; the planar adenine group should be normal to the plane of the paper

bonds. Regarding the adenine and ribose groups as rigid there are still many bonds in the molecule around which rotation might take place, to give a variety of conformations. It seems likely that more than one of these might exist in solution, and if so this would account for some of the difficulties experienced in preparing crystals.

Acknowledgements

I wish to thank Dr Olga Kennard and Dr David Watson for their great help in providing the lists of references to structure determinations used throughout this chapter.

References

1. Anderson, B., Hodgkin, D. C. and Viswamitra, M. A. (1970). *Nature (London)*, **225,** 233
2. Adams, M. J., Blundell, T. L., Dodson, E. J., Dodson, G. G., Vijagan, M., Baker, E. N., Harding, M. M., Hodgkin, D. C., Rimmer, B. and Sheat, S. (1969). *Nature (London)*, **224,** 491
3. Hoard, J. L. (1968). *Structural Chemistry and Molecular Biology*, ed. by Rich, A. and Davidson, N., 573. (San Francisco: W. H. Freeman and Company)
4. Sweet, R. M. and Dahl, L. F. (1970). *J. Amer. Chem. Soc.*, **92,** 5489
5. Kennard, O. and Watson, D. G. (ed.) (1970). *Molecular Structures and Dimensions*, Vol. 1, Bibliography 1935–1969. (Utrecht: Oosthoek)
6. Berman, H. M. (1970). *Acta Crystallogr.*, **B26,** 290
7. Fleischer, E. B. (1963). *J. Amer. Chem. Soc.*, **85,** 146
8. Hamor, T. A., Caughey, W. S. and Hoard, J. L. (1965). *J. Amer. Chem. Soc.*, **87,** 2305

9. Koenig, D. F. (1965). *Acta Crystallogr.*, **18,** 663
10. Hoard, J. L., Hamor, M. J., Hamor, T. A. and Caughey, W. S. (1965). *J. Amer. Chem. Soc.*, **87,** 2312
11. Timkovich, R., Tulinsky, A. (1969). *J. Amer. Chem. Soc.*, **91,** 4430
12. Collins, D. M. and Hoard, J. L. (1970). *J. Amer. Chem. Soc.*, **92,** 3761
13. Hoard, J. L., Cohen, G. H. and Glick, M. D. (1967). *J. Amer. Chem. Soc.*, **89,** 1992
14. Countryman, R., Collins, D. M. and Hoard, J. L. (1969). *J. Amer. Chem. Soc.*, **91,** 5166
15. Pettersen, R. C. (1969). *Acta Crystallogr.*, **B25,** 2527
16. Webb, L. E. and Fleischer, E. B. (1965). *J. Chem. Phys.*, **43,** 3100
17. Silvers, S. J. and Tulinsky, A. (1967). *J. Amer. Chem. Soc.*, **89,** 3331
18. Hoppe, W., Will, G., Gassmann, J. and Weichselgartner, H. (1969). *Z. Krist.*, **128,** 18
19. Gillespie, R. J. (1963). *J. Chem. Soc.*, 4679
20. Watson, H. C. (1969). *Progr. in Stereochem.*, **4,** 299
21. Dunitz, J. D., Mayer, E. F. (1965). *Proc. Roy. Soc. A*, **288,** 324
22. Dale, D., Hodgkin, D. C., Venkatasan, K. (1963). *Crystallography and Crystal Perfection*, ed. by Ramachandran, G. N., 237. (London: Academic Press)
23. Hodgkin, D. C,, Pickworth, J., Robertson, J. H., Prosen, R. J., Sparks, R. A. and Trueblood, K. N. (1959). *Proc. Roy. Soc. A*, **251,** 306
24. Nockolds, C. K., Ramaseshan, S., Hodgkin, D. C., Waters, T. N. M. and Waters, J. M. (1967). *Nature (London)*, **214,** 129
25. Moore, F. M., Willis, B. T. M. and Hodgkin, D. C. (1967). *Nature (London)*, **214,** 130
26. Brink-Shoemaker, C., Cruickshank, D. W. J., Hodgkin, D. C., Kamper, M. J. and Pilling, D. (1964). *Proc. Roy. Soc. A*, **278,** 1
27. Hodgkin, D. C., Lindsey, J., Mackay, M. and Trueblood, K. N. (1962). *Proc. Roy. Soc. A*, **266,** 475
28. Hodgkin, D. C., Lindsey, J., Sparks, R. A., Trueblood, K. N. and White, J. G. (1962). *Proc. Roy. Soc. A*, **266,** 494
29. White, J. G. (1962). *Proc. Roy. Soc. A*, **266,** 440
30. Lenhert, P. G. (1968). *Proc. Roy. Soc. A*, **303,** 45
31. Hodgkin, D. C. (1965). *Proc. Roy. Soc. A*, **288,** 294
32. Clark, G. L., Kay, W. I., Pipenberg, K. J. and Schieltz, N. C. (1949). *The Chemistry of Penicillin*, 367. (Princeton: Princeton University Press)
33. Structure Reports (1949). **12,** 424
34. Vaciago, A. (1960). *Atti. Accad. naz. Lincei. Rend. Classe. Sci. fis. mat. nat.*, **28,** 851
35. Abrahamsson, S., Hodgkin, D. C. and Maslen, E. N. (1963). *Biochem. J.*, **86,** 514
36. Hardcastle, G. A., Johnson, D. A., Panetta, C. A., Scott, A. I. and Sutherland, S. A. (1966). *J. Org. Chem.*, **31,** 897
37. James, M. N. G., Hall, D. and Hodgkin, D. C. (1968). *Nature (London)*, **220,** 168
38. Hodgkin, D. C. and Maslen, E. N. (1961). *Biochem. J.*, **79,** 393
39. Abraham, E. P. (1968). *Top. Pharm. Sci.*, **1,** 1
40. Agtarap, A., Chamberlain, J. W., Pinkerton, M. and Steinrauf, L. K. (1967). *J. Amer. Chem. Soc.*, **89,** 5737
41. Shiro, M. and Koyama, H. (1970). *J. Chem. Soc. B*, 243
42. Steinrauf, L. K., Pinkerton, M. and Chamberlain, J. W. (1968). *Biochem. Biophys. Res. Commun.*, **33,** 29
43. Johnson, S. M., Herrin, J., Lin, S. J. and Paul, I. C. (1970). *J. Amer. Chem. Soc.*, **92,** 4428
44. Kilbourn, B. T., Dunitz, J. D., Pioda, L. A. R. and Simon, W. (1969). *Helv. Chim. Acta*, **52,** 2573
45. Pinkerton, M., Steinrauf, L. K. and Dawkins, P. (1969). *Biochem. Biophys. Res. Commun.*, **35,** 512
46. Dobler, M., Dunitz, J. D. and Krajewski, J. (1969). *J. Mol. Biol.*, **42,** 603
47. Brufani, M., Fedeli, W., Giacomello, G. and Vaciago, A. (1964). *Experientia*, **20,** 339
48. Brufani, M., Fedeli, W., Giacomello, G. and Vaciago, A. (1967). *Experientia*, **23,** 508
49. Kamiya, K., Sugino, T., Wada, Y., Nishikawa, M. and Kishi, T. (1969). *Experientia*, **25**, 901
50. Donohue, J., Dunitz, J. D., Trueblood, K. N. and Webster, M. S. (1963). *J. Amer. Chem. Soc.*, **85,** 851
51. Cid-Dresdner, H. (1965). *Z. Krist.*, **121,** 170
52. Restivo, R. and Palenik, G. J. (1969). *Biochem. Biophys. Res. Commun.*, **36,** 621
53. van den Hende, J. H. (1965). *J. Amer. Chem. Soc.*, **87,** 929
54. Anderson, B., Hodgkin, D. C. and Viswamitra, M. A. (1970). *Nature (London)*, **225,** 233

55. James, M. N. G. and Watson, K. J. (1966). *J. Chem. Soc. C,* 1361
56. Truter, M. R. (1971). *Chem. in Brit.,* 203
57. Koyama, G., Iitaka, Y., Maeda, K. and Umezawa, H. (1967). *Tetrahedron Lett.,* 3587
58. Hughes, R. E., Muxfeldt, H., Tsai, C. and Stezowski, J. J. (1970). *J. Amer. Chem. Soc.,* **92,** 5267
59. Wei, C. H., Doherty, D. G. and Einstein, J. R. (1969). *Acta Crystallogr.,* **25A,** S192
60. Ueki, T., Ashida, T., Kakudo, M., Sasada, Y. and Katsube, Y. (1969). *Acta Crystallogr.,* **25B,** 1840
61. Low, B. W., Lovell, F. M. and Rudko, A. D. (1969). *Acta Crystallogr.,* **25A,** S188
62. Karle, I. L. and Karle, J. (1963). *Acta Cryst.,* **16,** 969
63. Groth, P. (1970). *Acta Chem. Scand.,* **24,** 780
64. Karle, I. L., Gibson, J. W. and Karle, J. (1970). *J. Amer. Chem. Soc.,* **92,** 3755
65. Konnert, J. and Karle, I. L. (1969). *J. Amer. Chem. Soc.,* **91,** 4888
66. Zalkin, A., Forrester, J. D. and Templeton, D. H. (1966). *J. Amer. Chem. Soc.,* **88,** 1810
67. Marsh, R. E. and Donohue, J. (1967). *Advan. Protein Chem.,* **22,** 235
68. Corey, R. B. and Pauling, L. (1953). *Proc. Roy. Soc. B,* **141,** 10
69. Ramachandran, G. N., Ramakrishnan, C. and Sasisekharan, V. (1963). *J. Mol. Biol.,* **7,** 95
70. Ramachandran, G. N. and Sasisekharan, V. (1968). *Advan. Protein Chem.,* **23,** 283
71. Scott, R. A. and Scheraga, H. A. (1966). *J. Chem. Phys.,* **45,** 2091
72. Edsall, J. T., Flory, P. J., Kendrew, J. C., Liquori, A. M., Nemethy, G., Ramachandran, G. N. and Scheraga, H. A. (1966). *J. Mol. Biol.,* **15,** 399
73. de Santis, P., Giglio, E., Liquori, A. M. and Ripamonti, A. (1965). *Nature (London),* **206,** 456
74. Marsh, R. E. and Glusker, J. P. (1961). *Acta Crystallogr.,* **14,** 1110
75. Koyama, G., Iitaka, Y., Maeda, K. and Umezawa, H. (1967). *Tetrahedron Lett.,* 3587
76. Hassall, C. H. and Thomas, W. A. (1971). *Chem. in Brit.,* 145
77. Bennett, I., Davidson, A. G. H., Harding, M. M. and Morelle, I. (1970). *Acta Crystallogr.,* **B26,** 1722
78. Harding, M. M. and Long, H. A. (1968). *Acta Crystallogr.,* **B24,** 1096
79. Hunt, D. J. and Rogers, D. (1964). *Biochem. J.,* **93,** 35C
80. Cooper, R. D. G., Demarco, P. V., Cheng, J. C. and Jones, N. D. (1969). *J. Amer. Chem. Soc.,* **91,** 1408
81. Kennard, O., Isaacs, N. W., Coppola, J. C., Kirby, A. J., Warren, S., Motherwell, W. D. S., Watson, D. G., Wampler, D. L., Chenery, D. H., Larson, A. C., Kerr, K. A. and Riva di Sanseverino, L. (1970). *Nature (London),* **225,** 333
82. Shefter, E., Barlow, M., Sparks, R. A. and Trueblood, K. N. (1969). *Acta Crystallogr.,* **25B,** 895
83. Shefter, E., Kotick, M. P. and Bardos, T. J. (1967). *J. Pharm. Sci.,* **56,** 1293
84. Shefter, E. and Kalman, T. I. (1968). *Biochem. Biophys. Res. Commun.,* **32,** 878
85. Berman, H. M. (1970). *Acta Crystallogr.,* **B26,** 290
86. Hybl, A., Rundle, R. E. and Williams, D. E. (1965). *J. Amer. Chem. Soc.,* **87,** 2779
87. Koyama, G., Iitaka, Y., Maeda, K. and Umezawa, H. (1968). *Tetrahedron Lett.,* 1875
88. Neidle, S., Rogers, D. and Hursthouse, M. B. (1968). *Tetrahedron Lett.,* 4725
89. Harris, D. R., McGeachin, S. G. and Mills, H. H. (1965). *Tetrahedron Lett.,* 279
90. Mechlinski, W., Schaffner, C. P., Ganis, P. and Avitabile, G. (1970). *Tetrahedron Lett.,* 3873
91. Sundaralingam, M. (1968). *Biopolymers,* **6,** 189
92. Sundaralingam, M. (1969). *Biopolymers,* **7,** 821
93. Berman, H. M., Chu, S. C. and Jeffrey, G. A. (1967). *Science,* **157,** 1576
94. Jeffrey, G. A. and Rosenstein, R. D. (1964). *Advan. Carbohydrate Chem.,* **19,** 7
95. Voet, D. and Rich, A. (1970). *Prog. Nucleic Acid Res. Molec. Biol.,* **10,** 183
96. Marchessault, R. H. and Sarko, A. (1967). *Advan. Carbohydrate Chem.,* **22,** 421
97. Langridge, R. and Sundaralingam, M. (1972). *Progr. in Nucleic Acid Research and Molecular Biology.* (to be published)
98. Brown, G. M. and Levy, H. A. (1963). *Science,* **141,** 920
99. Rees, D. A. and Scott, W. E. (1971). *J. Chem. Soc. B,* 469
100. Anderson, N. S., Campbell, J. W., Harding, M. M., Rees, D. A. and Samuel, J. W. B. (1969). *J. Mol. Biol.,* **45,** 85
101. Venkatasan, K., Dale, D., Hodgkin, D. C., Nockolds, C. E., Moore, F. H. and O'Connor, B. H. (1971). *Proc. Roy. Soc. A,* **323,** 455

6
Protein Crystallography

T. L. BLUNDELL and LOUISE N. JOHNSON
University of Oxford

6.1 INTRODUCTION

One of our great pleasures is to listen to the stories of the early days of protein crystallography. Stories of how pepsin crystals were carried from Sweden in a coat pocket to Cambridge for J. D. Bernal, who made the first x-ray pictures[1, 2]. Stories of Dorothy Crowfoot Hodgkin cycling back to her laboratory in Oxford late at night to check that her insulin crystals had really given a diffraction pattern. Each new x-ray picture caused great excitement. They showed that these large molecules were ordered in a crystal lattice if the crystals were kept wet. Nevertheless, little was understood of the nature of proteins or indeed of the methods by which their structures might be solved. Today we know that complex protein structures may be solved by x-ray analysis; we may now call protein crystallography a 'science'. According to Bertrand Russell, only 'definite knowledge belongs to science'. In 1934 it required fantastic vision – Russell might have called it 'faith' – to believe that the complicated patterns of x-ray intensities given by a protein crystal could possibly be transformed into an equally complex protein structure. In those days the analysis of small molecules of a few atoms was proving a formidable problem.

The eventual success of x-ray analyses of small molecules did not lead immediately to success with proteins. The x-ray analysis of proteins differs in many ways from the study of small molecules. The difference is partly quantitative; protein crystals might have as many as 5000 atoms in the crystallographic asymmetric unit. Consequently the number of positional and thermal parameters may exceed the number required for a small molecule by a factor of a hundred. This causes great difficulty in conventional Patterson, Fourier and least-squares refinement techniques. An atom with an impossibly high atomic weight would be needed for the heavy-atom method, and the probability of any sign relationships between individual structure factors would be very low. However, the difference is also qualitative. Disorder in the protein crystals usually does not permit a structure analysis at atomic resolution. Protein crystal data may extend to 1.5 Å resolution in certain optimum cases; often they will terminate closer to 3 Å resolution. For both reasons it is difficult to use structural information about the protein in the gradual improvement of phases before the electron density map becomes fully interpretable.

Dorothy Hodgkin recalls that, in 1939, J. M. Robertson, who had recently solved the structure of nickel phthalocyanine by the method of isomorphous replacement, suggested that she use the method to solve the structure of insulin by comparing the zinc and cadmium insulin crystals. However, the differences between the diffraction patterns of these two series were too small to give accurately measurable differences at that time. Many years elapsed before Perutz and his colleagues[3] in Sir Lawrence Bragg's Cambridge laboratory showed, in 1954, that the method of isomorphous replacement could be used effectively to determine the structure of a protein directly, without the need to make assumptions about the structure of the protein. Indeed, at the present time, the only proven technique for solving the phase problem in protein crystallography is the method of isomorphous replacement sometimes supplemented by the use of anomalous scattering effects.

Protein crystallography has been reviewed several times during the various stages in its development. Of the more recent reviews we refer the reader to Phillips'[4] most lucid account of the interpretation of x-ray diffraction data from protein crystals and the monograph of Holmes and Blow[5] for the methods employed in protein crystallography. The preparation of heavy-atom derivatives was described by Blake[6] in 1968 and the current state of protein crystallography reviewed in the same year by North and Phillips[7]. The results as they appeared during 1968–1971 have been summarised by Blake[8] and a descriptive account of the structure and function of proteins is given in Dickerson and Geiss[9]. An excellent account of the contribution of x-ray analysis to our present understanding of the mechanism of action of enzymes has been given by Blow and Steitz[10] and an illuminating review of the current state of the field has recently been published by Blake[228].

At the time of writing, the structures of 30 proteins have been determined at resolutions greater than 3.5 Å. These proteins are listed in Table 6.1. They include globins, redox proteins, enzymes and a hormone. The proteins range

Table 6.1 Proteins whose structures have been determined by x-ray analysis to a resolution of 3.5 Å or better

A. *Globins*	16. α-Chymotrypsin[37–41]
1. Myoglobin[11–14]	17. Carboxypeptidase[42–46]
2. Horse oxyhaemoglobin[15]	18. Papain[47, 48]
3. Horse deoxyhaemoglobin[16]	19. Subtilisin BPN′[49–51]
4. Human deoxyhaemoglobin[17]	20. Elastase[52–55]
5. Erythrocruorin[18–20]	21. Chymotrypsinogen[56, 57]
6. Glycera haemoglobin[21]	22. γ-Chymotrypsin[58–60]
7. Lamprey haemoglobin[21, 22]	23. Trypsin[61]
	24. Subtilisin Novo[62]
B. *Redox systems*	25. Staphylococcus nuclease[63, 64]
8. Rubredoxin[23, 24]	26. Lactate dehydrogenase[65]
9. Cytochrome C[25, 26]	27. Carbonic anhydrase[66]
10. Cytochrome b_5[27]	
11. High potential iron protein[28]	D. *Hormones*
12. Flavodoxin[29]	28. Insulin[67, 68]
C. *Enzymes*	E. *Others*
13. Lysozyme[30–32]	29. Trypsin inhibitor[69, 70]
14. Ribonuclease A[33]	30. Calcium binding protein from Carp muscle[71]
15. Ribonuclease S[34–36]	

in molecular weight of the asymmetric unit from *c.* 6000 to 52 000 daltons. Most are monomeric, two are tetramers and one is a hexamer.

The method used in these successful analyses proceeds in several well defined steps, all of which have posed new problems to the protein crystallographer. The solutions of these problems have represented some of the most important advances in x-ray crystallography over the past 20 years. The first steps in the method of isomorphous replacement are the preparation of large crystals of the native protein and of isomorphous derivatives containing electron dense atoms, the so called 'heavy-atoms'. Crystallisation of proteins has previously meant 'microcrystalline' to most biochemists and techniques of growing and keeping protein crystals of 1 mm in size have been developed. For example, the addition of organic solvents, in particular 2-methylpentane-2,4-diol, which is relatively non-volatile and miscible with water has proved very useful[72], and a number of proteins now solved at high resolution have been crystallised from organic solvents. Many proteins are available in very small quantities (the structure of rubredoxin was solved with as little as 15 mg of protein)[23] and in recent years there has been careful study of micromethods of crystallisation. Each laboratory has tended to develop its own techniques and often these are relevant only to the particular protein under study and the manual dexterity of the research workers involved. Of particular interest are the methods involving equilibrium dialysis using a microdiffusion cell[73] and modifications in which a second capillary is introduced to establish a concentration gradient of precipitant[74]. This has proved useful in the recent crystallisation of glutamine synthetase[75]. Micro-methods of the vapour-diffusion technique developed first for t-RNA have also been modified for use with proteins[76]. Protein crystallisation requires patient investigation of all parameters affecting solubility coupled with an understanding and a certain 'feel' for the molecule. This approach is well illustrated by the work of Lynen and his colleagues[77] on fatty acid synthetase, a multi-enzyme complex comprising seven proteins and a total molecular weight of 2.3×10^6.

The introduction of heavy atoms into crystals to give isomorphous derivatives was inevitably a 'trial and error' process in the early x-ray analyses of proteins. The complexity of protein structure and the intricate nature of the molecular packing in the crystals has meant that the rationalisation of the preparation of isomorphous derivatives has been difficult and rather unrewarding. Nevertheless, the accumulated experience of 30 successful structure analyses now allows some general conclusions which may be helpful in future work. The replacement of a metal in a metalloprotein has proved useful with several proteins containing either zinc or calcium, which are moderately strongly bound[42, 63, 66, 68]. Replacement of the metal ion by one of higher molecular weight and hence larger ionic radius, usually leads to occupation of a slightly different position[42, 63, 66]. The successful use of such derivatives depends critically on the precise positioning of the metal ion in the native protein, so that this may be used as a negative atom in the phase calculations. Heavy-atom labelled specific inhibitors have also played an important part in the preparation of isomorphous derivatives of several enzymes[37, 63, 66]. These have the advantage of specificity and allow the active site to be directly located once the electron density map is interpreted. However, interference by the inhibitor with this part of the molecule by reaction with the side-chains possibly causing a conformational change in this locality may restrict the use of this method. Other specific methods such as the introduction of new functional groups[78–82] or the replacement of amino acids and peptides[83, 84] have been widely studied but have yet to be useful in x-ray analysis. In contrast the direct binding of heavy-atom compounds to the protein or prosthetic group, traditionally considered to be much less specific than other methods, is generally more useful. Despite the complexity of the heavy-atom binding caused partly by chelation and dissociation of the heavy-atom compounds, it is now clear that some heavy atom reagents are more specific than was previously thought. The uranyl, plumbous, thallous and rare-earth cations tend to bind carboxylate groups; *p*-chloromercuribenzoate (PCMB) and other mercurials are good sulphydryl reagents[8]; $PtCl_4^{2-}$ is a good methionine reagent[85]; and mersalyl may be useful for imidazole groups[85], although its large size may lead to lack of isomorphism in many cases. Iodination of tyrosines has been widely studied but rarely used; it tends to give rise to a displacement of the tyrosyl ring position[49, 86]. Xenon[87] and mercuri-iodide[88, 89] bind to non-polar regions. However the complexity of mercuri-iodide binding to proteins emphasises the need for caution in the use of the above generalisations as a guide for successful exploitation of the different functional groups in the preparation of heavy-atom derivatives.

The next step requires techniques of measurement of large numbers of weak x-ray intensities. The great majority of protein structures have been solved using conventional fine-focus sealed tubes. However, greater resolution and greater intensity are required for small crystals with large unit cells ($a > 140$ Å). Rotating anode tubes have given an increase in brilliance of a factor of four over the sealed x-ray tubes[90], and focusing of the beam onto a smaller target area leads to a smaller divergence, but their unreliability and maintenance problems may prevent their routine use except in relatively large x-ray laboratories. The x-radiation emitted by an electron synchroton could be one hundred times more intense, and may make feasible an investigation of a wide range of proteins, protein complexes and subcellular structures which are of interest in molecular biology[91].

The accurate measurement of x-ray intensities of protein crystals was successfully achieved in most cases by diffractometers, and development of multi-counter machines has led to their more efficient use[93–95]. However, the availability of fully automatic micro-densitometers such as the rotating drum and flying spot densitometers[99–103] has led to the return to photographic methods in some laboratories. The use of the new densitometers in conjunction with rotation photographs and screenless precession photographs[92, 96–98] is likely to prove most effective for large proteins.

The need to determine the positions of the heavy atoms in the methods of isomorphous replacement has led to many thoughtful contributions to the theory of x-ray analysis. In particular the use of isomorphous differences in Patterson syntheses[4] and 'lack of closure' methods of refinement[104, 105] have been extensively studied. Also the use of anomalous scattering differences in combination with isomorphous differences has stimulated the derivation of expressions for the heavy atom contribution to the structure factor for a general reflection[106–110]. These expressions have also been used in the refinement of heavy-atom positions, and it is interesting that anomalous scattering measurements have been used to advantage in the determination of heavy atom positions of derivatives of ribonuclease[33], insulin[67, 68], nuclease[63], calcium binding protein[71], rubredoxin[23] and cytochrome b_5 [27].

The heavy-atom contributions of several isomorphous derivatives can be combined with the measured structure amplitudes of the native protein and derivatives[111, 112] and the anomalous scattering measurements of the derivatives to give information about the phases of the reflections. This stage in the analysis has led to many ideas about the use of isomorphous replacement and anomalous scattering data, and to the proper treatment of errors in these methods[113–116]. The phases are finally used to compute an electron density map. Recently attention has turned to the display of electron density maps. Methods involving half-silvered mirrors to enable direct comparison of the protein model and the electron density[117] have proved useful in most high-resolution studies, but methods using cathode-ray tubes are also being developed[118,119]. Direct methods of phase refinement including the use of the tangent formula have been tried out[120–123], but have been used in only one structure analysis, that of lamprey haemoglobin[21]. Real[124,125] and reciprocal[24] space-refinement techniques have proved themselves more useful. The recent x-ray analysis of the small protein, rubredoxin[23,24] at 1.5 Å resolution with a final agreement index of 0.132 shows that protein crystallography is now beginning to approach the limits of accuracy placed upon it by the quality and order of the protein crystals rather than the techniques used in the analysis.

In this review, space prevents us from a full discussion of the techniques which are proving useful in the different laboratories of protein crystallography. Instead we will confine ourselves to a description of the structural principles of proteins elucidated by this method which have led to a revolution in our ideas about the activity and function of these complex molecules.

6.2 STRUCTURES OF PROTEINS AS FOUND BY X-RAY ANALYSIS

In order to evaluate the structural principles of the 30 diverse protein molecules studied at high resolution by x-ray analysis, we have found it convenient to organise the discussion of structure in the following way:

(i) primary: the use of the x-ray method to identify amino-acid residues in the absence of complete sequence data.

(ii) secondary: study of the conformation of the polypeptide main chain and the nature of the hydrogen bond interactions between CO and NH groups of the peptides.

(iii) tertiary: study of the chain folding, which allows interactions between different side groups and so stabilises the globular structure.

(iv) quaternary: study of the symmetry relating and the interactions between subunits of proteins.

x-Ray analysis provides the only effective way of finding the detailed nature of the secondary, tertiary and quaternary structures of a protein. x-Ray analysis can contribute some information to the primary structure determination but a complete determination depends on chemical analysis.

6.2.1 Primary structure

There has been considerable discussion amongst protein crystallographers about the possibilities of correctly identifying different amino-acid residues from an electron density map. Most agree that at resolutions of $\leqslant 3$ Å, it is possible to identify an appreciable proportion of side chains and to narrow down the choice of the remainder to a few of the twenty possibilities. Sequences have been proposed for several proteins before the complete sequence determination had been achieved by chemical methods. These include lactate dehydrogenase[65], rubredoxin[23] and flavodoxin[29] where most of the sequence was completely unknown, and myoglobin[11], carboxypeptidase[42], lamprey haemoglobin[21] and carp calcium binding protein[71], where the sequences of a number of tryptic peptides were available but their order in the polypeptide chain had not been established. Consider the case of carboxypeptidase. A complete sequence was published by Lipscomb *et al.*[43] based on the x-ray data at 2 Å resolution and the sequences of nine fragments. In 1969 Neurath and his co-workers[126] published the full sequence, which confirmed the number of residues established as 307 in the x-ray studies. The sequenced fragments had been placed in the electron density of the polypeptide chain correctly. By comparison with the chemical sequence, it is now clear that 56 of the remaining 93 amino-acid residues (60%) were identified correctly[44]. It is instructive to consider what sort of mistakes may be made in this kind of analysis by considering the remaining 37 residues. Several valines and threonines were confused, and three leucines were incorrectly identified as asparagines. These are isosteric pairs and are easily confused. In fact, this was not the only type of confusion. Other leucines and a good percentage of the isoleucines and serines were also incorrectly identified. However, the misidentification of two of the three histidines and three of the four tyrosines was more surprising; these were not previously considered difficult to identify. The confusion of these residues with arginines in two cases and tryptophane once appears to indicate that pyrosyl rings may not always be as obvious as has been considered. In contrast four phenylalanines were correct; but identification here may have been aided by considering their non-polar environments. As the carboxypeptidase electron density map is clearly of very good quality in relation to others at similar resolution, the results of the study show that identification of the amino-acid residues by the x-ray method must be treated with caution. It is clear that the difficulty of identifying R groups is very dependent on their position in the protein molecule; those on the surface and at the chain termini will tend to be more disordered than the internal residues and will be more difficult to identify correctly. However, there is every indication that collaboration between the crystallographer and chemist may often provide the speediest solution of the primary structure. The carboxypeptidase and myoglobin analyses demonstrate that it is possible to place tryptic peptides whose sequences are known, and to detect the total number of amino acids in the chain. In the case of papain[47] the incorrect placing of the fragments by the chemists who were unable to obtain overlapping sequences was clearly recognised by the protein crystallographer. Drenth and his colleagues were able to rearrange these properly and tentatively identify thirteen missing residues.

6.2.2 Secondary structure

The discovery of the α-helix and β-pleated sheet structures by Pauling and Corey[127, 128] and their relation to the well known x-ray patterns from fibrous compounds were based on two assumptions; the first that peptide groups are planar and the second that hydrogen bonds would be important to the stability of such structures. The structures of myoglobin and haemoglobin seemed to confirm the expected occurrence of α-helix in globular protein structures, and perhaps led early workers to overestimate the importance of this secondary structure. The regularity of the α-helix also provided an attractive simplification in the conceptualisation of protein structure. Recently, the x-ray analyses of carboxypeptidase[42] subtilisin[19], lactate dehydrogenase[26], and carbonic anhydrase[27] have shown that pleated-sheet structures also may be important features in the structures of globular proteins. It is now clear that protein structures can contain variable percentages of α-helix, β-pleated sheet and, of course, irregular, non-repeating conformations. Although the α-helix and β-pleated sheet were suggested originally because they allow formation of good hydrogen-bonds, a number of workers (see the review by Ramachandran and Sasisekhran (1968))[129] have shown that these structures would also be predicted on the grounds of minimising unfavourable interatomic contacts. Conformations that are 'allowed' can be calculated using either a hard-sphere model for the atoms or more sophisticated potential functions. These can then be plotted in

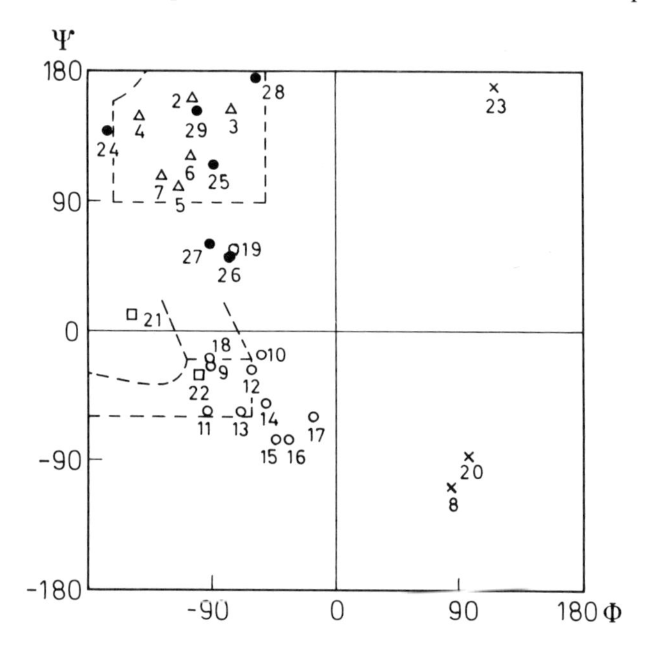

Figure 6.1 A 'Ramachandran plot' for the dihedral angles of the insulin B chain. The three residues with positive ϕ angles are glycines. The conformation is shown in Figure 6.2

terms of the ϕ and ψ angles that define them. The two largest allowed regions correspond to the right-handed helix, including the α-helix, and the β-pleated-sheet conformation. The so-called 'Ramachandran plot' is shown in Figure 6.1 for the insulin B chain, the structure of which is shown in Figure 6.2. It can be seen that the observed conformations correspond reasonably with the predicted, except for three residues which are glycines. As glycine has no β-carbon, the allowed regions for this residue are symmetrical about the origin of the Ramachandran plot[129].

Careful study of the helices found in globular proteins have shown that these tend to be rather distorted from idealised arrangements. Phillips and his colleagues[31] have found that the plane of the peptide group is often rotated in lysozyme so that the CO group is pointing slightly outwards from the helix axis. The variation in the extent of these effects excludes the possibility that they result from diffraction effects. The consequence of such a rotation is that each NH group does not point at the CO group four residues back along the chain, but between that and the carbonyl three residues back. The helices lie somewhere between an α-helix and a 3.0_{10} α-helix. Similar irregularities are observed in the eight helices found in carboxypeptidase.

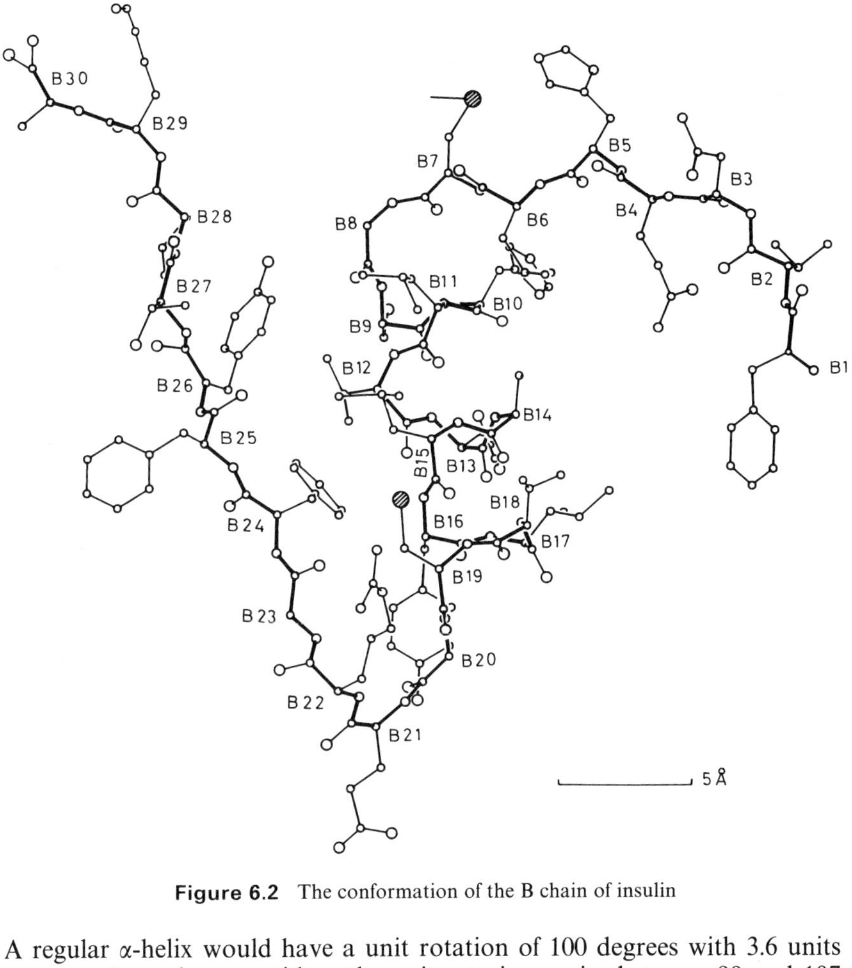

Figure 6.2 The conformation of the B chain of insulin

A regular α-helix would have a unit rotation of 100 degrees with 3.6 units per turn. In carboxypeptidase the unit rotation varies between 90 and 107 degrees and there are between 3.4 and 3.9 units per turn. In one length involving eleven residues, only one hydrogen bond exists. Similar distortions of α-helix are found in carbonic anhydrase[66] as shown in Figure 6.3. In many cases the terminal residues of the helix are most distorted forming a 3.0_{10} helix as in the helix 5 to 10 of lysozyme. On the other hand, the carboxy terminals of the helix F of both haemoglobin α-chain and myoglobin are opened out towards the 4.4_{16} (π) helix). Hydrogen bonds to the third nearest neighbour back along the chain are very common. The 3.0_{10} helix is an example of one possible type. Alternatively, the orientation of the second of the three peptide bonds can be reversed in a way that is only permitted for glycine[130]. This gives rise to a hairpin bend, and such a bend has been observed in three cases in chymotrypsin[40]. Two of these are associated with the active centre sequence

—cys—	met—	gly—	asp—	ser—	gly—	gly—
191	192	193	194	195	196	197

It is of interest that the glycines 193 and 196 are invariant in all known sequences of vertebrate serine proteases. The two invariant glycines of the

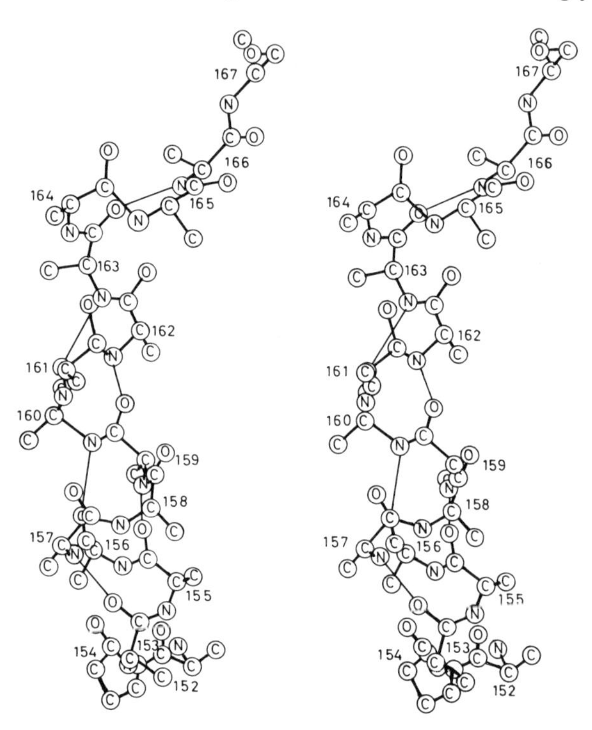

Figure 6.3 Stereoscopic diagram of a distorted helix showing the orientation of carbonyl groups away from the helix axis in carbonic anhydrase
(Reproduced by permission of Dr. A. Liljas and Dr. K. K. Kannan)

insulin B chain (Figure 6.2) are at similar bends; the third also has a positive ϕ angle but the chain here is twisted rather than bent. Similar conformations have also been observed in ferricytochrome C[25].

The first pleated sheet discovered in a globular protein was the antiparallel structure of lysozyme. Similar regions containing just two or three antiparallel chains have been found in ribonuclease[33], nuclease[63], papain[47] and lactate dehydrogenase[26]. However, a more extensive structure containing

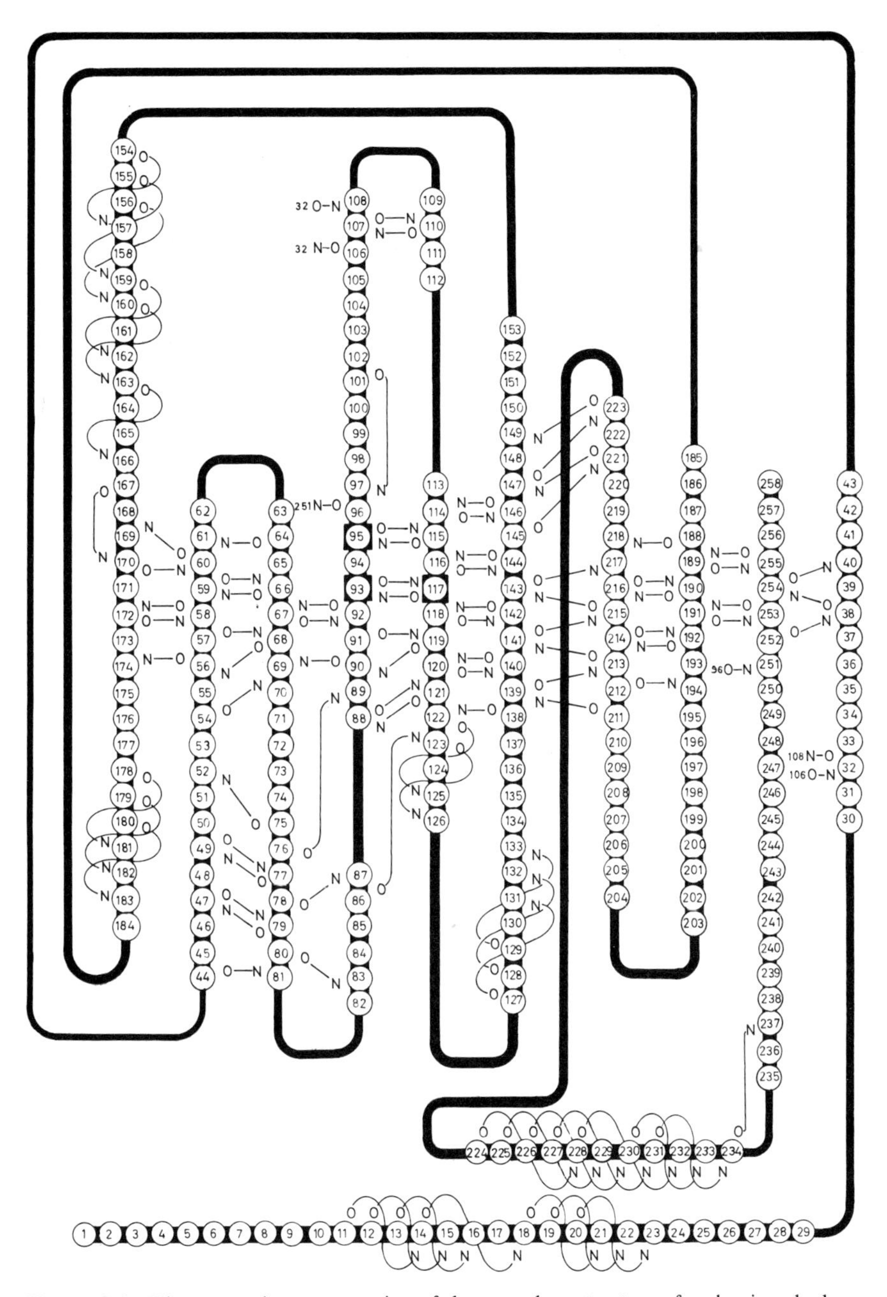

Figure 6.4 Diagrammatic representation of the secondary structure of carbonic anhydrase showing the extensive anti-parallel pleated sheet arrangement
(Reproduced by permission of Dr. A. Liljas and Dr. K. K. Kannan)

40% of the residues in ten mainly antiparallel strands has recently been discovered in carbonic anhydrase[66], the structure of which is shown in Figure 6.4. Chymotrypsin has a structure comprising two regions of antiparallel chains wrapped in very distorted cylinders[40], as shown diagrammatically in Figure 6.5. There are only very short regions where the full classical pleated-sheet structure is found. In contrast extensive parallel pleated sheets

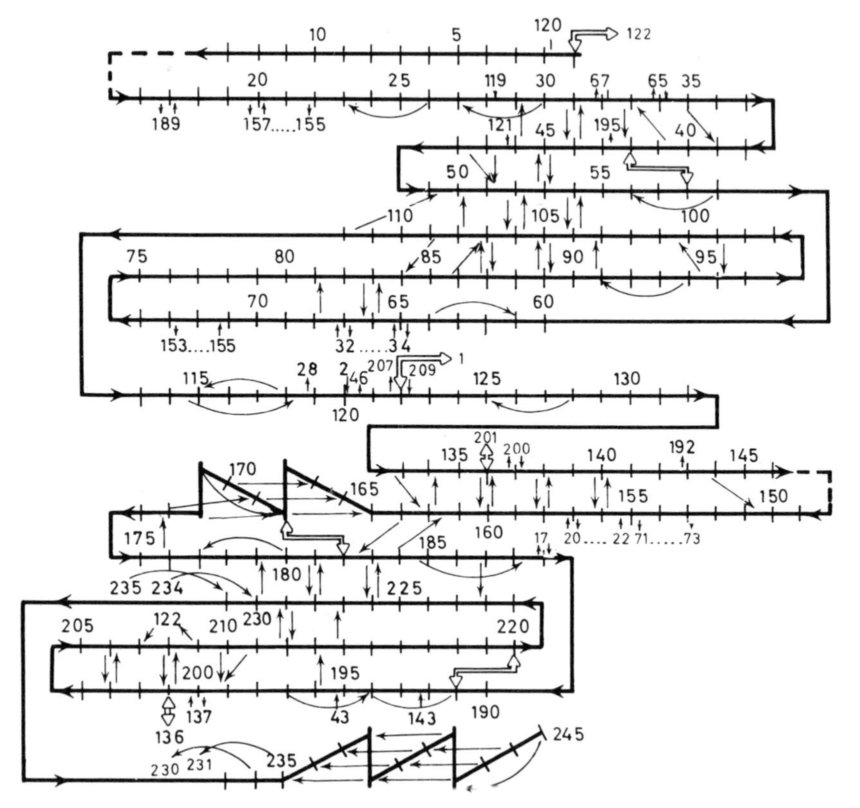

Hydrogen bonds between the main-chain peptide bonds in α-chymotrypsin. The black arrows (⟶) representing hydrogen bonds are drawn from amido to carbonyl group. The white arrows (⇔) represent disulphide bridges

Figure 6.5 Diagrammatic representation of the secondary structure of α-chymotrypsin showing the organisation in two halves
(From Birktoff, *et al*[40] by permission of the Royal Society)

involving five strands exist in both lactate dehydrogenase[65] and subtilisin[49]. In the latter case, 16 of 20 potential hydrogen bonds are formed. Carboxypeptidase[42] contains both types of pleated sheet; 8 strands give rise to 4 parallel and 3 antiparallel pairs in which there are 33 hydrogen bonds between the 45 residues involved. All of the sheet structures tend to be twisted; there is a 120 degrees twist from top to bottom in carboxypeptidase[42]. Most of the structures are also rather distorted from the idealised arrangements. For instance, the 3 antiparallel chains in nuclease[63] form 12 hydrogen bonds, but the structure is convex with a pronounced kink where two adjacent amino acids have their side chains on the same side.

6.2.3 Tertiary structure

The convoluted paths of the polypeptide chains of globular proteins defy generalisation. No protein is typical. The variation in extent and nature of the secondary structure, in size and in function of proteins has precluded standardisation in nature. Figures 6.6 and 6.7 show some examples of protein structures. Nevertheless, some structural features are shared by a limited number of molecules. Also the tertiary interactions of the side groups are presumably based on the same thermodynamic principles.

Several proteins, including the related serine proteases, chymotrypsin[37] and elastase[52] have a structure made up of a number of loops in which the

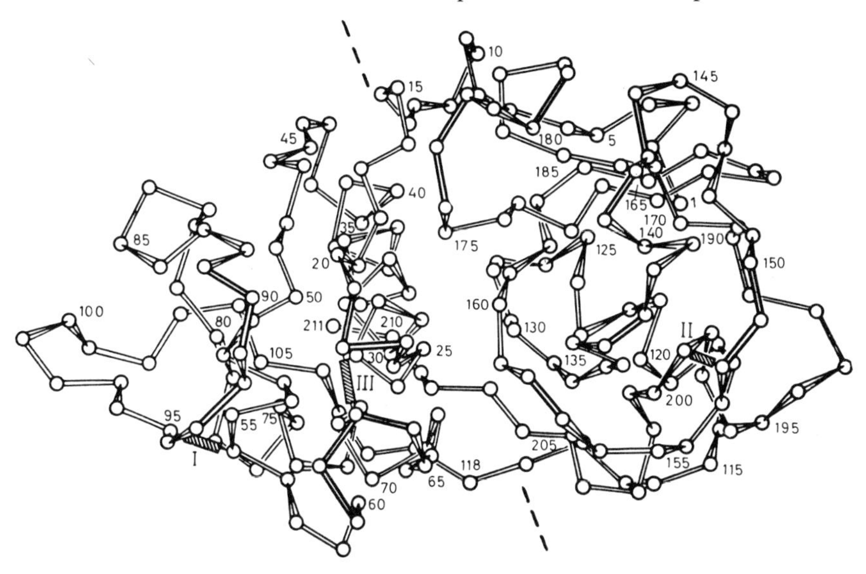

Figure 6.6 The path of the main chain of papain showing two locally organised regions (From Drenth, *et al.*[47] by permission of Macmillan Journals Limited)

polypeptide chain turns back on itself (see Figure 6.5). Often a number of antiparallel type hydrogen bonds are used in stabilising each loop, or alternatively they may be stabilised by disulphide bridges such as in the 'histidine', 'methionine' and 'active serine' loops of chymotrypsin. In many cases these loops are built up into the twisted-sheet structures described above. However, the chains are often not in sequential order. In elastase[52] and chymotrypsin[37], either the serine or the histidine loops must be threaded through pre-folded sections of chain depending on whether the protein is considered to fold from the *N* or *C* terminal ends. Similarly in carboxypeptidase[42] the final hydrogen bonds of the pleated sheet in residues 60–66 and 200–204 cannot be formed until 104–109 and 265–271 respectively have been folded into place. Blow and his colleagues[40] have remarked that on building chymotrypsin with a space-filling model, they initially assembled each loop separately. When these were placed together the packing forces gave rise to a tendency of the loops to spiral round each other in the way observed in chymotrypsin[40], carboxypeptidase[42] and subtilisin[49]. The assembled structure in subtilisin and carboxypeptidase forms a sheet through the centre of the molecule, which appears to be important in the stability of the globular structure.

Lengths of helix may also be important in stabilising the tertiary structure observed. In the small protein, insulin[67,68], the helix of the B chain residues 9–19 appears to be a spacer between the disulphide bridges, B7–A7, and B19–A20. In subtilisin[49] the eight helical segments are approximately parallel to each other, lying within ± 15 degrees of a common direction, and in fact, actually have the same *N*-terminal to *C*-terminal sense. In carboxypeptidase[42] six of the nine helices are on one side of the extensive sheet structure. This has the appearance of greater structural stability in comparison to the other side which contains many residues in random coil. This may be related to the large change of tertiary structure that occurs only in the latter part of the molecule on adding a substrate analogue.

Several authors have remarked on the tendency of the larger proteins to be organised in several distinct pieces. The three locally organised regions of the subtilisin polypeptide chain are ala-1 to gly-100, gly-100 to ala-176 and ala-176 to gln-275[49]. The folding of the polypeptide appears to be more straightforward if these regions attain structures approaching these tertiary arrangements before they are assembled to give the overall globular structure. A similar feature occurs in the papain[47]. Figure 6.6, which is apparently binuclear, residues 1–111 and 112–212 comprising the two halves. It has already been pointed out that α-chymotrypsin and elastase[52,54] can be considered as two cylinders of antiparallel sheet. Shotten and Watson[54] have remarked that organisation within the *N*-terminal half of the molecules, traced from the *N*-terminus to the middle of the chain, resembles that within the *C*-terminus half, traced from the *C*-terminus to the middle of the chain. The antiparallel chains form a twisted sheet which curves round to unite with itself to form the cylinder around which the chains run in an oblique fashion. The cylinders lie above each other and are oriented at right angles to one another. Apart from the obvious interest of these locally organised regions to the folding problem, it is also exciting to find that the active site in all these proteins is at point where the different pieces come together. Thus in subtilisin[49] the active his-64, and ser-221 are on different locally organised regions. Also the active his and ser loops of chymotrypsin[40] and elastase[54] are on separate cylinders. Finally, the active site of papain[47] is the region formed between the two halves. One can imagine that this arrangement may be a consequence of the evolution of these molecules from several simpler globular structures.

The convoluted folding of the polypeptide chains in the way discussed gives rise to a surprisingly compact globular structure. The shape varies from roughly spherical in subtilisin (dia. 42 Å)[49] to ellipsoidal in papain (dimensions: 50×37 Å)[48]. The packing of the amino-acid residues can be most easily considered by drawing circles representing the van der Waals radii of the atoms. However, the accessibility of a particular region to a solvent molecule can only be tested by consideration of the van der Waals radii of the solvent as well as the atoms of the protein. This has recently been discussed in an important paper by Lee and Richards[131]. The static accessibility contours are shown in Figure 6.7 for a section of ribonuclease. The arrow indicates a cavity inside the molecules large enough to accommodate a solvent molecule of radius 1.4 Å. Several such small cavities have been found in lysozyme and myoglobin. Two of these in lysozyme appear to contain water molecules. One in myoglobin is the non-polar cavity that can accept a xenon atom. Cavities probably disappear in the ribonucleases of certain other species where valines next to cavities are replaced by larger

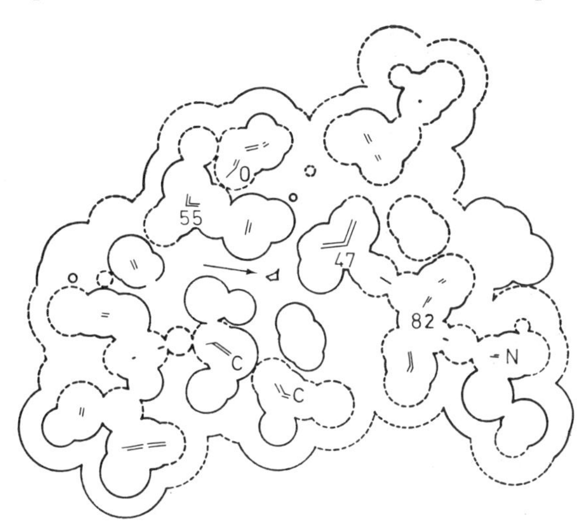

Figure 6.7 The static accessibility contours for some sections of ribonuclease
(From Lee, and Richards[131], by permission of Academic Press)

isoleucines or leucines. The accessibility per non-hydrogen atom of lysozyme and myoglobin are 72, 67 and 69 Å^2 respectively. This is a reflection of the compact overall shapes.

The internal core generally contains well ordered non-polar side-chains; inaccessibility of water to these side-chains is the basis of the so-called 'hydrophobic' forces. However, it is true that a protein cannot be described as 'hydrophobic' side-chains in: 'hydrophilic' out. Although most charged and other polar groups are on the surface, there are also many non-polar groups[7]. Lee and Richards[131] have found that on the average nitrogen and oxygen atoms are 3.5 times as accessible as carbon and sulphur atoms in a native protein. However, the figure for an extended chain is two times. Thus, the change in accessibility in going from an extended chain to threefolded conformation for these different atoms differs by a factor less than two. When the main chain is also considered the factor becomes even smaller. This sounds a note of caution in discussion on the importance of hydrophobic forces in a protein structure. The satisfactory pairing of buried polar groups, in particular the NH and CO of the main chain, is likely to be an important force in the precise organisation of the protein.

6.2.4 Quaternary structure

The structures of three proteins with subunits have been successfully studied at high resolution by x-ray techniques. Lactate dehydrogenase and haemoglobin each have four subunits, although two different subunits, the α- and β-chains, are present in haemoglobin. Zinc insulin on the other hand is a hexamer[67, 68], with two zinc atoms binding the six insulin molecules. X-ray crystallography can give information about the symmetry of these protein aggregates not only from the high-resolution electron-density maps but also either from the crystal symmetry or from diffraction intensities if the symmetry axes of the aggregate are non-crystallographic. The high-resolution electron density map gives the only precise description of the nature of the interactions between subunits.

Rossmann and his colleagues[65] have shown that the muscle form of lactate dehydrogenase of dogfish crystallises in the space group $F422$ with the protein tetramer at the intersection of three mutually perpendicular twofold axes. However, on diffusion of NAD into the crystals of the apo-enzyme there is a reduction of crystal symmetry, which has been interpreted as a small quaternary structural change such that only the twofold axis parallel to c is retained. However, addition of pyruvate gives a ternary complex, enzyme-NAD-pyruvate, of 222 symmetry which has yet a different space group, $P42_12$. This is a beautiful example of the ease with which preliminary crystallographic studies can give information about the relations between subunits. In a similar way the space groups of oxyhaemoglobin[15] and zinc insulin[67] have been used to demonstrate the twofold and threefold axes respectively.

Haemoglobin[15] and insulin[67] both have two subunits in the crystallographic asymmetric unit. The relative positions of these subunits can be determined by an analysis of the origin of the Patterson function. This is achieved by superposing the Patterson function upon itself, and rotating it in all possible ways in order to find the orientation in which there is good overlap in the region about the origin. Rossmann and Blow[132] have developed a reciprocal-space method for carrying out this procedure. The so-called 'rotation function' has been used in the studies of α-chymotrypsin[133], haemoglobin and insulin[134] and the results have since been confirmed by calculation of high-resolution electron density maps. Low-resolution studies have been reported for α-lactalbumin, aldolase and asparaginase. It is clear that the differences of conformation observed between the two crystallographically independent insulin molecules in the high-resolution study did not interfere with the successful identification of the approximate local axis. However, it would be clearly unwise to assume that the local axis found in low-resolution studies with the rotation function is always of exact symmetry.

The absence of a complete sequence in the case of lactate dehydrogenase makes a detailed description of the subunit interactions of this protein impossible. Rossmann and his colleagues[65] note that subunits related by the twofold axis parallel to c have three helices of each molecule in contact with each other. The *N*-terminal helix has contacts with a winglike feature of the subunit related by the *c*-axis in a surprising way. Subunits related by the two-fold axes involve contacts between the extensive parallel pleated sheets.

In the case of haemoglobin and insulin the sequences are known and a detailed analysis of the subunit interactions is possible. In haemoglobin[135] the contacts are of two different kinds. The contact $\alpha_1\beta_1$ is extensive and is made from 34 residues and about 110 atoms come within a distance of 4 Å of each other. The great majority of the interactions are non-polar. However, five hydrogen bonds seem to be quite important on the basis of certain abnormal human haemoglobins.

The contact $\alpha_1\beta_2$ is made up of 19 residues contributing about 80 atoms within a distance of 4 Å of each other. The contacts are again mainly non-polar. On the transition from oxyhaemoglobin to deoxyhaemoglobin movement in the contact $\alpha_1\beta_1$ is slight, with a relative displacement of atoms at the contact of *c*. 1 Å. On the other hand the movement of the contact $\alpha_1\beta_2$ is large. Relative to the α_1 subunit, the β_2 subunit rotates by 13.5 degrees about a screw axis and moves by 1.9 Å; relative displacements of atoms at the contact are as great as 5.7 Å in some cases. Perutz notes that the contact $\alpha_1\beta_2$ is not only smaller than $\alpha_1\beta_1$ but also smoother; it is constructed to allow the two subunits to slide past one another. No contacts between like chains are visible in the electron density maps, but Perutz considers that at low salt concentrations, certain salt bridges involving terminal residues may be possible. An internal cavity extends all the way along the molecular dyad axis, for a length of 50 Å. Perutz describes this as two boxes, each about 20 Å long, 8–10 Å wide and 25 Å deep; one separates the α-chains and the other separates the β-chains from each other. The two boxes are set one above the other with their 20 Å axes at right angles to the dyad and to each other, each box being open at the top and bottom. The internal cavity is lined with many polar groups. The oxyhaemoglobin tetramer resembles a spheroid with length 64 Å, width 55 Å and height 50 Å.

The main subunit interactions in zinc insulin hexamers are also between non-polar groups. The two crystallographically independent molecules of the asymmetric unit form a tightly bound dimer which is exceedingly stable in solution. Contacts between the two molecules of the dimer are between atoms of the insulin B chains only. Van der Waals contacts are made between like residues B24 and B25 phenylalanines and B12 valines in the region of the approximate dyad axis relating the molecules. There is no cavity around this local dyad. Apart from the non-polar interactions, the carboxyterminal residues of the B chains are held in an antiparallel arrangement by four hydrogen bonds between the NH and CO groups of residues B24 and B26 of each molecule. Distortions from exact twofold symmetry are widely distributed in the molecule and very significant deviations are observed in the locality of the dyad. For instance, one of the B25 phenylalanines lies on the approximate dyad axis. The simple assumptions of the Monod symmetry model appear to be inapplicable in the case of insulin.

The three insulin dimers are related by the crystallographic threefold axis to give a hexamer shown in Figure 6.8. This means that the dimers must also be related together by an approximate dyad axis. There are many non-polar contacts between the A chain as well as the B chain residues, but the dimers are less tightly bound to each other than the monomers. A cylindrical cavity lined with polar groups surrounds the threefold axis. Two zinc atoms lie on the threefold axis, each bound to three histidines, one from each insulin dimer. The zinc insulin hexamer has a 'dough-nut' shape of 35 Å height and 50 Å diameter. The hexamer has approximate 32 or D_3 symmetry.

It is very interesting that in the insulin and haemoglobin aggregates, the crystallographic rotation axes are surrounded by cavities while the approximate local axes have a very close packing of residues. It is tempting to

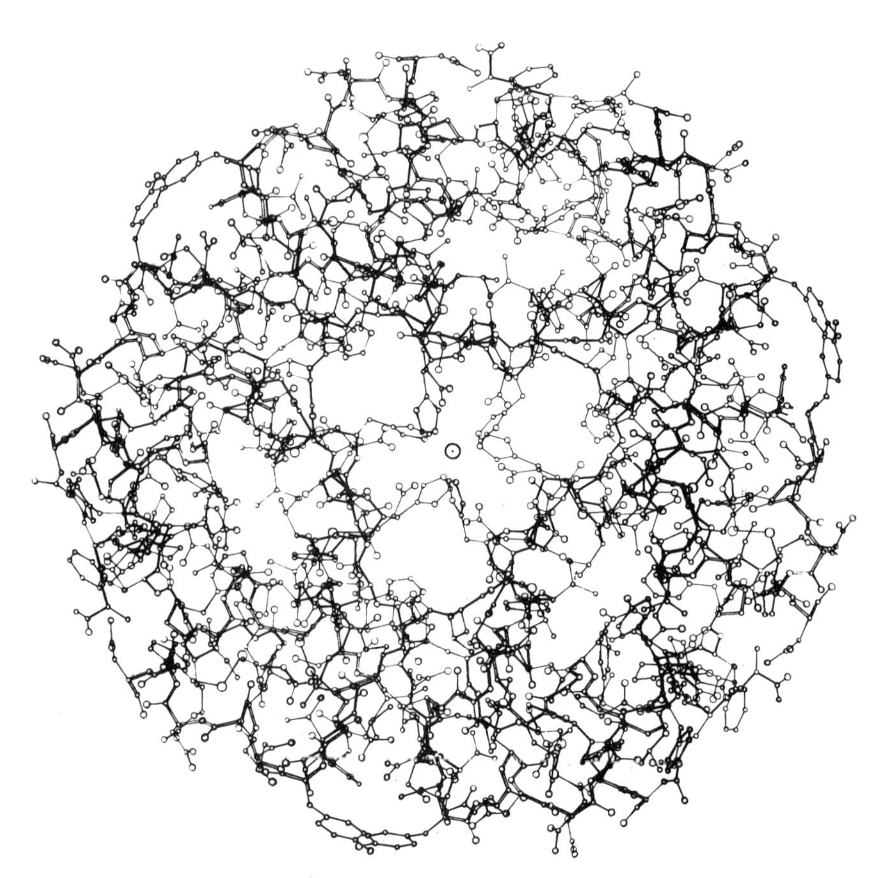

Figure 6.8 The zinc insulin hexamer viewed along the crystallographic threefold axis. The zinc atoms lie on the threefold axis and are each coordinated by three histidines (From Blundell, *et al.*[226], by permission of Macmillan Journals Limited)

suggest that the close packing of equivalent amino-acid side-chains in the vicinity of a local axis may preclude an exactly symmetrical arrangement. Indeed the close juxtaposition of two large groups, such as phenylalanines, may be best achieved, as observed in insulin, by breaking the symmetry and consequently gaining better van der Waals contacts.

The predominantly non-polar interactions between the subunits in these aggregates is consistent with the observations of van Holde that proteins with subunits tend to have a higher percentage of non-polar residues than monomeric molecules.

6.3 THE RELATIONSHIP BETWEEN STRUCTURE AND FUNCTION

The results of x-ray diffraction studies on protein molecules have revolutionised our fundamental understanding of the relationship between the structure and function of biological macromolecules. Once a protein structure has been solved the study of the association of small molecules with the protein may be accomplished relatively easily by means of difference Fourier synthesis based on the phases obtained for the native protein providing that the complexed structure is isomorphous with the native structure. Recently the difference Fourier method has been used to solve the structure of a small molecule bound to elastase[136]. The errors which are likely to arise in difference maps have been analysed[138, 139]. Difference Fouriers will also give indications of changes in conformation, although analysis of these shifts in terms of the conventional formulation[137] is often difficult, except in well defined cases[140], due to the lack of atomic resolution and large thermal vibrations of some of the side chains in protein electron density maps.

In this section we summarise the main contribution of x-ray crystallography to our present understanding of the activity of the molecules listed in Table 6.1 under the four headings, globins, redox systems, enzymes and hormones.

6.3.1. Globins

The available information to date on globin structures, which have been determined independently at high resolution, comprises whale myoglobin[11–14], monomeric insect haemoglobin from *chironomus thummi*[18–20], monomeric blood-worm haemoglobin from *glycera dibranchiata*[21], monomeric lamprey haemoglobin from *petromyzon marinus*[21, 22] and a series of publications on horse oxyhaemoglobin[15], horse deoxyhaemoglobin[16], human deoxyhaemoglobin[17] and a number of human mutant haemoglobins[162, 163].

The first outstanding feature, which has emerged from the structural studies on the globins, is the very great similarity in the overall conformation of these molecules to that of myoglobin (Figure 6.9). It appears that the 'myoglobin fold' has remained almost unchanged throughout evolution and it has been possible to rationalise the conservation of the tertiary structure in terms of the conservation of polar and non-polar residues[141]. In myoglobin the protein shields the haem from the solvent and provides a medium of low dielectric constant within which oxygenation is favoured and oxidisation discouraged, as indeed was anticipated by model studies[142]. However, under the conditions which are used for crystallisation of the globins the iron is readily oxidised from Fe^{II} to Fe^{III}.

The reversible combination of oxygen with the haem group of the globins results in a change in the electronic structure of the iron which is characterised by a transition from a high-spin paramagnetic state to a low-spin diamagnetic state. Myoglobin exhibits a normal hyperbolic oxygen-binding curve. For tetrameric haemoglobin, however, the oxygen affinity depends on the number of oxygen molecules already combined and the protein exhibits a characteristic sigmoid binding curve. From Perutz's latest work, published during 1970 and 1971[143–145], these cooperative effects in haemoglobin appear to be triggered by the small shifts in the iron atom relative to the porphyrin ring

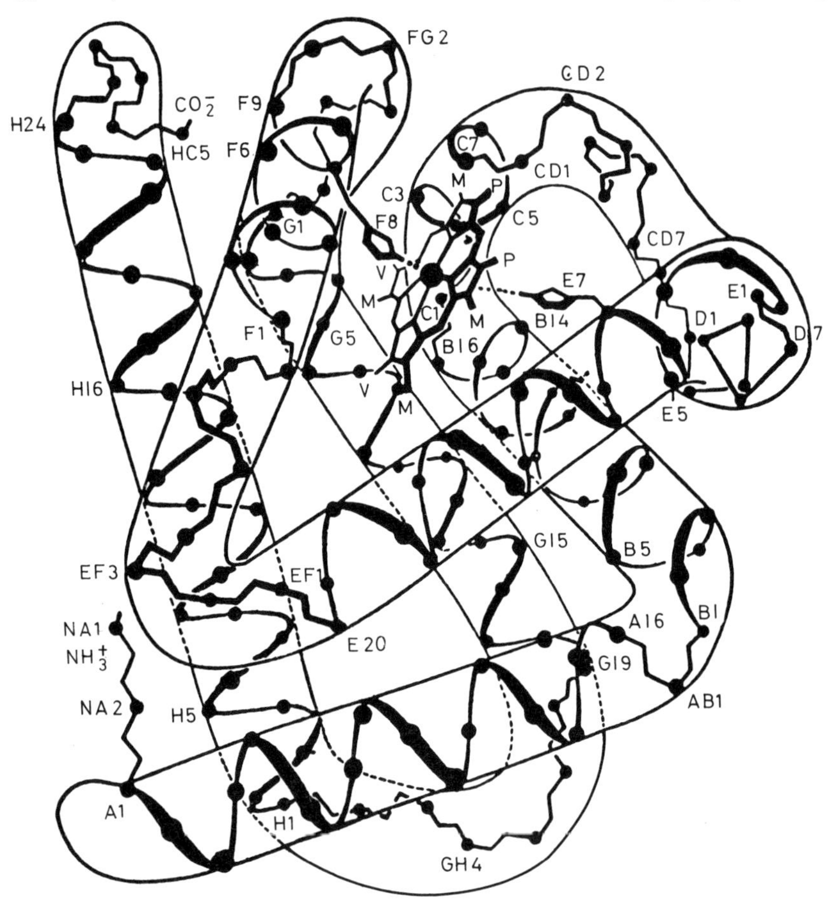

Figure 6.9 A diagrammatic sketch showing the course of the polypeptide chain in myoglobin. The helical regions are numbered A to H from the amino terminal to the carboxyl end. Nonhelical residues between helices are labelled AB, BC, etc., and nonhelical residues at the α-amino and carboxyl end are denoted as NA and HC respectively. The haem lies in a pocket between helices E and F
(From Dickerson, R. E., *The Proteins*, vol. 2, ed. by H. Neurath, by permission of Academic Press (copyright holders))

which occur on uptake of oxygen and the transmission of these shifts to other parts of the molecule, as previously suggested[146].

Small crystal structure studies have shown that in high spin Fe^{III} porphyrin compounds the iron is displaced out of the plane defined by the four porphine nitrogens by 0.4–0.5 Å[147]. The shape of the near planar haem group is slightly puckered. In the structure of the only low-spin Fe^{III} compound studied to date, the iron is in the plane of the haem and centred among the four nitrogens of the porphine[148], as predicted[147]. Porphine structures are notoriously difficult to crystallise and often contain complex disorder effects. As yet no crystal studies on Fe^{II} porphyrins have been published. The ionic radius of low-spin Fe^{III} is believed to be the same, within 0.02 Å, as the ionic radius of low-spin Fe^{III} and hence it is anticipated that the structures of the two porphyrin complexes will be similar. High-spin Fe^{II} has a somewhat greater ionic radius than high spin Fe^{III} and it is therefore expected that in high-spin Fe^{II} haem complex, the iron will be displaced out of the plane of the haem by rather more than the 0.4–0.5 Å observed for high-spin Fe^{III} porphyrin compounds[147].

The x-ray studies on myoglobin[11–14] and erythrocruorin[18–20] were carried out on the met-aqua form in which the iron is in the high-spin Fe^{III} state and the sixth coordination position is occupied by a water molecule. The distances of the iron from the plane of the four pyrrole nitrogens are given in Table 6.2

Table 6.2 The out-of-plane displacement of the iron atom from the porphyrin ring in myoglobin and haemoglobins. In each case the displacement is towards the proximal histidine. Distances are in Å.

Protein	Fe^{III}		Fe^{II}	
	High spin	*Low spin*	*High spin*	*Low spin*
Met-aqua-myoglobin*	0.45			
Deoxymyoglobin[149] *			0.45	
Oxymyoglobin[150] *				0.45
Azide-met-myoglobin[151] *		0.45		
Met-aqua-erythrocruorin[20]	0.3			
Deoxyerythrocruorin[20]			0.3	
Carbonmonoxyerythrocruorin[20]		0.05		
Cyano-met-lamprey haemoglobin[21, 22]		0.01		
Met-aqua-horse haemoglobin[143]	0.3			
Deoxyhaemoglobin[143]			0.75	

*Watson, H. C. (Private communication). *Note*: The original value of 0.3 Å quoted for the out-of-plane displacement of the iron in myoglobin referred to the distance of the iron from the best mean plane through the porphyrin and the iron. The figure of 0.45 Å represents the distance of the iron from the best mean plane of the porphyrin.

and these values are in reasonable agreement with those observed in small crystal structures. Three-dimensional difference Fourier transformations at high resolution for deoxymyoglobin[149] and deoxyerythrocruorin[20] have shown that in the deoxy-form there is a loss of the haem-bound water molecule, resulting in a 5-coordination state of the iron, and no indication of shifts or conformational changes in the haem group or in amino acids lining the haem pocket (Table 6.2). In erythrocruorin[20], the authors estimate that changes of 0.2 Å for light atoms and 0.1 Å for the iron would have been detected.

Several different forms of low-spin monomeric globins have been studied crystallographically. In the difference Fourier analysis of carbonmonoxyerythrocruorin[20], and in the direct determination of the structure of cyano-met-lamprey haemoglobin[21, 22] (both Fe^{III}; low spin), the iron was found to be within 0.05 Å of the plane of the four pyrrole nitrogens (Table 6.2). Likewise two-dimensional analysis of the monomeric haemoglobin from the annelid worm *glycera dibranchiata* indicated that shifts in the position of the iron and the haem accompany changes in the spin state of the iron[153, 154] although it is not yet possible to analyse these changes in detail. In summary, it appears that in each of the monomeric haemoglobins studied so far, there are indications of conformational changes in the iron and the porphyrin which accompany a change in the spin state of the iron and the magnitude and direction of these changes are in agreement with theoretical expectations. x-Ray analysis of two low-spin derivatives of myoglobin, azide myoglobin[151] (Fe^{III}; predominantly low spin) and oxymyoglobin[150] (Fe^{II}; low spin) however revealed no changes in conformation in the haem on change of spin state, although preliminary two dimensional studies on cyano-met-myoglobin and hydroxy-met-myoglobin[152] indicated that some change may occur with these two derivatives. It will be interesting to see if the absence of shifts in myoglobin are confirmed by structural studies on other myoglobins. A high-resolution map of seal myoglobin has recently been obtained (H. Scouloudi, private communication) but details are not yet available.

How do these results relate to the recent experimental results for tetrameric haemoglobins? The solution of the structure of haemoglobin at 2.8 Å resolution[15] showed that the four subunits are assembled into a compact tetrahedral structure with extensive areas of contact between the α_1 and β_1 subunits and the α_2 and β_1 subunits. Detailed comparison[155] of the electron density maps, determined independently, of human deoxy- and horse oxy-haemoglobin at 5.5 Å resolution, with the aid of a computer program, showed that the tertiary structure of the individual subunits are the same to within the limits that could be detected at low resolution but that there are large differences in quaternary structure. Analysis showed that the α_1 and β_1 subunits rotate almost in concert so that the $\alpha_1\beta_1$ contact undergoes only slight changes but the α_2 and β_1 subunits rotate in opposite directions leading to large relative movements of atoms at the $\alpha_2\beta_1$ contact of up to 6 Å. The solution of the structures of horse deoxyhaemoglobin[16] at 2.8 Å and human deoxyhaemoglobin[17] at 3.5 Å resolution has revealed some of the finer stereochemical details associated with these changes.

In horse deoxyhaemoglobin (Fe^{II}: high-spin), the iron atom is found to be *c*. 0.75 Å out of the plane of the haem towards the proximal histidine, in contrast to deoxymyoglobin and deoxyerythrocruorin (Table 6.2). Crystals which were originally thought to be oxyhaemoglobin are now known to be met-aqua-haemoglobin. In horse met-aqua-haemoglobin (Fe^{III}: high-spin) the iron is 0.3 Å from the plane of the haem (Table 6.2). The mean distance of the proximal histidine from the plane of the porphyrin also changes rather dramatically from *c*. 2.9 Å in deoxyhaemoglobin to 2.3 Å in met-haemoglobin[145]. These shifts in the iron and haem linked histidine give rise to changes elsewhere in the structure, and these have been further investigated in a chemically modified haemoglobin whose quaternary structure is locked in the oxy-form[156]. The modified haemoglobin has allowed the details of the tertiary structure changes to be probed in the absence of the quaternary structure changes. The results indicate that the binding of ligand results in a

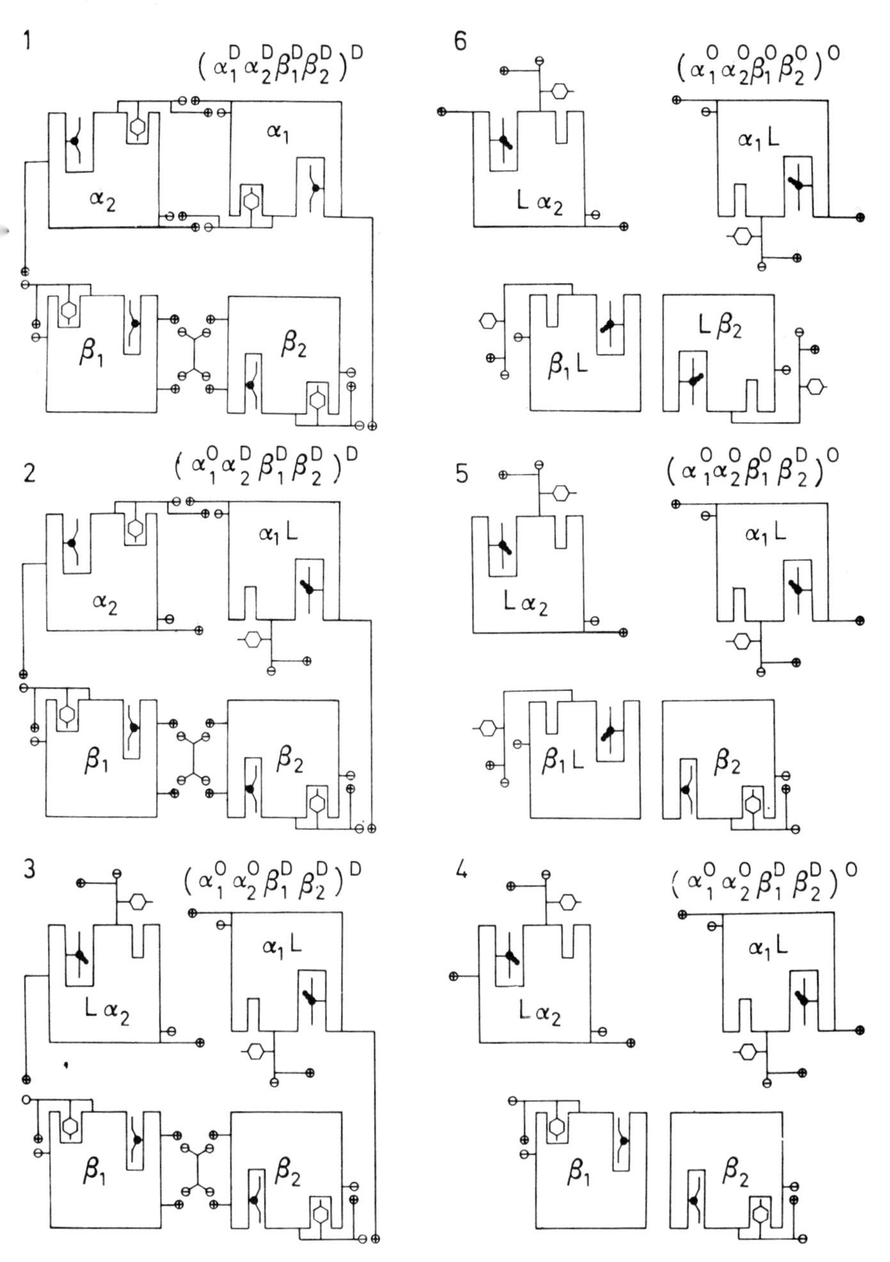

Figure 6.10 Diagrammatic sketch showing one possible sequence of steps in the reaction of haemoglobin with oxygen. 1. Deoxyhaemoglobin with all salt bridges intact and with one molecule of diphosphoglycerate clamped between the two β-chains. At steps 1–2 and 2–3 the α-chains are oxygenated. The tyrosine pockets are narrowed, the tyrosines expelled and the salt bridges with the partner α-chains are broken. In step 3–4 the quaternary structure clicks to the oxy-form, accompanied by expulsion of diphosphoglycerate, and breakage of the salt links between subunits α_1–β_2 and α_2–β_1. Note that the internal salt bridges of the chains are intact and their haems too narrow to admit ligands. At steps 4–5 and 5–6 the β subunits react in turn, accompanied by widening of the haem pockets, narrowing of the tyrosine pockets and rupture of the internal salt links from the *C*-terminal histidines to aspartates-94. *Note*: The change in quaternary structure could in fact take place at any stage of the reaction, and the co-operative mechanism is independent of the sequence in which the individual subunits react (From Perutz[143], by permission of Macmillan Journals Limited)

movement of helix F towards the centre of the molecule which narrows the pocket between it and helix H and the expulsion of the penultimate tyrosine of the α-chain from its pocket (Figure 6.9 and 6.10). Expulsion of the β-chain tyrosine is prevented by the chemical modification. Examination of the structure of deoxyhaemoglobin reveals that the penultimate tyrosines are firmly held by hydrogen bonds to the carbonyl oxygen of val-FG5 as in myoglobin[14] and recent chemical and crystallographic experiments[157–159] have indicated that this hydrogen bond is essential for the expression of co-operativity. In deoxy-haemoglobin the adjacent *C*-terminal residues are also firmly held in salt bridges, as indicated schematically in Figure 6.10. In met-aqua haemoglobin the *C*-terminal residues and the penultimate tyrosines are barely visible in the electron density map and these residues appear to be free to take up several positions. Perutz[143, 145] has suggested that the salt bridges form the main constraints which distinguish deoxy- from met- and, by inference, from oxy-haemoglobin.

The essence of the mechanism may be summarised as follows: the binding of a ligand to the iron in deoxyhaemoglobin results in a shift of the iron and the proximal histidine which, by a mechanism as yet to be elucidated, causes the expulsion of the penultimate tyrosine from its niche between helices F and H. The expelled tyrosine pulls the *C*-terminal residue with it, rupturing the salt links that had held the reacting subunit to its neighbours in the deoxy-form. The rupture of each salt bridge removes one of the constraints holding the molecule in the deoxy conformation and tips the equilibrium in favour of the oxy form. A more detailed diagrammatic sketch of the proposed possible steps is shown in Figure 6.10.

The theory is attractive in that it also accounts for the Bohr effect by postulating that the release of the Bohr protons on oxygen binding arises from the opening of the salt bridges made by the *C*-terminal residues[144]. From a series of elegant biochemical[160] and crystallographic[145] experiments, it has been possible to show that removal or blocking of the salt bridges leads to an increase in oxygen affinity, reduction in co-operativity and modification of the Bohr effect although the rupture of the salt bridges does not significantly perturb the structure.

A wide variety of mutant haemoglobins are known and have been characterised. Many of these involve single amino-acid changes and this natural supply of slightly different proteins has been exploited by the crystallographers. From studies by difference Fourier transformations at 3.5 Å resolution, Greer and Perutz[162, 163, 145] have been able to elucidate the stereochemical consequences of several mutations and to explain the changes in oxygen affinity in terms of Perutz's hypothesis. The good correlation between structural and functional abnormalities in these mutants and the ability to provide stereochemical interpretations of other abnormal haemoglobins[161] provides support for the biological significance of the crystallographic studies.

The rather large displacement of the iron from the porphyrin (Table 6.2), together with the salt links which characterise the structure of deoxyhaemoglobin support the results of chemical and spectroscopic studies[165, 166]. These have indicated that deoxy-haemoglobin is constrained in a way which lowers the affinities of the subunits for oxygen while the affinity of oxyhaemoglobin is essentially similar to that of the isolated subunits. Williams[164] has argued that the stereochemistry of the protein as a whole and the grouping of the porphyrin and the ligands to the iron are intimately connected so that changes in the metal can control the protein and changes in the protein can control the metal. This certainly seems to be true for haemoglobin.

The explanation for the stereochemistry of the co-operative effects in haemoglobin is based on the assumption that the structure of oxy-haemoglobin is closely similar to that of met-aqua-haemoglobin and certainly the determination of a low-spin Fe^{II} haemoglobin, such as carbonmonoxy-haemoglobin, will form one of the crucial tests of the hypothesis. This is especially important in view of recent n.m.r. experiments which have indicated that certain parts of the structure of met-haemoglobin may be more similar to deoxy- than to oxyhaemoglobin[167]. Although several details still remain to be worked out, Perutz's hypothesis accounts for many of the important properties of haemoglobin and its general consistency with all the crystallographic results combine to make it readily likeable.

6.3.2 Redox systems

The structures of five redox proteins are known to date. These include two haem proteins, cytochrome C[25, 26, 168, 169] and cytochrome b_5 [27], two non-haem iron proteins, rubredoxin[23, 24], and the high potential iron protein[28, 170], and one flavoprotein, flavodoxin[29, 171]. Each of these proteins is involved in an electron-transport system and in this sense an understanding of redox systems requires an understanding of multi-enzyme complexes and poses problems concerned with biological organisation in more complex systems than the individual proteins studied so far. Nevertheless, considerable structural information has emerged from these studies and represents the first important step towards a fuller understanding.

6.3.2.1 Cytochrome C and cytochrome b_5

The redox potentials of these two haem proteins are very different: (Cytochrome C; +0.260 V, Cytochrome b_5; +0.020 V) and this data suggests that the protein environment in cytochrome C destabilises the ferric state with respect to the ferrous state to a far greater extent than in cytochrome b_5. The high-resolution x-ray studies have shown that the structures of these two proteins and the coordination to the iron are indeed very different. In cytochrome C the haem group is located in a crevice tightly enveloped by hydrophobic groups with the propionic acid residues pointing towards the interior of the molecule[25]. In contrast to the globins, the haem is covalently linked to the protein by thioether linkages from the vinyl side-chains of the porphyrin to cysteinyl residues 14 and 17. The iron is coordinated to a histidine residue at the fifth coordination site and to a methionine residue at the sixth coordination site. As yet no detailed bond lengths are available. In cytochrome b_5 the haem is again located in a hydrophobic crevice but in this protein the propionic acid groups are on the surface[27]. The iron is coordinated to two histidine residues and these together with the four nitrogens of the porphine ring result in coordination of the iron by six nearly equivalent nitrogens.

In the structure of ferricytochrome C, Dickerson *et al.*[25] have observed two 'channels' filled with hydrophobic side-chains which lead to the right and left from the haem to the surface of the molecule. Each channel contains at least two aromatic rings in roughly parallel orientations and is surrounded by a cluster of positively charged lysines at the surface. Between the two positive regions there is a cluster of nine negatively charged acidic groups. The significance of these charge clusters is not yet known but it has been proposed that they are involved in the binding to other macromolecular complexes, including the cytochrome oxidase system.

The x-ray structure of tuna heart ferro-cyrochrome C has now been achieved at 2.45 Å resolution[26]. Comparison with the 2.8 Å map for tuna ferricytochrome C shows that the haem ligands remain the same in both oxidation states as suggested by n.m.r. experiments[172]. Although the overall conformations of the two molecules are similar, there are considerable changes in several parts of the flexible region of the molecule, the most dramatic of which involve residues 77–83 which form part of one of the evolutionary invariant regions of the molecule. In an attempt to elucidate the pathway by which an electron is transferred from the surface to the iron, Winfield[173] has proposed a free-radical mechanism which involves the haem-linked methionine sulphur and the phenolic oxygen of tyrosine 67. This mechanism has been discussed in terms of the structure of ferricytochrome C [169] and appears possible in the light of the new structural information.

6.3.2.2 Non-haem iron proteins

The structure of rubredoxin (M.W. 6000) at 1.4 Å has revealed that the single iron atom is bound to the sulphur atoms of four cysteine residues in a tetrahedral arrangement[24]. At 2.5 Å no deviations from regularity were observed[23] but at 1.4 Å resolution, where bond lengths of the iron–sulphur links could be determined to 0.04 Å, further examination revealed that one iron–sulphur bond was considerably shorter (1.97 Å) than the others (2.31, 2.33 and 2.39 Å) and that the angles at the iron vary from 118 to 102 degrees. The particular properties of a metal in a metal–protein system are likely to depend most critically on the small deviations from standard geometry caused by the protein environment[164]. It is salutary that in the case of rubredoxin such deviations could be determined at 1.4 Å resolution although not at 2.5 Å resolution.

The crystallographic studies on the high-potential iron protein commenced with a partially successful project aimed at locating the positions of the iron atoms using the anomalous components of the scattered x-rays[170]. The structure has now been determined at high resolution by heavy-atom isomorphous replacement methods[28]. The four iron atoms are arranged in a tetrahedral array with the four inorganic sulphur atoms forming a second concentric tetrahedron of similar dimensions and each sulphur atom making equivalent bonds to three iron atoms. Each iron is connected to the polypeptide chain by a cysteine sulphur link. The protein is so called because of its high redox potential of +0.35 V.

6.3.2.3 Flavodoxin

Until recently the flavoproteins had been neglected by the crystallographers. However, Ludwig and her colleagues[29] are now well advanced with studies on a flavodoxin purified from *Clostridium MP*, a FMN-containing flavoprotein of molecular weight 16 000 that can substitute for both plant and bacterial ferredoxins. All three oxidations states crystallise into the same space group but there are slight changes in cell dimensions and significant differences in diffracted intensities[171]. An electron density map of the semiquinone form has been calculated at 3.25 Å resolution and a provisional fitting of the polypeptide backbone has been accomplished, although the complete interpretation awaits sequence data and higher resolution. An independently determined map of the oxidised form has been calculated at 4 Å resolution.

6.3.3 Enzymes

Out of the 800 or so enzymes listed in the Enzyme Handbook[174] the structures of 14 enzymes have been determined in detail by x-ray crystallography (Table 6.1). Some of these are identical in amino-acid sequence and others are closely related. Although it is impossible to provide generalisations from such a small sample, which nevertheless contains considerable diversity, certain common features have emerged from the structural studies. We will summarise these features and then give a brief account of the individual enzymes.

The results of x-ray diffraction experiments have shown that the specificity of an enzyme towards its substrate may be elucidated fairly readily from binding studies on various inhibitors and substrate analogues. The complementarity observed between the enzyme and inhibitor in these studies provides striking confirmation of the lock-and-key analogy proposed by Fischer[175] in 1894. Active sites appear to be located either in a cleft on the enzyme surface which separates two halves of the molecule as in lysozyme[176], ribonuclease[34], papain[47], and lactate dehydrogenase[65] or in a depression or cavity as in chymotrypsin[177] and other serine proteinases[52, 60, 61, 178], carboxypeptidase[45], staphylococcus nuclease[63] and carbonic anhydrase[66]. The substrate is held in place through a number of specific hydrogen bonds and non-polar contacts. Contacts not only involve interactions between the substrate and the side chains of the protein but also involve specific hydrogen bonds to the main-chain peptide groups, as for example in the lysozyme interaction with the acetamido side-chain of the oligosaccharide substrate[176] and the subtilisin and γ-chymotrypsin interactions with the chloromethylketone substrate analogues[60, 178]. In these cases the specificity is uniquely determined by the folding of the protein. Many of the enzymes described in the following sections act on polymeric substrates and contain on their surfaces multiple sites of slightly differing specificity. In the serine proteinases, carboxypeptidase and ribonuclease the most specific site for inhibitor binding is immediately adjacent to the bond cleaved. In lysozyme and papain this site is one removed from the site of action.

The time required for a set of x-ray intensity measurements is of the order of 2–4 days and this vastly exceeds the turnover number of all enzymes (30 moles of substrate per minute in the case of lysozyme). It is therefore not possible to use x-ray methods to study the *productive* binding of substrates. Three possibilities are then open to the crystallographer. (i) The enzyme can be chemically modified so that it is catalytically inactive but still binds substrate. Although a number of elegant specific chemical modifications of active site groups have been achieved, which have provided information on the importance of these groups in the catalytic process, such modifications usually result in some alteration in the structural details of the binding properties. (ii) The substrate can be modified so that it still binds to the enzyme but is not cleaved. Perhaps the prettiest example of this type of approach is that given by ribonuclease-S binding studies with a phosphonate dinucleotide in which the phosphate ester linkage is replaced by a CH_2 group[36]. However, synthesis of equivalent compounds for other enzymes poses formidable problems for the organic chemist. (iii) In the most popular and universal approach, the crystallographer combines the results of inhibitor binding studies and the results of chemical experiments with a careful examination of the model and attempts to deduce the most likely mode of substrate binding from model-building studies. It is assumed that no further conformational changes take place in the enzyme on substrate binding beyond those observed on inhibitor binding. This approach was pioneered by Phillips in his work on lysozyme where the geometry of the active site cleft provided stringent stereochemical limitations so that only one binding mode, the correct one, appeared plausible[32, 179].

The structural studies on the active sites of enzymes have provided a framework within which it is possible to investigate proposals for mechanistic pathways. As a result of chemical and crystallographic experiments the mechanism of action of two enzymes, lysozyme[176, 180, 181] and ribonuclease[36], are known in detail and the essential features of the mechanisms of chymotrypsin[39] and the other serine proteinases and of carboxypeptidase[45] are also understood. In lysozyme the rate enhancement is effected by distortion of substrate and general acid–nucleophilic attack by two acid groups; in ribonuclease, the precise orientation of two histidine residues with respect to the substrate enables them to act, one as an acid the other as a base, in the hydrolysis of phosphate ester linkages in ribonucleic acids; in chymotrypsin and related enzymes hydrolysis appears to be brought about by nucleophilic attack combined with general acid catalysis, in which 'electronic strain'[10] from the enzyme results in increased reactivity of the catalytic groups; in carboxypeptidase, hydrolysis is effected by a zinc-carbonyl mechanism together with general acid hydrolysis by a tyrosine residue whose position is determined by an 'induced-fit' mechanism[182] and either general base or nucleophilic attack by a nearby carboxyl. These enzymes provide examples of many of the theories which have been put forward to account for enzyme action and to some extent they all illustrate a single catalytic principle which transcends their individual idiosyncracies. This principle, which was foreseen and proposed in various forms by many authors[10, 164, 183–186], suggests that the folding of the protein as a whole provides a reversible energy store through which certain regions of the protein may be activated either by steric effects or through electronic effects. The binding energy gained by interaction of substrate may then be used to further enhance the reactivity of these groups and/or to activate the substrate. In either case both the specificity and the overall conformation of the enzyme are critically involved in providing the correct orientation and activation of the catalytic groups at the active site. It is from an understanding of these structural principles that it is possible to propose, although not entirely explain, mechanisms for the origin of rate enhancement.

The recent results of protein crystallography have provided a stimulus for many new biochemical and chemical experiments and several of the specific findings of the structural studies can be tested by a variety of methods, including n.m.r. In general, there is an excellent correlation between the solution and crystallographic studies.

6.3.3.1 Lysozyme

Studies of hen egg-white lysozyme provided the first example of the way in which the results of protein crystallography may be used to elucidate the specificity and mechanism of enzymes. The structure[30–32] and proposals for the catalytic mechanism[146, 179, 180] have been recently reviewed[181, 187, 188].

A schematic diagram of the proposed binding of a hexasaccharide substrate which is based on inhibitor binding studies and on model-building experiments is shown in Figure 6.11. The hexasaccharide occupies six sites labelled A,B,C,D,E,F. In order to relieve overcrowding between the enzyme and sugar at site D, it was proposed that the sugar residue in this site was distorted. In the crystallographic hypothesis, hydrolysis occurs between sites D and E through general acid catalysis by glu-35 leading to the formation of a carbonium ion intermediate which is favoured and stabilised both by the nearby negative charge of asp-52 and by the distortion of sugar residue D to the half-chair conformation. Subsequent chemical experiments in a number of laboratories[181, 187, 188] have supported the crystallographic proposals for the specificity of lysozyme and in particular these studies have provided evidence for the distortion of the sugar in site D and the existence of sites E and F which had been previously deduced by model-building studies alone. The stereochemistry of site D has also been explored in more recent crystallographic studies[189]. Similar studies with γ-gluconolactone have illustrated the difficulties involved in predicting binding modes even when considerable structural information is available[189].

Knowledge of the crystal structure has greatly facilitated more recent chemical experiments aimed at elucidating the catalytic mechanism as well as substrate specificity. Chemical studies indicated that oxidation of lysozyme with iodine results in complete inactivation[190]. Subsequent crystallographic work[191] at high resolution revealed that iodine catalyses the oxidation of trp–108 to form an oxindole–ester with the nearby carboxyl side-chain of

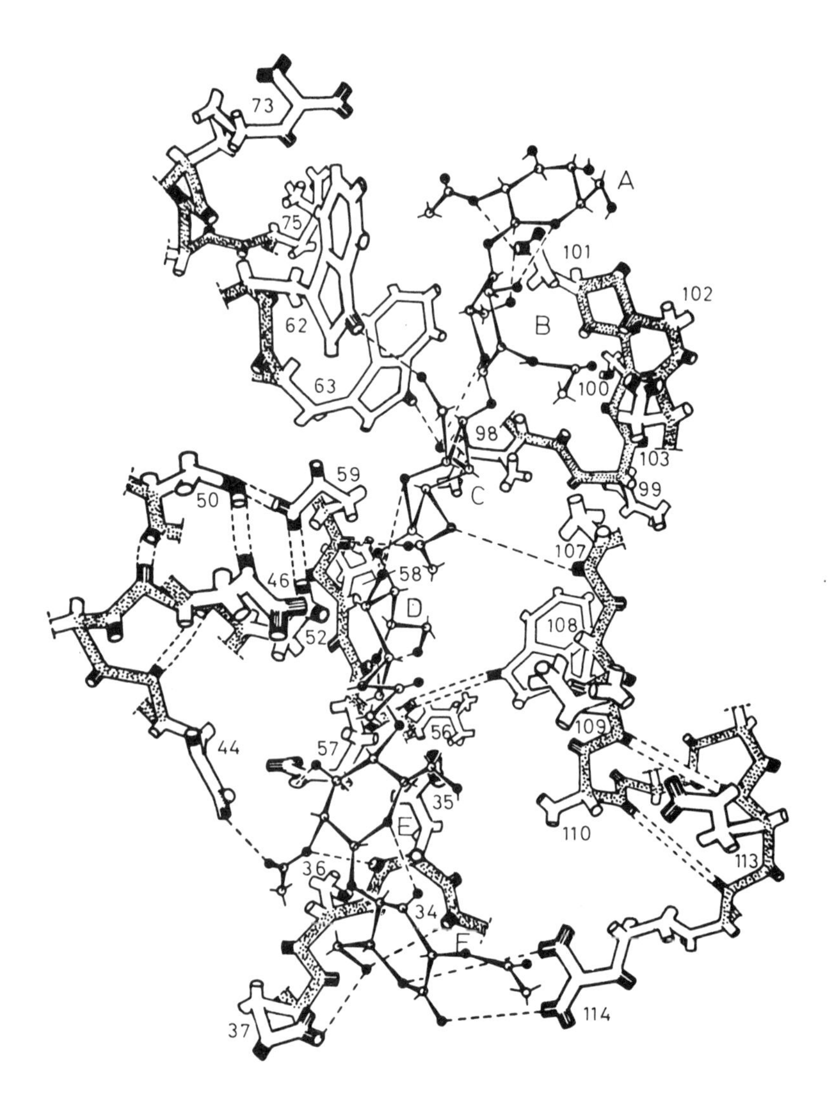

Figure 6.11 A diagrammatic representation of the atomic arrangement in the lysozyme molecule in the neighbourhood of the cleft with a hexa-*N*-acetylchitohexose shown bound to the enzyme. The main polypeptide chain is shown speckled, the NH and CO are indicated by line and full shading respectively. Sugar residues A, B and C are as observed in the binding of tri-*N*-acetylchitotriose. Residues D, E, and F occupy positions inferred from model building. The linkage hydrolysed by the action of the enzyme is between residues D and E
(From Phillips, D. C. (1967). *Proc. Nat. Acad. Soc.*, **57**, 493, by permission of National Academy of Sciences)

the active-site residue glu-35. The modification of glu-35 and the small movement of trp-108, which is involved in binding substrate, account for the loss of activity. Other solution studies have involved the synthesis of model substrates for kinetic studies, the measurement of the deuterium isotope effect[192], the design of new reagents for modification of carboxyl groups [193–195], the synthesis of a possible transition-state analogue[196], and the increasing use of n.m.r. to probe the microenvironment of substrate and inhibitor groups on the enzyme[188]. It is significant that many of these studies could not have been accomplished five years ago nor could meaningful experiments have been designed in the absence of the crystallographic data. They illustrate the fruitful interaction which is now taking place between crystallographers, chemists and biochemists and at the same time demonstrate the difficulties encountered in an attempt to account quantitatively for catalysis. This still remains to be achieved for lysozyme.

6.3.3.2 Serine proteinases

The structures of four mammalian serine proteinases have been obtained by high resolution x-ray studies: α-chymotrypsin at 2 Å[37–41], γ-chymotrypsin at 2.8 Å[58–60], elastase at 3.5 Å[52–55] and trypsin at 2.7 Å[61] and the structure of one of the zymogen precursors chymotrypsinogen A has been obtained at 2.5 Å[56, 57]. γ-chymotrypsin is identical to α-chymotrypsin in primary sequence and is defined operationally as the high-pH crystalline form of chymotrypsin. In addition the structures of two bacterial serine proteinases have also been determined: Subtilisin BPN′ at 2.5 Å[49–51] and Subtilisin Novo at 2.8 Å[62]. All of these enzymes are characterised by their possession of a uniquely reactive serine residue, ser-195 in chymotrypsin, which reacts specifically with diisopropyl fluorophosphate, resulting in loss of enzymatic activity. The amino-acid sequences of chymotrypsin, elastase and trypsin show a high proportion of amino acids which are either identical or closely similar and these enzymes may be classed as a homologous series which have diverged from a common ancestral gene. The amino-acid sequence of the subtilisins is unrelated to that of chymotrypsin. The structure and function of these proteins are reviewed in reference 197.

In 1966 from the standpoint of biochemical experiments, chymotrypsin formed one of the most thoroughly studied enzymes and yet the x-ray results have demonstrated stereochemical features which were entirely unsuspected. The first striking feature which emerged from the structural studies of α-chymotrypsin was that the α-amino group of ile-16, whose formation during the activation process of the zymogen is essential for activity of the enzyme, forms an ion pair with asp-194 in a region of the molecule which is shielded from solvent[37]. Initially it was thought that this ionic interaction might help to stabilise the active site serine-195 in its correct orientation, but the recent results with chymotrypsinogen (discussed later) suggest that this is not so. In elastase a similar internal ion pair is made between val-16 and asp-194 but it does not appear to be essential for activity against small substrates[55].

The second striking feature involves residue ser-195, the active-site serine.

It is placed so that the hydroxyl group is within hydrogen-bonding distance of the $N^{\varepsilon 2}$ atom of his-57 which in turn is hydrogen bonded through its $N^{\delta 1}$ atom to the $O^{\delta 2}$ of asp-102 [39]. The $O^{\delta 1}$ atom of asp-102 is also hydrogen bonded to the hydroxyl of ser-214 and to the peptide nitrogen of his-57 [41] (Figure 6.12). His-57 and ser-195 are on the surface of the molecule but asp-102 is shielded from the solvent by a number of non-polar residues. It has been proposed that this hydrogen-bonding arrangement acts as a charge relay system whereby the effect of burying the negatively charged carboxylate group of asp-102 is partially compensated by the hydrogen bonds and results in a transfer of the charge to the serine at the surface of the molecule.

A series of crystallographic studies on the binding of virtual substrates,

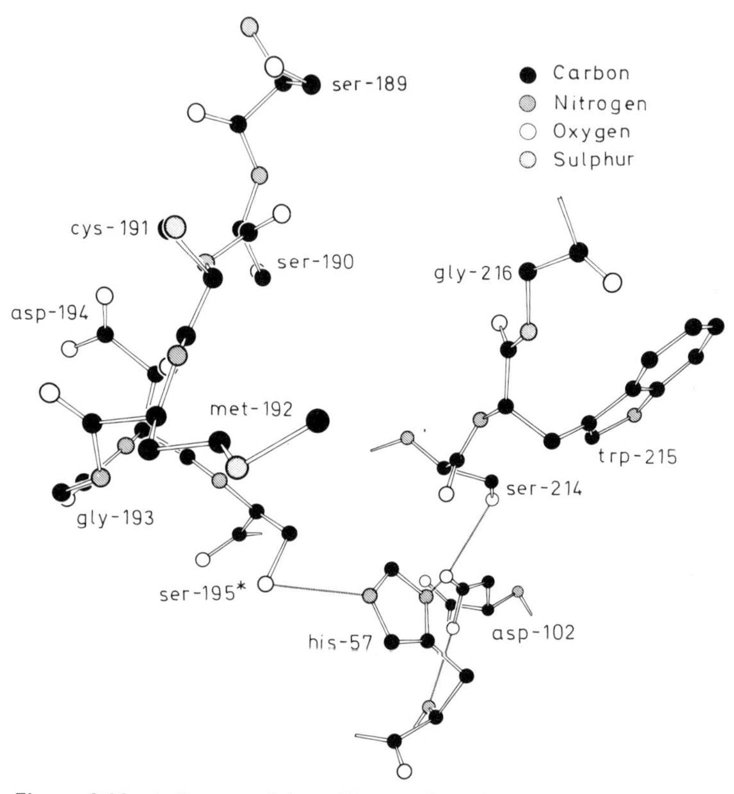

Figure 6.12 A diagram of the residues at the active site of α-chymotrypsin showing the hydrogen bonding scheme from ser-214 to asp-102 to his-57 to ser-195
(Reproduced by permission of Dr. J. Kraut)

such as formyl-1-tryptophan, have revealed the nature of the specificity site and have shown how the sensitive bond of a real substrate might be orientated[177]. The aromatic side-chain of the inhibitor is positioned in a hydrophobic pocket adjacent to the active-site residues which accounts for the specificity of the enzyme for peptide bonds following aromatic side-chains.

Based on the crystallographic data and the chemical evidence for the existence of a covalent acyl-enzyme intermediate in the reaction pathway[198], Blow *et al.*[39] have proposed the following mechanism for the catalysis (Figure 6.13). The charge relay system results in a partial withdrawal of the proton from the side chain of ser-195 towards his-57. During catalysis this

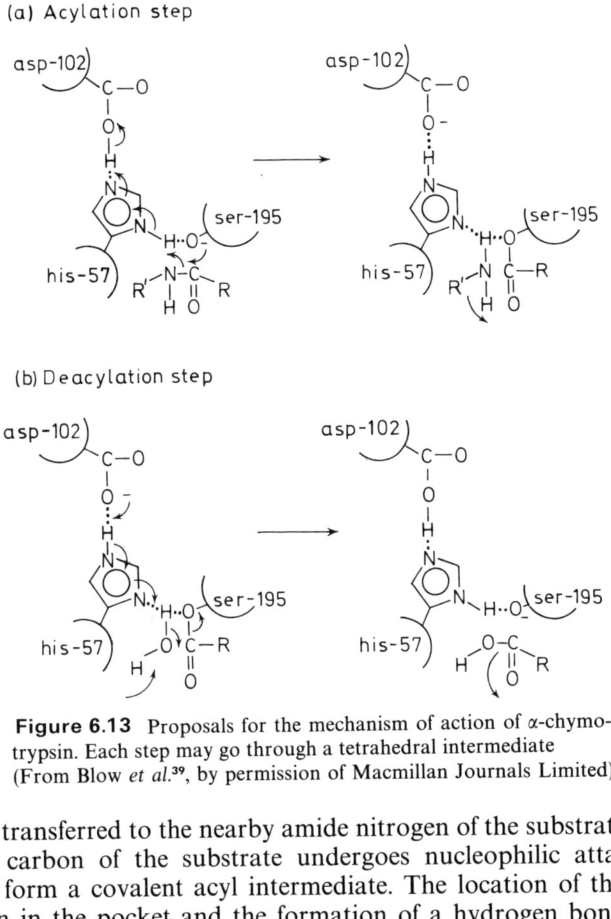

(a) Acylation step

asp-102 C—O O H N N his-57 H··O ser-195 R'—N—C—R H O

(b) Deacylation step

asp-102 C—O O H N N his-57 H··O ser-195 O C—R H O

Figure 6.13 Proposals for the mechanism of action of α-chymotrypsin. Each step may go through a tetrahedral intermediate (From Blow *et al.*[39], by permission of Macmillan Journals Limited)

proton is transferred to the nearby amide nitrogen of the substrate while the carbonyl carbon of the substrate undergoes nucleophilic attack by the serine to form a covalent acyl intermediate. The location of the aromatic side chain in the pocket and the formation of a hydrogen bond from the carbonyl of ser-214 to the amide nitrogen appear to be the most important features of the specificity. Binding of the substrate results in desolvation of ser-195 and may also contribute to the increased nucleophilicity of this group[199]. As yet the crystallography is not sufficiently precise to determine if the substrate is bound in a high-energy state. Deacylation is presumed to be the reverse of the acylation step, with water in the position previously occupied by the leaving group, as identified in the crystallographic studies of the indoleacryloyl-acyl-enzyme complex[140].

The crystallographic studies of α-chymotrypsin were performed at pH 4 at which chymotrypsin is inactive. The effectiveness of the charge relay system in increasing the nucleophilicity of ser-195 will depend critically on pH. It is therefore encouraging that studies on *γ-chymotrypsin* show no significant changes in the conformation of these atoms between pH 6 and 10 (D. R. Davies quoted in reference 200). Similarly in the structure analysis of the homologous enzyme, *elastase*[136] no significant changes in conformation of the charge relay system were observed on changing the pH from 5 to 8.5, the pH for optimum activity. These results, together with the observations of high catalytic activity of crystalline native elastase[136], suggest that the relative orientations of the groups in the charge relay system observed in these crystal structures relate to the active conformation of the enzyme and are not an artefact of crystallisation.

The major differences between the structures of chymotrypsin and elastase concern the substrate binding pocket. In elastase the replacement of two glycine residues by valine and threonine prevents the access of aromatic substrates and accounts for the specificity of elastase towards small aliphatic residues such as alanine[52, 53, 136]. The specificity of a third homologous enzyme, *trypsin*, towards basic residues is also neatly explained. The residue equivalent to ser-189, which is at the back of the binding pocket in chymotrypsin, is aspartic acid in trypsin. Model-building studies have shown this group could make a plausible salt link with the ε-amino or guanidinium group of a basic substrate which is otherwise maintained in the same orientation with respect to ser-195 as a substrate in chymotrypsin[177]. The x-ray analysis of an inhibited trypsin at 2.7 Å has confirmed the relative orientation of these groups to be the same as in chymotrypsin[58–60]. These studies illustrate most beautifully the conservation in the pancreatic serine proteinases of a common primary and tertiary structure within which divergent evolutionary changes have provided differing specificities.

The two bacterial serine proteinases *subtilisin BPN′* and *Novo* have been shown to have identical amino-acid sequences[201, 202]. In spite of very markedly different crystallisation conditions, the structures of the subtilisin BPN′ [49–51] and subtilisin Novo[62] are almost identical. In the active-site region the essential serine (which is residue 221 in subtilisin) is hydrogen bonded to his-64 which in turn is hydrogen bonded to asp-32 (Figure 6.14). The arrangements of these atoms bear a striking similarity to the charge relay system observed for chymotrypsin, although the structures and amino-acid sequences of the bacterial serine proteinases bear no relationship to those of the mammalian enzymes. In subtilisin asp-32 appears to be somewhat more exposed to the solvent than the corresponding residue asp-102 in chymotrypsin. In both structures the aspartate interacts with other amino acids (Figures 6.12 and 6.14).

In view of the apparent importance of the charge relay system in these enzymes it was surprising to find that from the high-resolution x-ray studies on the inactive precursor of chymotrypsin, *chymotrypsinogen*, the spatial arrangement of the three residues ser-195, his-57, asp-102 is almost identical to that in the catalytically active enzyme[56]. The overall conformation of the zymogen is very similar to that of α-chymotrypsin (mean square displacement of all α carbon atoms is 1.8 Å) but detailed comparison of the two structures reveal some of the essential changes which take place on activation. Chymotrypsinogen shows low affinity for chymotrypsin substrates and in the zymogen structure the active-site pocket is blocked. Kraut and his co-workers[57] have been careful to stress that their interpretation of chymotrypsinogen at 2.5 Å is still preliminary and small changes in the active site such as a small rotation of his-57 and the blocking of the active-site cavity may account for the inactivity of the zymogen. Nevertheless it is surprising that the catalytic residues, in contrast with the specificity cavity, are essentially correctly orientated in the zymogen. Further it is not at all clear why the simple act of cleaving the exterior peptide bond between residues 15 and 16 should cause the conformational changes observed.

In an attempt to evaluate the contribution of the charge relay system to catalysis Henderson *et al.*[200, 203] have studied a derivative chymotrypsin in

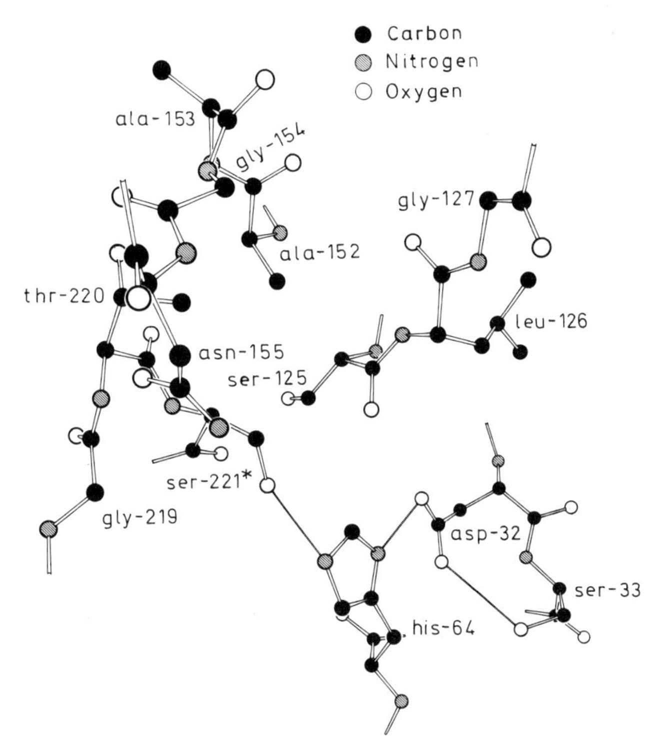

Figure 6.14 A diagram of the residues at the active site of subtilisin BPN′ from the equivalent view shown for chymotrypsin. Note the hydrogen bonding scheme from ser-33 to asp-32 to his-64 ser-221
(Reproduced by permission of Dr. J. Kraut)

which the active site his-57 is methylated at N^{ε}. The modified enzyme binds substrate as well as the native protein, although both acylation and deacylation are slower. No gross structural changes were observed in the crystal structure[200]. It appears that interruption of the hydrogen-bonded quartet ser-195–his-57–asp-102–ser-214 results in a decrease in rate by three to four orders of magnitude but does not completely inactive the enzyme.

Mechanistic and crystallographic studies on these enzymes were, until recently, restricted to the study of model compounds. More recently the subsite specificity of these enzymes has been probed with larger ologopeptide substrates. Shotton *et al.*[136] have studied the binding of a number of di- and tri-peptides to elastase and confirmed the existence of a series of subsites. In particular, there appear to be indications of a conformational change when a fourth subsite is filled which may have implications for the correct positioning of the substrate with respect to ser-195.

The subsite interactions observed in elastase are different from those observed for γ-chymotrypsin by Segal *et al.*[59, 60] for the binding of chloromethylketone peptide substrate analogues which are covalently attached to his-57. Although there is a lack of isomorphism between the inhibited

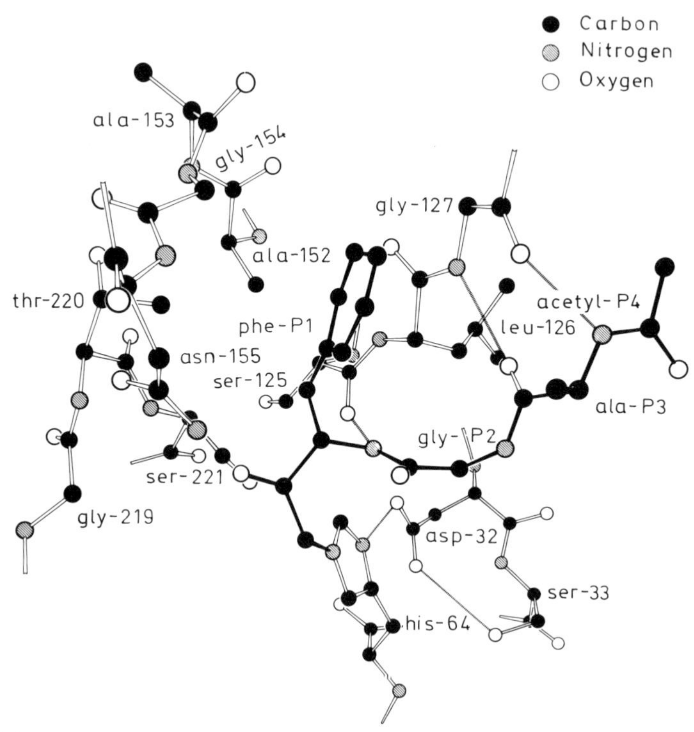

Figure 6.15 A diagram of the active site region of subtilisin with the inhibitor acetyl-L-ala-gly-L-phe-methylene bound to $N^{\varepsilon 2}$ of his-64 (Reproduced by permission of Dr. J. Kraut)

derivatives and the native γ-chymotrypsin the inhibited derivatives were found to be isomorphous with each other. Segal *et al.* concluded from model-building interpretations based on difference Fourier transformation between pairs of inhibited chymotrypsins and from comparisons with α-chymotrypsin that the peptide substrate analogue bound to the enzyme in an anti-parallel β-pleated-sheet fashion to a length of extended backbone chain containing residues ser-214, trp-215 and gly-216. An almost equivalent binding mode of peptide substrates has been reported for subtilisin BPN′ by use of a similar range of chloro-ketones[204] (Figure 6.15) where the main chain residues involved are ser-125, leu-126 and gly-127. It appears that not only are the charge relay systems in subtilisin and chymotrypsin similar but also certain aspects of their substrate binding.

The structure of pancreatic trypsin inhibitor has been determined at 2.5 Å resolution[69], and model-building studies suggest that the inhibitor can be held firmly to the enzyme by a number of contacts which include a similar antiparallel β-structure as proposed for substrate analogue binding to γ-chymotrypsin and subtilisin[70].

6.3.3.3 *Papain*

The structural studies on the plant proteolytic enzyme papain[47, 48] have confirmed the proximity of the essential cysteine and a histidine residue[205]. Under the conditions of crystallisation the enzyme is inactive[206] and this has hampered detailed studies on inhibitor binding, although some work has been done[207]. On the basis of the model, and further kinetic and chemical work, a plausible mechanism has been proposed[208, 209] which takes into account the specificity of papain for aromatic groups in the site one removed from the bond cleaved[210]. Confirmation of these proposals awaits further structural studies.

6.3.3.4 *Carboxy-peptidase*

The x-ray analysis of the zinc containing enzyme carboxy-peptidase at 2.0 Å resolution[42–46] and of the enzyme complexed with glycyl-tyrosine, a slowly hydrolysed substrate, has led to the successful identification of the active-site residues. These data have enabled the number of possible catalytic mechanisms to be limited to two which cannot be distinguished on the basis of x-ray evidence. In both these mechanisms the zinc atom acts as a Lewis acid polarising the carbonyl group of the susceptible peptide bond on the substrate. The zinc is chelated to the enzyme through the side chains of two histidine residues and a glutamic acid residue. On binding of the inhibitor (and presumably the substrate), a tyrosine residue was observed to shift 12 Å via a rotation about its C_α—C_β bond from an external position in the native enzyme to a position in which it can participate as a general acid in the catalytic reaction. These conformational changes provide a clear example of the induced-fit mechanism[182] in which the specific binding of the substrate triggers off a change in the structure which results in the correct orientation of the catalytic groups. Carboxy-peptidase is only 0.33% as active in the crystal as in solution[211]. It would be interesting to know if the decrease in activity can be attributed to restricted mobility in the crystal or to some other effects.

6.3.3.5 *Carbonic anhydrase*

The high resolution results[66] on this so-called 'impossible' enzyme[185] have only recently become available. The zinc atom is situated at the bottom of a

15 Å cavity and is bound to the enzyme by three histidine residues. At pH 8.5 in the crystal structure, the fourth zinc ligand appears to be a water molecule and the geometry at the zinc is that of a distorted tetrahedron with the largest deviation from regularity of 20 degrees. The active-site cavity appears to be filled with ordered solvent and this solvent is displaced on binding sulphonamide inhibitors. No details are available as yet concerning the origins of the unusual catalytic properties although several proposals have been made which focus on the ordered water structure[185, 212]

6.3.3.6 *Ribonuclease*

The crystallographic results on ribonuclease A [33] and ribonuclease S [34–36] (a modified enzyme in which the peptide bond joining residues 20 and 21 has been cleaved) have shown that the conformation of the two molecules are very similar and have confirmed the proximity of the two histidines, his-12 and his-119 at the active site, as originally proposed by chemical studies (see review by Barnard[213]). Further analysis on the binding of a number of nucleotides and substrate analogues with ribonuclease S (Figure 6.16) have provided an explanation for the specificity of this enzyme towards pyridine nucleotides[36], and have shown that his-119 probably occupies several alternate positions in the native enzyme[35, 36]. Richards and Wyckoff[36] have summarised this crystallographic data with respect to the various proposals for the mechanism of action of ribonuclease and the recent chemical studies. They show that although the x-ray data cannot as yet provide conclusive evidence, the mechanism originally proposed by Rabin and his colleagues[214, 215], which involves the two histidines, one as an acid the other as a base, is compatible with almost all the data. A third residue lys-41 has also been implicated from chemical studies in the activity. The x-ray results show that although lys-41 is near the phosphate binding site, it is not in contact with it. The precise role of this group is still not clear.

6.3.3.7 *Nuclease of staphylococcus Aureus*

The crystallographic and extensive biochemical studies of this enzyme have been reviewed recently[64, 216]. The structure determination at 2 Å resolution was carried out on Ca^{2+}-nuclease inhibited with thymidine 3′,5′-diphosphate[63]. The inhibitor is located in a large pocket and appears to be bound in a markedly specific and rigid manner. The calcium ion, which is essential for activity is located 4.7 Å from the phosphate group. This distance is probably too great to allow any strong interaction and is somewhat similar to the position of the positively charged lys-41 in ribonuclease. Although the crystallographers are able to make some suggestions as to the mechanism of action of nuclease, definitive statements have been deferred until all the chemical and crystallographic results can be merged.

6.3.3.8 *Dehydrogenases*

The structural studies on three dehydrogenases are well advanced. The crystal structure of lactate dehydrogenase has been solved at 2.8 Å resolution and now extended to 2.5 Å[65, 217], a 3 Å map has recently been calculated for malate dehydrogenase[218] and a high-resolution map for alcohol dehydrogenase appears iminent[219, 220]. In addition a low-resolution study on the binary complex of LDH with the co-enzyme NAD[221, 222] and a preliminary map of the ternary complex[223–225] at 3.0 Å resolution has been calculated.

Although the amino-acid sequence of LDH has not yet been completed, Rossmann and his colleagues[65] have been able to interpret the map of the native enzyme in terms of the overall folding of the polypeptide chain. In particular, a dodecapeptide of known sequence which contains an essential

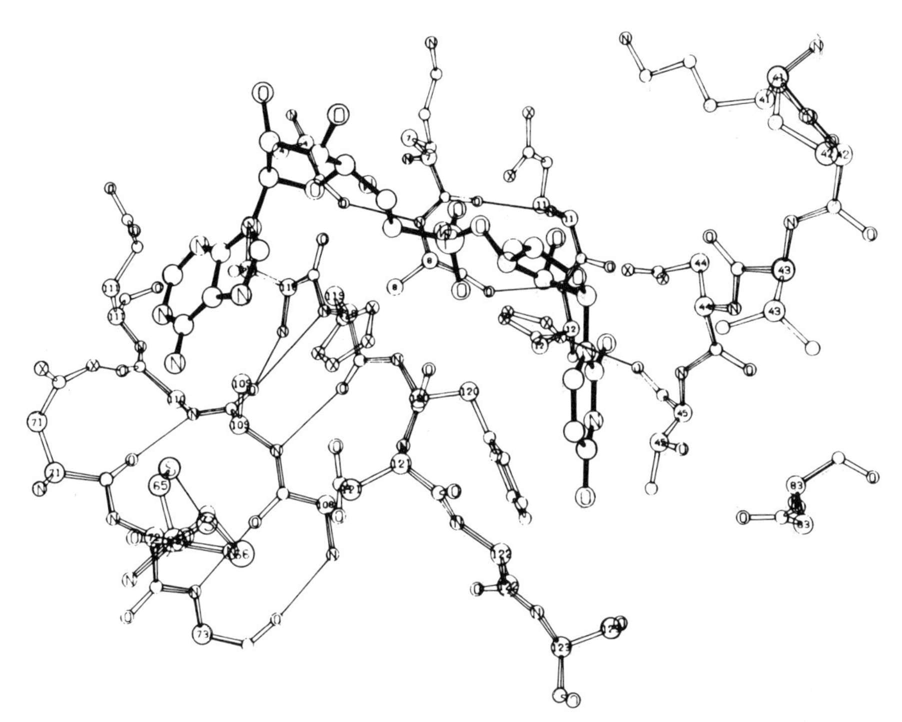

Figure 6.16 A drawing of the dinucleotide phosphonate, uridine-3′-phosphonate-(5′)-adenine, bound to ribonuclease-S from an analysis of a difference electron-density map at 2 Å resolution. Histidine-119 is in a position where it is forced to be by the adenosine moiety. The CH_2 which is bound to the phosphorus is close to his-119 and some further adjustments may be necessary. The rather large separation of lys-41 from the phosphate appears to eliminate the possibility of a strong interaction between these groups
(From Richards, and Wyckoff[36], by permission of Academic Press (copyright holders))

thiol group has been fitted to the electron density and has been shown to be close to the edge of a cleft which divides the two halves of the globular subunit. The nicotinamide end of the co-enzyme binding site is situated within this cleft.

It seems likely that the structural studies on these enzymes will result in mechanistic proposals for this particular class of enzyme in the near future. It will be especially interesting to understand the nature of the quaternary structure changes which take place between the native enzyme, the binary complex and the ternary complex (discussed in Section 6.2) which, as far as we are aware, are not co-operative but appear to be an integral part of the dehydrogenase mechanism.

6.3.4 Hormones

For the protein crystallographer, hormones open new areas of investigation. They pose questions concerning the relation of the structure of the protein hormone to the different aspects of the rather complicated physiology-activation, storage, secretion, circulation in the blood, and finally action at a receptor in the target tissue. However, two problems must be faced by the protein crystallographer who is interested in hormone action. First, is the structure of a protein hormone, ill-defined and variable in solution, assuming a different structure on interaction with the receptor? Secondly, in the absence of an isolated receptor, how can the protein crystallographer begin to relate the structure to the activity?

x-Ray analysis at high resolution has been achieved with only one protein hormone, insulin[67, 226]. The insulin monomer has two polypeptide chains, the A and the B chain, held together by disulphide bridges, as shown in Figure 6.17. It aggregates as dimers and hexamers as described in Section 6.2.4. The monomer has a compact core of non-polar residues conserved in the twenty sequenced insulins. Chemical studies designed to examine the availability of different functional groups, whether they are buried or exposed, and also to study the geometric relations between parts of the molecule are in general agreement with the x-ray model[68]. These observations indicate that the general tertiary structure found in porcine insulin is common to insulins of other mammals and the fishes, and is conserved in solution[68]. These are of course features which are common to enzyme structures, but may not be true of smaller polypeptide hormones such as glucagon and ACTH.

Insulin is synthesised as a single-chain precursor, proinsulin, in which A1 and B30 are connected by a peptide of *c.* 30 residues in most species. It was of great interest to find that in the x-ray model the residues A1 and B30 were separated by only 10 Å [68]. This implied that not all the length of the connecting peptide was important to the folding of the protein, and this is consistent with high species variability of the connecting peptide.

After cleavage of the connecting peptide, insulin is stored in granules which are usually crystalline and contain zinc. They have clearly defined shapes often resembling the rhombs of the insulin crystals used in the x-ray analysis. Under the electron microscope granules of some species appear to have repeat distances which correspond to the diameter of the zinc insulin hexamer shown in Figure 6.8. This suggests that the structure of the granules resembles insulin crystals[68].

Identification of the active region has been approached by considering the natural variations of different insulins and chemically modified insulins in relation to their biological activity. Only a small surface region of the molecule is invariant. This includes B25 phenylalanine, A1 glycine, A5 glutamine, A19 tyrosine and A21 asparagine. The x-ray model shows these residues to be

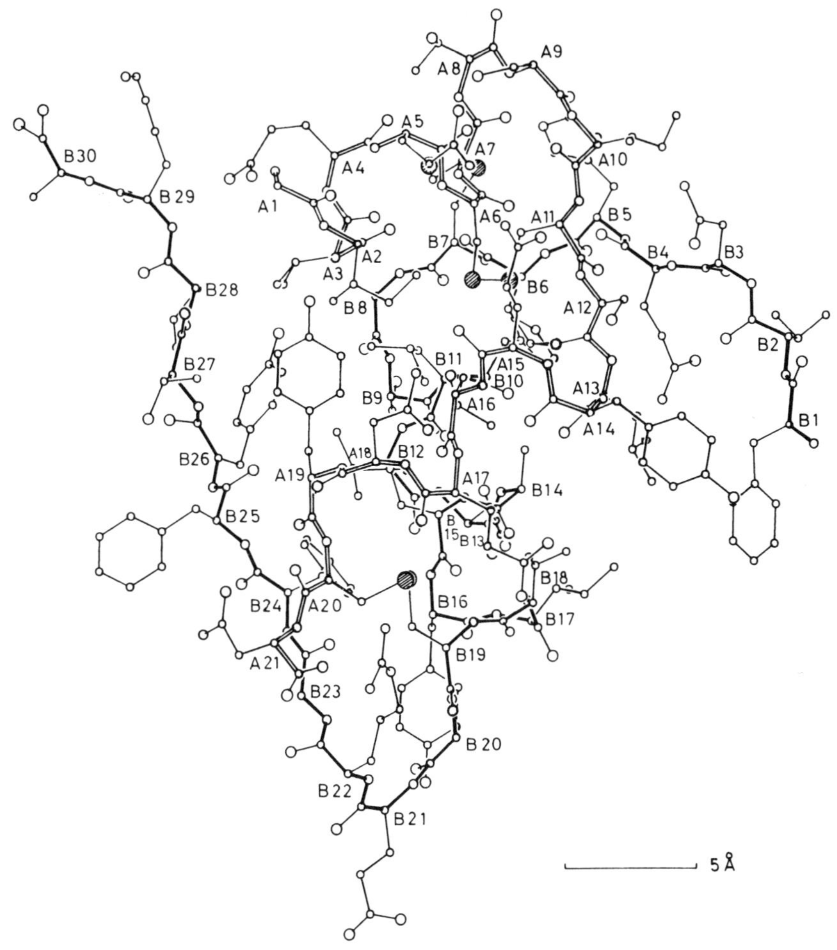

Figure 6.17 The structure of the insulin monomer
(From Blundell, *et al.*[226], by permission of Macmillan Journals Limited)

clustered together as may be seen from Figure 6.17. The biological activity is very susceptible to chemical modification of these residues, and also depends critically on the conservation of the general three-dimensional arrangement of the protein hormone structure, and this region in particular. For these reasons, these residues are thought to be involved in action at the receptor[68].

6.4 CONCLUDING REMARKS

Several new protein crystal structures have already been determined at low resolution[227, 228] and it seems likely that many of these will emerge at high resolution in the not too distant future. Part of the present trend in the selection of new protein structure problems is illustrated by the concerted attempt to solve the structures of the enzymes in the glycolytic pathway[65, 229–231].

From these studies it is hoped to gain a more detailed understanding of the molecular basis of this particular aspect of metabolism.

Many of the more interesting biological molecules are large (M.W.> 100 000) or aggregate together to form multi-enzyme complexes. Calculations indicate that the heavy-atom isomorphous replacement method is likely to be applicable for crystals which contain up to 100 000 daltons per asymmetric unit[227]. The largest structure for which successful derivatives have been obtained to high resolution is probably malate dehydrogenase (M.W. 74 000)[218] and considerable difficulty has been experienced with other large proteins. It seems likely that information obtained on the gross structure of the molecule, as seen in the electron microscope, may well facilitate the preliminary interpretation of Patterson and Fourier syntheses. An encouraging start along these lines has been made with catalase[232], γ-Gl immunoglobulin[233, 234] and the lipoyl transsuccinylase component of the γ-ketoglutarate dehydrogenase complex[235, 236].

It is anticipated that future x-ray studies on biological molecules will provide similar fundamental changes in our understanding of more complex systems as those which have already taken place in our understanding of globins and enzymes.

In Oppenheimer's phrase, 'the future is likely to be only more radical and not less, only more strange and not more familiar, and it will have its own new insights for the inquiring mind'.

Acknowledgements

We are most grateful to all our colleagues who generously sent us their manuscripts in advance of publication and supplied us with copies of figures. It is also a great pleasure to acknowledge Professor D. C. Phillips for his encouragement in the preparation of this article.

References

1. Bernal, J. D. and Crowfoot, D. C. (1934). *Nature (London)*, **133**, 794
2. Hodgkin, D. C. and Riley, D. P. (1968), in *Structural Chemistry and Molecular Biology*, 15. (San Francisco: Freeman)
3. Green, D. W., Ingram, V. M. and Perutz, M. F. (1954). *Proc. Roy. Soc.*, **A225**, 287
4. Phillips, D. C. (1966). *Advan. in Research by Diffraction Methods*, **2**, 75
5. Holmes, K. C. and Blow, D. M. (1966). *The Use of X-ray Diffraction in the Study of Protein and Nucleic Acid Structure*. (New York: Interscience)
6. Blake, C. C. F. (1969, 1970, 1971). *Specialist Periodical Reports, Amino Acids, Peptides and Proteins*, Vols. 1, 2, and 3. (London: The Chemical Society)
7. North, A. C. T. and Phillips, D. C. (1968). *Progress in Biophysics*, 1.
8. Blake, C. C. F. (1968). *Advan. in Protein Chemistry*, **23**, 59
9. Dickerson, R. E. and Geiss, I. (1969). *Structure and Function of Proteins*. (New York: Harper Row)
10. Blow, D. M. and Steitz, T. A. (1970). *Ann. Rev. Biochem.*, **39**, 63
11. Kendrew, J. C., Dickerson, R. E., Strandberg, B. E., Hart, R. G., Davies, D. R., Phillips, D. C. and Shore, V. C. (1960). *Nature (London)*, **185**, 442
12. Kendrew, J. C., Watson, H. C., Strandberg, B. E., Dickerson, R. E., Phillips, D. C. and Shore, V. C. (1961). *Nature (London)*, **190**, 666

13. Kendrew, J. C. (1962). *Brookhaven Symposia on Biology,* **15,** 216
14. Kendrew, J. C. and Watson, H. C. (1969). *Progress in Stereochemistry,* **4,** 299
15. Perutz, M. F., Muirhead, H., Cox, J. M. and Goaman, L. G. C. (1968). *Nature (London),* **219,** 131
16. Bolton, W. and Perutz, M. F. (1970). *Nature (London),* **228,** 551
17. Muirhead, H. and Green, J. (1970). *Nature (London),* **228,** 516
18. Huber, R., Epp, O. and Formanek, H. (1969). *Naturwissenschaften,* **56,** 362
19. Huber, R., Epp, O. and Formanek, H. (1969). *J. Mol. Biol.,* **42,** 591
20. Huber, R., Epp, O. and Formanek, H. (1970). *J. Mol. Biol.,* **52,** 349
21. Love, W. E., Hendrickson, W. A., Klock, P. A., Lattmann, E. E., Padlan, E. A. and Ward, K. B. (1971). *Cold Spring Harbour Symposium,* **36,** 349
22. Hendrickson, W. A. and Love, W. E. (1971). *Nature (London),* in the press
23. Herriott, J. R., Sieker, L. C., Jensen, L. H. and Lovenberg, W. (1970). *J. Mol. Biol.,* **50,** 391
24. Watenpaugh, K., Sieker, L. C., Herriott, J. R. and Jensen, L. H. (1971). *Cold Spring Harbor Symposium,* **36,** 359
25. Dickerson, R. E., Takano, T., Eisenberg, D., Kallai, O. B., Samson, L., Cooper, A. and Margoliash, E. (1971). *J. Biol. Chem.,* **246,** 1511
26. Takano, T., Swanson, R., Kallai, O. B. and Dickerson, R. E. (1971). *Cold Spring Harbor Symposium,* **36,** 397
27. Mathews, F. S., Argos, P. and Levine, M. (1971). *Cold Spring Harbor Symposium,* **36,** 387
28. Carter, C. W., Freen, S. T., Xuong, N. H., Alden, R. A. and Kraut, J. (1971). *Cold Spring Harbor Symposium,* **36,** 381
29. Ludwig, M. L., Anderson, R. D., Apgar, P. A., Burnett, R. M., LeQuesne, M. E. and Mayhew, S. G. (1971). *Cold Spring Harbor Symposium,* **36,** 369
30. Blake, C. C. F., Koenig, D. F., Mair, G. A., North, A. C. T., Phillips, D. C. and Sarma, V. R. (1965). *Nature (London),* **206,** 757
31. Blake, C. C. F., Mair, G. A., North, A. C. T., Phillips, D. C. and Sarma, V. R. (1967). *Proc. Roy. Soc.,* **B167,** 365
32. Phillips, D. C. (1967). *Proc. Nat. Acad. Sci., Washington,* **57,** 493
33. Kartha, G., Bello, J. and Harker, D. (1967). *Nature (London),* **213,** 862
34. Wyckoff, H. W., Hardman, K. D., Allewell, N. M., Inagami, T., Tsernoglou, D., Johnson, L. N. and Richards, F. M. (1967). *J. Biol. Chem.,* **242,** 3984
35. Wyckoff, H. W., Tsernoglou, D., Hanson, A. W., Knox, J. R., Lee, B. and Richards, F. M. (1970). *J. Biol. Chem.,* **245,** 305
36. Richards, F. M. and Wyckoff, H. W. (1971). *The Enzymes,* Vol. 4, 3rd edn., ed. by Boyer, P. D. (New York: Academic Press)
37. Mathews, B. W., Sigler, P. B., Henderson, R. and Blow, D. M. (1967). *Nature (London),* **214,** 652
38. Sigler, P. B., Blow, D. M., Mathews, B. W. and Henderson, R. (1968). *J. Mol. Biol.,* **35,** 143
39. Blow, D. M., Birktoft, J. J. and Hartley, B. S. (1969). *Nature (London),* **222,** 337
40. Birktoft, J. J., Blow, D. M., Henderson, R. and Steitz, T. A. (1970). *Phil. Trans. Roy. Soc. London,* **B257,** 67
41. Blow, D. M. (1971), in *The Enzymes,* Vol. 3, 3rd edition, (ed. by Boyer, P. D.), 185. (New: York: Academic Press)
42. Reeke, G. N., Hartsuck, J. A., Ludwig, M. L., Quiocho, F. A., Steitz, T. A. and Lipscomb, W. N. (1967). *Proc. Nat. Acad. Sci., Washington,* **58,** 220
43. Lipscomb, W. N., Hartsuck, J. A., Reeke, G. N., Quiocho, F. A., Bethge, P. H., Ludwig, M. L. Steitz, T. A., Muirhead, H. and Coppola, J. C. (1968). *Brookhaven Symposium in Biology,* **21,** 24
44. Lipscomb, W. N., Hartsuck, J. A., Quiocho, F. A. and Reeke, G. N. (1969). *Proc. Nat. Acad. Sci., Washington,* **64,** 28
45. Lipscomb, W. N., Reeke, G. N., Hartsuck, J. A., Quiocho, F. A. and Bethge, P. H. (1970). *Phil. Trans. Roy. Soc., London,* **B257,** 177
46. Hartsuck, J. A. and Lipscomb, W. N. (1971), in *The Enzymes,* Vol. 3, 3rd edition, (ed. by Boyer, P. D.), 1. (New York: Academic Press)
47. Drenth, J., Jansonius, J. N., Koekoek, R., Swen, H. H. and Wolthus, B. G. (1968). *Nature (London),* **218,** 929

48. Drenth, J., Jansonius, J. N., Koekoek, R. and Wolthus, B. G. (1971). *The Enzymes,* Vol. 3, 3rd, edn., (ed. by Boyer, P. D.), 484. (New York: Academic Press)
49. Wright, C. S., Alden, R. A. and Kraut, J. (1969). *Nature (London),* **221,** 233
50. Alden, R. A., Wright, C. S. and Kraut, J. (1970). *Phil. Trans. Roy. Soc., London,* **B257,** 119
51. Kraut, J. (1971), in *The Enzymes,* Vol. 3, 3rd edn., (ed. by Boyer, P. D.), 547. (New York: Academic Press)
52. Shotton, D. M. and Watson, H. C. (1970). *Phil. Trans. Roy. Soc., London,* **B257,** 111
53. Watson, H. C., Shotton, D. M., Cox, J. M., Muirhead, H. (1970). *Nature (London),* **225,** 806
54. Shotton, D. M. and Watson, H. C. (1970). *Nature (London),* **225,** 811
55. Hartley, B. S. and Shotton, D. M. (1971), in *The Enzymes,* Vol. 3, 3rd edn., (ed. by Boyer, P. D.), 323. (New York: Academic Press)
56. Freer, S. T., Kraut, J., Robertus, J. D., Wright, H. T. and Xuong, N. H. (1970). *Biochemistry,* **9,** 1997
57. Kraut, J. (1971), in *The Enzymes,* Vol. 3, 3rd edn., (ed. by Boyer, P. D.), 165. (New York: Academic Press)
58. Davies, D. R., Cohen, G. H., Silverton, E. W., Braxton, H. P., Mathews, B. W. (1969). *Acta Crystallogr.,* **A35,** 5182
59. Segal, D. M., Powers, J. C., Cohen, G. H., Davies, D. R. and Wilcox, P. E. (1971).
60. Segal, D. M., Cohen, G. H. and Davies, D. R. (1971). *Cold Spring Harbor Symposium,* **36,** 85
61. Stroud, R. M., Kay, L. M. and Dickerson, R. E. (1971). *Cold Spring Harbor Symposium,* **34, 36,** 125
62. Drenth, J., Hol, W. G. J., Jansonius, J. N. and Koekoek, R. (1971). *Cold Spring Harbor Symposium,* **34, 36,** 107
63. Arnone, A., Bier, C. J., Cotton, F. A., Day, V. W., Hazen, E. E., Richardson, D. C., Richardson, J. S. and Yonath, A. (1971). *J. Biol. Chem.,* **246,** 2302
64. Cotton, F. A. and Hazen, E. E. (1971), in *The Enzymes,* Vol. 4, 3rd edn., (ed. by Boyer, P. D.). (New York: Academic Press)
65. Adams, M. J., Ford, G. C., Koekoek, R., Lentz, P. J., McPherson, A., Rossmann, M. G., Smiley, I. E., Schevitz, R. W. and Wonacott, A. J. (1970). *Nature (London),* **227,** 1998
66. Kannan, K. K., Liljas, A., Waara, I., Bergsten, P. C., Lovgren, S., Bengtsson, U., Carlbom, U., Fridborg, K., Jarup, L., Petef, M. and Strandberg, B. (1971). *Cold Spring Harbor Symposium,* **36,** 221
67. Adams, M. J., Blundell, T. L., Dodson, E. J., Dodson, G. G., Vijayan, M., Baker, E. N., Harding, M. M., Hodgkin, D. C., Rimmer, B. and Sheet, S. (1969). *Nature (London),* **224,** 491
68. Blundell, T. L., Hodgkin, D. C., Dodson, E. J., Dodson, G. G. and Vijayan, M. (1971). *Recent Progr. Horm. Res.,* **27,** 1
69. Huber, R., Kukla, D., Ruhlmann, A., Epp, O. and Formanek, H. (1970). *Naturwissenschaften,* **57,** 389
70. Kukla, D., Ruhlmann, A., Steigeman, W. and Huber, R. (1971). *Cold Spring Harbor Symposium,* **36,** 141
71. Kretsinger, R. H., Nockolds, C. E., Coffee, D. J. and Bradshaw, R. A. (1971). *Cold Spring Harbor Symposium,* **36,** 217
72. Zeppezauer, M. (1971). *Advan. in Enzymology,* **22,** 253
73. Zeppezauer, E., Soderburg, B. O., Branden, C. I., Akeson, A. and Theorell, H. (1967). *Acta Chem. Scand.,* **21,** 1099
74. Weber, B. H. and Goodkin, P. E. (1970). *Arch. Biochem. Biophys.,* **141,** 489
75. Eisenberg, D., Heidner, E. G., Goodkin, P., Dastoor, M., Weber, B. H., Wedler, F. and Bell, J. D. (1971). *Cold Spring Harbor Symposium,* **36,** 291
76. Davies, D. R. and Segal, D. M. (1971). *Methods in Enzymology,* **22,** 266
77. Oesterhelt, D., Bauer, H. and Lynen, F. (1969). *Proc. Nat. Acad. Sci., Washington,* **63,** 1377
78. Benesch, R. and Benesch, R. E. (1956). *J. Amer. Chem. Soc.,* **78,** 1597
79. Shall, S. and Barnard, E. A. (1969). *J. Mol. Biol.,* **41,** 237
80. Avey, H. P. and Shall, S. (1969). *J. Mol. Biol.,* **43,** 341
81. Bernsek, W. and Richards, F. M. (1968). *J. Biol. Chem.,* **243,** 4267
82. Sokolovsky, M., Riordan, J. F. and Vallee, B. L. (1967). *B. B. Res. Commun.,* **27,** 20

83. Borras, F. and Offord, R. E. (1971). Private communication
84. Brandenburg, D. (1971). Private communication
85. Dickerson, R. E., Eisenberg, D., Warnum, J. and Kopka, M. L. (1969). *J. Mol. Biol.*, **45,** 77
86. Kretsinger, R. H. (1968). *J. Mol. Biol.*, **31,** 315
87. Schoenborn, B. P., Watson, H. C. and Kendrew, J. C. (1965). *Nature (London)*, **207,** 28
88. Kretsinger, R. H., Watson, H. C. and Kendrew, J. C. (1968). *J. Mol. Biol.*, **31,** 305
89. Scouloudi, H. (1960). *Proc. Roy. Soc., London*, **A258,** 81
90. Parratt, L. J. (1959). *Rev. Sci. Instr.*, **30,** 297
91. Rosenbaum, G., Holmes, K. C. and Witz, J. (1971). *Nature (London)*, **230,** 434
92. Perutz, M. F. (1949). *Proc. Roy. Soc., London*, **A195,** 474
93. Phillips, D. C. (1964). *J. Sci. Instr.*, **41,** 123
94. Arndt, U. W., North, A. C. T. and Phillips, D. C. (1964). *J. Sci. Instr.*, **41,** 421
95. Banner, D. W. and Phillips, D. C. (1971). Private communication
96. Milledge, H. J. (1966). *Acta Crystallogr.*, **A21,** 220
97. Xuong, N., Kraut, J., Seely, O., Freer, S. T. and Wright, C. S. (1968). *Acta Crystallogr.*, **B24,** 289
98. Arndt, U. W. (1968). *Acta Crystallogr.*, **B24,** 1355
99. Abrahamsson, S. (1966). *Acta Crystallogr.*, **A21,** 213
100. Arndt, U. W., Crowther, R. A. and Mallett, J. F. W. (1968). *J. Sci. Instr.*, **1,** 510
101. Xuong, N. (1969). *J. Sci. Instr.*, **2,** 485
102. Werner, P. E. (1970). *Acta Crystallogr.*, **A26,** 489
103. Nockolds, C. F. and Kretsinger, R. H. (1970). *J. Sci. Instr.*, **3,** 842
104. Dickerson, R. E., Kendrew, J. C. and Strandberg, B. E. (1961). *Acta Crystallogr.*, **14,** 1188
105. Dickerson, R. E., Weinzierl, J. E. and Palmer, R. A. (1968). *Acta Crystallogr.*, **B24,** 997
106. Kartha, G. and Parthasarathy, R. (1965). *Acta Crystallogr.*, **18,** 745
107. Matthews, B. W. (1965). *Acta Crystallogr.*, **20,** 320
108. Harding, M. M. (1962). *D.Phil. Thesis*, Oxford University
109. Singh, A. K. and Ramaseshan, S. (1966). *Acta Crystallogr.*, **21,** 279
110. Kartha, G. (1965). *Acta Crystallogr.*, **19,** 883
111. Bragg, W. L. and Perutz, M. F. (1954). *Proc. Roy. Soc.*, **A225,** 315
112. Blow, D. M. (1958). *Proc. Roy. Soc.*, **A247,** 302
113. Blow, D. M. and Crick, F. H. C. (1959). *Acta Crystallogr.*, **12,** 794
114. Blow, D. M. and Rossmann, M. G. (1961). *Acta Crystallogr.*, **14,** 1195
115. North, A. C. T. (1965). *Acta Crystallogr.*, **18,** 212
116. Matthews, B. W. (1966). *Acta Crystallogr.*, **20,** 82
117. Richards, F. M. (1968). *J. Mol. Biol.*, **37,** 225
118. Barry C. D. and North, A. C. T. (1971). *Cold Spring Harbor Symposium*, **36,**
119. Levinthal, C. (1966). *Scientific American*, **214,** June, p. 42
120. Coulter, C. L. (1965). *J. Mol. Biol.*, **12,** 292
121. Weinzierl, J. E., Eisenberg, D. and Dickerson, R. E. (1969). *Acta Crystallogr.*, **B25,** 380
122. Reeke, G. N. and Lipscomb, W. N. (1969). *Acta Crystallogr.*, **B25,** 2614
123. Main, P. and Rossmann, M. G. (1966). *Acta Crystallogr.*, **21,** 67
124. Diamond, R. (1965). *Acta Crystallogr.*, **19,** 774
125. Diamond, R. (1966). *Acta Crystallogr.*, **21,** 253
126. Bradshaw, R. A., Ericsson, L. H., Walsh, K. A. and Neurath, H. (1969). *Proc. Nat. Acad. Sci., Washington*, **63,** 1389
127. Pauling, L. and Corey, R. B. (1951). *Proc. Nat. Acad. Sci., Washington*, **37,** 729
128. Pauling, L., Corey, R. B. and Branson, H. R. (1951). *Proc. Nat. Acad. Sci., Washington*, **37,** 205
129. Ramachandran, G. N. and Sasisekharan, V. (1968). *Advan. in Protein Chemistry*, **23,** 283
130. Venkatachalam, C. M. (1968). *Biopolymers*, **6,** 1425
131. Lee, B. and Richards, F. M. (1971). *J. Mol. Biol.*, **55,** 379
132. Rossmann, M. G. and Blow, D. M. (1962). *Acta Crystallogr.*, **15,** 24
133. Blow, D. M., Rossmann, M. G. and Jeffrey, B. A. (1964). *J. Mol. Biol.*, **8,** 65
134. Dodson, E., Harding, M. M., Hodgkin, D. C. and Rossmann, M. G. (1966). *J. Mol. Biol.*, **16,** 227
135. Perutz, M. F. (1969). *Proc. Roy. Soc., London*, **B173,** 113
136. Shotton, D. M., White, N. J., Campbell, J. C. and Watson, H. C. (1971). *Cold Spring Harbor Symposium*, **36,** 91

137. Lipson, H. and Cochran, W. (1966). *The Determination of Crystal Structures,* 3rd edn., 319. (London: Bell and Sons)
138. Beddell, C. R. (1970). *D.Phil. Thesis,* Oxford University
139. Henderson, R. and Moffat, J. K. (1971). *Acta. Crystallogr.,* **B27,** 1414
140. Henderson, R. (1970). *J. Mol. Biol.,* **54,** 341
141. Perutz, M. F., Kendrew, J. C. and Watson, H. C. (1965). *J. Mol. Biol.,* **13,** 669
142. Nakahara, A. and Wang, J. H. (1958). *J. Amer. Chem. Soc.,* **80,** 6526
143. Perutz, M. F. (1970). *Nature (London),* **228,** 726
144. Perutz, M. F. (1970). *Nature (London),* **228,** 734
145. Perutz, M. F. and Ten-Eyck, L. F. (1971). *Cold Spring Harbor Symposium,* **36,** 295
146. Williams, R. J. P. (1961). *Fed. Proc.,* **20,** No. 3 supplement 10, p.5
147. Hoard, J. L. (1968). *Structural Chemistry and Molecular Biology* (ed. by Rich, A. and Davidson, N.). (San Francisco: Freeman and Sons)
148. Countryman, R., Collins, D. M. and Hoard, J. L. (1969). *J. Amer. Chem. Soc.,* **91,** 5166
149. Nobbs, C. L., Watson, H. C. and Kendrew, J. C. (1966). *Nature (London),* **209,** 339
150. Watson, H. C. and Nobbs, C. L. (1968). *Biochemie der Sauerstoffs,* 19, Colloquium Gesellschaft fur Biologische Chemie. (Berlin: Springer Verlag)
151. Stryer, L., Kendrew, J. C. and Watson, H. C. (1964). *J. Mol. Biol.,* **8,** 96
152. Watson, H. C. and Chance, B. (1966). *Hemes and Heme Proteins* (ed. by Chance, B., Estabrook, R. W. and Yonetani, T.), Johnson Research Foundation. (New York: Academic Press)
153. Padlan, E. A. and Love, W. E. (1968). *Nature (London),* **220,** 376
154. Padlan, E. A. and Love, W. E. (1969). *Acta Crystallogr.,* **A25,** 5187
155. Muirhead, H., Cox, J. M., Mazzarella, L. and Perutz, M. F. (1967). *J. Mol. Biol.,* **28,** 117
156. Simon, S. R., Konigsberg, W. H., Bolton, W. and Perutz, M. F. (1967). *J. Mol. Biol.,* **28,** 451
157. Simon, S. R., Arndt, P. J. and Konigsberg, W. H. (1971). *J. Mol. Biol.,* **58,** 69
158. Moffat, J. K. (1971). *J. Mol. Biol.,* **58,** 79
159. Moffat, J. K., Simon, S. R. and Konigsberg, W. H. (1971). *J. Mol. Biol.,* **58,** 89
160. Kilmarten, J. W. and Hewitt, J. A. (1971). *Cold Spring Harbor Symposium,* **36,** 311 press
161. Morimoto, H., Lehman, H. and Perutz, M. F. (1971). *Nature (London),* **232,** 408
162. Greer, J. (1971). *Cold Spring Harbor Symposium,* **36,** 315
163. Greer, J. (1971). *J. Mol. Biol.,* **59,** 99, 107
164. Williams, R. J. P. (1971). *Cold Spring Harbor Symposium,* **36,** 53
165. Brunori, M., Noble, R. W., Antonini, E. and Wyman, J. (1966). *J. Biol. Chem.,* **241,** 523
166. Brunori, M., Antonini, E., Wyman, J. and Anderson, S. R. (1968). *J. Mol. Biol.,* **34,** 357
167. Raftery, M. A., Huestis, W. H. and Millett, F. (1971). *Cold Spring Harbor Symposium,* **34,** in the press
168. Margoliash, E., Fitch, W. M. and Dickerson, R. E. (1968). *Brookhaven Symposium in Biology,* **21,** 259
169. Dickerson, R. E., Takano, T., Kallai, O. B. and Samson, O. (1971). *Werner–Gren Foundation Symposium on Oxidative Enzymes,* in the press
170. Kraut, J., Strahs, G. and Freer, S. T. (1968). *Structural Chemistry and Molecular Biology* (ed. by Rich, A. and Davidson, N.), 55. (New York: Freeman)
171. Ludwig, M. L., Anderson, R. D., Mayhew, S. G. and Massey, V. (1969). *J. Biol. Chem.,* **244,** 6047
172. Redfield, A. G. and Gupta, R. D. (1971). *Cold Spring Harbor Symposium,* **36,** 405
173. Winfield, M. E. (1965). *J. Mol. Biol.,* **12,** 600
174. Barnum, T. E. (1970). *The Enzyme Handbook.* (Berlin: Springer-Verlag)
175. Fischer, E. (1894). *Ber. Deutsch. Chem. Ges.,* **27,** 2985
176. Blake, C. C. F., Johnson, L. N., Mair, G. A., North, A. C. T., Phillips, D. C. and Sarma, V. R. (1967). *Proc. Roy. Soc.,* **B167,** 378
177. Steitz, T. A., Henderson, R. and Blow, D. M. (1969). *J. Mol. Biol.,* **46,** 337
178. Kraut, J., Robertus, J. D., Birktoft, J. J. and Alden, R. A. (1971). *Cold Spring Harbor Symposium,* **36,** 117
179. Phillips, D. C. (1966). *Scientific American,* **215,** 78
180. Vernon, C. A. (1967). *Proc. Roy. Soc.,* **B167,** 389
181. Johnson, L. N., Phillips, D. C. and Rupley, J. A. (1968). *Brookhaven Symposium in Biology,* **21,** 120

182. Koshland, D. E. (1963). *Cold Spring Harbor Symposium,* **28,** 473
183. Lumry, R. (1959). *The Enzymes,* Vol. 1, 2nd edn. (ed. by Boyer, P. D., Lardy, H. and Myrbach, K.), 157. (New York: Academic Press)
184. Jenks, W. P. (1966). *Current Aspects of Biochemical Energetics* (ed. by Kaplan, N. and Kennedy, E. P.), 273. (New York: Academic Press)
185. Wang, J. H. (1968). *Science,* **161,** 328
186. Vallee, B. C. and Williams, R. J. P. (1968). *Proc. Nat. Acad. Sci., Washington,* **59,** 498
187. Chipman, D. M. and Sharon, N. (1968). *Science,* **165,** 454
188. Dahlquist, F. W. and Raftery, M. A. (1969). *Fortschr. Chem. Org. Naturstoffe,* **27,** 340
189. Beddell, C. R., Moult, J. and Phillips, D. C. (1970). *Ciba Foundation Symposium on Molecular Properties of Drug Receptors* (ed. by Porter, R. and O'Connor, M.), 85. (London: Churchill)
190. Hartdegen, F. J. and Rupley, J. A. (1967). *J. Amer. Chem. Soc.,* **89,** 1743
191. Beddell, C. R. and Blake, C. C. F. (1970). *Chemical Reactivity and Biological Role of Functional Groups in Enzymes,* Biochemistry Society Symposium, **31,** 157. (London: Academic Press)
192. Dahlquist, F. W., Rand-Meir, T. and Raftery, M. A. (1968). *Proc. Nat. Acad. Sci., Washington,* **61,** 1994
193. Lin, T. Y. and Koshland, D. E. (1969). *J. Biol. Chem.,* **244,** 505
194. Parsons, S. M., Jao, L., Dahlquist, F. W., Borders, C. F., Groff, T., Racs, J. and Raftery, M. A. (1969). *Biochemistry,* **8,** 700
195. Thomas, E. W., McKelvey, J. F. and Sharon, N. (1969). *Nature (London),* **222,** 485
196. Secemski, I. I. and Lienhard, G. E. (1971). *Cold Spring Harbor Symposium,* **36,** 45
197. Boyer, P. D. (ed.) (1971). *The Enzymes,* Vol. 3, 3rd edn. (New York: Academic Press)
198. Cunningham, L. (1965). *Comparative Biochemistry,* **16,** 85
199. Doonan, S., Vernon, C. A. and Banks, B. E. C. (1970). *Progr. in Biophys. and Molecular Biology,* **20,** 249
200. Henderson, R., Blow, D. M., Hess, G. D. and Wright, C. S. (1971). *Cold Spring Harbor Symposium,* **36,** 63
201. Olaitan, S. A., Delange, R. J. and Smith, E. (1968). *J. Biol. Chem.,* **243,** 5296
202. Robertus, J. D., Alden, R. A. and Kraut, J. (1971). *Biochem. Biophys. Res. Commun.,* **42,** 334
203. Henderson, R. (1971). *Biochem. J.,* **124,** 13
204. Kraut, J., Robertus, J. D., Birktoft, J. J. and Alden, R. A. (1971). *Cold Spring Harbor Symposium,* **36,** 117
205. Hussain, S. S. and Lowe, G. (1968). *Biochem. J.,* **108,** 861
206. Drenth, J., Hol, W. G. J., Visser, J. W. E. and Sluyterman, L. A. E. (1968). *J. Mol. Biol.,* **34,** 369
207. Wolthers, B. G., Drenth, J., Jansonius, J. N., Koekock, R. and Swen, M. M. (1970). *Proc. Int. Symp. Structure Function Relationships of Proteolytic Enzymes,* 272. (Copenhagen: Munksgaard)
208. Lowe, G. (1970). *Phil. Trans. Roy. Soc., London,* **B257,** 237
209. Lowe, G. and Yuthavong, Y. (1971). *Biochem. J.,* **124,** 107
210. Schechter, J. and Berger, A. (1968). *Biochem. Biophys. Res. Commun.,* **32,** 898
211. Quiocho, F. A. and Richards, F. M. (1966). *Biochemistry,* **5,** 4062
212. Khalifa, R. G. (1971). *J. Biol. Chem.,* **246,** 2561
213. Barnard, E. A. (1969). *Ann. Rev. Biochem.,* **38,** 677
214. Deavin, A., Mathias, A. P. and Rabin, B. R. (1960). *Nature (London),* **211,** 252
215. Findley, D., Herries, D. G., Mathias, A. P., Rabin, B. R. and Ross, C. A. (1962). *Biochem. J.,* **85,** 152
216. Anfinsen, C. B., Cuatrocassus, P. and Taniuchi, H. (1971). *The Enzymes,* Vol. 4, 3rd edn. (ed. by Boyer, P. D.). (New York: Academic Press)
217. Adams, M. J., Ford, G. C., Koeboek, R., Lentz, P. J., McPherson, A., Rossmann, M. G., Smailey, I. E., Schevitz, R. W. and Wonacott, A. J. (1970). *Nature (London),* **227,** 1098
218. Tsernoglou, D., Banaszak, L. J. and Hill, E. (1971). *Cold Spring Harbor Symposium,* **36,** 171
219. Branden, C. I., Zeppezauer, E., Soederberg, B. O., Boiwe, T., Nordstroem, B., Soderlund, G., Zeppezauer, H., Werner, P. E. and Akeson, A. (1970). *Werner–Gren Symposium on Structure and Function of Oxidation Reduction Enzymes, Stockholm* (ed. by Akeson, A. and Ehrenberg, A.), in the press

220. Branden, C. I., Zeppezauer, F., Boiwe, T., Soderlund, G., Soderberg, B. O. and Nordstroem, B. (1970). *Pyridine Nucleotide Dependent Dehydrogenases* (ed. by Bund, H.), 129. (Berlin:Springer)
221. Adams, M. J., Haas, D. J., Jeffrey, B. A., McPherson, A., Mermall, L., Rossmann, M. G., Schevitz, R. W. and Wonacott, A. J. (1969). *J. Mol. Biol.*, **41,** 159
222. Adams, M. J., McPherson, A., Rossmann, M. G., Schevitz, R. W. and Wonacott, A. J. (1970). *J. Mol. Biol.*, **51,** 31
223. Leberman, R., Smiley, L. E., Haas, D. J. and Rossmann, M. G. (1969). *J. Mol. Biol.*, **46,** 217
224. Smiley, L. E., Koekoek, R., Adams, M. J. and Rossmann, M. G. (1971). *J. Mol. Biol.*, **55,** 467
225. Adams, M. J., Buehner, M., Ford, G. C., Hackert, M., Lentz, P. J., McPherson, A., Rossmann, M. G., Schevitz, R. W. and Smiley, L. E. (1971). *Cold Spring Harbor Symposium*, **36,** 179
226. Blundell, T. L., Cutfield, J. F., Dodson, G. G., Dodson, E. J., Hodgkin, D. C., Mercola, D. A. and Vijayan, M. (1971). *Nature (London)*, **231,** 506
227. Eisenberg, D. (1970). *The Enzymes*, Vol. 1, 3rd edn. (ed. by Boyer, P. D.). (New York: Academic Press)
228. Blake, C. C. F. (1971). *Progr. in Biophys.*, in the press
229. Banner, D. W., Bloomer, A. C., Petsko, G. A. and Phillips, D. C. (1971). *Cold Spring Harbor Symposium*, **36,** 151
230. Campbell, J. W., Duee, E., Hodgson, G., Mercer, W. D., Stammers, D. K., Wendell, P. L., Muirhead, H. and Watson, H. C. (1971). *Cold Spring Harbor Symposium*, **36,** 165
231. Steitz, T. A. (1971). *J. Mol. Biol.*, in the press
232. Vainshtein, B. K., Barynin, V. V. and Gurskaya, G. V. (1967). *Krisstallographiya*, **12,** 750
233. Sarma, V. R., Silverton, E. W., Davies, D. R. and Terry, W. D. (1971). *J. Biol. Chem.*, **246,** 3753
234. Labaw, L. W. and Davies, D. R. (1971). *J. Biol. Chem.*, **246,** 3760
235. De Rosier, D. J., Oliver, M. and Read, L. J. (1971). *Proc. Nat. Acad. Sci., Washington*, **68,** 1135
236. De Rosier, D. J. (1971). *Cold Spring Harbor Symposium*, **36,** 199

7
Application of Direct Methods in X-Ray Crystallography

J. KARLE and ISABELLA L. KARLE
U.S. Naval Research Laboratory, Washington

7.1 INTRODUCTION

There has been a great proliferation of investigations employing direct methods of x-ray crystallography since the middle 1960s owing to the fact that the newly developed procedures have afforded a rapid and generally quite reliable method for structure determination. Many types of structures are now readily accessible to study, which heretofore were either quite difficult to solve or not solvable at all. Particularly difficult structures have been ones consisting of almost equal atoms crystallising in the non-centrosymmetric space groups. Many substances of interest to organic and biological chemists fall in this class. A large number of such structures have by now been determined by direct methods, and in many instances their molecular formulae were initially unknown.

Progress in structure determination has also been facilitated in recent years by the rapid developments in the field of computers and the increasing use of automatic diffractometers for data collection. As a consequence, x-ray structure analysis has become a significant analytical tool which is finding increasing application in areas such as organic structural analysis, replacing classical chemical methods. x-Ray structure analysis may be readily integrated with a chemical or biological research programme, not as a separate entity but as a tool to be used conjointly with other aspects of the programme, e.g. to enlighten an intermediate step of a synthesis or an analysis. On the molecular level, x-ray analysis can be employed to obtain the structural formula, stereoconfiguration, conformation, bond lengths and angles, charge distributions and average vibrational amplitudes. Intermolecular information concerning hydrogen bonding, molecular complexes, charge-transfer complexes, coordinations and clathrate formation is also provided.

Direct methods of structure determination are designed to overcome the phase problem associated with crystal-structure analysis. In the usual diffraction experiment, the amplitudes of the scattered waves are obtained readily from the measured x-ray intensities whereas their phases are not. It is necessary to know both the amplitude and phase associated with the scattered rays in order to make an immediate calculation of the electron density in a crystal. The maxima of the electron density function locate the positions of the atoms in the unit cell of a crystal.

The phase problem has been overcome in the past by many means such as the use of the Patterson function[1] which represents the distribution of interatomic vectors in a crystal. The Patterson function is particularly useful when the structure contains a small number of heavy atoms. Methods have been developed for employing the known positions of the heavy atoms derived from the Patterson function to obtain initial, approximate phases from which the entire structure can be ultimately developed. Other types of heavy-atom techniques involve the use of isomorphous replacement and anomalous dispersion which can be used for phase determination under special circumstances. Isomorphous replacement plays an important role in current research on protein structures. In contrast, direct methods of phase determination are readily applicable to equal atom structures as well as those containing heavy atoms. Direct methods owe their existence to the remarkable fact that it is possible to compute the required unknown phases from the measured x-ray scattering intensities. This can be seen from a simple mathematical description of the problem which will now be presented.

The electron density function for a crystal, whose maxima locate the positions of the atoms in a unit cell, may be represented by a three-dimensional Fourier series,

$$\rho(\vec{r}) = V^{-1}\sum_{\vec{h}=-\infty}^{\infty} F_{\vec{h}} \exp(-2\pi i\vec{h}\cdot\vec{r}) \qquad (7.1)$$

where the coefficients

$$F_{\vec{h}} = |F_{\vec{h}}| \exp(i\phi_{\vec{h}}) \tag{7.2}$$

are the crystal structure factors whose magnitudes are proportional to the square roots of the corresponding measured x-ray intensities. The quantities $|F_{\vec{h}}|$ and $\phi_{\vec{h}}$ represent the amplitude and phase, respectively, of the x-ray wave scattered by crystal planes which may be labelled by the vector $\vec{h} = (h,k,l)$, where the components of $\vec{h}$ assume integer values and are known as the Miller indices. The $|F_{\vec{h}}|$ represent the sets of four numbers which are obtained from an x-ray diffraction pattern, each set representing the amplitude and the three associated Miller indices derivable from the measurement of the intensity of an x-ray reflection and its position in the diffraction pattern. The vector $\vec{r}$ in equation (7.1) is the position vector of any point in the unit cell of volume V. Since the phases, $\phi_{\vec{h}}$, are not obtainable from a diffraction experiment, the electron density given by equation (7.1) cannot be immediately computed. Only the magnitudes $|F_{\vec{h}}|$ are ordinarily obtainable. However, as a consequence of the fact that the electron-distribution functions associated with the individual atoms in crystals are known to a good approximation, it is possible to compute the required phases from the measured x-ray intensities. The direct methods of crystal structure analysis are based on such a calculation, since, once the phases have been computed, the electron density in the unit cell of a crystal and therefore the structure may be obtained directly from equation (7.1).

The crystal structure factor, $F_{\vec{h}}$, may be expressed in terms of the electron density distribution, $\rho(\vec{r})$, by means of a Fourier inversion of equation (7.1). The integral expression for the Fourier inversion may be replaced by a sum of the contributions from each of the N discrete atoms in the unit cell which comprise the electron density, obtaining

$$|F_{\vec{h}}| \exp(i\phi_{\vec{h}}) = \sum_{j=1}^{N} f_{j\vec{h}} \exp(2\pi i\vec{h}\cdot\vec{r}_j) \tag{7.3}$$

where $f_{j\vec{h}}$ is the atomic scattering factor for the jth atom in a unit cell containing N atoms and $\vec{r}_j$ is its position. Equations (7.3) form a set of simultaneous equations, one for each independent reflection labelled by $\vec{h}$. In writing these equations, an exponential damping term which represents the effect of the vibrational motion has been omitted. The damping may be eliminated in the treatment of the intensity data and it is assumed that this adjustment has been made. The unknown quantities in equations (7.3) are the phases ϕ_h and the atomic positions $\vec{r}_j$, whereas the known quantities are the structure-factor magnitudes $|F_{\vec{h}}|$ obtained from experiment and the atomic scattering factors $f_{j\vec{h}}$ which have been calculated and tabulated for free atoms. The quantities $|F_{\vec{h}}|$ and $f_{j\vec{h}}$ are generally known within an uncertainty that does not exceed 5–10%. Since there are complex expressions on both sides of equations (7.3), there are actually two equations for each $\vec{h}$, one for the real part and one for the imaginary part. For moderate size structures, i.e. structures containing no more than a few hundred atoms per unit cell, it is usually possible to collect data from the full sphere of scattering for x-ray radiation from a copper target. Under such circumstances the simultaneous equations (7.3) are greatly overdetermined by the number of data from which the values of the $|F_{\vec{h}}|$ are derived. The overdeterminacy for each atomic coordinate amounts to a factor of *c*. 50 for centrosymmetric crystals and *c*. 25 for non-centrosymmetric ones. Because of the high degree of overdeterminacy, fairly simple phase determining relations exist which can be employed with high reliability, despite the fact that the simultaneous equations given by equations (7.3) form a very complicated system. As a practical matter, it is at present often more readily possible to employ phase-determining relations to evaluate phases and compute the structure from equation (7.1) than to obtain the atomic positions directly, for example, from an analysis of the interatomic vectors as given by a Patterson function.

An underlying feature which facilitated the development of simple phase-determining relationships and determines their character is the non-negativity of the electron density function $\rho(\vec{r})$. Phase determining relations were initially obtained in the form of inequalities whose validity depended upon the non-negativity of $\rho(\vec{r})$. The first inequalities were derived by Harker and Kasper[2] demonstrating that phases could be defined in terms of the measured x-ray intensities. The practicality of the inequalities was illustrated by the structure determination of orthorhombic decaborane by Kasper, Lucht and Harker[3]. The criterion of the non-negativity of the electron density function was employed by Karle and Hauptman[4] as a basis for obtaining a complete set of inequalities and led to an explicit form for the main phase-determining relations. Further application of probability theory has resulted in reliability measures for the phase-determining relations. Based on these mathematical developments, procedures for direct phase determination have been devised having general applicability to the centrosymmetric and non-centrosymmetric space groups. The relations, reliability measures and procedures will be discussed in the next section.

In a previous review article[5], applications of direct phase determination to a wide variety of problems were described. By this time, groups of related compounds have been investigated and can be discussed under special topics such as, for example, photorearrangement products and polypeptide conformations. Such topics and a variety of additional investigations will be reviewed in this article.

Owing to the great proliferation of investigations employing direct methods of analysis, this review will not be all-inclusive. A choice has been made in terms of the general interest in the topics discussed with emphasis on the most difficult type of structure problem, the non-centrosymmetric crystal possessing equal or almost equal atoms.

7.2 RELATIONS AND PROCEDURE FOR PHASE DETERMINATION

It has been found quite useful for phase determination to alter the experimental x-ray intensities to represent the scattering intensities which would have been obtained from motionless point scatterers rather than from vibrating atoms with spatially distributed electron densities. This is accomplished by making use of statistical properties of the $|F_{\vec{h}}|^2$ determined by means of the probability theory of Wilson[6]. As a result it is possible to define a normalised structure factor[7] $E_{\vec{h}}$ in terms of $F_{\vec{h}}$ corrected for vibrational motion having the property that $\langle |E_{\vec{h}}|^2\rangle_{\vec{h}} = 1$,

$$E_{\vec{h}} = F_{\vec{h}}/[(m_2^0 + m_0^2)s_{2,\vec{h}}/n]^{\frac{1}{2}} \tag{7.4}$$

where

$$s_{2,\vec{h}} = \sum_{j=1}^{N} f_{j\vec{h}}^2, \tag{7.5}$$

n is the number of symmetry related atomic positions in the unit cell and

$$m_i^k = \int_0^1\int_0^1\int_0^1 \xi^i\eta^k \,\mathrm{d}x\,\mathrm{d}y\,\mathrm{d}z \tag{7.6}$$

where $\xi = \xi(x,y,z;\vec{h})$ and $\eta = \eta(x,y,z;\vec{h})$ are the real and imaginary parts, respectively, of the trigonometric portion of the structure factor as defined, for example, in the International Tables for X-ray Crystallography[8] for the space group of interest. The quantity $s_{2,\vec{h}}$, defined in terms of the atomic scattering factors, corrects the $F_{\vec{h}}$ for the effect of the spatial distribution of electron density. The normalised structure factor magnitudes $|E_{\vec{h}}|$ are the end products of the data reduction procedure and are the quantities which are used to represent the scattering amplitudes from a structure consisting of motionless point atoms. The phase-determining formulae and their associated probability measures are defined in terms of the $E_{\vec{h}}$ whose phases correspond to those for the $F_{\vec{h}}$.

The phase-determining formulae currently used can be described by the terms 'main' and 'auxiliary'. The bulk of the phase determination is carried out by means of the main formulae and the auxiliary formulae afford additional and often very useful phase information. The main phase-determining relation for centrosymmetric crystals is the sigma-2 relation of Hauptman and Karle[7],

$$sE_{\vec{h}} \simeq s\sum_{\vec{k}} E_{\vec{k}}E_{\vec{h}-\vec{k}}, \tag{7.7}$$

where s means 'sign of'. A plus sign corresponds to a phase of zero and a minus sign corresponds to a phase of π, the only two phase values which reflections from centrosymmetric crystals can possess. For non-centrosymmetric crystals, the main phase-determining formulae are the 'sum of angles' formula[4, 9],

$$\phi_{\vec{h}} \simeq \langle \phi_{\vec{k}} + \phi_{\vec{h}-\vec{k}} \rangle_{\vec{k}_r} \tag{7.8}$$

and the 'tangent formula'[10],

$$\tan \phi_{\vec{h}} \simeq \frac{\sum_{\vec{k}} |E_{\vec{k}}E_{\vec{h}-\vec{k}}| \sin(\phi_{\vec{k}} + \phi_{\vec{h}-\vec{k}})}{\sum_{\vec{k}} |E_{\vec{k}}E_{\vec{h}-\vec{k}}| \cos(\phi_{\vec{k}} + \phi_{\vec{h}-\vec{k}})}. \tag{7.9}$$

The symbol $\vec{k}_r$ implies that the $\vec{k}$ vectors entering into the average are restricted to those associated only with large magnitudes for $|E_{\vec{k}}|$ and $|E_{\vec{h}-\vec{k}}|$.

It is apparent from an examination of equations (7.7), (7.8) and (7.9) that a certain number of phases are required to be known in order to employ these formulae for determining additional ones. Experience has shown that only a relatively few phases are needed to initiate a phase determination and carry it through, providing that the initial phases are chosen to satisfy a simple set of criteria and close attention is paid to reliability measures as the phase determination proceeds. Auxiliary formulae can provide additional phase information.

Initial phase information is available in the form of values for phases which must be specified in order to fix the origin in the crystal. In certain crystals an additional specification is required to fix the enantiomorph and/or reference frame. Generally, the values of a few additional phases are required to carry through a phase determination and these may be assigned values by means of symbols. The remaining phases may be determined in terms of the specified phases and the ones with unknown symbols. Since few symbols are required, the number of ambiguous alternatives which must be considered at the end of a phase determination is small and can be handled by modern computing facilities. There are only two possible values for a symbol for centrosymmetric crystals. More alternatives need to be considered for non-centrosymmetric crystals since the phases may have values anywhere between $-\pi$ and π. In actual practice it is only necessary to consider values at intervals. The procedure for phase determination based on symbolic phases is called the symbolic addition procedure[11].

Usually less than four symbols are required to carry through a phase determination and there are several ways in which they may be evaluated before it is necessary to compute a Fourier series for the remaining ambiguous alternatives. In the course of the phase determination, relations among the symbols often develop. Auxiliary formulae may also be employed to determine the values of the symbols. Examples of such formulae are known as sigma-1, sigma-3 and $B_{3,0}$ and have been defined for all the space groups in a previous review article[12]. Modifications of $B_{3,0}$ have been given by Hauptman[13] and Karle[14]. Additional chemical and packing considerations can also be used to reduce the number of symbols. The number of alternatives which need to be considered can be further limited by measures of the internal consistency of the phase-determining procedure, e.g. in terms of the number of violations of the phase-determining formulae. Whereas it is often not true that the most internally consistent set of phases (besides the one with all phases equal to zero) is the proper one, the correct answer is generally to be found among the first few most consistent possibilities. All of these restrictions applied to the original set of unknown symbols leave few, if any, alternatives to be considered.

In the application of the phase-determining procedure, new phases are obtained in a stepwise fashion. It is evidently important to have a measure of the reliability of each determination. The probability measure to be associated with equation (7.7) is[7,15]

$$P_+(\vec{h}) \simeq \tfrac{1}{2} + \tfrac{1}{2} \tanh \sigma_3 \sigma_2^{-\frac{3}{2}} \, | E_{\vec{h}} | \sum_{\vec{k}} E_{\vec{k}} E_{\vec{h}-\vec{k}} \tag{7.10}$$

where $P_+(\vec{h})$ represents the probability that the sign of $E_{\vec{h}}$, as determined by equation (7.7), is positive and

$$\sigma_n = \sum_{j=1}^{N} Z_j^n \tag{7.11}$$

where Z_j is the atomic number of the jth atom in a unit cell containing N atoms.

A measure of reliability which can be associated with equations (7.8) and (7.9) is the variance of the value of the phase angle $\phi_{\vec{h}}$ obtained by means of these equations. On the assumption that the values of the phases used to determine $\phi_{\vec{h}}$ are known accurately, the formula for the variance, V, is[9, 11]

$$V = \frac{\pi^2}{3} + [I_0(\alpha)]^{-1} \sum_{n=1}^{\infty} \frac{I_{2n}(\alpha)}{n^2} - 4[I_0(\alpha)]^{-1} \sum_{n=0}^{\infty} \frac{I_{2n+1}(\alpha)}{(2n+1)^2} \tag{7.12}$$

where I_n is the Bessel function of imaginary argument of order n,

$$\alpha = \{[\sum_{\vec{k}} \kappa(\vec{h},\vec{k}) \cos(\phi_{\vec{k}} + \phi_{\vec{h}-\vec{k}})]^2 + [\sum_{\vec{k}} \kappa(\vec{h},\vec{k}) \sin(\phi_{\vec{k}} + \phi_{\vec{h}-\vec{k}})]^2\}^{\frac{1}{2}} \tag{7.13}$$

and

$$\kappa(\vec{h},\vec{k}) = 2\sigma_3\sigma_2^{-\frac{3}{2}} \,|\, E_{\vec{h}} E_{\vec{k}} E_{\vec{h}-\vec{k}} \,|\,. \tag{7.14}$$

The expressions (7.10) and (7.12) indicate that those contributors to the phase-determining relations associated with the largest normalised structure factor magnitudes $|E|$ give the most reliable phase indications. It is therefore a matter of considerable importance to initiate a phase determination employing the largest $|E|$ values. In the initial stages of a phase determination there may be only a single contributor to equations (7.7) or (7.8). Nevertheless it is usually possible to proceed with $P_+(\vec{h}) \geqslant 0.98$ for centrosymmetric crystals and $V \leqslant 0.5$ for non-centrosymmetric ones. The great overdeterminacy of the data, and therefore its relative extensiveness, provides the opportunity to find a path of phase determination among the largest structure-factor magnitudes which satisfies these rather restrictive reliability criteria.

For specifying the origin in a crystal, phase values are assigned to an appropriate set of magnitudes $|E_{\vec{h}}|$ as determined by the nature of the components of $\vec{h}$ and the particular space group under consideration. This subject is covered by the theory of semi-invariants and there are tables to facilitate the origin assignment for any space group[16]. A semi-invariant is a linear combination of phases whose values or magnitudes, depending upon the space group involved, are independent of the choice of permissible origin, but are determined by the x-ray intensities, once a functional form has been chosen for the structure factor. Many space groups permit more than one functional form for the structure factor and the one usually chosen is one which is found in the International Tables for X-ray Crystallography[8].

In choosing a suitable set of $|E_{\vec{h}}|$ for origin assignment it is important to consider the extent to which a particular $\vec{h}$ enters into the relationships required by the right sides of equations (7.7)–(7.9). An additional specification is required for most non-centrosymmetric space groups. Its purpose is to distinguish one of two possible enantiomorphs when they are distinct and/or a reference frame when the frames are distinct. This is readily achieved by assigning a sign to the magnitude of the imaginary part of a suitable $|E_{\vec{h}}|$, once the origin in the crystal has been fixed. The magnitude of the particular $|E_{\vec{h}}|$ should be large and the phase should differ significantly in value from 0 or π. Two structures related to one another by a reflection through the origin are enantiomorphs. Enantiomorphs are not distinct in crystals having centres of symmetry or symmetry planes.

Once suitable phase specifications have been made, equations (7.7) or (7.8) are employed to proceed with the phase determination. New phases are obtained in a stepwise fashion with close attention paid to the reliability measures, equations (7.10) or (7.12). If, in the course of the phase determination, it is not possible to proceed without violating a restriction concerning an acceptable value for the reliability measure, an additional phase value is assigned by means of an unknown symbol. A suitable $|E_{\vec{h}}|$ of large magnitude is chosen having a vector $\vec{h}$ which interacts often with other normalised structure factors of large magnitude to make significant contributions to the phase-determining formulae. In the beginning of the procedure there may be few or only one contributor to equations (7.7) or (7.8). As the determination proceeds, there are generally more and more contributors to equations (7.7) or (7.8), since the newly determined phases are used to determine still others. For centrosymmetric crystals, this process is continued until *c.* 15 or 20 phases per atom in the asymmetric unit are obtained. The phases are defined in terms of the originally specified values and any remaining unknown symbols. Since the symbols have 2 possible values, for p symbols there would be at most 2^p possible maps to consider. Fourier maps are then computed employing equation (7.1) with the $E_{\vec{h}}$ replacing the $F_{\vec{h}}$ as Fourier coefficients, E-maps[17]. E-maps have the property of increasing the resolution of the atomic positions over that obtained from maps computed from the $F_{\vec{h}}$. The map representing the correct structure is generally recognised on the basis that the maxima make good chemical sense in terms of bond lengths and angles. Ultimately the structure is refined by least-squares procedures. The agreement between the structure-factor magnitudes computed from the refined structure and the measured structure-factor magnitudes affords a further confirmation of the correctness of the structure.

Phase determination for non-centrosymmetric structures proceeds in a fashion which is similar to that for centrosymmetric ones. However, in the symbolic addition procedure, equation (7.8) is used with specified phase values and symbols until *c.* 50 phases are determined. Alternative values are then assigned to the remaining symbols and the phase determination is continued for each of the alternatives by use of the tangent formula, equation (7.9). The calculation is usually carried out until *c.* 20–30 phases per atom in the asymmetric unit are obtained. E-maps are then computed and the one representing the chemically sensible structure is chosen for refinement.

With non-centrosymmetric crystals it often happens that the E-map may reveal only a portion of the structure. The tangent formula can be employed to develop a partial structure into a complete one[18]. Phases and amplitudes are computed from the partial structure by use of a formula such as equation (7.3). A comparison with the measured structure-factor magnitudes is then made and the computed phase is retained if its associated amplitude indicates that the partial structure makes a significant contribution to the measured magnitude and if it is associated with a large $|E|$. The basic set of phases, so obtained, may be expanded further by means of the tangent formula equation (7.9)) and a new E-map may be computed. More and more of the structure continues to appear as the process is repeated and finally the entire structure will appear. The source of the partial structure, of course, need not be the procedure for phase determination. Partial structures may be obtained from the Patterson function, for example, in the location of heavy atoms for the heavy-atom method[19] or oriented fragments from vector search methods[20, 21].

In using the symbolic addition procedure, it is possible to readily carry out the initial stages of the phase determination by hand. In fact, the use of symbols greatly facilitates the bookkeeping associated with the ambiguous alternatives. There are many variations of the symbolic addition procedure in use and they mainly concern strategies which permit the immediate use of computers in order to fully automate the procedure. The minimal use of symbols and the increased use of alternative numerical values for certain of the phases facilitates the application of the tangent formula (equation (7.9)) at the start of the phase determination. Procedures which are largely based on the use of numerical phases make use of the auxiliary phase-determining formulae such as sigma-1, sigma-3 and $B_{3,0}$ in the beginning of the phase determination. If alternative numerical values for phases are used exclusively as a basic set for carrying through a phase determination, their number is limited by the number of alternatives which it is convenient to handle with a computer. Probability measures and rejection criteria play an important role in limiting the number of E-maps which need to be considered.

Many computer programs have been written for carrying out phase determination. Some of the people who have prepared phase-determining programs which follow the symbolic addition procedure closely are S. R. Hall and F. R. Ahmed of the National Research Council, Ottawa, R. B. K. Dewar and E. B. Fleischer of the Illinois Institute of Technology in Chicago and the University of California, respectively, A. Bednowitz of the International Business Machines Research Center, Yorktown Heights, N.Y., W. D. S. Motherwell and A. C. Larson of the University Chemical Laboratory, Cambridge, England and Los Alamos Scientific Laboratory, New Mexico, respectively, and P. Gum, S. Brenner, R. Gilardi, J. Konnert and J. Karle of the Naval Research Laboratory, Washington, D.C. Similar programs have been prepared by P. T. Beurskens of the Laboratory for Crystal Chemistry, Niemegen, and by J. M. Stewart, R. V. Chastain, E. G. Boonstra, H. L. Ammon, J. R. Holden and C. Dickinson of the University of Maryland, College Park. Programs which use few or no symbols, but rather develop sets of numerical phases with the tangent formula have been written by S. R. Hall of the National Research Council, Ottawa, M. G. Drew of Reading University, England, P. Main, M. M. Woolfson and G. Germain of York University, England, and Louvain University, Belgium, C. Tsai and J. Donohue of the University of Pennsylvania in collaboration with W. E. Streib of Indiana University and by W. Hoppe, J. Gassmann and K. Zechmeister of the Institut für Eiweiss und Lederforschung, Munich.

7.3 IRRADIATION PRODUCTS OF NUCLEIC ACID BASES

Radiation damage to nucleic acids has broad implications in living systems and is a subject which has been receiving increased attention. Of particular concern are the reactions forming photoproducts which are apparently either irreversible or very difficult to reverse, resulting, for example, in permanent damage to hereditary material. It has been noted[22] that pyrimidines are much more sensitive to photochemical change than purines and that ribose and deoxyribose are essentially insensitive to light of wavelengths greater than 2300 Å. The vulnerable site in pyrimidines is the double bond in the six-membered ring. Thymine, uracil and their derivatives have often been used as starting materials for investigations of radiation damage. The products which are obtained have been identified as substances which are also formed by the action of radiation on DNA or RNA.

Several types of irradiation reactions have been observed for nucleic acid bases. Among them are the formation of cyclobutyl-type dimers as described by Wang[23–26] and by Beukers and Berends[27,28], adducts[29], and polymers[30], by the action of u.v. radiation (2537 Å) on frozen aqueous solutions. Crystals have been obtained for each of these types of products and elucidation of the structures has confirmed their molecular formulae and established their stereoconfigurations.

The four possible types of cyclobutyl isomers formed by the dimerisation of thymine and uracil are shown in Figure 7.1. The *cis-syn* type is the most

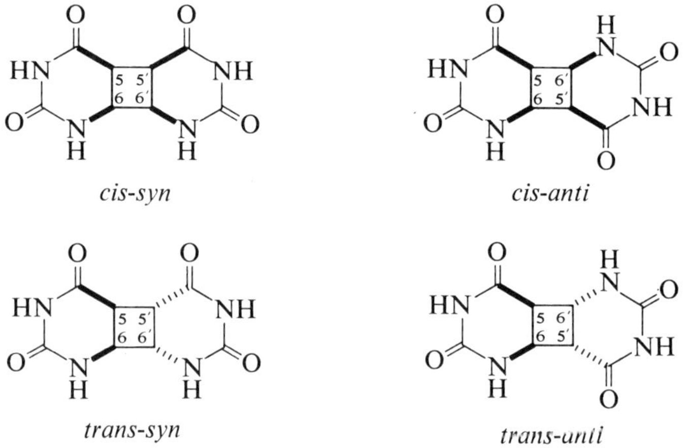

cis-syn *cis-anti*

trans-syn *trans-anti*

Figure 7.1 The four possible cyclobutyl type dimers

prevalent in photochemical reactions[26,31] while the *trans-syn* occurs in such small quantities that crystals suitable for x-ray analysis have not been isolated yet. Structure analyses have been made on the *cis-syn* dimers of uracil by Adman, Gordon and Jensen[32], of 6-methyl uracil, crystallising with one molecule of water, by Gibson and Karle[33], and of dimethyl thymine by Camerman and Camerman[34]. In each of these *cis-syn* isomers, the molecule assumes a cup shape. The four-membered rings are markedly puckered with dihedral angles of 155 to 162 degrees and the six-membered rings have a saturated C(5)—C(6) bond. As a consequence, the rings are no longer planar and the deviations from planarity are somewhat different in each of the molecules studied. A common feature of the non-planar configuration is that C(6) usually occurs out of the plane of the ring. A *cis-syn* internal dimer (1) formed by u.v. action on 1,1′-trimethylenebisthymine has been investigated by Leonard *et al.*[35]. It differed in conformation from the above-mentioned *cis-syn* dimers in that the four-membered ring is virtually planar, a constraint on the conformation probably imposed by the trimethylene bridge.

Structures of the *cis-anti* dimers of 1,3-dimethylthymine were determined by Camerman, Weinblum and Nyburg[36], of uracil by Konnert and Karle[37],

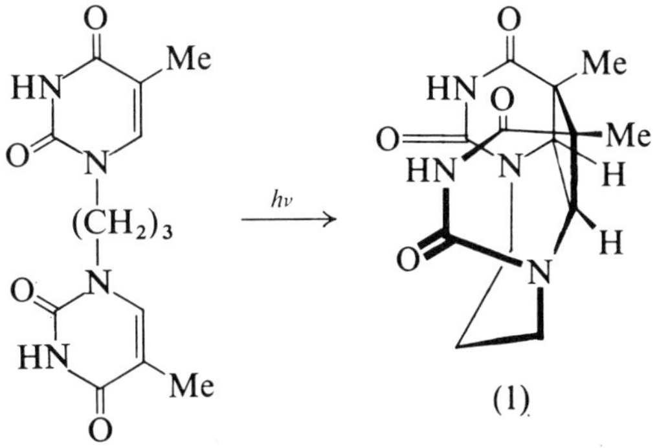

(1)

and of 6-methyl uracil by Flippen[38]. These dimers crystallised in space groups *Cc*, $P\bar{1}$ and *C*2/*c*, respectively. The dimer of 6-methyl uracil possesses a twofold axis in the molecule as required in this case by the space-group symmetry. The other two molecules very nearly contain a twofold axis. The major difference in the structures of the three *cis-anti* dimers is in the configuration of the cyclobutyl ring. In the first two molecules, the rings are puckered with dihedral angles of 154 and 150 degrees, respectively, whereas in the dimer of 6-methyl uracil the ring is virtually planar even though there are no apparent constraints.

Trans-anti dimers of thymine[39, 40] and 1-methylthymine[41] crystallise with the centres of symmetry of the molecules coincident with centres of symmetry in the crystal lattices. In these cases, the four-membered rings are planar.

Although conformations of the cyclobutyl rings and the six-membered rings can be different, and although extensive hydrogen bonding of the type NH···O=C occurs in some of the crystals and none at all exists in others, the bond lengths for comparable bonds in the thymine and uracil moieties in all the cyclobutyl dimers are nearly the same and are comparable to the values in dihydrothymine[42] and dihydrouracil[43].

The cyclobutyl-type dimers revert to the monomeric state upon exposure to additional radiation. In fact, in three of the crystal-structure determinations of *cis-syn* dimers from thymine derivatives[34, 35, 44], the x-rays partially dissociated the dimers into monomers in the crystalline state. A different kind of thymine–thymine product has been isolated from ultraviolet-irradiated DNA and from frozen aqueous solutions of thymine[29] which does not appear to revert to the monomeric state upon additional irradiation and may represent permanent damage when it occurs in the DNA molecule. A crystal-structure analysis of the adduct[45] has confirmed the molecular formula and established the stereoconfiguration to be (2).

The formation of the adduct may proceed through an intermediate[29]:

(2)

Transfer of an oxygen atom from one thymine moiety to another creates a 5-hydroxydihydrothymine moiety, ring I. However, the Me group on C(5) is in an axial position in contrast to dihydrothymine[42] and dihydrothymidine[46] where the Me group is equatorial.

A trimer of thymine is another photoproduct which has been isolated from irradiated frozen aqueous solutions of thymine[30]. An *E*-map was computed for this crystal[47] with 458 reflections whose phases were determined by the symbolic addition procedure, Figure 7.2. The 29 peaks corresponded to the

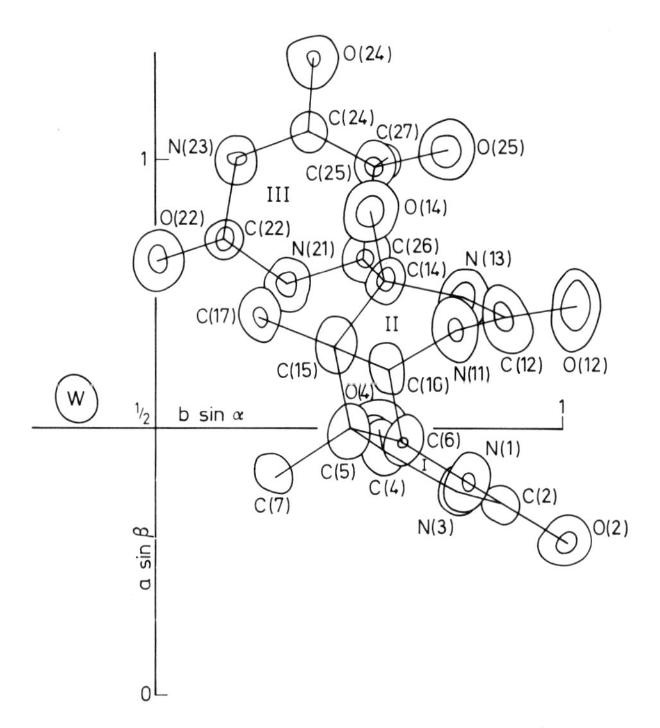

Figure 7.2 The structure of a hydrated thymine trimer[47]. The sections are from a three-dimensional *E*-map computed with 458 reflections whose phases were determined directly from the structure factor magnitudes. The space group is $P\bar{1}$

C, N and O atoms of the trimer (3) and to the oxygen atom of a water of crystallisation. Rings I and II resemble the *cis-syn* cyclobutyl dimer whereas rings II and III resemble the thymine–thymine adduct with an additional OH and H moiety on ring II. Two internal hydrogen bonds (C=O...HN and OH...O) increase the rigidity of the molecule. Additional irradiation of the

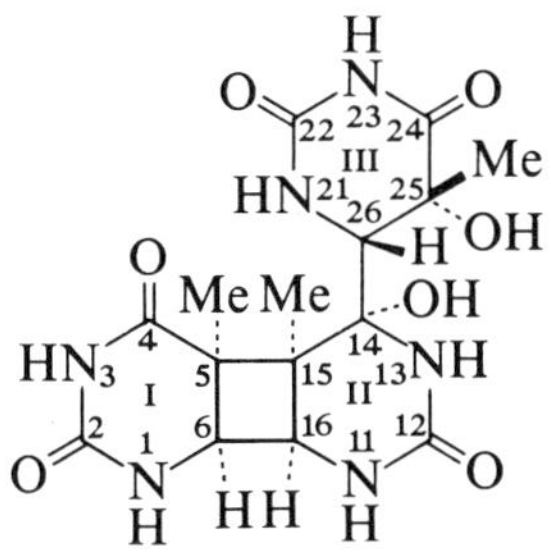

(3)

molecule causes the cyclobutyl ring to rupture and the end products are thymine and the thymine–thymine adduct.

A phototetramer of nucleic acid bases (4) has been prepared[48] by the u.v. irradiation of the dehydrated uracil–thymine adduct, analogous to a dehydrated thymine–thymine adduct:

$h\nu$ H_2O Me_2SO_4

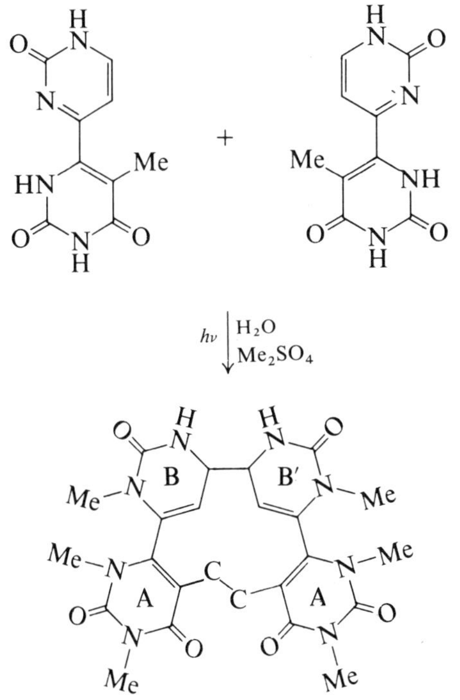

(4)

The crystal structure analysis of the tetramer[49], established the nature of the molecule. It showed that the head-to-head and tail-to-tail dimerisation of the uracil–thymine adduct was accomplished by the formation of a dimethylene bridge with the Me groups on C(5) in rings A and A′ and by a *trans* junction between atoms C(6′) of rings B and B′. The 12-membered ring of carbon atoms created by the dimerisation has four double bonds with two *cis* and two *trans* conformations. Although not required by the space group, the molecule very nearly contains a twofold axis of rotation.

The action of u.v. light on thymine and similar molecules can be summarised in the following diagram:

$$\text{thymine} \overset{h\nu}{\rightleftharpoons} \textit{cis-syn}\ \text{dimer} + \text{other dimers}$$

$$\text{thymine} \xrightarrow{h\nu} \text{adduct} \xrightarrow[h\nu,\ \text{MeOH}]{-H_2O} \text{tetramer}$$

$$\text{thymine} \xrightarrow{h\nu} \text{trimer} \xrightarrow[-H_2O]{h\nu} \text{adduct}$$

The photoreversibility of the dimer and the relative stability of the adduct may play a significant role in the photobiology of the nucleic acids.

7.4 NUCLEIC ACID DERIVATIVES

Nucleotides are the basic chemical constituents of DNA and RNA. In addition, these materials and their derivatives, such as nucleosides and other modifications, play an important role in the biochemistry of living organisms. The β-nucleosides are the more abundantly occurring ones. They are present in coenzymes and are the sole constituents of nucleic acids. The α-nucleosides are constituents of certain coenzymes such as vitamin B_{12} coenzyme and the DPN–DPNH system.

It is of considerable interest to investigate the three dimensional structure of these substances in order to develop relationships between structure and function. There is also an interest in building models of large molecules such as DNA from the structural information obtained about their component units. Since a variety of conformations appears to be readily accessible to some of the components, a significant degree of arbitrariness in model building may be inevitable. The following discussion will contain examples of the high degree of variation, especially in the sugar moiety.

Some examples of adenosine derivatives which have been recently investigated with direct methods of x-ray structure analysis are adenosine triphosphate (5) by Kennard *et al.*[50], 3′-*O*-acetyladenosine (6) by Rao and Sundaralingam[51], α-D-2′-amino-2′-deoxyadenosine monohydrate (7) by Rohrer and Sundaralingam[52], and 2′-*O*-tetrahydropyranyladenosine (8) by Kennard *et al.*[53]. The important molecule, adenosine triphosphate (ATP) is a coenzyme which plays a major role in energy conversion in living systems. For example, more than half of the food energy derived from oxidation in the human body is converted to the free energy of ATP. It, in turn, supplies the energy for processes such as muscular contraction.

On introducing the tetrahydropyranyl group into adenosine to form (8), diastereoisomers were formed. These differ in the configuration of the C(2″) of the tetrahydropyranyl group. The structure of the more *laevo* compound

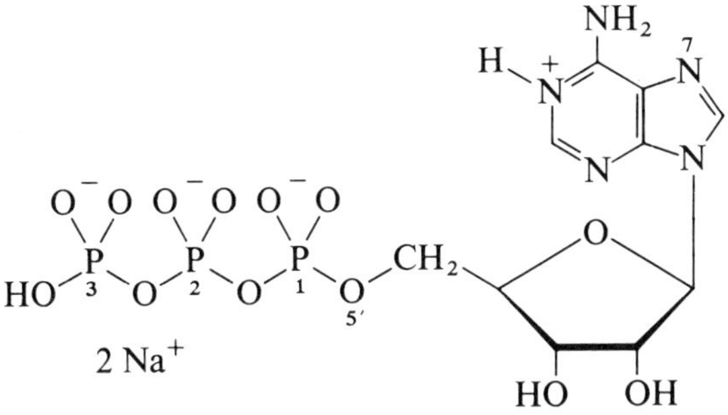

(5)

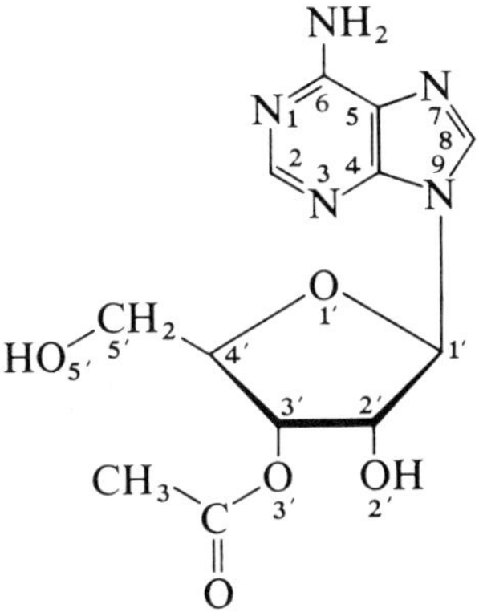

(6)

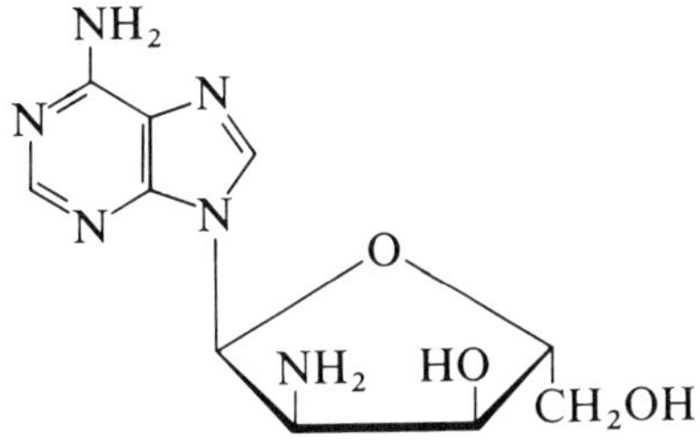

(7)

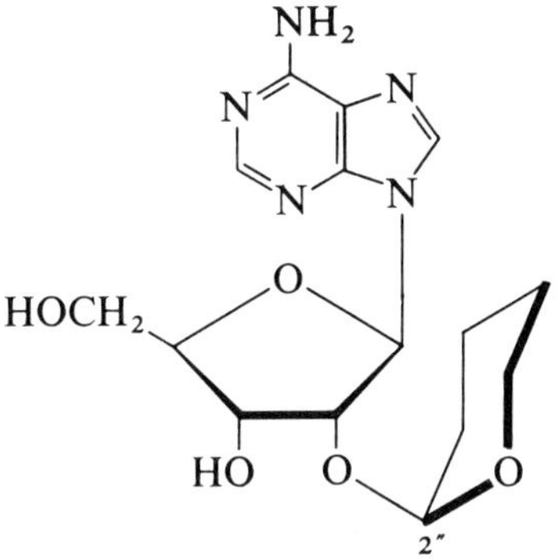

(8)

was investigated and the *S*-configuration was confirmed (Figure 7.3). The relationship among the ring systems in adenosine may be seen from Figure 7.3. The purine ring system is seen to be essentially planar and the ribose ring is rotated approximately perpendicular to the purine group. Unlike inorganic triphosphates the triphosphate chain in ATP is not extended, but rather curls back toward the purine group. The four adenosine derivatives discussed here have bond distances and bond angles for their common parts which are quite comparable. However, variations are found in the conformation of the ribose ring and in the angle of orientation of the ribose group with respect to the purine group expressed in terms of the glycosidic torsional angle. The conformation of the ribose ring has been usually found to have the envelope conformation with C(2′) or C(3′) out of the plane of the other four atoms. It was reported that in (6) the ribose ring is in the envelope conformation with a C(2′) *endo* pucker of 0.56 Å. In contrast both (7) and (8) have a ribose ring in the half-chair conformation with C(2′) in an *endo* pucker of −0.15 and −0.29 Å, respectively, and C(3′) in an *exo* pucker of +0.34 Å for each with respect to the plane of the remaining atoms in the ring.

A nucleotide is formed from a nucleoside by attachment of a phosphate group to O(5′) similar to the attachment of the triphosphate in ATP. The

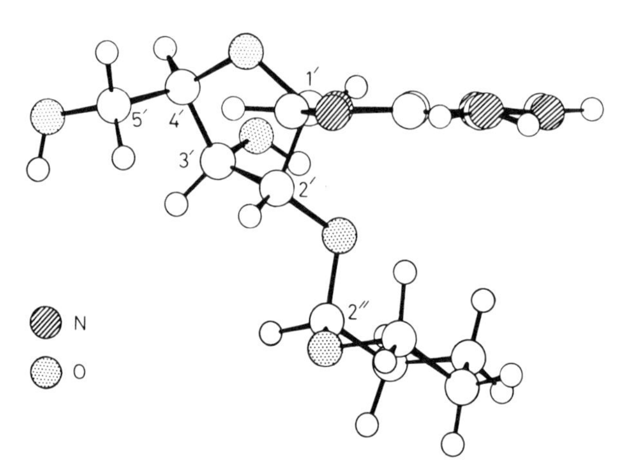

Figure 7.3 2′-*O*-Tetrahydropyranyladenosine. View parallel to the plane of the adenine ring system[53]

orientation of O(5′) with respect to the ribose ring is therefore of interest. The orientation may be defined in terms of the rotation of O(5′) about the C(4′)—C(5′) bond with respect to the O(1′)—C(4′) and C(3′)—C(4′) bonds. The conformation in (6) is *gauche-gauche*, in (7) it is *gauche-trans* and in (8) it is *trans-gauche*.

The glycosidic torsional angle is a measure of the rotation of the sugar unit with respect to the base about the bond connecting the two. It was reported that (6) is the *syn* conformation with the torsional angle $\chi = -140$ degrees. The *syn* conformation in (6) is thought[51] to be a consequence of the fact that the 5′-hydroxy bond is rotated from its usual conformation so as to form an intramolecular hydrogen bond, Figure 7.4, with N(3) of the base (2.77 Å). In contrast to (6), the conformation for (7) is *anti* with $\chi = -60$ degrees. The latter conformation has been found in several α-nucleosides[54]. In (8) the conformation is *anti* with $\chi = +44.8$ degrees and the preliminary report on ATP indicates that the conformation is also *anti*.

The structure of two pyrimidine nucleosides and a nucleotide having unusual structural properties have been investigated by direct methods. They are dihydrothymidine (9) investigated by Konnert, Karle and Karle[46], α-pseudouridine (10) by Rohrer and Sundaralingam[55] and the triethylammonium salt of cyclic uridine-3′,5′-phosphate (11) by Coulter[56, 57]. The molecule α-pseudouridine is the isomer of the naturally occurring β-pseudouridine which is found as a minor constituent of transfer RNA. Linkage of

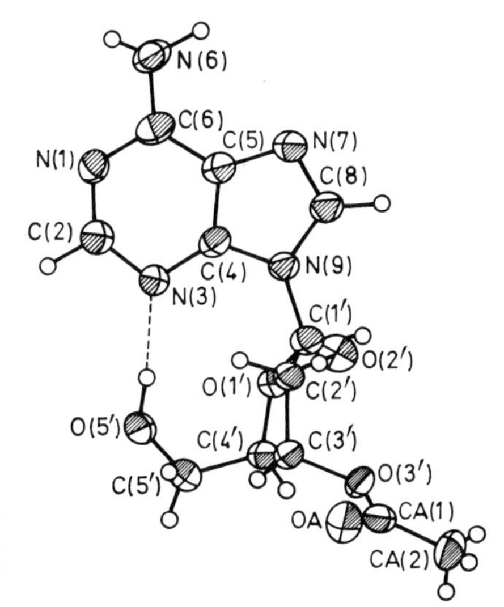

Figure 7.4 3′-*O*-Acetyladenosine. Conformation of the molecule with the internal hydrogen bond and ellipsoids of thermal vibrations illustrated
(From Rao and Sundaralingam[51], by courtesy of The American Chemical Society)

(9)

(10)

(11)

the sugar to the base in (10) is made by means of a bond between C(1′) and C(5) instead of between C(1′) and N(1). This accounts for the prefix *pseudo*. The structures of α-pseudouridine (10) and the cyclic nucleotide (11) were determined by finding the location of some of the atoms by means of the Patterson function[1] and developing the partial structure to completion by means of the tangent formula[18] (equation (7.9)).

The conformations of the ribose moiety in (9), (10) and (11) are considerably different. In (9), the ribose ring assumes an unusual conformation in which atom O(1′) is 0.42 Å out of the plane and is *endo* to C(5′). By contrast, in thymidine atom C(3′) is out of the plane by 0.57 Å and is *exo*[58]. In α-pseudouridine, C(2′) is out of the plane by 0.57 Å and is *exo*. In the cyclic nucleotide (11), C(3′) is *endo*. The three molecules have glycosidic torsion angles in the *anti* range.

It was found in the crystal of dihydrothymidine that atom O(5′) was disordered, occupying two different positions. Both the *gauche–gauche* and the *gauche–trans* conformations about the C(4′)–C(5′) bond are represented with the latter being favoured by a ratio of 2:1. For α-pseudouridine the conformation is *gauche-trans* similar to that for (7) and for the cyclic nucleotide it is *trans–gauche*, constrained by the spatial requirements for forming the cyclic phosphate.

7.5 PHOTO- AND CHEMICAL REARRANGEMENTS

Rearrangement reactions often produce quite unusual products. This has been observed to be especially true in the field of photo-rearrangements. The changes in chemistry and in configuration may be so drastic that the usual methods of chemical and physical analysis may not suffice for characterising the reaction products and establishing their structural formula. Given a sufficiently good crystal of a reaction product, the recent development in x-ray crystal structure analysis have made it possible to gain such information even in those instances when the substance of interest consists only of atoms of similar atomic number and its empirical formula is not known.

Some photo rearrangement reactions are a part of the normal physiology of living systems and are reversible. An example of considerable interest involves the substance 11-*cis*-retinal[59] which is the prosthetic group of the visual pigment and is its light-absorbing part. It is thought that the light absorption causes a photochemical transformation of the 11-*cis*-retinal to the all *trans*-isomer. This isomerisation is followed by processes described by Wald *et al.*[60] which lead to the phenomenon of sight.

A challenging structural problem arises with respect to 11-*cis*-retinal. It had been thought that the configuration of this isomer would resemble (12) with the possibility of relief from steric interference at the circled H atom and Me group by a rotation of 30–50 degrees about the single bonds adjacent to the C(11)=C(12) double bond[61]. The *trans* form was expected to have the extended configuration with a planar side-chain. The precise configurations of both of these materials were established by Gilardi *et al.*[62] employing x-ray analysis and are illustrated in Figure 7.5. The *trans* form has the expected configuration in the chain. The *cis* form assumes an unanticipated twisted configuration. The main feature is a rotation of the Me group attached to C(13) around the C(12)–C(13) bond by *c.* 135 degrees from the planar position illustrated in (12). The chain from C(6) to C(13) is essentially in one plane and the remaining atoms including C(13) are in another plane. In terms of bond distances, the alternation in values between *c.* 1.36 and 1.48 Å along the chain indicates that the conjugation is not disturbed despite the fact the C(13)═C(14) double bond is rotated by *c.* 135 degrees from the usual position

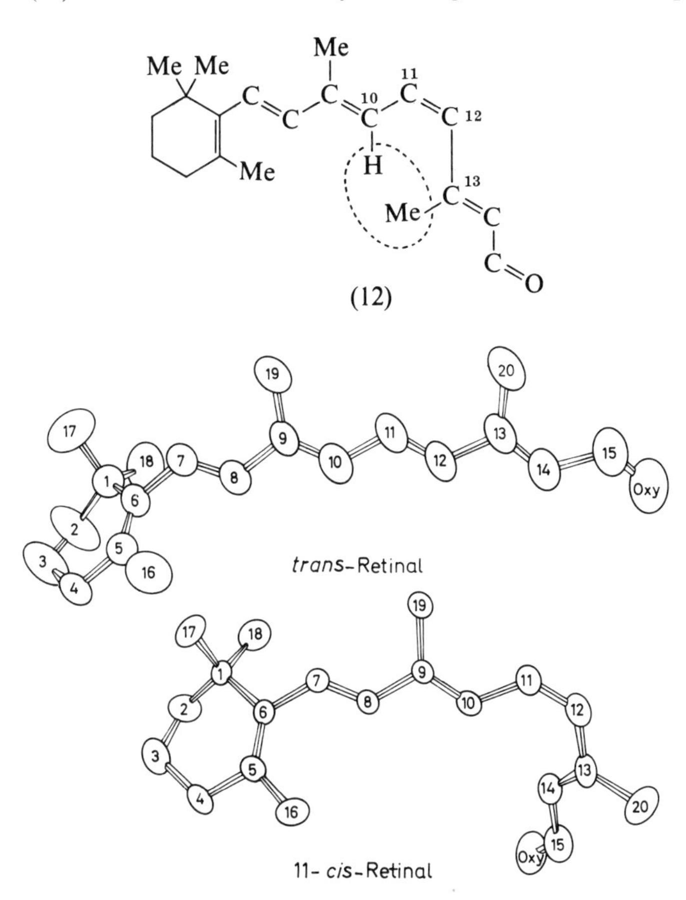

Figure 7.5 Configurations of 11-*cis*-retinal and *trans*-retinal as determined by x-ray analysis
(From Gilardi *et al.*[62], by courtesy of Macmillan Journals Limited)

in a conjugated system. The double bond in the ring is approximately *cis* to the first double bond in the chain.

Considerable chemical changes have been observed in many systems irradiated with u.v. light. A series of interesting studies have been carried out on pharmacodynamic amines. These investigations concern the type of problem which can offer the greatest difficulty in structure analysis, namely, a material of unknown molecular formula consisting of atoms having similar atomic numbers and crystallising in a non-centrosymmetric space group. The materials which were subjected to u.v. light were *N*-chloroacetyl-*p*-*O*-methyl-L-tyrosine (13), *N*-chloroacetyl-3,4-dimethoxyphenethylamine (15) and *N*-chloroacetylmescaline (17). The initial motivation for studying these materials was an interest in the possibility of ring closure with the elimination of chlorine[63]. This indeed occurs. However additional reactions of a complex nature take place as indicated by the following equations:

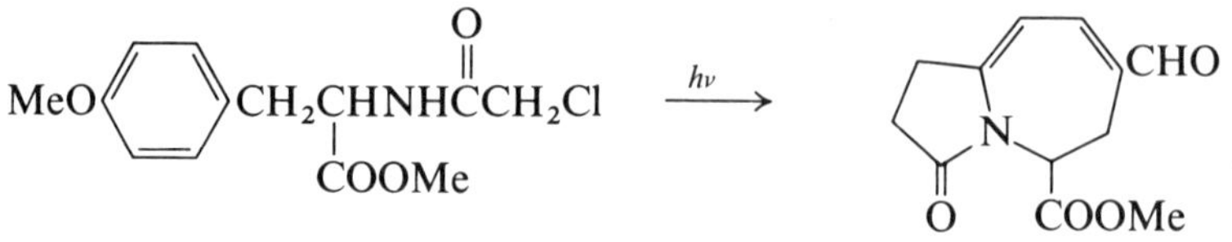

MeO ... $CH_2CHNHCCH_2Cl$ (O above C; COOMe below CH) — $h\nu$ → CHO, N, O, COOMe

(13) (14)

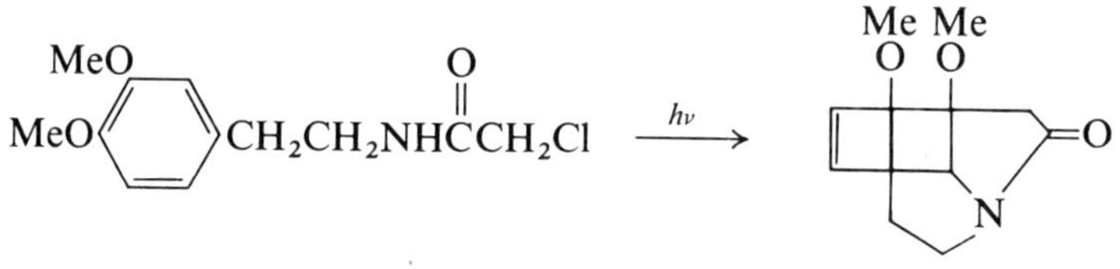

MeO, MeO ... $CH_2CH_2NHCCH_2Cl$ (O above C) — $h\nu$ → Me Me, O O, =O, N

(15) (16)

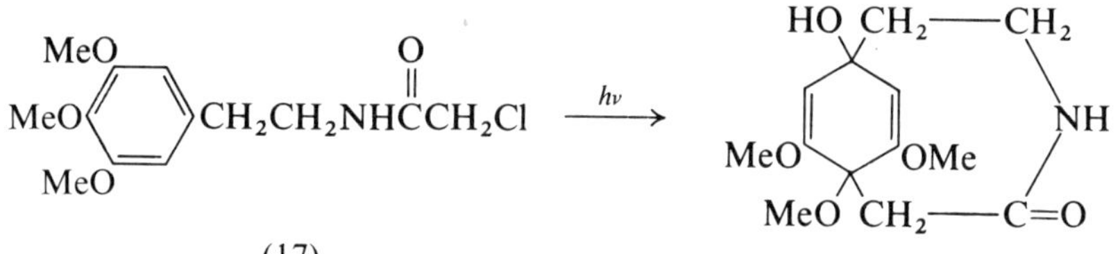

MeO, MeO, MeO ... $CH_2CH_2NHCCH_2Cl$ (O above C) — $h\nu$ → HO CH_2—CH_2, NH, MeO, OMe, MeO CH_2—C=O

(17) (18)

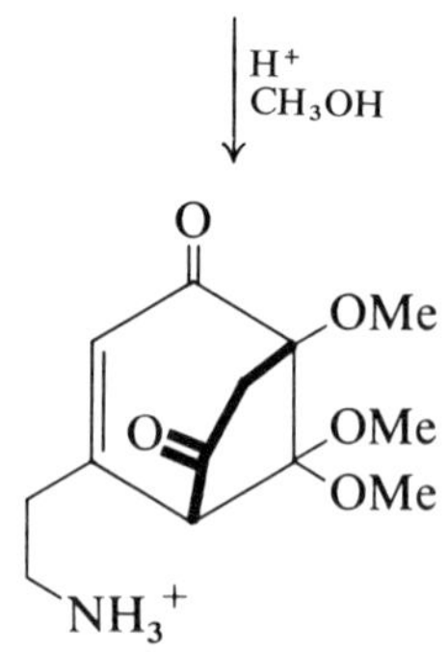

↓ H^+, CH_3OH

O, OMe, O, OMe, OMe, NH_3^+

(19)

Each of the products was identified by an x-ray structure analysis. As often occurs with non-centrosymmetric space groups, the initial E-map in the study of (14) revealed only a portion of the structure[64], Figure 7.6. Part (a) shows the six largest peaks in the first map. The five atoms represented by open circles were used to obtain initial phases for the tangent formula. The use of the tangent formula produced the ten large peaks shown in Part (b). A repetition of the procedure produced 14 atoms in Part (c), 16 atoms in Part (d) and 17 atoms in Part (e) (dotted lines refer to the last atom obtained). The dashed circles represent extraneous peaks whose heights corresponded to those of some of the weaker correct peaks. Extraneous peaks are eliminated by chemical considerations, but can also be readily eliminated by use of

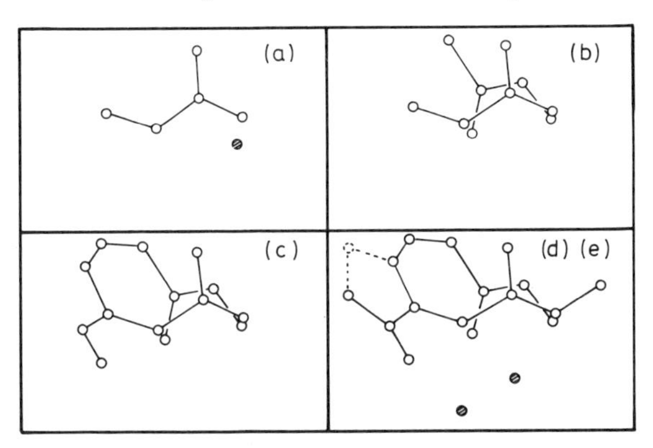

Figure 7.6 Development of a partial structure for (14)[64] to completion by use of the tangent formula (equation (7.9)) (From Karle *et al.*[64], by courtesy of the International Union of Crystallography)

more terms in computing the E-maps. The number of data used in the maps corresponding to the steps in Figure 7.6 was about one quarter of the total number.

The structures of (16)[65,66] and of (18)[67,68] were also obtained by the development of partial structures by means of the tangent formula. Despite the fact that there is a considerable degree of similarity among the starting materials (13), (15) and (17), there is a marked dissimilarity among the products (14), (16) and (18). The treatment of (18) with HCl for about 12–16 hours induced a chemical rearrangement shown in (19)[67]. The structural formula of (19), which was again established by x-ray analysis, is quite different from (18).

Other examples of photon-induced internal rearrangements occur in the strained aromatic system (20) of *anti*-[2,2]-paracyclophane. In the presence of oxygen the major reaction product is (21), whereas in the absence of oxygen the major products is (22)[69,70]. The reaction products (21) and (22) were confirmed by x-ray analysis[70]. The features of special interest are the interior six-membered rings, which must of necessity have the boat conformation, and the cyclobutane rings in (22). The latter were found to be highly puckered with dihedral angles of 124 degrees. This can be compared with bicyclo-[1,1,1]-pentane which was shown by electron diffraction[71] to

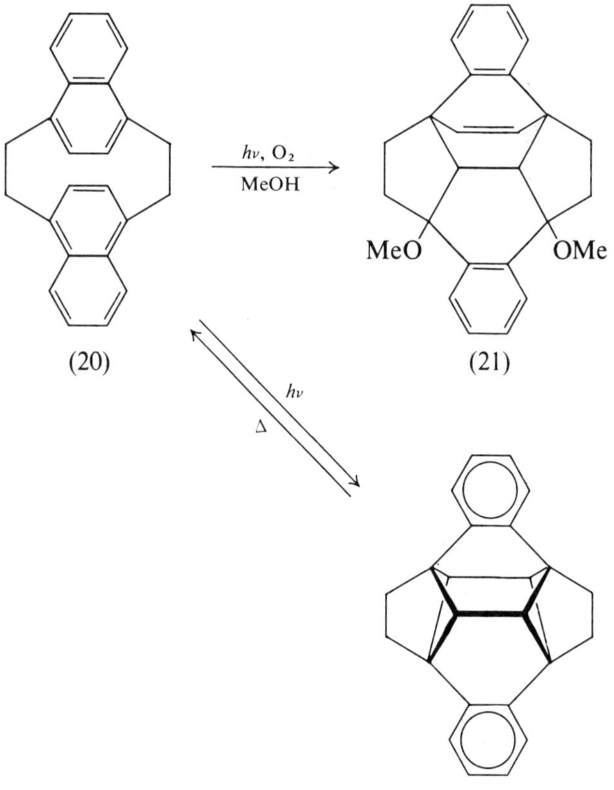

hν, O₂
MeOH

MeO OMe

(20) (21)

hν
Δ

(22)

have D_{3h} symmetry and therefore the dihedral angles must by 120 degrees for the four-membered rings. When unconstrained, cyclobutane rings are much flatter.

Another type of photosensitised autoxidation has been reported[72] from the irradiation of furano-paracylophane (23) in the presence of oxygen. One of the three products was characterised by x-ray analysis. The *E*-map contained the superposition of two molecules from a disordered structure. It was unravelled to yield the configuration shown in (24).

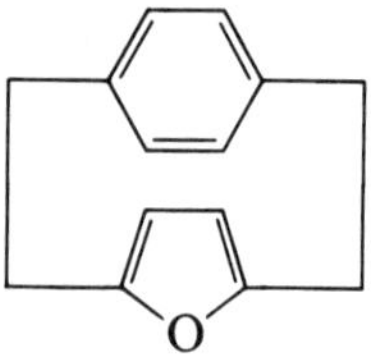

(23) 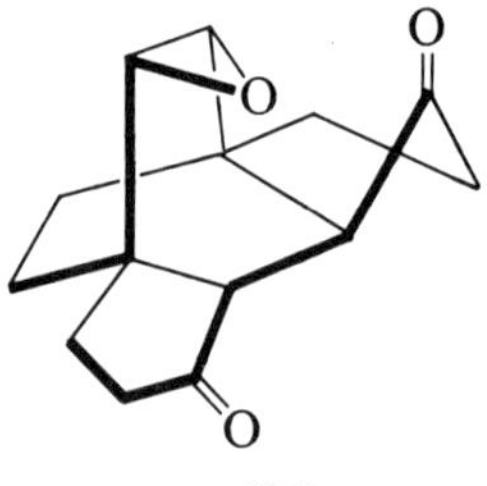 (24)

Similarly, unusual rearrangements have been observed to take place solely under the influence of chemical reagents. For example, the DPN–DPNH system can be described by

$$\text{(pyridinium ring, } N^{+}\text{–R, } CONH_2\text{)} + 2\,H \rightleftharpoons \text{(1,4-dihydropyridine ring, } N\text{–R, H H, } CONH_2\text{)} + H^{+}$$

where R is ribose pyrophosphate adenosine. This system is involved in biological oxidation-reduction, acting as coenzymes. If R is replaced on the right side of the equation by a methyl group and the compound is treated with acid[73], a dimeric product is formed which has been shown by x-ray structure analysis to have the complex ring system shown in Figure 7.7. Apparently the double bonds have opened to form the saturated cage compound.

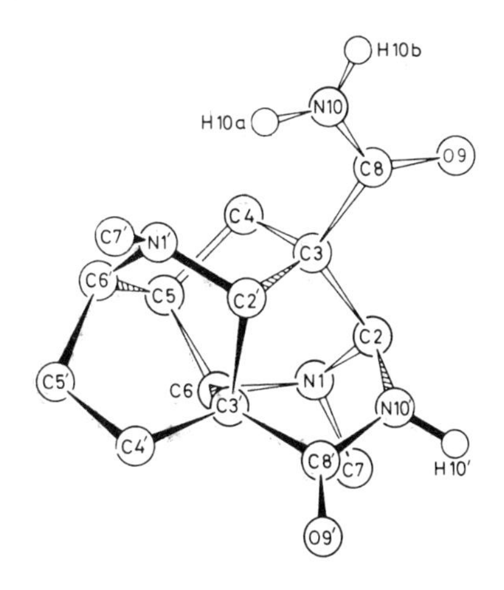

Figure 7.7 Dimerisation product from 1-methyl-1, 4-dihydronicotinamide
(From Ammon and Jensen[73], by courtesy of the International Union of Crystallography)

A quite unexpected product has been found from a high temperature treatment of α-phenyl-cinnamoyl chloride (25). The resultant product formed a red glass which was recrystallised from glacial acetic acid to form red–orange prismatic crystals. A structure determination showed that the molecule had the structural formula shown in (26)[74].

H COCl

(25)

$SOCl_2$, SO_2Cl_2
170–200°C

H
O Ph O

(26)

By means of the x-ray diffraction techniques, the structural formulae and configurations of entirely unanticipated products can be determined readily. This makes it possible to consider an investigation of a series of related photo or chemical rearrangement reactions, e.g. structures (13)–(19) and greatly facilitates an understanding of the reaction mechanisms. Reaction mechanisms have been proposed for several of the reactions which have been described here and may be found in the literature cited earlier.

7.6 CONFORMATIONS OF AMINO ACIDS AND POLYPEPTIDES

One of the striking results of structural investigations of protein molecules is that a variety of proteins performing the same function will have quite comparable three-dimensional shapes even though there is little correspondence in the amino-acid sequences. This is exemplified by the hemoglobins and myoglobins studied to date in which the amino-acid correspondences may be as low as 10 per cent. Nevertheless, the helical backbones in all of them assume comparable configurations. The amino acid proline plays a key role in interrupting the helical pattern, causing the backbone to turn. This characteristic is a consequence of the structure of the proline molecule which contains the imino group in a 5-membered ring. It will be seen that the ring can assume a variety of conformations and therefore its effect on the helix in a particular protein is not readily predictable.

Proline (27), principally found in collagen, has the envelope conformation[75] with the γ-carbon atom displaced by *c.* 0.6 Å from the plane formed by the nitrogen atom and the other three carbon atoms of the ring. The displacement of the γ-carbon atom is *cis* to the carboxyl group. In 3-hydroxyproline (28), found in scleroproteins and keratins, the γ-carbon atom is displaced by *c.* 0.4 Å from the plane of the other four atoms, but is *trans* to the carboxyl group[76]. By contrast, in 3,4-dihydroxyproline (29), occurring in the cell walls of diatoms, the β-carbon atom is displaced 0.60 Å from the plane of the other four atoms in the ring[77, 78]. In two additional investigations the conformations of the five-membered rings were found to be the same, but differed from each of the three above. *cis*-3,4-Methylene-L-proline (30) is a toxic constituent of horse chestnuts. The *trans*-isomer (31), also toxic, was prepared synthetically. In both molecules it is the nitrogen atom which is displaced by 0.40 Å from the plane of the four carbon atoms in the ring[79].

In the polypeptides containing proline which have been investigated to date both the β- and the γ-carbon atoms have been found displaced from the plane formed by the other atoms in the prolyl residue. The γ-carbon atom has been found to be displaced in tosyl-prolyl-hydroxyproline[80], leucyl-prolyl-glycine[81] and *p*-bromocarbobenzoxy-gly-L-pro-L-leu-gly(OH)[82]. The β-carbon atom is displaced in the cyclic dipeptide ┌L-pro-L-leu┐[83].

The amino acid arginine exists as a zwitterion in the crystalline state[84] with the extra proton on the guanidyl group rather than on the α-amino group, as is characteristic for α-amino acids. The nitrogen atom in the

Natural prolines

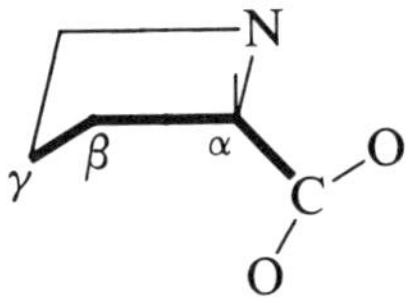

(27)

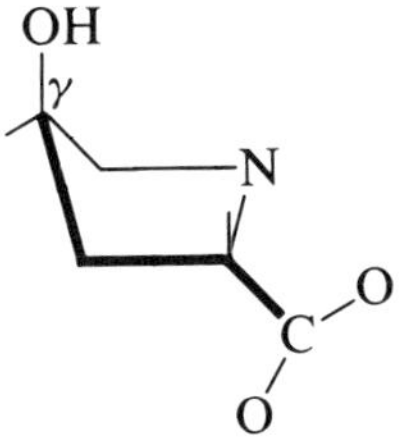

(28)

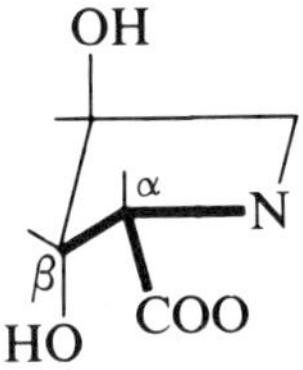

(29)

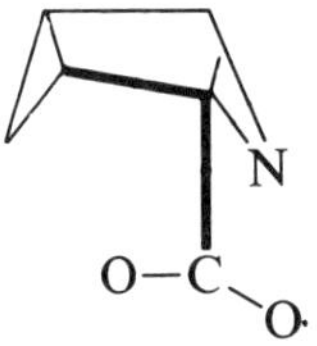

(30)
cis
natural

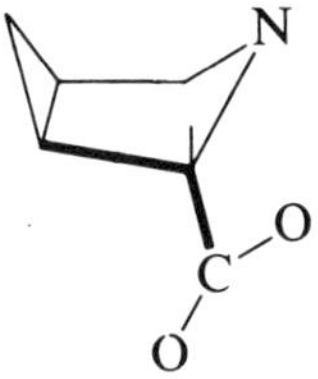

(31)
trans
synthetic

α-amino group has a tetrahedral configuration with the fourth position accepting a hydrogen bond from a guanidyl group of another molecule.

Two molecules of L-*N*-acetylhistidine monohydrate crystallise in a unit cell of space group *P*1. Materials crystallising in space group *P*1 are rare. The phase determination for this space group proceeded in the usual way for non-centrosymmetric crystals. The symbolic addition procedure led to an *E*-map showing a partial structure which was developed to a complete structure by means of the tangent formula. The two molecules of L-*N*-

acetylhistidine monohydrate in the unit cell have different conformations as shown in Figure 7.8[85]. Both nitrogen atoms in the imadazole ring are protonated. This is comparable to the protonation in arginine which occurs in the guanidyl group rather than on the nitrogen atom in the α-amino group.

Another unusual space group for a simple polypeptide is *I*4, the one in which L-alanyl-L-alanine was found to crystallise[86]. The packing of the molecules and the hydrogen bonding is illustrated in Figure 7.9.

Other investigations of amino acids and derivatives which have employed direct methods of structure analysis concern DL-valine[87], DL-*N*-chloroacetylalanine[88], the dipeptide L-alanylglycine[89], DL-acetylleucine *N*-methylamide[90], and 1,4-cyclohexadiene-1-glycine[91]. In the latter compound, the 1,4-cyclohexadiene ring is planar.

Cyclic polypeptides occur in nature and they have a number of important physiological properties. Significant among these is antibiotic behaviour. The diketopiperazine ring represented by cyclic diglycine occurs widely and is contained, for example, in the molecule of cyclic serine which is an effective agent against a tuberculosis bacillus. In considering the conformations of cyclic polypeptides, it is useful to categorise certain of the structural features of peptide groups and residues. A portion of a polypeptide chain is illustrated in (32) showing a peptide unit enclosed within the solid curve and a residue enclosed within the dotted curve. Peptide groups are found to be essentially planar and usually assume the *trans*-conformation (33), as is illustrated in L-alanylglycine[89], Figure 7.10.

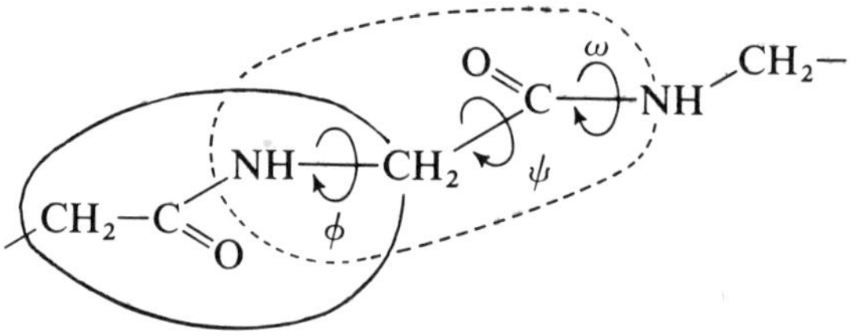

$CH_2-C(=O)-NH-CH_2-C(=O)-NH-CH_2-$ (rotations ϕ, ψ, ω)

(32)

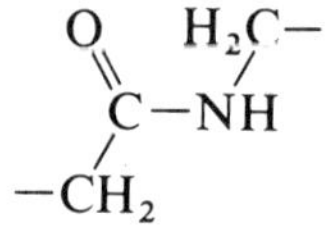

trans

(33)

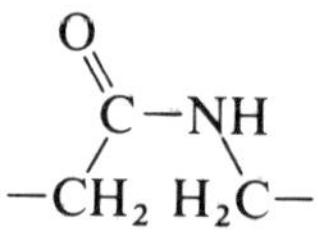

cis

(34)

The simplest examples of cyclic polypeptides are the cyclic dipeptides forming a diketopiperazine ring. In order to close the ring, it is necessary for the peptide groups to assume the *cis*-conformation (34). The investigation of the isomers, cyclic D-ala-L-ala and cyclic L-ala-L-ala[92], has shown that in the D L-isomer, the ring is nearly planar and the methyl groups are axial whereas in the L,L-isomer, the ring is in the twist-boat conformation and the methyl groups are closer to the equatorial positions. In the cyclic dipeptide L-pro-L-leu[83] the ring is found to be in the boat conformation and the

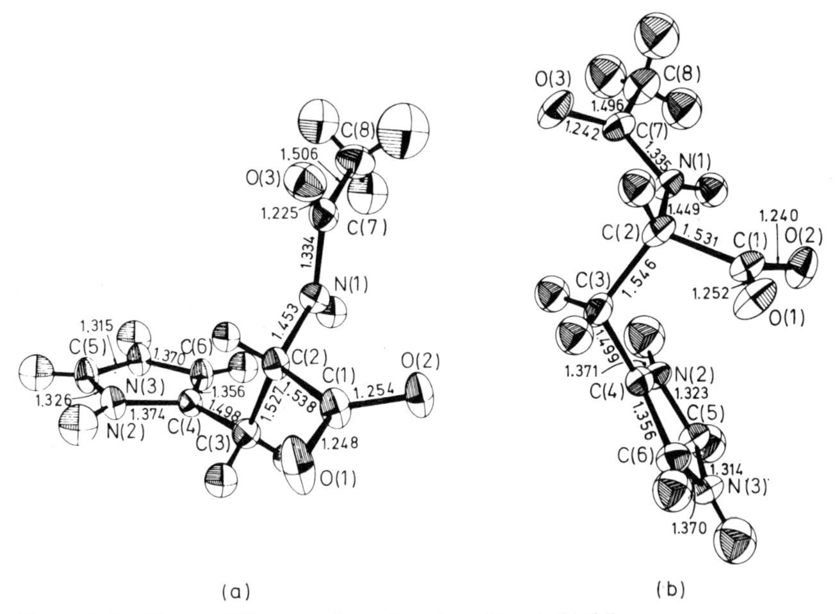

Figure 7.8 The two different conformations for L-*N*-acetylhistidine
(From Kistenmacher *et al.*[85], by courtesy of the International Union of Crystallography)

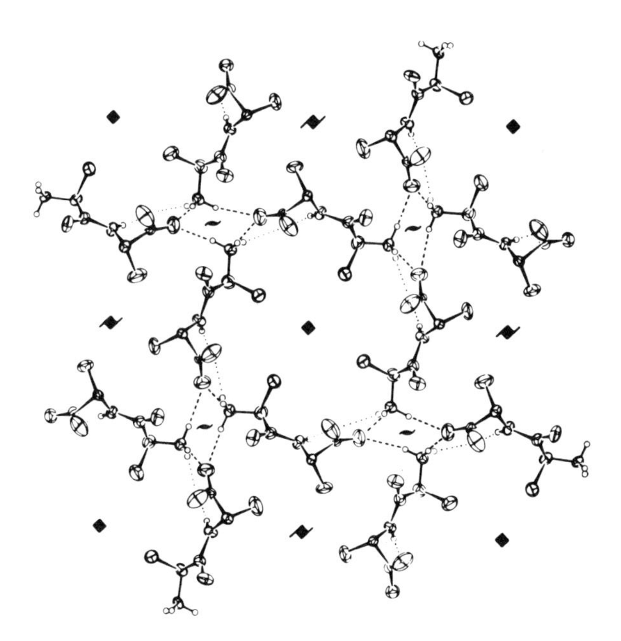

Figure 7.9 A view of the packing along the tetragonal axis of L-alanyl-L-alanine. The dotted and dashed lines indicate the hydrogen bonding scheme
(From Fletterick, Tsai and Hughes[86], by courtesy of The American Chemical Society)

dimethylethyl side chain of the leucyl moiety is equatorial and fully extended. As pointed out previously, the β-carbon atom is displaced from the plane of the other atoms in the prolyl group.

One group of naturally occurring cyclodepsipeptides has the general formula given in (35). Depsipeptides consist of alternating α-amino and

Me_2CH R

[CHCO—NCHCO]

D Me L

O O $]_n$

L D

COCHN—COCH

R_1 Me $CHMe_2$

(35)

α-hydroxy acid moieties. Certain cyclohexadepsipeptides have significant antibiotic behaviour. In particular enniatin A and enniatin B are cyclohexadepsipeptides with $n = 2$ in (35) and $R = R_1$ = CHMeEt for enniatin A and $CHMe_2$ for enniatin B. The conformation of the cyclotetradepsipeptide ⌐D-hyiv-L-Meileu-D-hyiv-L¬Meleu- has been investigated by x-ray analysis[93].

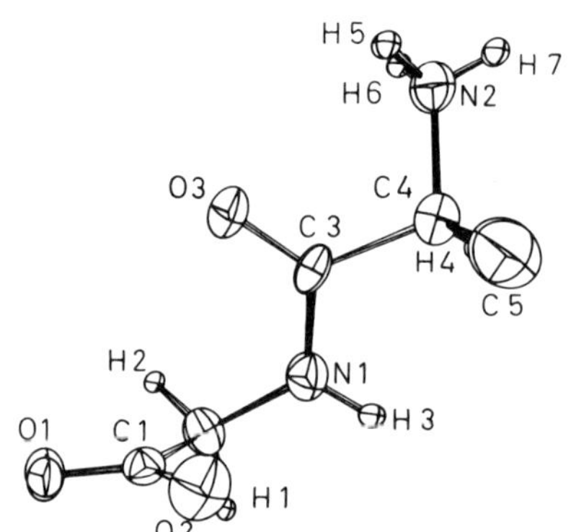

Figure 7.10 L-alanylgycine (From Koch and Germain[89], by courtesy of the International Union of Crystallography)

This molecule is represented by (35), when $n = 1$, R = CHMeEt and $R_1 = CH_2CHMe_2$ and its conformation is shown in Figure 7.11. It was found that the two α-hydroxy units are in the *trans*-conformation, which is comparable to (33) with O replacing NH. The peptide units are again in the unusual *cis*-conformation (34). All the carbonyl groups occur on one side of the average plane of the ring and are directed away from it, while all the hydrocarbon side chains occur on the opposite side of the average plane of the ring in the fully extended form. The conformation angles illustrated in Figure 7.11 follow the convention of Edsall *et al.*[94]. Their values are listed in Table 7.1.

The cyclic polypeptide cyclohexaglycyl crystallises in space group $P\bar{1}$ with eight molecules in the unit cell having four different conformations[95, 96],

Figure 7.12. Conformer (a), the only one with internal hydrogen bonds, NH···O=C, occurs four times, conformer (b) occurs twice and conformers (c) and (d) occur once each. In the four types of conformers, the peptide groups are planar and in the *trans*-conformation. Apparently a ring consisting of six peptide units is large enough to accomplish closure without requiring peptide

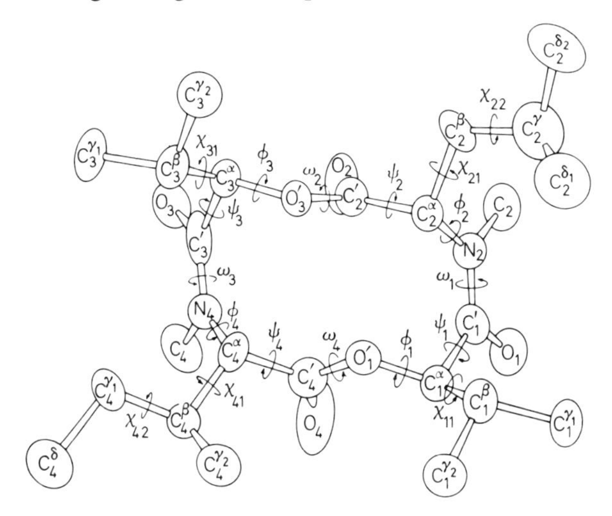

Figure 7.11 The configuration of –D-hyiv-L-Meileu-D-hyiv-L-Meleu– (From Konnert and Karle[93], by courtesy of The American Chemical Society)

Table 7.1 Standard conformational parameters

j	ϕ_j	ψ_j	ω_j	χ_{j1}	χ_{j2}
1	−88.7	−143.7	−177.5	−175.2	
2	77.9	−33.7	−2.6	−55.5	−50.9
3	−114.1	−141.2	−173.5	−172.9	
4	72.4	−59.2	0.3	−66.5	174.9

groups to assume a *cis*-conformation, as was found in the cyclic dipeptides and the cyclic tetradepsipeptide. In order to form a cyclic polypeptide, there are angular twists at the junctions of the planar peptide groups causing the residues (32) to assume various conformations. For the four conformers of cyclohexaglycyl, the approximate arrangements for the residues are:

(a) *cis, cis, trans, cis, cis, trans*
(b) *cis, cis, skew, cis, skew, skew*
(c) *cis, cis, skew, cis, cis, skew*
(d) *cis, skew, skew, cis, skew, skew*

The dihedral angles between the planes of adjacent peptide units range from 90–116 degrees.

A cyclic hexapeptide[96] which differs from cyclohexaglycyl by only two methyl groups, ⌐4-gly-2D-ala⌐, crystallises in space group $P2_12_12_1$ with only

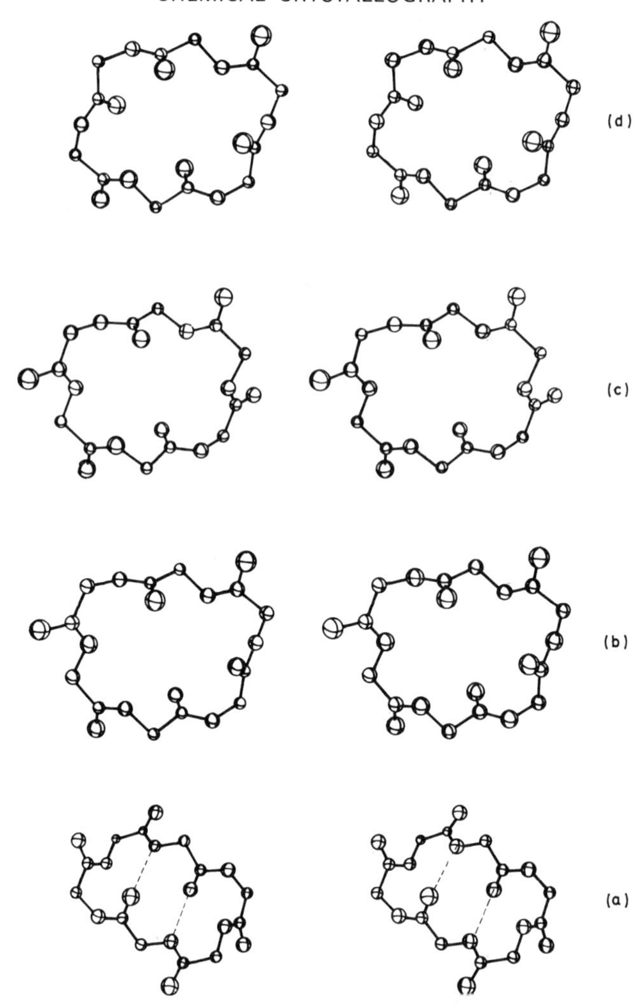

Figure 7.12 A stereodiagram of four different conformers which exist in one unit cell of cyclohexaglycyl·$\frac{1}{2}H_2O$. The different size spheres depict C, N and O atoms in order of size[95, 96]

Table 7.2 Conformational angles for *cyclic* gly-gly-D-ala-D-ala-gly-gly

	$j = 1$	$j = 2$	$j = 3$	$j = 4$	$j = 5$	$j = 6$
ϕ_j	73°32′	−78°54′	−114°20′	−49°05′	74°39′	109°52′
ψ_j	−163°50′	−6°01′	−164°47′	148°53′	11°38′	164°54′
ω_j	2°16′	1°54′	2°06′	4°06′	−3°40′	−8°19′

one molecule per asymmetric unit. This molecule resembles conformer (a) of cyclohexaglycyl, the most frequently occurring one, and is shown in Figure 7.13. It too contains two internal hydrogen bonds, NH...O=C. The conformation angles are listed in Table 7.2. The information contained in

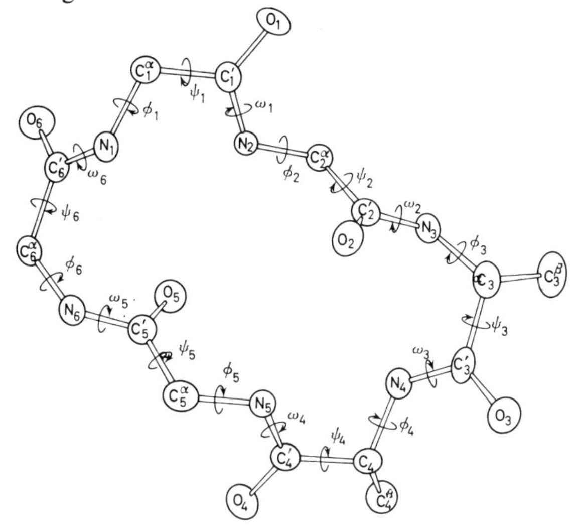

Figure 7.13 Configuration of ⌐gly-gly-gly-gly-D-ala-D-ala¬ and labelling of rotational angles
(From Karle *et al.*[96], by courtesy of The American Chemical Society)

Tables 7.1 and 7.2 is relevant to the formulation of minimum energy criteria[97] in attempts to predict the conformation of cyclic polypeptides.

7.7 LARGE RINGS

In the previous section several large ring systems, occurring in polypeptides and depsipeptides, were discussed. Other types of substances having large rings present similar conformational problems generated by steric hindrance and other intramolecular forces.

Conformational isomers, as exhibited in cyclohexaglycyl, also occur in a cyclic dimer of Nylon 66 (36)[98]. The four molecules in the unit cell of space group $P2_1/c$ consist of pairs of two conformational isomers, Figure 7.14. The

```
     O              O
     ||             ||
HN—C—(CH2)4—C—NH
 |                   |
(CH2)6            (CH2)6
 |                   |
HN—C—(CH2)4—C—NH
     ||             ||
     O              O
```

(36)

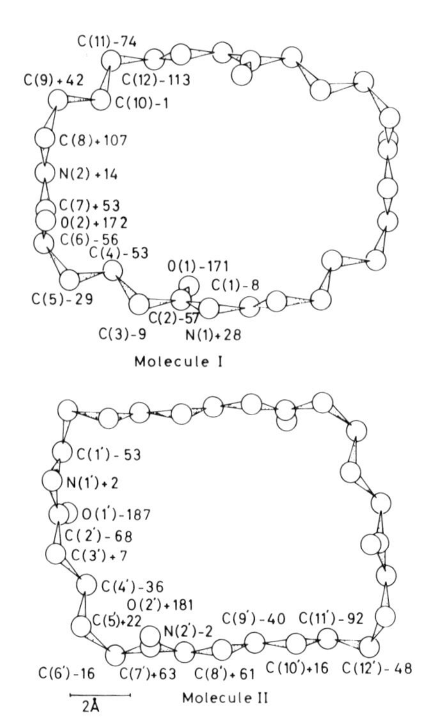

Figure 7.14 The two conformational isomers of a cyclic dimer of Nylon 66: The numbers indicate the perpendicular distances from average least-squares planes in $Å \times 10^2$
(From Northolt[98], by courtesy of the International Union of Crystallography)

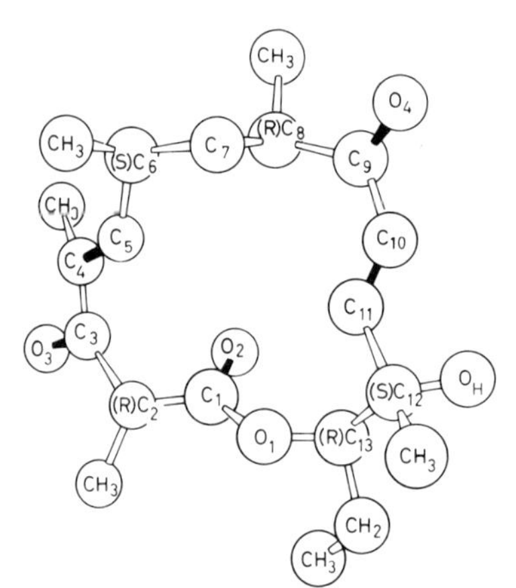

Figure 7.15 A perspective view of the kromycin molecule
(From Hughes *et al.*[99], by courtesy of The American Chemical Society)

molecules have 28-membered rings which are approximately rectangular in shape and they lie on symmetry centres. Hydrogen bonds, NH...O=C, link individual molecules together forming sheets. Northolt[98] points out that the conformers of the dimers are formed from a relatively small number of planar elements, each consisting of a planar amide group, a planar sequence of methylene groups, or both. This approximates the planar zigzag conformation which characterises a linear polymer.

The configuration of double bonds in rings is a matter of structural interest since they may be fixed in the *cis* or *trans* positions, and occasionally may even be skewed. For example, kromycin possesses a 14-membered ring (37) containing two double bonds. The spatial configuration of the ring[99] is shown in Figure 7.15. It may be seen that the double bonds C(4)=C(5) and C(10)=C(11) are in the *trans*-configuration in the ring.

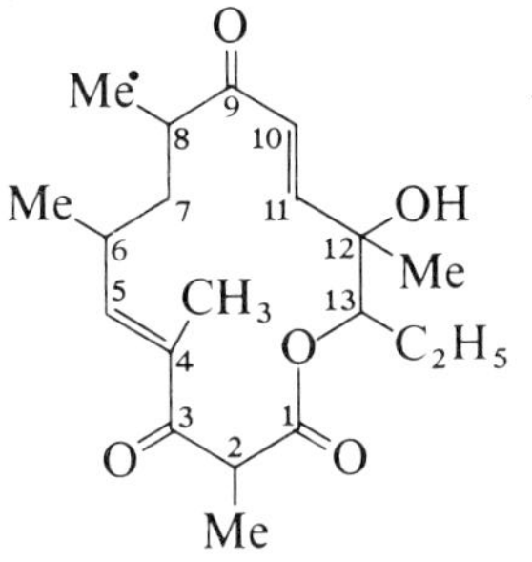

(37)

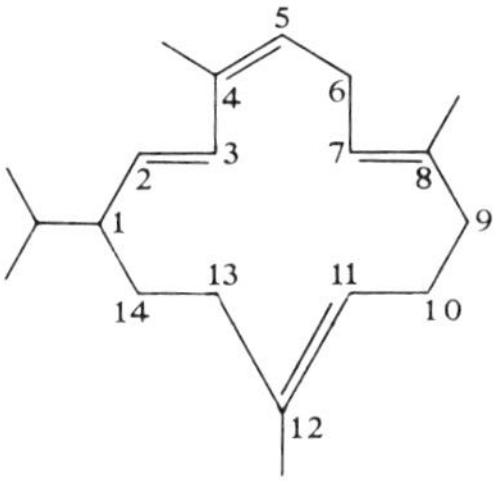

(38)

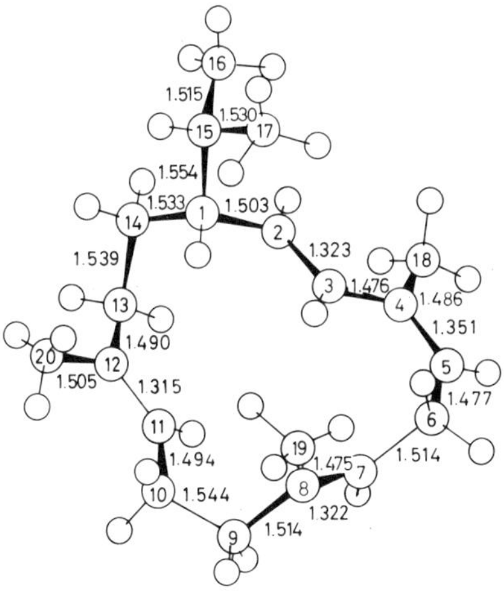

Figure 7.16 The configuration of the cembrene molecule with the C—C bond lengths indicated
(From Drew *et al.*[100], by courtesy of the International Union of Crystallography)

Cembrene is a diterpene possessing a 14-membered ring composed entirely of carbon atoms (38). Three of the four double bonds in the ring have been determined to be in the *trans*-configuration and the one at C(4)=C(5) is in the *cis*-configuration[100], Figure 7.16. In both kromycin and cembrene the rings are significantly puckered, but have planar segments.

The deep purple hydrocarbon $C_{16}H_{16}$, [16]-annulene, occurs with alternating single and double bonds. N.M.R. studies at low temperature (−110 °C) and higher temperatures imply that there are two conformational isomers. The x-ray analysis carried out at 4 °C revealed the presence of only one conformer[101]. It was found that the configuration of the double bonds alternated between *cis* and *trans* around the ring which is somewhat non-planar with atoms displaced as much as 0.57 Å from an average plane. The nonplanarity of the ring increases the separation between the four hydrogen atoms which point roughly toward the centre of the ring, Figure 7.17. The

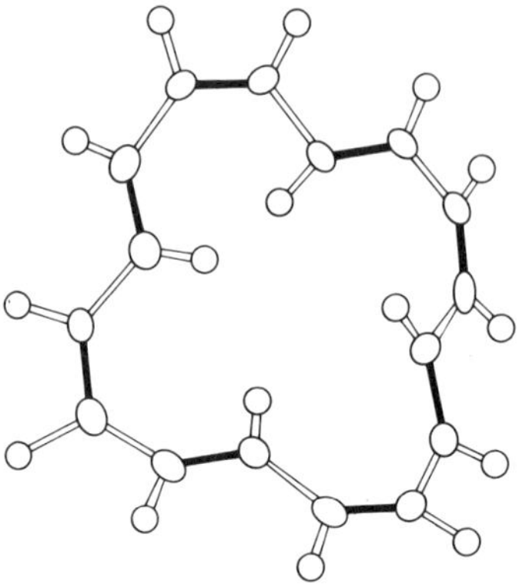

Figure 7.17 View of the [16]-annulene molecule. Two of the inner H atoms are on one side of the mean plane of the molecule and the other two inner H atoms are on the other side[101]

values for the bond distances in the ring are close to those found in conjugated systems.

The diketo tautomer (39) of 1,6-dihydroxy-[10]-annulene is a light brown material which possesses three double bonds in the ring as a consequence of the tautomerisation. An x-ray analysis has shown that the tautomer is in the *cis,trans,cis*-configuration[102]. The double bond at C(4)=C(5) has a somewhat skewed *trans*-configuration, Figure 7.18, since the torsional angle is 160 degrees instead of 180 degrees. Distortion about the two *cis* double bonds is very small.

Protopine (40) and the related cryptopine are alkaloids which occur in

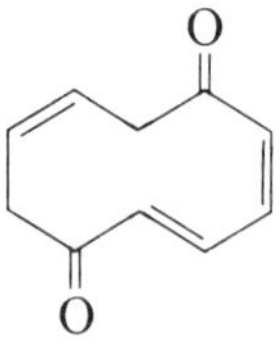

(39)

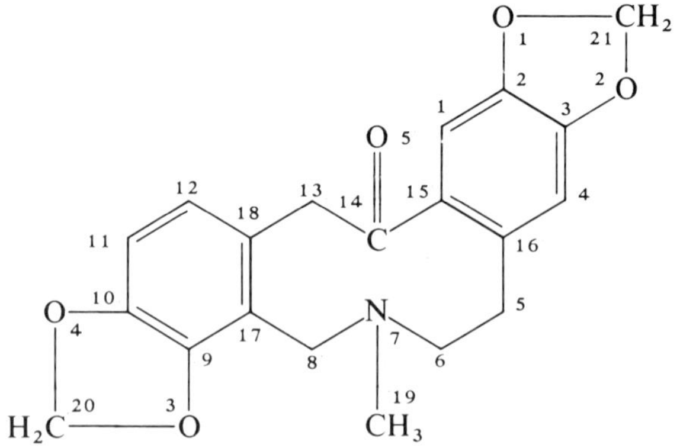

Protopine

(40)

opium. Both molecules contain a 10-atom ring which possesses a short non-bonded distance between C(14) and N(7), 2.555 and 2.581 Å for protopine and cryptopine, respectively[103, 104]. The bonding for N(7) is oriented so that the fourth position in a tetrahedral configuration corresponds to the relatively short non-bonded distance to C(14). It has been found to be fairly easy to form a bond across C(14) and N(7) to make two 6-membered rings. An example

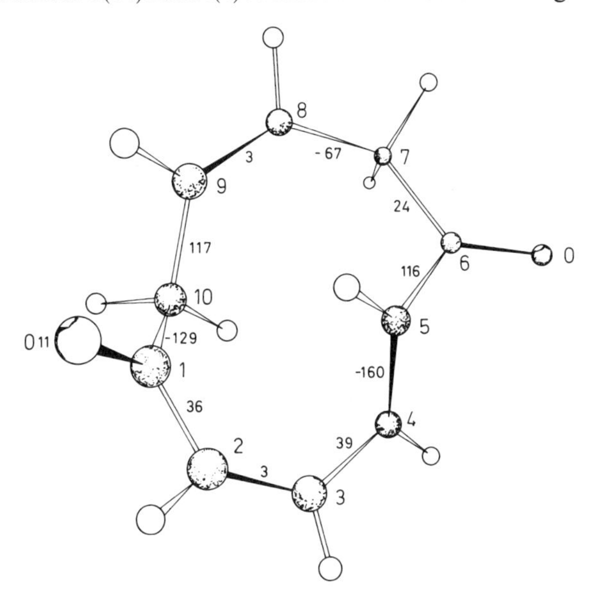

Figure 7.18 The configuration of the diketo tautomer of 1,6-dihydroxy-(10)-annulene showing the torsional angles about the C—C bonds[102]

of an *E*-map, the Fourier map obtained at the end of an initial phase determination, is shown for cryptopine in Figure 7.19. The calculation was made from 449 x-ray reflections for which phases were determined. The spatial configuration of the molecule is immediately apparent from such a map.

7.8 STEROIDS

The steroids comprise a large group of physiologically active substances and may be considered to be derivatives of the saturated hydrocarbon (41). In several instances, the character of the physiological activity appears to be

(41)

well correlated with the molecular shape. For example, *trans*-junctions between the rings or appropriately placed double bonds cause the molecule to flatten and extend. Cholesterol and the sex hormones are typical examples. *cis*-Junctions between the rings cause the molecule to curl up. Cardioactive steroids such as digitoxigenin have an A/B and a C/D *cis*-junction. Occasionally the steroid nucleus may vary significantly from (41) in that the C ring

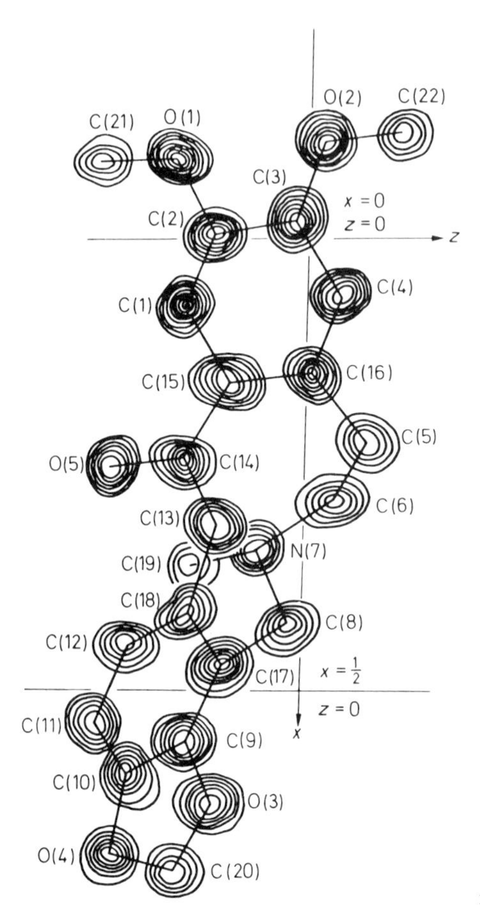

Figure 7.19 Composite drawing of the cryptopine molecule from an *E*-map computed with 449 reflections whose phases were determined directly from the structure factor magnitudes
(From Hall and Ahmed[104], by courtesy of the International Union of Crystallography)

may be 5-membered and the D ring 6-membered. Some steroids are alkaloids as well and, although they are relatively rare, they can be found in both animal and plant sources.

Digitoxigenin, Figure 7.20, occurring as a glycoside, is one of the active components of digitalis, a drug which is extensively used in heart therapy. Another cardioactive material is strophanthidin which can be obtained from the seeds of the African dogbane tree. The glycosides of digitoxigenin and strophanthidin have been used extensively in the past as active ingredients in arrow poisons. A structure investigation of strophanthidin[105] showed that the molecule crystallises in space group $P2_1$ with two molecules in the asymmetric unit of the unit cell. Drawings of the two molecules are shown in Figure 7.20. It is seen that there are only minor differences in the shapes of the two independent molecules. The crystal structure is characterised by extensive networks of hydrogen bonding, intramolecular, intermolecular and inter-asymmetric unit. However, although similar units are present, the crystal of

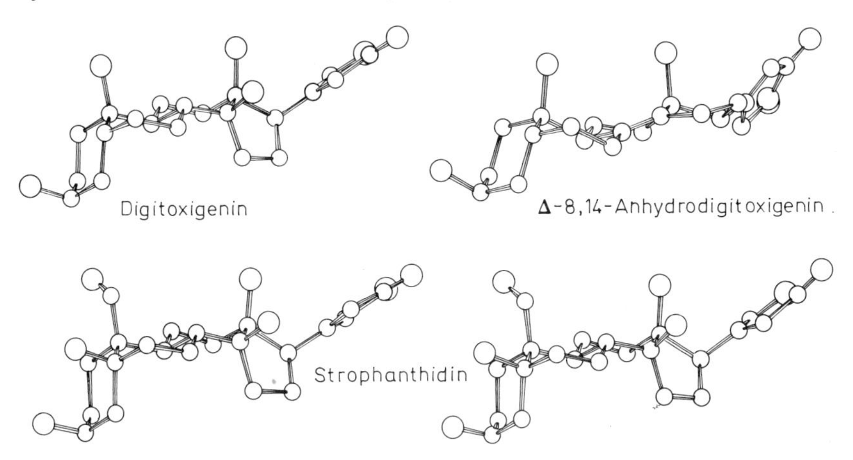

Figure 7.20 A comparison of the molecular configurations of digitoxigenin[106], Δ-8,14-anhydigitoxigenin[107] and the two independent molecules of strophanthidin[105]

strophanthidin did not show the head-to-tail hydrogen bonding characteristic of digitoxigenin[106]. It is apparent from Figure 7.20 that the molecular shapes of strophanthidin and digitoxigenin are almost identical. Each has a *cis*-junction between the A and B and C and D rings.

The molecule Δ-8,14-anhydrodigitoxigenin (42) is a dehydration product of digitoxigenin in which HOH is removed from the 8,14 positions. The resulting double bond has the effect of flattening the molecule in the vicinity of the C/D rings[107], Figure 7.20. It is of interest to note that this flattened molecule does not have cardioactive properties.

A substance closely related to the cardioactive steroids isolated by Tokuyama, Daly and Witkop[108] is batrachotoxin (43), the most potent

(42)

(43)

venom known. Batrachotoxin occurs in the skin of a brightly coloured Columbian frog, *Phyllobates Aurotaenia.* It is used by Columbian Indians as an arrow tip poison and has the worthwhile property of being digestible. However, it is quite lethal if it enters the blood stream directly, acting on the heart and the neuro-muscular system. x-Ray analysis played a significant role in establishing the molecular formula and stereoconfiguration of batrachotoxin. The structure analysis[109] was performed on a *p*-bromobenzoate derivative of batrachotoxinin A, a congener of batrachotoxin. The investigation was carried out on a crystalline needle which had a cross-section 0.03×0.05 mm. The number of scattering data from the small crystal was rather limited and the heavy atom, Br, was located on an axis in the unit cell giving rise to a fourfold ambiguity in the phase determination. It was possible to locate some light atoms in addition to the bromine atom and, using this as a partial structure, the entire structure was developed by means of the tangent formula (equation (7.9)). The stereoconfiguration of batrachotoxinin A, shown in Figure 7.21 has structural features in common with other steroidal cardioactive drugs, e.g. A/B and C/D *cis* ring junctions.

The hormonal steroids, in contrast, assume a relatively flat and extended configuration. An example is estriol (44), an estrogen which crystallises with

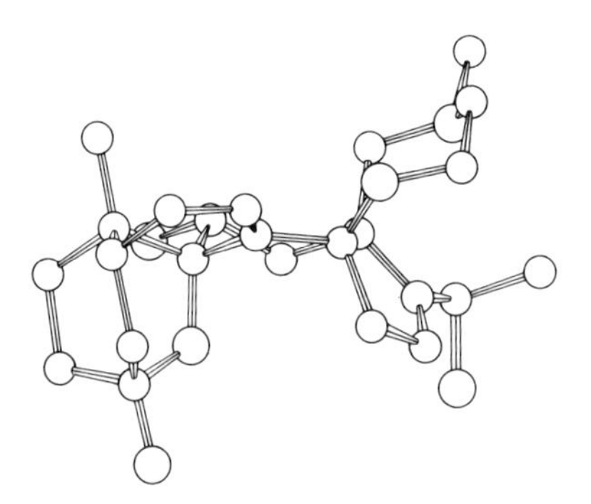

Figure 7.21 The stereoconfiguration of batrachotoxinin A
(From Karle and Karle[109], by courtesy of the International Union of Crystallography)

two molecules in the asymmetric unit[110]. The two molecules are not identical in that the B ring has a half-chair conformation in one and a twist-boat conformation in the other. The differences may be correlated with differences in the hydrogen bonding and the molecular environments.

An example of a synthetic steroid in which the B ring is a 5-membered ring is DL-9β-B-norestrone methyl ether (45). The molecule is unusual in that a *cis*-

(44)

(45)

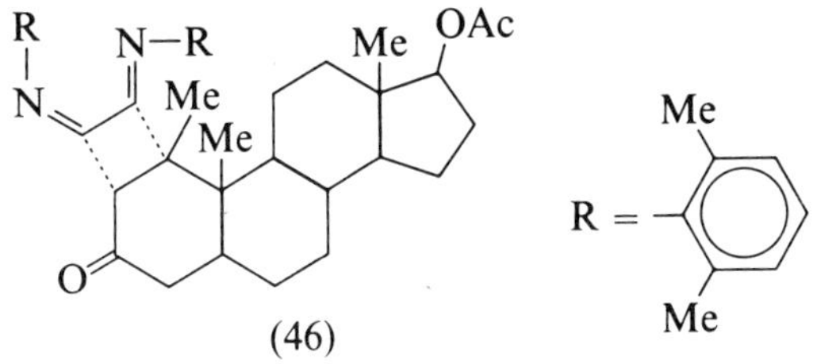

(46)

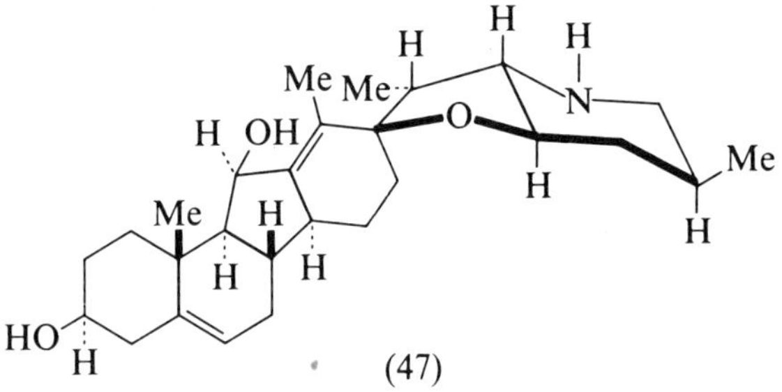

(47)

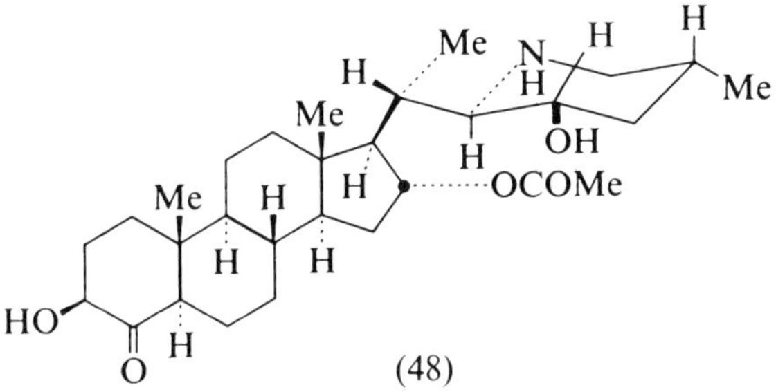

(48)

junction occurs between the B and C rings[111] causing the molecule to bend sharply at the B/C junction.

The nature of the addition product (46) between an isocyanide and a steroidal α,β unsaturated ketone was established by an x-ray structure analysis[112]. The addition took place at the 1,2-bond instead of at the ketone group, as had been previously supposed.

The determination of the structural formulae for batrachotoxinin A and batrachotoxin (43) revealed that these substances are steroidal alkaloids. Examples of steroidal alkaloids arising from plant rather than animal sources are veratrobasine (47) occurring in *Veratrum album*, the European hellebore, and solaphyllidine (48) which may be extracted from the leaves and green berries of *Solanum hypomalacophyllum Bitter*, a tree native to the Venezuelan Andes. The structure investigation of veratrobasine[113] established the empirical formula as well as the structural formula and stereoconfiguration. It is seen in (47) that veratrobasine has the unusual C-nor-D-homo steroid ring system. In the case of solaphyllidine, the empirical formula was known at the outset of the structure investigation. However, the analysis established the structural formula and stereoconfiguration[114, 115].

7.9 ANTIBIOTICS, SUGARS AND ADDITIONAL NATURAL PRODUCTS

Actinomycin C_1(D) (49) is an antibiotic which is composed of a chromophore and two cyclic polypeptides. It binds to double helical DNA and inhibits RNA synthesis. The structure of an actinomycin C_1(D)-deoxyguanosine complex was solved by Sobell and co-workers[116] by means of the development of a partial structure by recycling on the tangent formula[18] with data from the

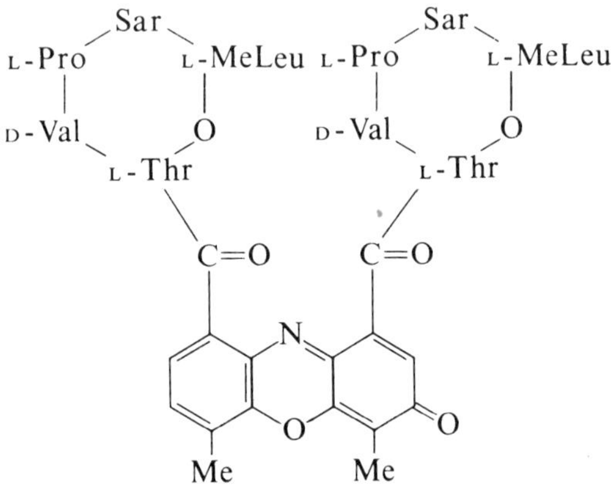

(49)

native complex. The development of the structure was cross-correlated with an analysis of the Patterson function[1] by means of vector search techniques of Nordman and Nakatsu[21] with data from a brominated derivative. It was found that models based on the structure determination helped to clarify the binding of actinomycin to DNA. The phenoxazone ring system of actinomycin slips into the DNA helix, while deoxyguanosine residues of the DNA interact with both cyclic peptides of the actinomycin through hydrogen bonds. The upper part of Figure 7.22 shows the phenoxazone ring system bordered by two deoxyguanosine residues which are attached to the cyclic peptides by means of the hydrogen bonds indicated by the horizontal dotted lines. The complex has twofold symmetry independent of the space-group symmetry which correlates with the twofold symmetry of DNA. The good fit of the actinomycin to the DNA helix correlates the structural properties of the repressor with its biological activity.

Myxin (50) is a fairly recently discovered antibiotic which is related to iodinin[117]. The structural formula for myxin was established by Hanson[118] by means of x-ray analysis, thereby distinguishing between two possibilities that had been proposed.

Daunomycin (51) has been found to interfere with nucleic acid metabolism in both mammalian and bacterial cells. It was found in the crystal that the

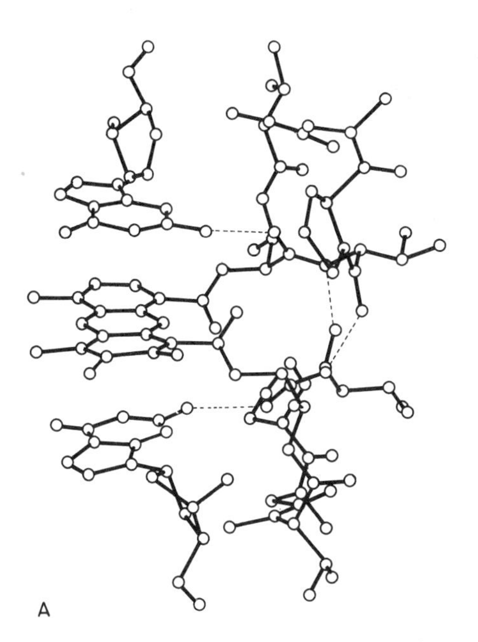

Figure 7.22 The configuration of the actinomycin–deoxyguanosine complex[116] and a space-filling model of this complex
(From Sobell *et al.*[116], by courtesy of Macmillan Journals Limited)

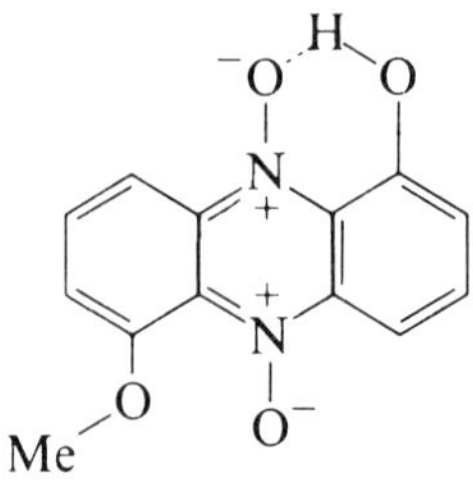

(50)

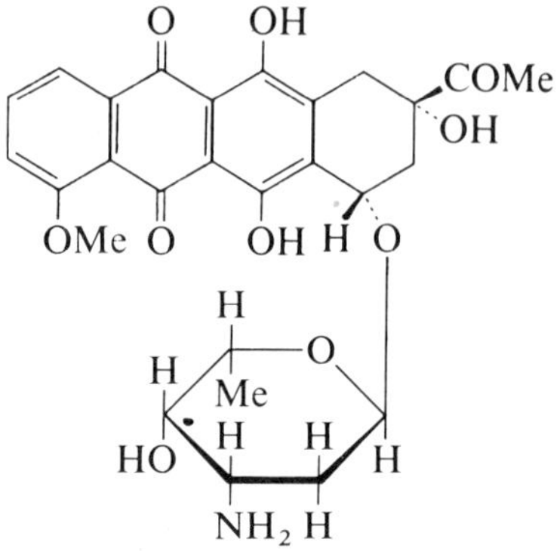

(51)

sugar moiety is approximately perpendicular to the tetracyclic moiety[119]. The tetracyclic part of daunomycin and the tricyclic part of myxin are flat like the chromophore of actinomycin and each of the molecules is capable of forming hydrogen bonds. Perhaps the biological activity of myxin and daunomycin may be accounted for in a manner similar to that proposed for actinomycin.

Showdomycin is an antibiotic which has been isolated from *Streptomyces showdoensis*. It is closely related chemically to the nucleoside uridine. The difference is that in showdomycin (52) one NH group is missing from the ring

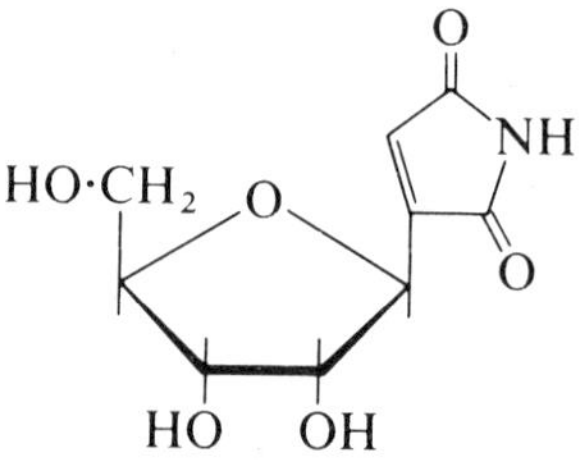

(52)

of the base. The furanose ring has the envelope configuration with C(2′) 0.58 Å out of the plane of the other four atoms in the ring[120], cf. Section 7.4 concerning nucleic acid derivatives. C(2′) lies on the same side of the ring as C(5′) (endo pucker). The furanose plane makes an angle of 46 degrees with the plane of the pyrroline ring and the torsion angle about the glycosidic bond is 129.5 degrees. This is the unusual *syn*-conformation associated with an intramolecular hydrogen bond. The C(5′)—O(5′) bond has the *gauche-gauche* conformation.

Cephalosporins are a different type of antibiotic. They inhibit the synthesis of bacterial cell walls in a fashion similar to that of the penicillins. The active Δ^3-cephalosporins (53) and the inactive Δ^2-cephalosporins (54) are related structurally to the penicillins (55). Ease of hydrolysis of the lactam amide bond, N(5)—C(8), has been correlated with the biological activity.

```
     H     H
     :     :
RCONH—C(6)—C(5)—S(1)        Me
      |     |       \     /
     C(7)—N(4)—C(3)—C(2)
     //           :      \
    O           CO2H  H   Me
```

penicillins

(55)

```
      H     H
      :     :             S(1)        H
RCONH—C(7)—C(6)—S(1)—C(2)
      |     |           |      H
     C(8)—N(5)—C(4)=C(3)—CH2R
     //          |
    O          CO2H
```

Δ^3-cephalosporins

(53)

```
      H     H
      :     :         S(1)      H
RCONH—C(7)—C(6)—          —C(2)
      |     |                ||
     C(8)—N(5)—C(4)—C(3)—CH2R'
     //          :   \
    O          CO2H   H
```

Δ^2-cephalosporins

(54)

A structure investigation of two active cephalosporins, cephaloridine hydrochloride monohydrate and cephaloglycine and one inactive cephalosporin, phenoxymethyl-Δ^2-desacetoxyl cephalosporin was carried out by Sweet and Dahl[121] in order to correlate structural differences with biological activity. Pyramidal character occurs in the β-lactam nitrogen atom of penicillin and the two active Δ^3-cephalosporins. The nitrogen atom is 0.22 Å out of the plane of the three attached atoms in cephaloglycine and 0.4 Å out of the plane in penicillin. In the non-active Δ^2-cephalosporin, it is only 0.06 Å out of the plane. The position of the double bond in the 6-membered ring may also influence the ease of hydrolysis of the lactam amide bond.

Direct methods of structure analysis have facilitated the study of the conformation of carbohydrate molecules. As a consequence of the asymmetric carbon atoms which occur in sugar molecules, these almost equal atom structures can be expected to crystallise in non-centrosymmetric space groups except when they occur as racemates. An example is the investigation by Kim and Jeffrey[122] of the conformation of xylitol, Figure 7.23, which

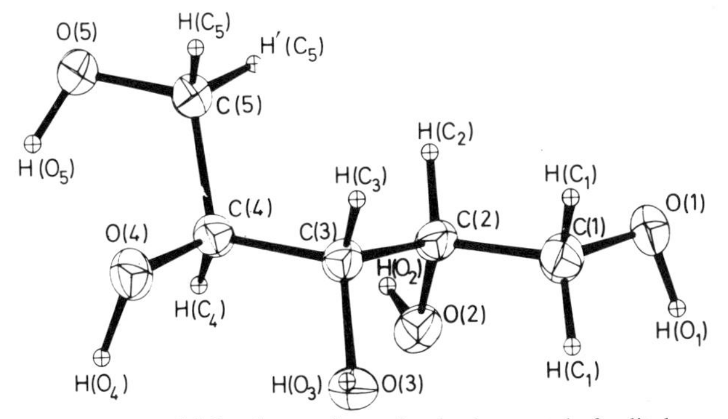

Figure 7.23 Molecular conformation in the crystal of xylitol (From Kim and Jeffrey[122], by courtesy of the International Union of Crystallography)

crystallises in space group $P2_12_12_1$. The conformation is the same as that for its stereoisomer ribitol[123], except for an interchange of O(3) and H on C(3) and the orientation of O(1). They are said to have the *meso*-configuration because of a mirror through C(3) in the symmetrical conformation. However, as seen in Figure 7.23, the molecules do not have a mirror plane. The molecule of xylitol has a planar zig-zag chain from O(1) to O(4). The O(1) atom in ribitol is rotated differently around the C(1)—C(2) bond. It is not apparent why this difference in orientation occurs. In xylitol, as is characteristic of polyol structures, each hydroxyl group is both a donor and acceptor in hydrogen bonding.

Coriose is a natural 3-heptulose. There were four possible structural formulae which could not be readily distinguished, α- or β-pyranose or α- or β-furanose, thus motivating an x-ray structure analysis. Coriose was found to occur in the novel α-furanose form[124], Figure 7.24, with three adjoining *cis*

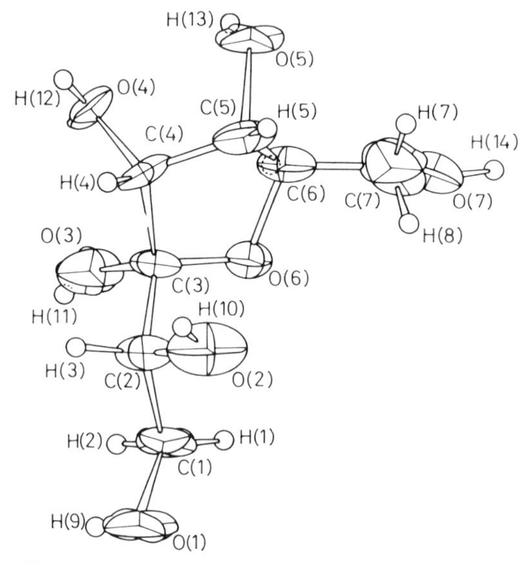

Figure 7.24 Molecular conformation of coriose (From Taga *et al.*[124], by courtesy of the International Union of Crystallography)

hydroxyl groups. The foranose ring is in the envelope conformation with C(5) out of the plane by 0.62 Å. This is the first case in which a furanose structure has been found in a crystal of a monosaccharide, although such a structure has been found among sugar derivatives, e.g. nucleosides as discussed in Section 7.4. All previously analysed monosaccharides have a pyranose form. An example of a pyranose sugar is methyl α-D-mannopyranoside[125] in which the 6-membered ring has the usual chair conformation.

Raffinose (56) is a naturally occurring trisaccharide composed of galactose, glucose and fructose. An x-ray structure investigation by Berman[126] showed that the galactose and glucose portions have pyranose ring configurations in the chair form and the fructose moiety has the puckered furanose ring configuration with C(3) and C(4) out of the plane. Raffinose possesses no melibiose sucrose

(galactose) (glucose) (fructose)

(56)

intramolecular hydrogen bonding. However, the molecules are arranged in two hydrogen-bonded helices per unit cell. Within each helix are 6 water molecules, each of which makes 2–4 hydrogen bonds with the raffinose molecules.

Methyl β-cellobioside (Figure 7.25) was chosen by Ham and Williams[127] for investigation because it could serve as a model compound for cellulose. It was found that the compound is composed of two glucopyranose rings

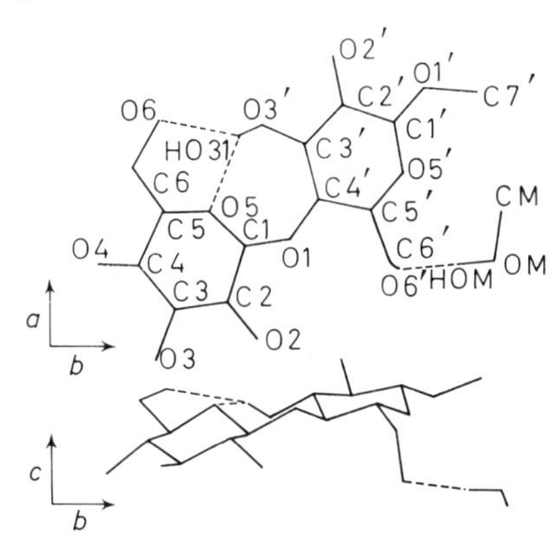

Figure 7.25 Two views of methyl-β-cellobioside·CH_3OH. The dotted lines indicate two intramolecular and one intermolecular hydrogen bonds
(From Ham and Williams[127], by courtesy of the International Union of Crystallography)

and the configuration at C(1) and C(1′) is β. All substituents on the pyranose rings are in the equatorial position. Two intramolecular hydrogen bonds occur.

Other types of natural products can be very detrimental to living systems. An example is a sesquiterpenoid from the yellow star thistle, solstitialin (57). On ingestion, horses develop a great difficulty in chewing and drinking and eventually die. The molecular structure of solstitialin was elucidated by

(57)

Thiessen and Hope[128] by performing an x-ray crystal-structure analysis. The absolute configuration of the molecule was determined from the anomalous scattering of the oxygen atoms.

X-ray crystal structure analysis has been employed to establish the structural formulae of the diterpenoid iso-eremolactone[129] (58) and the alkaloid bellendine[130] (59). The conformation of salts of hydroxy acids,

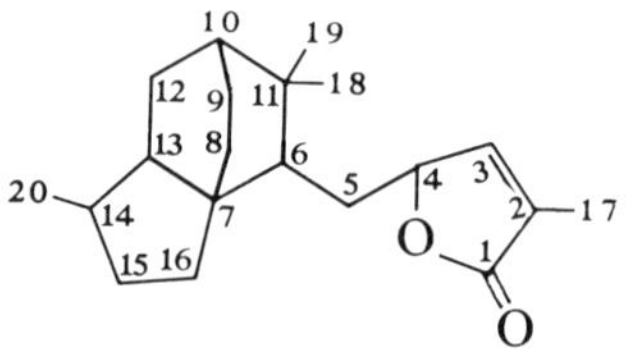

(58)

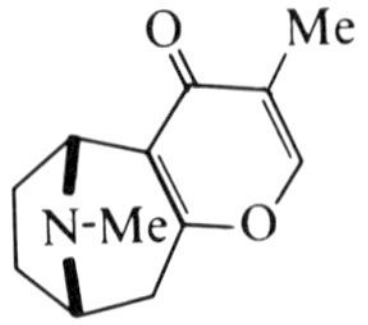

(59)

lithium ammonium hydrogen citrate monohydrate[131] and lithium glycolate[132], have also been established by direct methods of x-ray analysis.

7.10 COMPLEXES, CAGES AND CLATHRATES

Molecular attraction results in a variety of configurations ranging from complexes in which adjoining molecules maintain their structural integrity to adducts in which chemical bonds are formed. Cages may result from adduct formation or may be produced by extensive hydrogen bonding. When the space in a cage is large enough to trap other molecules, the host is called a clathrate, forming a complex with the enclosed molecule.

An example of a complex, apparently held together by charge transfer, is *s*-trinitrobenzene coupled with *s*-triaminobenzene. The complex is highly coloured, forming purplish black needles, which suggests a high degree of charge transfer. The arrangement of the molecules in the unit cell is shown in Figure 7.26[133]. The two different molecules interleave, are essentially planar and the distance between planes is 3.24 Å, a relatively short distance associated with the charge transfer.

It has been suggested that the biological behaviour of serotonin may be explained on the basis of its behaviour as an electron donor which may interact with electron acceptors in living systems. Serotonin is known to form complexes with flavins, nicotinamide derivatives and aromatic coenzymes. The serotonin–picrate complex (60) occurs in the form of red crystals. The approximately planar molecules stack on top of one another in a fashion similar to the previous charge-transfer complex[134]. The distance between planes is 3.3–3.4 Å leading to short intermolecular distances between ring atoms, as found in other acceptor–donor ring complexes. The complex occurs in the form of picrate anions and serotonin cations. In a serotonin–creatinine sulphate complex, in contrast, the different components are not stacked, but are connected together by hydrogen bonding[135]. This complex shows no evidence of charge transfer and is not highly coloured.

A complex has been formed in which one of the components is the adduct anthracene–tetracyanoethylene[136] (61) a cage structure. The complex consists

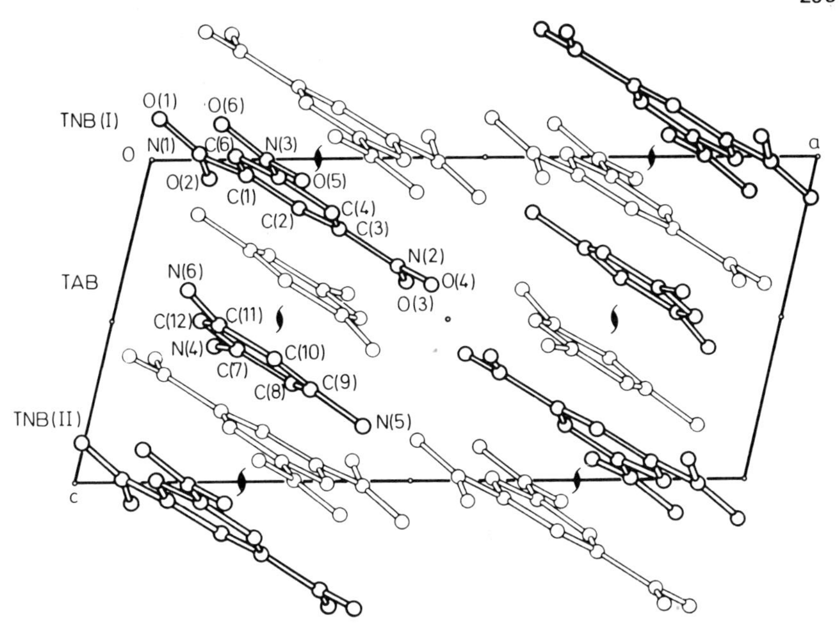

Figure 7.26 The layered structure of the 1:1 complex of *s*-trinitrobenzene and *s*-triaminobenzene
(From Iwasaki and Saito[133], by courtesy of the International Union of Crystallography)

(60)

(61)

of eight anthracene–tetracyanoethylene adducts, two methylene chloride molecules and one tetracyanoethylene molecule crystallising in a triclinic cell, space group $P\bar{1}$. The anthracene–TCNE adduct contrasts with a complex of naphthalene and TCNE (62) in which the two latter molecules are planar and stacked in such a way[137] as to facilitate a Diels–Alder reaction. However, the reaction does not proceed under the conditions of crystallisation. Adduct formation does proceed under these conditions when the naphthalene molecule is replaced with anthracene.

A cage compound, cedrone (63), has been formed by the dimerisation of trimethylphloroglucinol. The configuration was established by Beisler and Silverton[138] by x-ray analysis after several structures were proposed for the

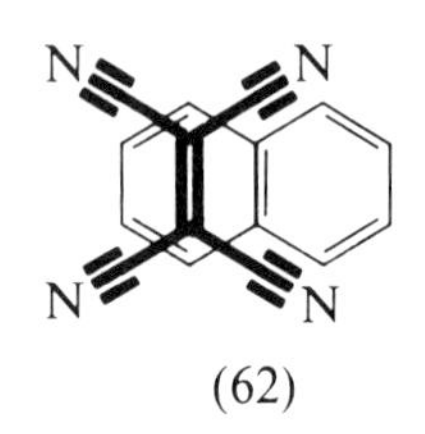

(62)

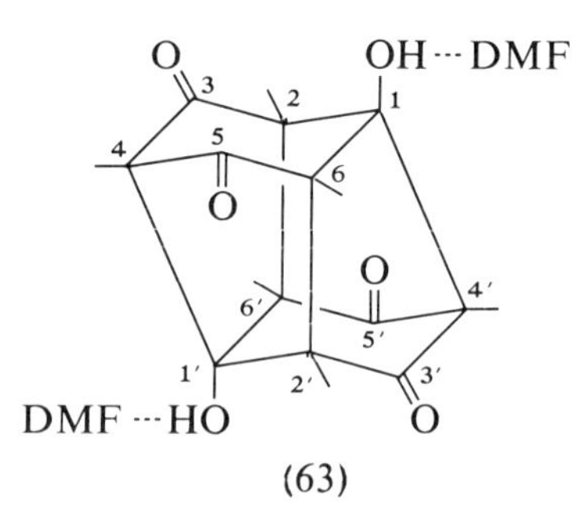

(63)

dimer, none of which were cages. The molecule has three six-membered rings, four five-membered rings and a centre of symmetry.

A heterocyclic compound was synthesised from the interaction of two molecules of ethylenediamine and three of glyoxal. Two possible configurations consistent with chemical properties and physical measurements are shown in (64) and (65). A crystal structure analysis[139, 140] showed that (64) was the correct structure. The molecule possesses three 6-membered and two 7-membered rings in the boat configuration and two 5-membered rings in the envelope configuration.

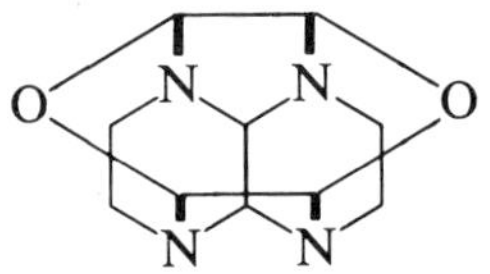

(64)

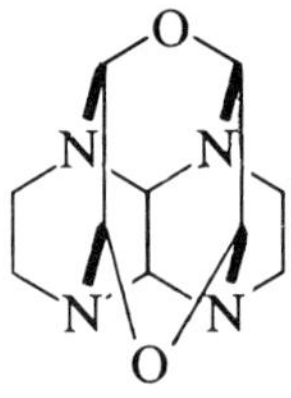

(65)

Another cage compound was prepared by the following reaction:

in MeOH
at 25°C
for 45 h

+ $C_{16}H_{10}OCl_4$
(66)

$3H_2$ Pd–C
AcONa
in MeOH

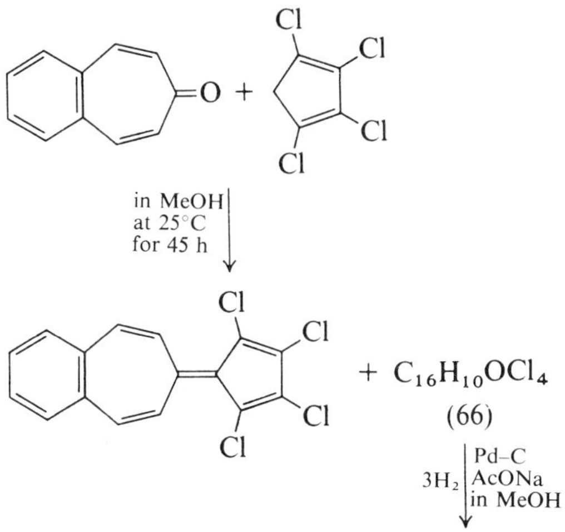

$C_{16}H_{14}OCl_2$
(67)

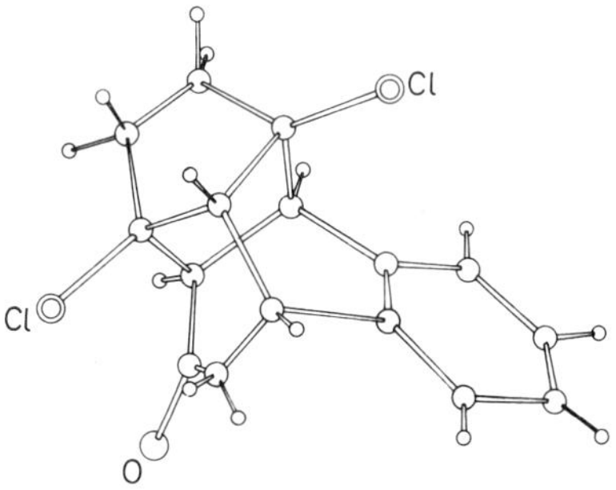

Figure 7.27 Configuration of the $C_{16}H_{14}OCl_2$ molecule
(From Shimanouchi and Sasada[141], by courtesy of the International Union of Crystallography)

The products (66) and (67) were of unknown molecular structure. The configuration of (67) was determined by Shimanouchi and Sasada[141] by x-ray analysis and is shown in Figure 7.27. From the configuration of (67) it was possible to establish the molecular configuration of (66).

A pentacyclononane (68) is produced from a cage compound by the following reaction:

OH^-

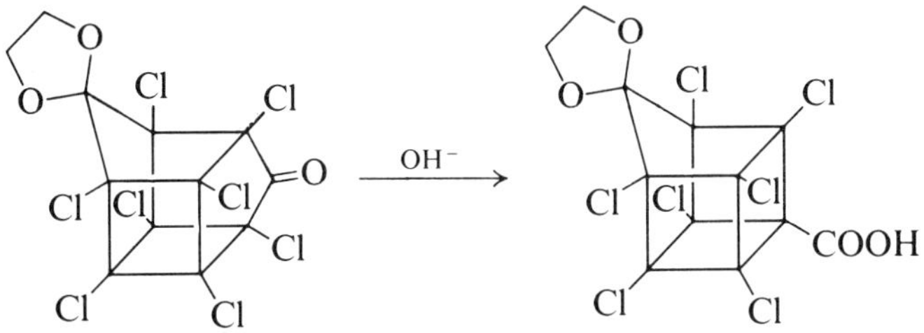

(68)

The objectives of an x-ray structure analysis were to locate the position of the carboxyl group in (68) and determine the shape of the condensed polyring carbon cage. It was found by Okaya[142] that the acid portion of the molecule makes a dimer by hydrogen bonding about a centre of symmetry. The cage consists of two planar and two puckered cyclobutane rings and two cyclopentane rings in the envelope conformation. In a study of a similar ring system (69) occurring in a cyclopentadecane, it was found[143] that the cage

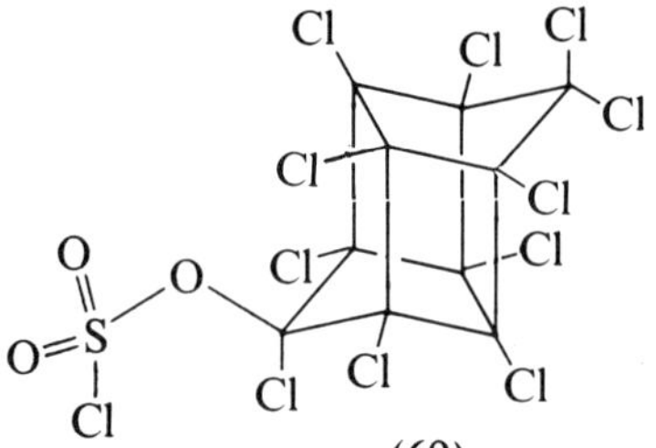

(69)

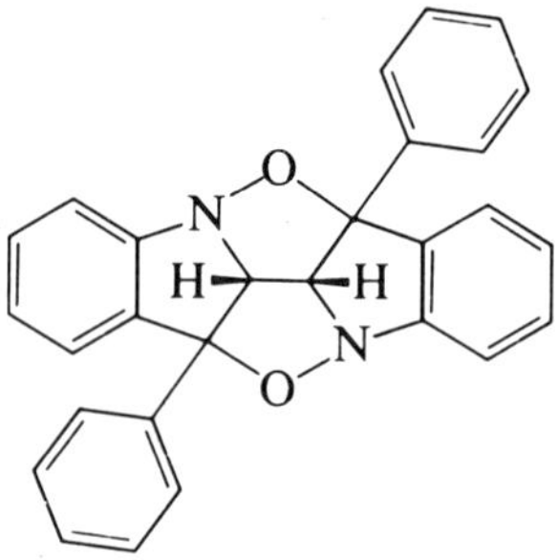

(70)

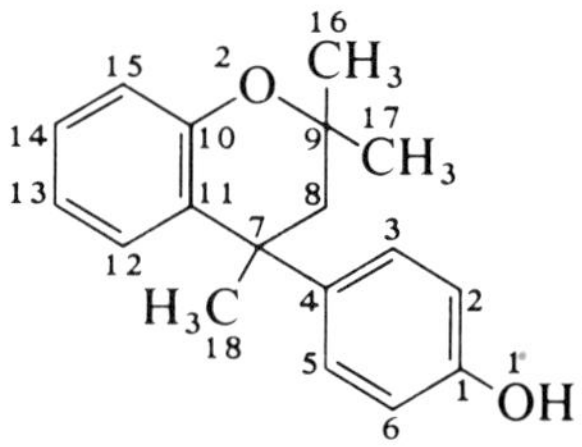

(71)

consists of two puckered cyclobutane rings and four cyclopentane rings in the envelope conformation.

A crystalline product is obtained from the reaction of phenylacetylene and nitrosobenzene[144]. An x-ray analysis identified the material to be a heterocyclic cage compound (70) of molecular formula $C_{28}H_{20}O_2N_2$.

Clathrates are cage structures which entrap guest molecules. A clathrate structure is formed from six molecules of Dianin's compound (71) held together by hydrogen bonding, Figure 7.28. A great variety of crystalline inclusion compounds may be formed by the entrapment of solvent molecules in the course of crystallisation. Structure analyses have been performed on compounds containing ethyl alcohol, chloroform[145] and heptanol[146] The cage was found to be unusually large, having a height of 11 Å and an approximate hour-glass shape with greatest diameters of 6.3 Å between van der Waals contacts. The occupancy of ethanol is approximately two molecules per cage. With chloroform, the occupancy is on the average one molecule per cage. Two orientations were found for the chloroform, one arranged so as to be consistent with the threefold symmetry of both the cage and the chloroform. The other violates this symmetry and projects a chlorine atom

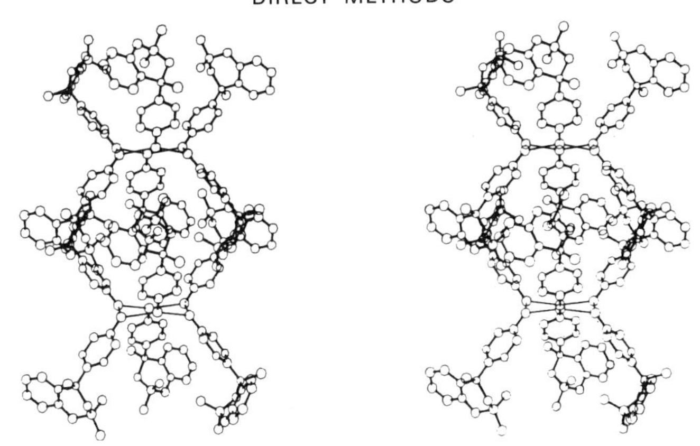

Figure 7.28 Stereodiagram of the stacking of two hydrogen bonded complexes to form the closed cavities in the crystal of Dianins compound
(From Flippen *et al.*[145], by courtesy of The American Chemical Society)

into the narrow central portion of the cage. The length of the usual extended configuration of n-heptanol is too great to fit into the cage. However, it was found that the molecule is trapped with an occupancy of *c*. 95%. Both ends of the n-heptanol molecule assume the unusual *gauche* conformation in order to be accommodated. To date, the entrapment of an n-octanol molecule has not been successful.

7.11 DIVERSE ORGANIC MATERIALS

There are continuing problems concerning the identification of the structural formula and conformation of various compounds. Often new compounds are obtained in minute amount, producing perhaps one small crystal. These substances may nevertheless be of great importance owing to their profound effect in small quantity, as may occur, for example, in biological systems. Such a substance was an unknown contaminant in the fat of animal feed producing hydropericardium in chickens. About 5 μg of the material can kill a chicken. It brought significant losses in 1957 to poultry farmers in the United States. A structure analysis by Cantrell, Webb and Mabis[147] of a single crystal of the contaminant identified the substance to be 1,2,3,7,8,9-hexachlorodibenzo-*p*-dioxin (72).

A triterpane of unknown structural formula, occurring as a petroleum distillate, crystallises in space group $P2_12_12_1$ with two molecules in the asymmetric unit. Thus, the structural problem involves the placement of 60

(72)

carbon atoms. A structure analysis by Smith, Fowell and Melsom[148] identified the substance as 18α-oleanane (73).

An unusual molecule having three fused 4-membered rings was prepared by the dimerisation of benzocyclobutadiene with a nickel tetracarbonyl catalyst, Figure 7.29. It was found by Barnett and Davis[149] that each of the

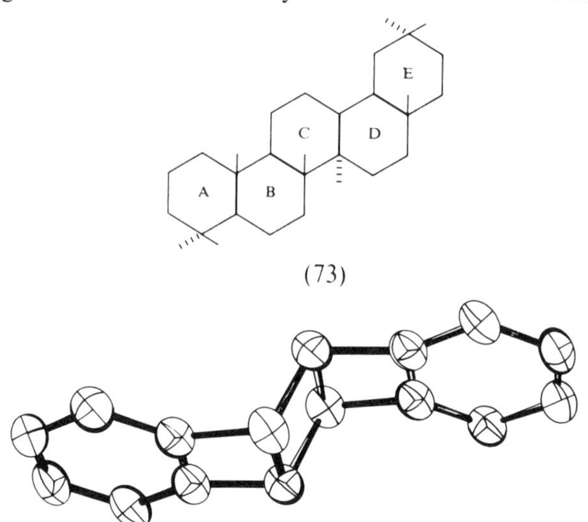

(73)

Figure 7.29 The *anti*- configuration of 3,4:7,8-dibenzotricyclo [4.2.0.0 ^{2,5}]-octa-3,7-diene
(From Barnett and Davis[149], by courtesy of the International Union of Crystallography)

4-membered rings is planar, although some angles within the rings differ by as much as 5 degrees from the average value of 90 degrees. This contrasts with the puckered conformation for the cyclobutane ring found in many other molecules.

A comparable fused ring system was prepared by the photodimerisation of 1,4-epoxy-1,4-dihydronaphthalene, Figure 7.30. The molecule crystallised in the non-centrosymmetric space group $Pca2_1$, although it was found to possess a centre of symmetry within experimental error by Bordner, Stanford and Dickerson[150]. It follows that the central 4-membered ring must be planar.

A derivative (74) of *cis*-metacyclophane possesses an oxygen bridge which causes the molecule to deviate significantly from planarity. The investigation

(74)

of the detailed configuration by Hanson and Huml[151] showed that deviations from otherwise planar ring systems were distributed throughout the molecule. The oxygen bridge possesses the small C—O—C angle of 99.7 degrees.

Triacetylsphingosine was chosen as a model by O'Connell and Pascher[152] for an investigation of the structure and function of lipids in biological

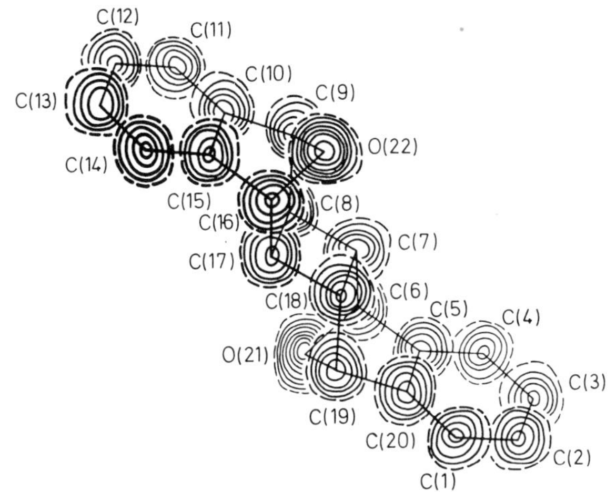

Figure 7.30 Photodimer of 1,4-epoxy-1,4-dihydronaphthalene
(From Bordner *et al.*[150], by courtesy of the International Union of Crystallography)

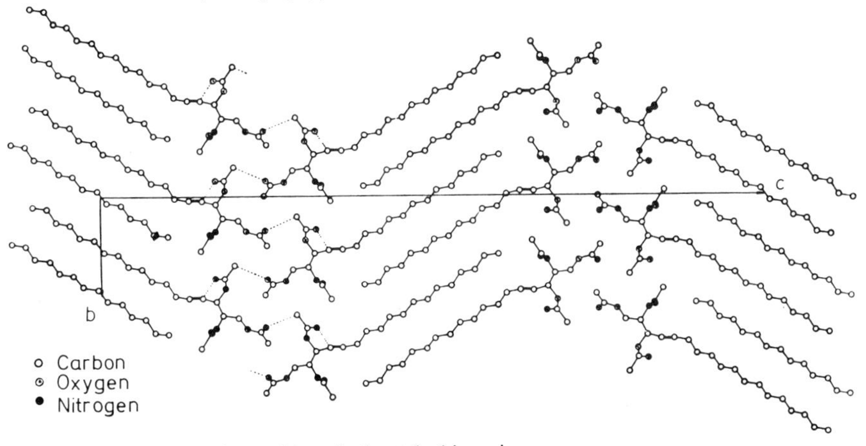

Figure 7.31 Molecular packing of triacetylsphingosine
(From O'Connell and Pascher[152], by courtesy of the International Union of Crystallography)

membranes. The molecules pack in an interleaved fashion, alternating head-to-tail and tail-to-head, Figure 7.31. The polar ends of the molecules form intermolecular hydrogen bonds in a continuous chain.

Nitrilotriacetic acid is widely used in complexometric titrations. An investigation of its structure showed the molecule to exist as a zwitterion

(75) in the crystal[153]. Both C—O distances in the carboxyl ion are 1.25 Å, whereas in the other two carboxyl groups the C=O distance is near 1.20 Å and the C—OH distance is near 1.30 Å. It is of interest that the ionisation is restricted to one specific carboxyl group, rather than being distributed amongst the three of them.

$$\begin{array}{c} COO^{-} \\ | \\ CH_2 \\ | \\ HOOC-CH_2-N-CH_2-COOH \\ | \\ H^{+} \end{array}$$

(75)

Single-crystal neutron diffraction data may be readily used with direct methods of structure analysis. A minor difference in the data reduction procedure arises from the fact that neutron scattering occurs only from the nuclei and therefore, in obtaining normalised structure factors it is only necessary to correct for the thermal motion and not the electron distributions around the atoms. A special problem arises in neutron scattering data when a significant part of the scattering derives from atoms having negative scattering factors, e.g. hydrogen. This special problem has been overcome by a special calculation[154] in which the experimental data are converted to values which would have been obtained if all the atoms scattered positively (with the squares of their scattering factors). An application of this procedure has been made to the neutron diffraction investigation of glycolic acid (hydroxyacetic acid) by Ellison, Johnson and Levy[155]. This substance crystallises with two molecules in the asymmetric unit of space group $P2_1/c$. The hydrogen bonding is different from that often found with carboxyl groups. Instead of forming dimers through the sharing of hydrogen bonds between two carboxyl groups, the hydrogen bonding is formed between the carbonyl group

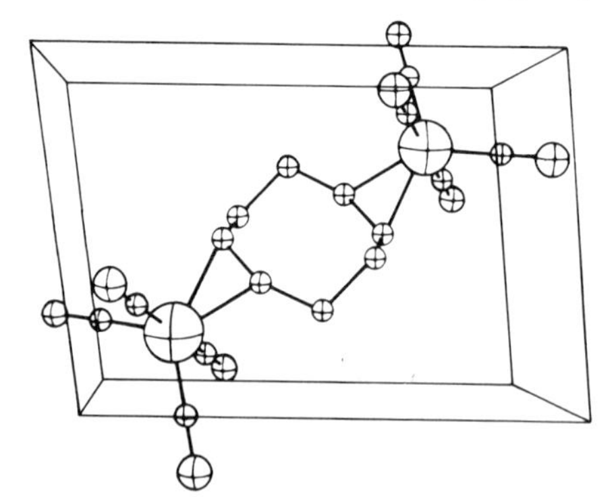

Figure 7.32 Contents of one unit cell of 1,5-cyclo-octadiene bis(tetracarbonyl iron)
(From Krüger[156], by courtesy of Elsevier Publishing Company)

of one molecule and the α-hydroxyl group of another molecule, building a three-dimensional network.

An example of an organometallic structure, investigated by Krüger[156], is 1,5-cyclo-octadiene bis(tetracarbonyl iron). One molecule crystallises in a unit cell in space group $P\bar{1}$, Figure 7.32. The cyclo-octadiene ring, whose centre coincides with a crystallographic centre of symmetry, assumes the chair conformation. The iron atom is linked to the double bond which increases in value from the usual 1.34 to 1.40 Å. Coordination about the iron atom is almost an ideal trigonal-bipyramid with two carbonyl groups in the axial directions and two other carbonyl groups along with the linkage to the organic moiety in the equatorial position.

Some examples of other interesting organic structures, natural products and organometallic compounds solved by direct methods are listed below. The detailed description of the results is not possible in a brief review, although these studies represent additional illustrations of the wide range of worthwhile applications which have been facilitated by direct methods of structure analysis. The topics covered concern identification, configuration and structural correlations with biological activity: $[TiCl(C_5H_7O_2)_2]_2O$ [157] octahydroxycyclobutane[158], tetracyanocyclobutane[159], *trans*-bicyclo[4.2.0] octyl-1-3,5-dinitrobenzoate[160], *anti*-head-to-head photodimer of 1,1-dimethyl-2(1*H*)-naphthalenone[161], an aminoborane, $C_{26}H_{38}N_2B_2Cl_2$ [162], cyclopentenylidenephosphorane[163] a lactam, $C_{19}H_{21}O_2N$ [164], 2-acetyl-3-indazolinone [165], α-(2-hydroxy-3,5-dibromobenzylidene)-γ-butyrolactone [166], a betaine, $C_{31}H_{25}N_3O_4PCl$ [167], Al_2O(2-methyl-8-quinolinolato)$_4$ [168], $Me_5Al_2NPh_2$ [169], $Bz^-Li^+N(C_2H_4)_3N$ [170], $Zn[S_2CN(CH_3)_2]$ [171], 2-keto-3-ethoxybutyraldehyde bis(thiosemicarbazone) [172], methylparathion[173], α-2-ethyl-5-methyl-3,3-diphenyltetrahydrofuran [174], cyclazocine [175], diphenylhydantoin and diazapam [176], 7-chloro-5-(2,4-dichlorophenyl)-4,5-dihydro-1,4-dimethyl-3*H*-1,4-benzodiazepin-2-one [177], *O*-di-t-butylquinoxaline [178], 3-(*N*-benzyl-*N*-methylaminoethyl)-2-norbornanol [179], hydrazone derivative of 2,2′-di-(1,4-naphthoquinone) [180], β-fumaric acid [181], pyridoxine hydrochloride [182], ε-aminocaproic acid [183], azulene-1,3-dipropionic acid [184], *N*-α-naphthyl-1,2,3,6-tetrahydrophthalamic acid [185], p-nitroperoxybenzoic acid [186], DL-2-methyl-7-oxodecanoic acid [187], 4,5-dioxo-2-thioxo-1,3-dithiolan (β-modification) [188], β-chloroethyl triptycene [189], 2-(2-pyridylmethyldithio) benzoic acid [190], pentachloroethoxycodide [191], cocarboxylase [192], C_8F_{12}, saturated dimer of hexafluorobutadiene [193], jamine [194], panamine [195], 6-hydroxycrinamine [196], reserpine [197].

7.12 INORGANIC MATERIALS

The presence of a small number of heavy atoms in a unit cell facilitates the structure analysis. Their location is usually readily obtained from a Patterson synthesis and simple procedures are then available for developing the complete structure. If direct methods are employed, such structure determinations can proceed with unusually high probability measures. However, many inorganic substances of interest possess relatively large numbers of heavy atoms. In such cases, it may be very difficult to obtain the locations of the heavy atoms in the unit cell by an analysis of the Patterson function. Under such circumstances, direct methods can expedite the analysis.

A material with many heavy atoms is the salt $(NH_4)_6[H_4Co_2Mo_{10}O_{38}]\cdot 7H_2O$ which crystallises in *Pc*. It was found by Evans and Showell[198] that the heteropolyion has the configuration shown in Figure 7.33. The two cobalt atoms are in the centre of the ion and each has an octahedral coordination to six oxygen atoms. Each oxygen atom participating in the octahedral coordination of the cobalt atoms is coordinated to two or three molybdenum atoms. Each of the ten molybdenum atoms is surrounded by six oxygen atoms in octahedral coordination.

Polyions containing boron assume a variety of configurations in mineral structures. The polyanion $[B_{14}O_{20}(OH)_6]^{4-}$ has been found in the mineral

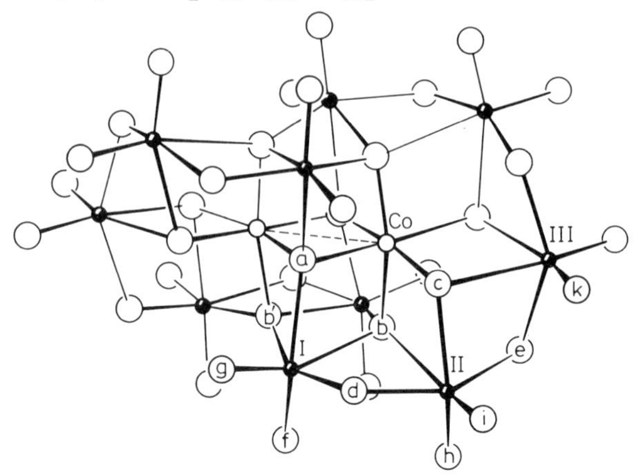

Figure 7.33 The decamolybdodicobaltate(III) ion, $[H_4Co_2Mo_{10}O_{38}]^{6-}$
(From Evans and Showell[198], by courtesy of the American Chemical Society)

strontioginorite by Konnert, Clark and Christ[199]. This polyion (76) is composed of two tunellite-like moieties, composed of three B—O tetrahedra and three B—O triangles connected at corners, which share O(11). In addition, it has a side chain consisting of two B—O triangles.

Another type of polyion containing boron, described by Merlino and Sartori[200], occurs in the mineral larderellite. The basic structural unit of $[B_5O_7(OH)_2]^{-1}$ is the double ring, Figure 7.34, composed of four B—O

B_4, O_6, O_7, OH_9, B_1, B_2, O_5, O_1, O_2, B_{10}, B_6, B_3, B_5, O_{17}, O_{18}, OH_8, O_4, O_3, O_{11}, O_{20}, O_{22}, OH_{24}, B_7, B_8, B_{13}, B_{14}, O_{10}, O_{16}, O_{12}, O_{13}, OH_{21}, OH_{23}, B_{12}, B_9, B_{11}, $_{19}HO$, O_{15}, O_{14}, O_{26}, O_{25}

(76)

triangles and one B—O tetrahedron. The two units shown in Figure 7.34 are a part of an infinite zig-zag chain where the next two segments are joined at O(1) and O(1″). The chains are linked into infinite sheets by hydrogen bonds to ammonium ions and water molecules.

The crystal structures of many silicate minerals have been determined by x-ray analysis. It is therefore of interest that new groupings of silicate tetrahedra continue to be discovered. Ardennite is a rare arsenic–vanadium silicate which was found by Donnay and Allmann[201] to contain a new Si_3O_{10} group, possessing $2mm$ symmetry and consisting of three tetrahedra, each of which has an apex pointing in the same direction instead of alternating. A low value of 126 degrees occurs for the Si—O—Si angle.

Soon after the discovery of the Si_3O_{10} group in ardennite, a similar group $[(Al,Si)_2SiO_{10}]$ was found in the mineral kornerupine by Moore and Bennett[202]. The symbol (Al,Si) implies that Al can replace Si in the two end tetrahedra in a disordered fashion. Kornerupine is a complex mineral whose composition and chemistry were considerably clarified by the crystal structure determination.

In low chalcocite there are 48 units of Cu_2S in the unit cell of space group $P2_1/c$. The structure[203] is composed of sheets of sulphur atoms in hexagonal close packing. Alternate sheets contain copper atoms at the centres of triangles similar to that found in high chalcocite. The remaining copper atoms occupy triangular sites between the sheets of sulphur atoms.

The crystal structure of washing soda[204], $Na_2CO_3{\cdot}1OH_2O$, is characterised by the presence of two ions arranged in a distorted sodium chloride type of structure. The cation $[Na_2(H_2O)_{10}]^{2+}$ consists of two sodium-water

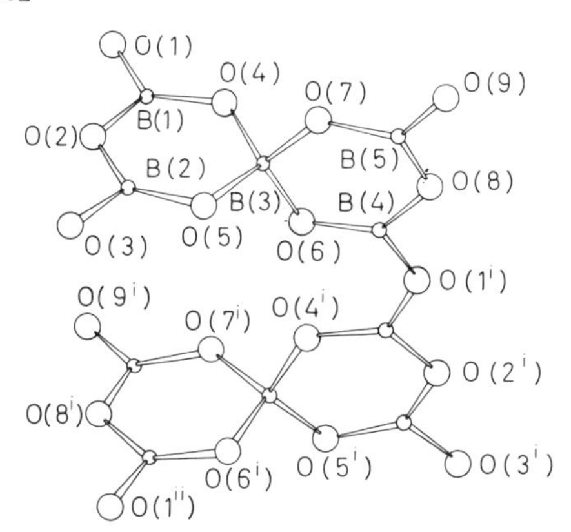

Figure 7.34 Two units of an infinite chain formed by the $[B_5O_7(OH)_2]_n^{-n}$ polyion in larderellite
(From Merlino and Sartori[200], by courtesy of the International Union of Crystallography)

octahedra which share an edge. The anion is CO_3^{2-}, which occurs with rotational disorder.

The final two materials to be discussed, $PaOCl_2$ and Ga_2Mg, afford examples of the facility with which chemical analysis in addition to structure determination can be performed by direct methods. This facility is valuable since the nature of the cell contents is often only approximately known for minerals and intermetallic compounds. In an investigation of $PaOCl_2$ [205], it was expected from chemical preparation that the substance under study would be $PaCl_4$. Density measurements indicated that there would be eight

Pa atoms per unit cell. However the *E*-map computed from the normalised structure factors showed that there were twelve Pa atoms in the unit cell. It was then readily possible to identify the chlorine and the oxygen atoms. $PaOCl_2$ crystallises in infinite polymeric chains $(Pa_3O_3Cl_6)_n$, Figure 7.35.

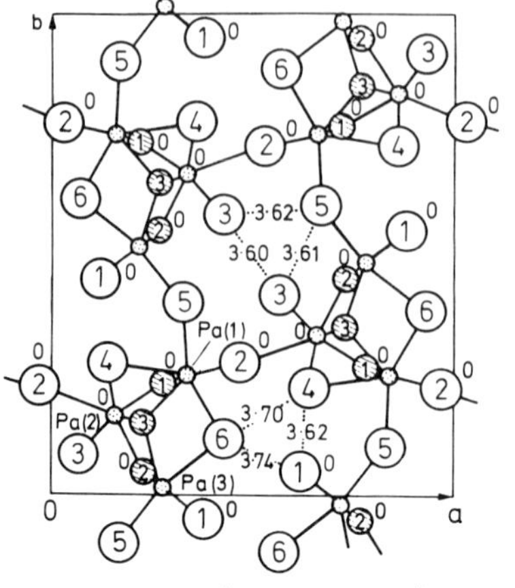

Figure 7.35 The structure of $PaOCl_2$ which occurs as infinite polymeric chains $(Pa_3O_3Cl_6)_n$, perpendicular to the projection shown
(From Dodge *et al.*[205], by courtesy of the International Union of Crystallography)

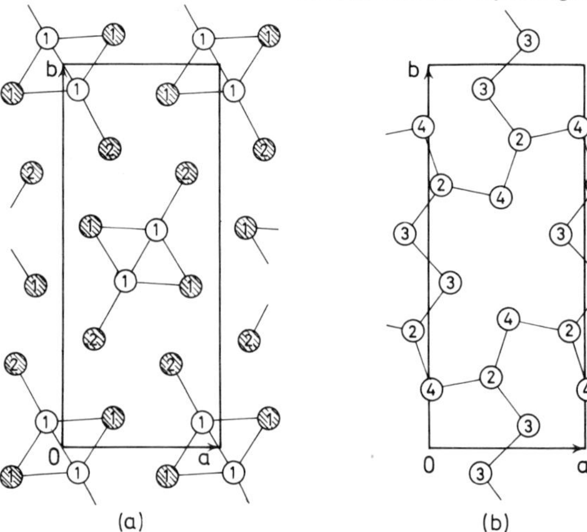

Figure 7.36 Two types of planar sheets which alternate in the Ga_2Mg structure. Layer (*a*), occurring at $z = 0$, contains both Ga and Mg atoms whereas layer (*b*), at $z = \frac{1}{2}$, contains only Ga atoms. Mg atoms are shaded; numbers refer to independent atoms
(From Smith *et al.*[206], by courtesy of the International Union of Crystallography)

In this fairly complicated structure the three independent Pa atoms have coordination numbers of 7, 8 and 9.

A structure investigation[206] was carried out on a crystal thought to be $ReCl_5$. The determination revealed however that the substance was Ga_2Mg. From the relative weights of the peaks in the Fourier map of the structure, it was suspected that Ga and Mg were present. This was then confirmed by comparison with a known sample. The structure can be described in terms of two planar sheets. One sheet contains Ga and Mg atoms and the other contains only Ga atoms, Figure 7.36. The layers stack so that the atoms of one layer intercollate with the interstices of adjoining layers. An unusual feature of the structure is the occurrence of two holes at $0,0,\frac{1}{2}$ and $\frac{1}{2},\frac{1}{2},\frac{1}{2}$ that are almost large enough to accommodate another Ga atom.

7.13 CONCLUDING REMARKS

Direct methods of crystal structure analysis have been applied to a variety of structural problems involving both organic and inorganic materials. The methods have met with success where the chemical formula or structural formula was unknown, heavy or moderately heavy atoms were absent in non-centrosymmetric structures and also where low space-group symmetries were involved. As a general rule the phase determination appears to be facilitated by the presence of high space-group symmetry and of some moderately heavy atoms. However, these special structural features are not required and do not pertain to the majority of structures described in this article. There have been many examples of essentially equal atom non-centrosymmetric structures which were readily solved.

x-Ray structure analysis has developed as a useful tool in chemistry. It can be used, for example, to identify and characterise natural products, reaction products and rearrangements. As such, it facilitates the development of reaction mechanisms. Classes of compounds can be studied with a view toward correlating structure with chemical and physical behaviour. Of particular interest are problems which arise in biological systems.

In view of the large amount of structural information inherent in scattering data, it is legitimate to look forward to further developments in analytic procedures which can increase the facility with which more and more complex systems may be investigated.

References

1. Patterson, A. L. (1935). *Phys. Rev.*, **46,** 372; *Z. Kristallogr.*, **90,** 517
2. Harker, D. and Kasper, J. S. (1948). *Acta Crystallogr.*, **1,** 70
3. Kasper, J. S., Lucht, C. M. and Harker, D. (1950). *Acta Crystallogr.*, **3,** 436
4. Karle, J. and Hauptman, H. (1950). *Acta Crystallogr.*, **3,** 181
5. Karle, J. (1969). *Advan. in Chemical Physics,* Vol. 16, 131. (New York–London: Interscience)
6. Wilson, A. J. C. (1949). *Acta Crystallogr.*, **2,** 318
7. Hauptman, H. and Karle, J. (1953). *American Crystallographic Association Monograph No. 3.* (Pittsburgh: Polycrystal Book Service)
8. *International Tables for X-ray Crystallography,* Vol. I (1952). (Birmingham: The Kynoch Press)
9. Cochran, W. (1955). *Acta Crystallogr.*, **8,** 473
10. Karle, J. and Hauptman, H. (1956). *Acta Crystallogr.*, **9,** 635
11. Karle, J. and Karle, I. L. (1966). *Acta Crystallogr.*, **21,** 849
12. Reference 5, p. 157
13. Hauptman, H., Fisher, J., Hancock, H. and Norton, D. (1969). *Acta Crystallogr.*, **B25,** 811
14. Karle, J. (1970). *Acta Crystallogr.*, **B26,** 1614

15. Woolfson, M. M. (1954). *Acta Crystallogr.*, **7,** 61
16. Karle, J. (1970). *Crystallographic Computing,* 13. (Copenhagen: Munksgaard)
17. Karle, I. L., Hauptman, H., Karle, J. and Wing, A. B. (1958). *Acta Cryst.*, **11,** 257
18. Karle, J. (1968). *Acta Crystallogr.,* **B24,** 182
19. Robertson, J. M. and Woodward, J. (1937). *J. Chem. Soc.*, 219; (1940). *J. Chem. Soc.*, 36
20. Hoppe, W. (1957). *Z. Elektrochemie,* **61,** 1076
21. Nordman, C. E. and Nakatsu, K. (1963). *J. Amer. Chem. Soc.,* **85,** 353
22. Smith, K. C. and Hanawalt, P. C. (1969). *Molecular Photobiology, Inactivation and Recovery,* 58. (New York: Academic Press)
23. Wang, S. Y. (1960). *Nature (London),* **188,** 844
24. Wang, S. Y. (1961). *Nature (London),* **190,** 690
25. Wang, S. Y. (1964). *Photochem. Photobiol.*, **3,** 395
26. Wang, S. Y. (1965). *Federation Proc.,* **24,** S-71
27. Beukers, R. and Berends, W. (1960). *Biochim. Biophys. Acta,* **41,** 550
28. Beukers, R. and Berends, W. (1961). *Biochim. Biophys. Acta,* **49,** 181
29. Varghese, A. J. and Wang, S. Y. (1968). *Science,* **160,** 186
30. Varghese, A. J. and Wang, S. Y. (1968). *Biochem. Biophys. Res. Commun.*, **33,** 102
31. Wang, S. Y. (1963). *Nature (London),* **200,** 879
32. Adman, E., Gordon, M. P. and Jensen, L. H. (1968). *Chem. Commun.*, 1019
33. Gibson, J. and Karle, I. L. (1971). *J. Cryst. and Mol. Struc.*, **1,** 115
34. Camerman, N. and Camerman, A. (1968). *Science,* **160,** 1451
35. Leonard, N. J., Golankiewicz, K., McCredie, R. S., Johnson, S. M. and Paul, I. C. (1969). *J. Amer. Chem. Soc.*, **91,** 5855
36. Camerman, N., Weinblum, D. and Nyburg, S. C. (1969). *J. Amer. Chem. Soc.*, **91,** 982
37. Konnert, J. and Karle, I. L. (1971). *J. Cryst. and Mol. Struc.,* **1,** 107
38. Flippen, J. L. Private communication
39. Camerman, N., Nyburg, S. C. and Weinblum, D. (1967). *Tetrahedron Lett.*, 4127
40. Camerman, N. and Nyburg, S. C. (1968). *Acta Crystallogr.*, **B25,** 388
41. Einstein, J. R., Hosszu, J. L., Longworth, J. W., Rahn, R. O. and Wei, C. H. (1967). *Chem. Commun.*, 1063
42. Furberg, S. and Jensen, L. H. (1968). *J. Amer. Chem. Soc.*, **90,** 470
43. Rohrer, D. C. and Sundaralingam, M. (1970). *Acta Crystallogr.*, **B26,** 546
44. Wei, C. H. and Einstein, J. R. (1968). *Abstr. of the Amer. Cryst. Assoc.*, Buffalo, N.Y., 102
45. Karle, I. L. (1969). *Acta Crystallogr.*, **B25,** 2119
46. Konnert, J., Karle, I. L. and Karle, J. (1970). *Acta Crystallogr.*, **B26,** 770
47. Flippen, J. L. and Karle, I. L. (1971). *J. Amer. Chem. Soc.*, **93,** 2762
48. Wang, S. Y. and Rhoades, D. F. (1971). *J. Amer. Chem. Soc.*, **93,** 2554
49. Flippen, J. L., Gilardi, R. D., Karle, I. L., Rhoades, D. F. and Wang, S. Y. (1971). *J. Amer. Chem. Soc.*, **93,** 2556
50. Kennard, O., Isaacs, N. W., Coppola, J. C., Kirby, A. J., Warren, S., Motherwell, W. D. S., Watson, D. G., Wampler, D. L., Chenery, D. H., Larson, A. C., Kerr, K. A. and Riva de Sanseverino, L. (1970). *Nature (London),* **225,** 333
51. Rao, S. T. and Sundaralingam, M. (1970). *J. Amer. Chem. Soc.*, **92,** 4963
52. Rohrer, D. C. and Sundaralingam, M. (1970). *J. Amer. Chem. Soc.*, **92,** 4956
53. Kennard, O., Larson, A. C., Rees, C. B., Coppola, J. C., Motherwell, W. D. S. and Watson, D. G. (1971). *J. Chem. Soc.*, in the press
54. Sundaralingam, M. (1969). *Biopolymers,* **7,** 821
55. Rohrer, D. C. and Sundaralingam, M. (1970). *J. Amer. Chem. Soc.*, **92,** 4950
56. Coulter, C. L. (1969). *Acta Crystallogr.*, **B25,** 2055
57. Coulter, C. L. (1968). *Science,* **159,** 888
58. Young, D. W., Tollin, P. and Wilson, H. R. (1969). *Acta Crystallogr.*, **B25,** 1423
59. Oroshnik, W., Brown, P. K., Hubbard, R. and Wald, G. (1956). *Proc. Nat. Acad. Sci., U.S.A.*, **42,** 578
60. Wald, G., Brown, P. K. and Gibbons, I. R. (1963). *J. Opt. Soc. Amer.*, **53,** 20
61. Nash, H. A. (1969). *J. Theoret. Biol.*, **22,** 314
62. Gilardi, R. D., Sperling, W., Karle, I. L. and Karle, J. (1971). *Nature (London),* **232,** 187
63. Yonemitsu, O., Cerutti, P. and Witkop, B. (1966). *J. Amer. Chem. Soc.*, **88,** 3941
64. Karle, I. L., Karle, J. and Estlin, J. A. (1967). *Acta Crystallogr.*. **23,** 494
65. Yonemitsu, O., Okuno, Y., Kanaoka, Y., Karle, I. L. and Witkop, B. (1968). *J. Amer. Chem. Soc.*, **90,** 6522

66. Karle, I. L., Gibson, J. W. and Karle, J. (1969). *Acta Crystallogr.*, **B25,** 2034
67. Yonemitsu, O., Nakai, H., Kanaoka, Y., Karle, I. L. and Witkop, B. (1970). *J. Amer. Chem. Soc.*, **92,** 5691
68. Karle, I. L. and Karle, J. (1970). *Acta Crystallogr.*, **B26,** 1276
69. Wasserman, H. H. and Keehn, P. M. (1967). *J. Amer. Chem. Soc.*, **89,** 2770
70. Fratini, A. V. (1968). *J. Amer. Chem. Soc.*, **90,** 1688
71. Chiang, J. F. and Bauer, S. H. (1970). *J. Amer. Chem. Soc.*, **92,** 1614
72. Wasserman, H. H., Doumaux, A. R. and Davis, R. E. (1966). *J. Amer. Chem. Soc.*, **88,** 4517
73. Ammon, H. L. and Jensen, L. H. (1967). *Acta Crystallogr.*, **23,** 805
74. Bednowitz, A. L., Hamilton, W. C., Brown, R., Donaruma, L. G., Southwick, P. L., Kropf, R. and Stanfield, R. A. (1968). *J. Amer. Chem. Soc.*, **90,** 291
75. Kayushina, R. L. and Vainstein, B. K. (1966). *Soviet Physics-Crystallography,* **10,** 698
76. Donohue, J. and Trueblood, K. N. (1952). *Acta Crystallogr.*, **5,** 414, 419
77. Karle, I. L., Daly, J. W. and Witkop, B. (1969). *Science,* **164,** 1401
78. Karle, I. L. (1970). *Acta Crystallogr.*, **B26,** 765
79. Fujimoto, Y., Irreverre, F., Karle, J. M., Karle, I. L. and Witkop, B. (1971). *J. Amer. Chem. Soc.*, **93,** 3471
80. Fridrichsons, J. and Mathieson, A. McL. (1962). *Acta Crystallogr.*, **15,** 569
81. Leung, Y. C. and Marsh, R. E. (1958). *Acta Crystallogr.*, **11,** 17
82. Ueki, T., Ashida, T., Kakudo, M., Sasada, Y. and Katsube, Y. (1969). *Acta Crystallogr.*, **B25,** 1840
83. Karle, I. L. (1972). *J. Amer. Chem. Soc.*, **94,** 81
84. Karle, I. L. and Karle, J. (1964). *Acta Crystallogr.*, **17,** 835
85. Kistenmacher, T. J., Hunt, D. J. and Marsh, R. E. (1971). *Acta Crystallogr.*, in the press
86. Fletterick, R. J., Tsai, C. and Hughes, R. E. (1971). *J. Phys. Chem.*, **75,** 918
87. Mallikarjunan, M. and Rao, S. T. (1969). *Acta Crystallogr.*, **B25,** 296
88. Cole, F. E. (1970). *Acta Crystallogr.*, **B26,** 622
89. Koch, M. H. J. and Germain, G. (1970). *Acta Crystallogr.*, **B26,** 410
90. Ichikawa, T. and Iitaka, Y. (1969). *Acta Crystallogr.*, **B25,** 1824
91. Jandacek, R. J. and Simonsen, S. H. (1969). *J. Amer. Chem. Soc.*, **91,** 6663
92. Sletten, E. (1970). *J. Amer. Chem. Soc.*, **92,** 172
93. Konnert, J. and Karle, I. L. (1969). *J. Amer. Chem. Soc.*, **91,** 4888
94. Edsall, J. T., Flory, P. J., Kendrew, J. C., Liquori, M. A., Nemetny, G., Ramachandran, G. N. and Scheraga, H. A. (1966). *J. Mol. Biol.*, **15,** 399
95. Karle, I. L. and Karle, J. (1963). *Acta Crystallogr.*, **16,** 969
96. Karle, I. L., Gibson, J. W. and Karle, J. (1970). *J. Amer. Chem. Soc.*, **92,** 3755
97. See, e.g. Momany, F. A., Vanderkooi, G., Tuttle, R. W., Scheraga, H. A. (1969). *Biochemistry,* **8,** 744
98. Northolt, M. G. (1970). *Acta Crystallogr.*, **B26,** 240
99. Hughes, R. E., Muxfeldt, H., Tsai, C. and Stezowski, J. J. (1970). *J. Amer. Chem. Soc.*, **92,** 5267
100. Drew, M. G. B., Templeton, D. H. and Zalkin, A. (1969). *Acta Crystallogr.*, **B25,** 261
101. Johnson, S. M. and Paul, I. C. (1968). *J. Amer. Chem. Soc.*, **90,** 6556
102. Kennard, O., Wampler, D. L., Coppola, J. C., Motherwell, W. D. S., Watson, D. G. and Larson, A. C. *Acta Crystallogr.*, in the press
103. Hall, S. R. and Ahmed, F. R. (1968). *Acta Crystallogr.*, **B24,** 337
104. Hall, S. R. and Ahmed, F. R. (1968). *Acta Crystallogr.*, **B24,** 346
105. Flippen, J. L. and Gilardi, R. D. *Abstr. C-4, Amer. Cryst. Assoc. Meeting,* Ames, Iowa, p. 33
106. Karle, I. L. and Karle, J. (1969). *Acta Crystallogr.*, **B25,** 434
107. Gilardi, R. D. and Karle, I. L. (1970). *Acta Crystallogr.*, **B26,** 207
108. Tokuyama, T., Daly, J. and Witkop, B. (1969). *J. Amer. Chem. Soc.*, **91,** 3931
109. Karle, I. L. and Karle, J. (1969). *Acta Crystallogr.*, **B25,** 428
110. Cooper, A., Norton, D. A. and Hauptman, H. (1969). *Acta Crystallogr.*, **B25,** 814
111. Hanson, J. C. and Nordman, C. E. (1970). *Abstr. L-9, Am. Cryst. Assoc. Meeting,* Ottawa, Canada, p. 73
112. Thiessen, W. E. (1970). *Abstr. G-3, Am. Cryst. Assoc. Meeting,* New Orleans, La., p. 40
113. Reeke, G. N., Jr., Vincent, R. L. and Lipscomb, W. N. (1968). *J. Amer. Chem. Soc.*, **90,** 1663

114. Karle, I. L. (1970). *Acta Crystallogr.*, **B26**, 1639
115. Usubillaga, A., Seelkopf, A., Karle, I. L., Daly, J. W. and Witkop, B. (1970). *J. Amer. Chem. Soc.*, **92**, 700
116. Sobell, H. M., Jain, S. C., Sakore, T. D. and Nordman, C. E. (1971). *Nature New Biology*, **231**, 200
117. Hanson, A. W. and Huml, K. (1969). *Acta Crystallogr.*, **B25**, 1766
118. Hanson, A. W. (1968). *Acta Crystallogr.*, **B24**, 1084
119. Angiuli, R., Arcamone, F., Foresti, E., Isaacs, N. W., Kennard, O., Motherwell, W. D. S., Riva di Sanseverino, L. and Wampler, D. L. (1971). *Nature New Biology*, **234**, 78
120. Tsukuda, Y. and Koyama, H. (1970), *J. Chem. Soc. B*, 1709
121. Sweet, R. M. and Dahl, L. F. (1970). *J. Amer. Chem. Soc.*, **92**, 5489
122. Kim, H. S. and Jeffrey, G. A. (1969). *Acta Crystallogr.*, **B25**, 2607
123. Kim, H. S., Jeffrey, G. A. and Rosenstein, R. D. (1969). *Acta Crystallogr.*, **B25**, 2223
124. Taga, T., Osaki, K. and Okuda, T. (1970). *Acta Crystallogr.*, **B26**, 991
125. Gatehouse, B. M. and Poppleton, B. J. (1970). *Acta Crystallogr.*, **B26**, 1761
126. Berman, H. M. (1970). *Acta Crystallogr.*, **B26**, 290
127. Ham, J. T. and Williams, D. G. (1970). *Acta Crystallogr.*, **B26**, 1373
128. Thiessen, W. E. and Hope, H. (1970). *Acta Crystallogr.*, **B26**, 554
129. Oh, Y. L. and Maslen, E. N. (1968). *Acta Crystallogr.*, **B24**, 883
130. Motherwell, W. D. S., Isaacs, N. W., Kennard, O., Bick, I. R. C., Brenner, J. B. and Gillard, J., to be published
131. Gabe, E. J. and Taylor, M. R. (1966). *Acta Crystallogr.*, **21**, 418
132. Gabe, E. J., Glusker, J. P., Minkin, J. A. and Patterson, A. L. (1967). *Acta Crystallogr.*, **22**, 366
133. Iwasaki, F. and Saito, Y. (1970). *Acta Crystallogr.*, **B26**, 251
134. Bugg, C. E. and Thewalt, U. (1970). *Science*, **170**, 852
135. Karle, I. L., Dragonette, K. S. and Brenner, S. A. (1965). *Acta Crystallogr.*, **19**, 713
136. Karle, I. L. and Fratini, A. V. (1970). *Acta Crystallogr.*, **B26**, 596
137. Williams, R. M. and Wallwork, S. C. (1967). *Acta Crystallogr.*, **22**, 899
138. Beisler, J. A. and Silverton, J. V. (1971). *Acta Crystallogr.*, in the press
139. Edwards, J. M., Weiss, U., Gilardi, R. D. and Karle, I. L. (1968). *Chem. Commun.*, 1649
140. Gilardi, R. D. (1971). *Acta Crystallogr.*, in the press
141. Shimanouchi, H. and Sasada, Y. (1970). *Acta Crystallogr.*, **B26**, 563
142. Okaya, Y. (1969). *Acta Crystallogr.*, **B25**, 1882
143. Okaya, Y. and Bednowitz, A. (1967). *Acta Crystallogr.*, **22**, 111
144. Iball, J., Motherwell, W. D., Pollock, J. J. S. and Tedder, J. M. (1968). *Chem. Commun.*, 365
145. Flippen, J. L., Karle, J. and Karle, I. L. (1970). *J. Amer. Chem. Soc.*, **92**, 3749
146. Flippen, J. L. and Karle, J. (1971). *J. Phys. Chem.*, **75**, 3566
147. Cantrell, J. S., Webb, N. C. and Mabis, A. J. (1969). *Acta Crystallogr.*, **B25**, 150
148. Smith, G. W., Fowell, D. T. and Melsom, B. G. (1970). *Nature (London)*, **228**, 355
149. Barnett, B. L. and Davis, R. E. (1970). *Acta Crystallogr.*, **B26**, 1026
150. Bordner, J., Stanford, R. H. and Dickerson, R. E. (1970). *Acta Crystallogr.*, **B26**, 1166
151. Hanson, A. W. and Huml, K. (1969). *Acta Crystallogr.*, **B25**, 2310
152. O'Connell, A. M. and Pascher, I. (1969). *Acta Crystallogr.*, **B25**, 2553
153. Stanford, R. H., Jr. (1967). *Acta Crystallogr.*, **23**, 825
154. Karle, J. (1966). *Acta Crystallogr.*, **20**, 881
155. Ellison, R. D., Johnson, C. K. and Levy, H. A. (1971). *Acta Crystallogr.*, **B27**, 333
156. Krüger, C. (1970). *J. Organometallic Chem.*, **22**, 697
157. Watenpaugh, K. and Caughlan, C. N. (1967). *Inorg. Chem.*, **6**, 963
158. Bock, C. M. (1968). *J. Amer. Chem. Soc.*, **90**, 2748
159. Greenberg, B. and Post, B. (1968). *Acta Crystallogr.*, **B24**, 918
160. Barnett, B. L. and Davis, R. E. (1970). *Acta Crystallogr.*, **B26**, 326
161. Carnduff, J., Iball, J., Leppard, D. G. and Low, J. N. (1969). *Chem. Commun.*, 1218
162. Tsai, C. C. and Streib, W. E. (1970). *Acta Crystallogr.*, **B26**, 835
163. Waite, N. E., Tebby, J. C., Ward, R. S., Shaw, M. A., Williams, D. H., Motherwell, W. D. S., Kennard, O. and Coppola, J. C., in the press
164. Birnbaum, K. B. (1970). *Acta Crystallogr.*, **B26**, 722
165. Smith, D. L. and Barrett, E. K. (1969). *Acta Crystallogr.*, **B25**, 2355
166. Koenig, D. F., Chiu, C. C., Krebs, B. and Walter, R. (1969). *Acta Crystallogr.*, **B25**, 1211

167. Huisgen, R., Brunn, E., Gilardi, R. D. and Karle, I. L. (1969). *J. Amer. Chem. Soc.*, **91,** 7766
168. Kushi, Y. and Fernando, Q. (1970). *J. Amer. Chem. Soc.*, **92,** 91
169. Magnuson, V. R. and Stucky, G. D. (1968). *J. Amer. Chem. Soc.*, **90,** 3269
170. Patterman, S. P., Karle, I. L. and Stucky, G. D. (1970). *J. Amer. Chem. Soc.*, **92,** 1150
171. Klug, H. P. (1966). *Acta Crystallogr.*, **21,** 536
172. Taylor, M. R., Gabe, E. J., Glusker, J. P., Minkin, J. A. and Patterson, A. L. (1966). *J. Amer. Chem. Soc.*, **88,** 1845
173. Bally, R. (1970). *Acta Crystallogr.*, **B26,** 477
174. Singh, P. and Ahmed, F. R. (1969). *Acta Crystallogr.*, **B25,** 2401
175. Karle, I. L., Gilardi, R. D., Fratini, A. V. and Karle, J. (1969). *Acta Crystallogr.*, **B25,** 1469
176. Camerman, A. and Camerman, N. (1970). *Science*, **168,** 1457
177. Karle, J. and Karle, I. L. (1967). *J. Amer. Chem. Soc.*, **89,** 804
178. Visser, G. J., Vos, A., de Groot, A. and Wynberg, H. (1968). *J. Amer. Chem. Soc.*, **90,** 3253
179. Fratini, A. V., Britts, K. and Karle, I. L. (1967). *J. Phys. Chem.*, **71,** 2482
180. Ammon, H. L. and Gibson, R. E. (1970). *Acta Crystallogr.*, **B26,** 1690
181. Bednowitz, A. L. and Post, B. (1966). *Acta Crystallogr.*, **21,** 566
182. Hanic, F. (1966). *Acta Crystallogr.*, **21,** 332
183. Bodor, G., Bednowitz, A. L. and Post, B. (1967). *Acta Crystallogr.*, **23,** 482
184. Ammon, H. L. and Sundaralingam, M. (1966). *J. Amer. Chem. Soc.*, **88,** 4794
185. Mornon, J. P. (1967). *Acta Crystallogr.*, **23,** 367
186. Kim, H. S., Chu, S. C. and Jeffrey, G. A. (1970). *Acta Crystallogr.*, **B26,** 896
187. O'Connell, A. M. (1968). *Acta Crystallogr.*, **B24,** 1399
188. Krebs, B. and Koenig, D. F. (1969). *Acta Crystallogr.*, **B25,** 1022
189. Karle, I. L. and Estlin, J. A. (1969). *Z. Krist.*, **128,** 371
190. Karle, J., Karle, I. L. and Mitchell, D. (1969). *Acta Crystallogr.*, **B25,** 866
191. Karle, I. L. and Karle, J. (1969). *Acta Crystallogr.*, **B25,** 1097
192. Karle, I. L. and Britts, K. (1966). *Acta Crystallogr.*, **20,** 118
193. Karle, I. L., Karle, J., Owen, T. B. and Hoard, J. L. (1965). *Acta Crystallogr.*, **18,** 345
194. Karle, I. L. and Karle, J. (1964). *Acta Crystallogr.*, **17,** 1356
195. Karle, I. L. and Karle, J. (1966). *Acta Crystallogr.*, **21,** 860
196. Karle, J., Estlin, J. A. and Karle, I. L. (1967). *J. Amer. Chem. Soc.*, **89,** 6510
197. Karle, I. L. and Karle, J. (1968). *Acta Crystallogr.*, **B24,** 81
198. Evans, H. T., Jr. and Showell, J. S. (1969). *J. Amer. Chem. Soc.*, **91,** 6881
199. Konnert, J. A., Clark, J. R. and Christ, C. L. (1970). *Amer. Mineralogist*, **55,** 1911
200. Merlino, S. and Sartori, F. (1969). *Acta Crystallogr.*, **B25,** 2264
201. Donnay, G. and Allmann, R. (1968). *Acta Crystallogr.*, **B24,** 845
202. Moore, P. B. and Bennett, J. M. (1968). *Science*, **159,** 524
203. Evans, H. T., Jr. (1968). *Abstr., Geological Soc. of Amer.*, p. 92
204. Taga, T. (1969). *Acta Crystallogr.*, **B25,** 2656
205. Dodge, R. P., Smith, G. S., Johnson, Q. and Elson, R. E. (1968). *Acta Crystallogr.*, **B24,** 304
206. Smith, G. S., Mucker, K. F., Johnson, Q. and Wood, D. H. (1969). *Acta Crystallogr.*, **B25,** 549

8
Structures of Natural Products

A. McL. MATHIESON

Division of Chemical Physics, CSIRO, Clayton, Victoria, Australia

8.1 INTRODUCTION

In a work of this form, particularly in the introductory volume, it is clearly impossible to attempt any approximations to a complete coverage of all the natural product structures which have been analysed by diffraction methods. Some selection is necessary. Our aim here is not to show how the structures were analysed, that is dealt with elsewhere in this series, but to indicate the extent to which diffraction methods (mainly x-rays, in practice) have interacted with the subject of natural product studies by outlining the range of information contributed. For this purpose, we will limit our survey to terpenes, including steroids, using the contributions to these groups to exemplify the impact on natural products studies as a whole.

Since 1966, when the author reviewed[1] this field, the trends visible then have accelerated. Automatic diffractometers, and latterly automatic densitometers, have become more widely used and access to computers more readily available, so that the potential for structure analysis evident earlier has been converted more and more into an actuality. As a result, the question of selection of structures to be analysed no longer rests with the feasibility of successful analysis but rather whether the analysis will contribute significant useful information to the chemical problem(s) on hand, thus warranting the effort involved. From this viewpoint, it is evident that the x-ray method constitutes a physical technique which, despite the tedium of data collection and handling compared with other techniques, nevertheless has compensating, if not unique, features in its favour, namely the details of atomic positions, which encourage its use in this particular field of study, not only by crystallographers but increasingly by organic chemists themselves. By contrast with the quasi-chemical physical techniques, u.v., i.r., n.m.r., o.r.d. etc., which focus attention on specific chemical groups, the x-ray method becomes relatively more valuable and rewarding where the structure studied is larger and more complex.

One aspect of the diffraction technique of special interest to the natural product chemist is the establishment of a structure on an absolute basis. This capability is of particular assistance in defining the chirality of compounds used as references in optical rotatory techniques and hence of aid in elaborating rules (and detecting anomalies) for specific chromophores. For natural products, use of the Bijvoet technique has expanded considerably in the last few years.

The literature search for this review has been greatly simplified by the generous assistance received from the Crystallographic Data Centre at Cambridge University, U.K., which provided up-to-date lists of the relevant papers from their files. Lists up to 1969 are available in print[2].

In order to conserve textual space, an attempt has been made to integrate figures and text by incorporating structural information in some of the figures. Also to conserve space, reference has been made to figures in an earlier work on the subject by the author[1]. In such cases, the figures are designated by the letter P.

It is necessary to stress that this review can merely skim over the subject. The mass of numerical data concerning atomic positions and hence bond lengths, bond angles, dihedral angles and other conformational details on chemical groups which is now on tape in the Crystallographic Data Centre is so extensive that it is only a matter of time before this store of information is more fully exploited to extract useful generalisations concerning specific atomic groupings. Earlier broad generalisations, e.g. on the peptide group, on lactone and ester groups, on dienes and other dissymetric groups, have been derived from a limited range of x-ray analyses. They have proved immensely valuable and stimulating in chemistry and biochemistry. With a more extensive range of data, the natural bounds of these generalisations can be more readily defined and further, more subtle generalisations may be evolved. Of the compounds which provide a wide range and variety of atomic groupings, natural products constitute an important source of such information on functional groups and conformational details of interest to chemists.

8.2 TERPENES

8.2.1 Monoterpenes

An early study in this group was that of the linear monoterpene, geranylamine, as the hydrochloride[3]. Analysis of (−)-bromodihydroumbellulone[4]

(1) involved two crystallographically independent molecules whose conformations were nearly identical, atom 1 being slightly out-of-plane to different extents, both rather small. Of menthyl derivatives, a number have been studied. That of (−)-menthyl(−)-*p*-iodobenzenesulphinate[5] used the absolute configuration of the menthyl component to establish that of the sulphinate. Menthyl trimethylammonium iodide[6] was revealed as having a six-membered ring with a distorted boat conformation with the methyl substituent equatorial and the isopropyl and trimethylammonium substituents axial. The crystal structure of (−)-menthyl-*p*-bromophenylglyoxalate[7] involves four crystallographically independent molecules showing slight differences in their relative conformational detail. The ester groups appear to have a preferred conformation in relation to the saturated ring systems as deduced earlier[8]. The study of 4*R*,8*R*-*p*-menth-1-en-9-ol *p*-iodobenzoate[9] (2) which

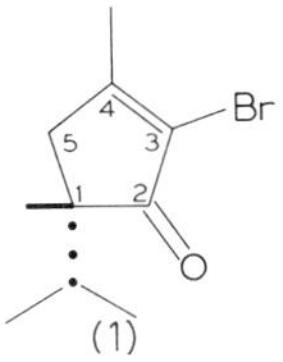

(1) 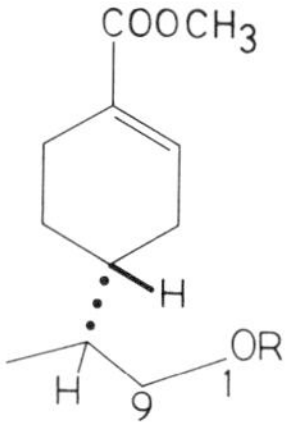 (2) 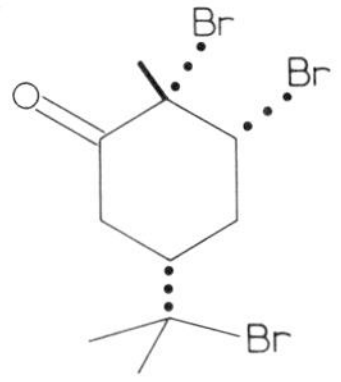 (3)

involved two independent molecules differing only in rotation about C(9)–O(1) led to a revision of the absolute stereochemistry of (+)-juvabione. In 2,4-dibromomenthone[10], the ring has the expected chair form. In (+)-*cis*-carvone tribromide[11] (isomerised from carvone tribromide with HBr and acetic acid at 0 °C) (3), the ring bromines are 2 axial, 3 equatorial.

Because of the possibility that marked differences in solvolysis rates depending on methylene bridgehead substitution might be associated with dimensional differences, a number of bornane derivatives have been studied. Both *syn*-[12] and *anti*-[13] 9-benzonorbornenyl *p*-bromobenzenesulphonate have been analysed but no significant dimensional difference was revealed. The bridgehead angle is 96 degrees and C(1) and C(4) are both slightly displaced from the plane of the benzene ring. The molecular dimensions of these molecules are very similar to those of other norbornene nuclei, *anti*-7-norbornenyl *p*-bromobenzoate[14] and a related structure, *anti*-8-tricyclo(3,2,1,$0^{2,4}$)-octyl *p*-bromobenzenesulphonate[15] and to norboramide[16]. The effect of substitution at an ethylene bridgehead was explored with 3-(*N*-benzyl-*N*-methylaminomethyl)-2-norbornanol[17]. The dihedral angle between the two planes of the six-membered boat-shaped ring is 114 degrees, other molecular dimensions being similar to those of 2-*endo*-phenyl-2-*endo*-norbornanol[18]. While the majority of studies indicate that the norbornane nucleus has twofold symmetry, the analysis of *exo*-2-tosylnorbornanol[19] suggests steric factors arising from substitution at position 2 and/or 3 can cause the cage to twist. Norbornadiene-2,3-dicarboxylic acid[20] involves a strong hydrogen bond between the carboxyl groups both of which lie in a plane of symmetry. A potassium salt of this compound has also been studied[21]. A closely related structure, 5-norbornen-2,3-*endo*-dicarboxylic acid anhydride[22] (4) has a conformation similar to previous norbornene structures (above) and also to cyclopentadiene dimer[23]. Calculations made on the geometry of these ring systems[24] are of interest.

Although not strictly natural products, the closely related compounds, 1-bi(norbornane), 1-bi(apocamphane) and 1-bi(adamantane) have been subjected to careful analyses[25]. A comparison of the dimensions of the central bond (C(1)–C(1′)) is of interest.

Three fenchone derivatives have been analysed: 2-bromo-2-nitrofenchone[26], 2-bromo-6-(n-dimethylaminomethyl)-fenchone[27] and 6-bromo-isofenchone[28]. In general, the molecules retain their symmetry about the median plane despite 'strain' introduced by substituents.

The early work on camphor[29] (5, R = Me) did not establish its absolute configuration but this was rectified by later workers[30, 31]. The latter report refers also to work on camphoroxime hydrobromide[32] which also established the absolute chirality of camphor. The structures of (+)-10-bromo-2-chloro-2-nitrosocamphor[33] and (−)-2-bromo-2-nitrocamphine[34] indicate that the presence of a relatively bulky group *cis* to the CMe_2 bridge causes some distortion of the molecular framework, the two groups bending from their ideal positions. In establishing the structure of retusamine, the absolute structure of α′-bromo-(+)-camphor-*trans*-π-sulphonate[35] (5; R = $-CH_2 \cdot SO_3^-$) was also derived.

In some of the crystal structures discussed above, disorder has occurred due mainly to the compact, symmetrical shapes of these molecules.

Two related structures, 3-(*N*-methylaminomethyl) pinane hydrobromide[36] (6) and 3-(*N*-dimethylaminomethyl) pinene-2(10) hydrobromide[37] indicate a

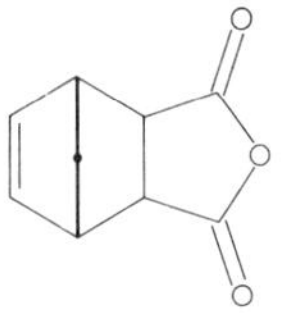

(4)

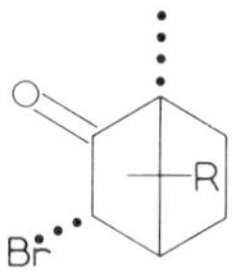

(5)

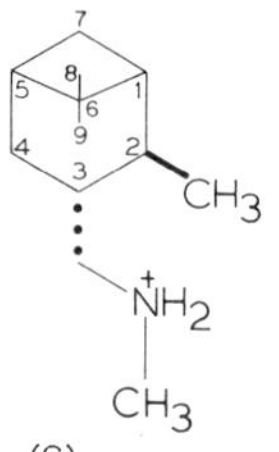

(6)

flattening of the six-membered ring due to interaction between H of C(3) and of C(9); the dihedral angle between planes 1,2,4,5 and 2,3,4 is 161 degrees. The reaction of thionyl chloride with (±)-10β-pinane-2,3α-diol yields two isomers, one of which, (±)-10β-pinane-2,3α-diol cyclic sulphite[38], has been analysed. In 3β-chloro-6,6-dimethylnorpinan-2-one[39], atom 3 lies on the same side as 8 due to the β-halogen substitution at 3.

The isomeric structures, iridomyrmecin[40] and isoiridomyrmecin[41] (P57) have both been analysed. In both cases, ring B is in a skew boat form with the lactone group planar. In the former, ring A is *endo* to ring B and, *exo* in the latter. In monotropein rubidium salt[42] (7), ring A is planar and ring B in a half-chair form while in the loganin derivative, loganin penta-acetate monomethyl ether bromide[43] (8), ring A appears to be in an envelope form and ring B in a boat form. These four compounds have a considerable number of common configurational features.

In anemonin[44] (P58), the lactone rings are in the *trans* configuration, the cyclobutane ring being bent with a dihedral angle of 152 degrees.

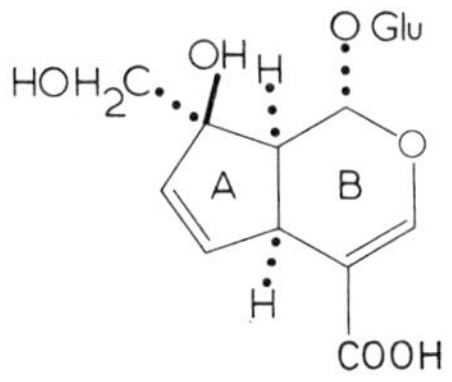

HOH$_2$C OH H O Glu A B O H COOH

(7)

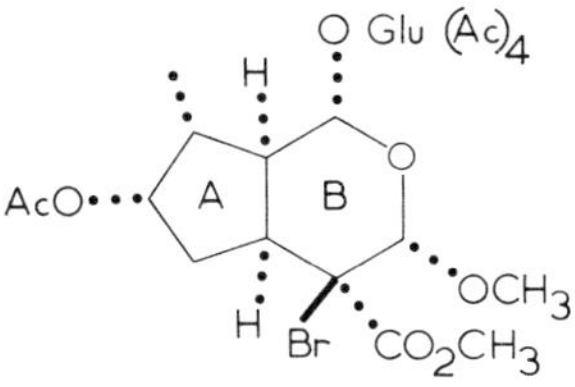

O Glu (Ac)$_4$ H AcO A B O OCH$_3$ H Br CO$_2$CH$_3$

(8)

8.2.2 Sesquiterpenes

In retro-β-ionylidene acetyl-p-bromanilide[45] (9), the chain of the β-ionylidene component is *cis* at the C(8)—C(9) double bond, in contrast to most carotenoids and vitamin-related compounds[46]. The conformation of laurinterol acetate[47] (10) is determined by intra-molecular overcrowding effects; ring B

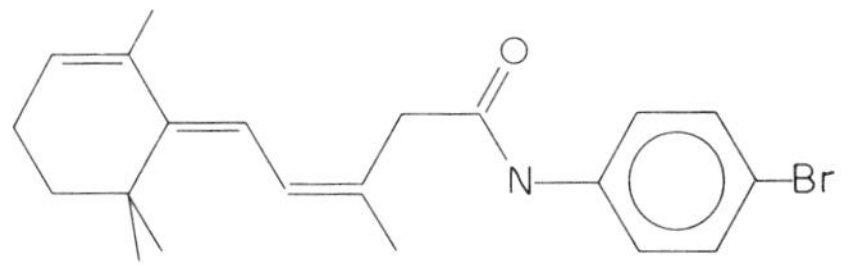

O N Br

(9)

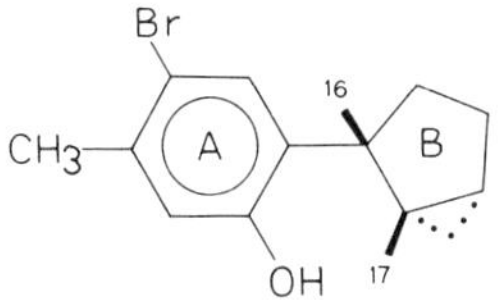

Br CH$_3$ A 16 B OH 17

(10)

has a boat form to avoid eclipsing C(16) and C(17). Ring A in dihydrofomannosin p-bromobenzoylurethane[48] (11) is also under considerable strain, the double bond substituents are displaced from their ideal positions and atoms 1,4,5,8 are noticeably non-planar. The lactone group is approximately planar but with atoms 4 and 5 displaced 0.56 and 0.80 Å out-of-plane, respectively, The A ring approximates to a slightly skewed boat. Ring B is not planar but has a dihedral angle 457/478 = 158 degrees. Two sesquiterpene derivatives with spiro ring junctions are acorone p-bromophenylsulphonyl hydrazone[49] (P82) and acorenone-B 4-iodo-2-nitrophenylhydrazone[50] (12).

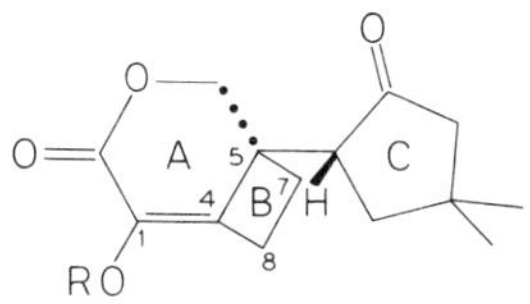

O O O A 5 C 4 B 7 H 1 8 RO

(11)

O 3 A B

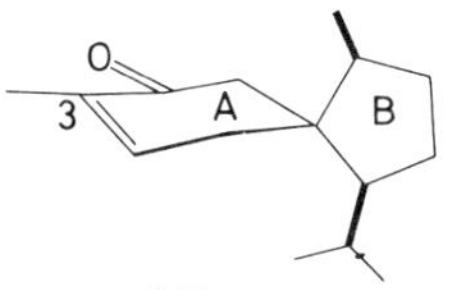

(12)

In the former, the 6-ring is in the chair form with the substituent at 3 axial. Final refinements indicate that the 5-ring approximates more closely to an envelope form. Because solution studies were interpreted as having the substituent at 3 in the equatorial rather than the axial position, caution in the extrapolation of solid-state studies to solution chemistry has been advised[51].

A number of 10-membered ring compounds have been studied. Germacratriene[52] (13) indicates the transannular interaction between double bonds. A C_{12} structure, pregeijerine[53] (14) shows a similar disposition; the minimum approach distances between double bonds are 2.91 and 3.13 Å. These two structures, together with those of shiromodial acetate *p*-bromobenzoate[53a] (15) and elephantol *p*-bromobenzoate[53a, 54]

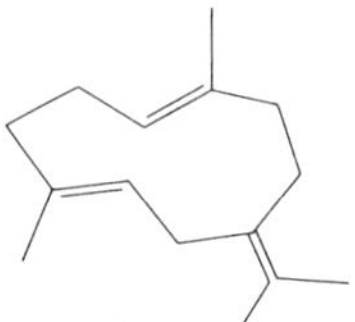

(13) 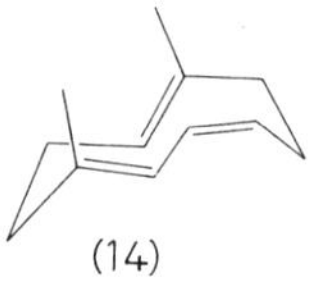 (14) 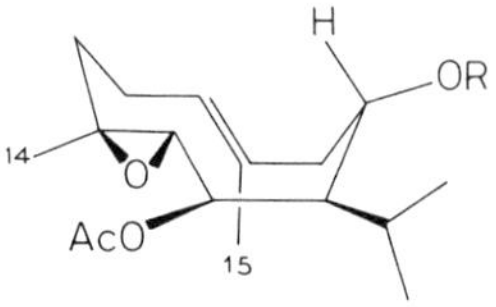 (15)

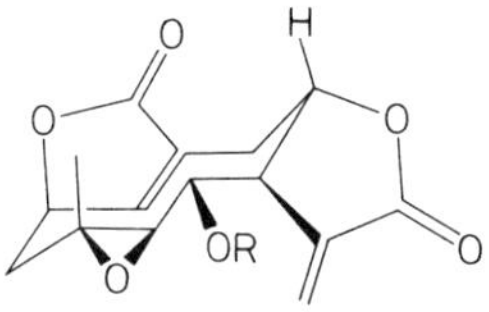

(16) 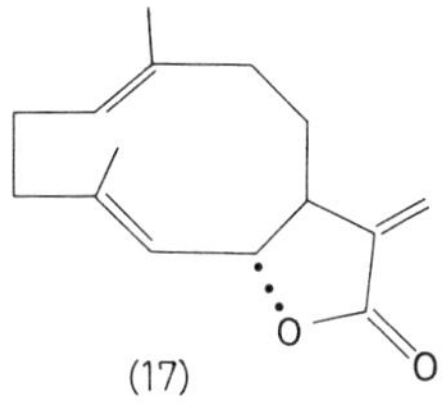 (17)

(16) throw light on the relation of most eudasmane sesquiterpenes to the precursor *trans*-farnesyl pyrophosphate. Costunolide[55] (17) also displays a similar mutual relation of the 4,5 and 1,10 double bonds, while vernolide iodobenzoate[56] (18) has substituent dispositions similar to those in elephantol. Humulene[57] (P59) was the first of this group of large ring structures to be studied revealing the double bonds as all *trans*. While the uncomplexed double bond is closely planar, those complexed with Ag^+ are slightly twisted. The structure of humulene bromhydrin[58] (19) indicates the attack on the 1,2 double bond in humulene is *cis*, but it is *trans* at the other two. The 8-ring is in a boat-chair conformation distorted from C_s symmetry by ring fusion. The 4-ring involves a dihedral angle of 151 degrees. Appropriate treatment of

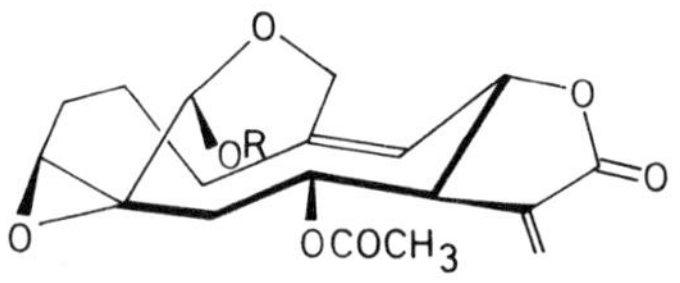

(18) 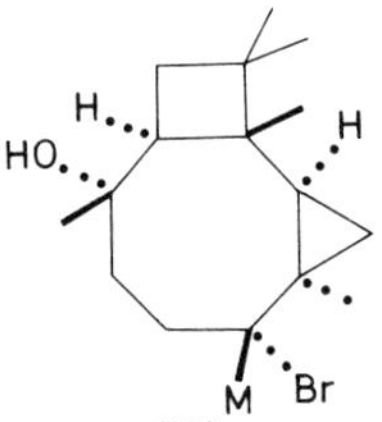 (19)

humulene yields the compound called α-caryophyllene alcohol[59] (20) in which ring A has a normal chair form with OH axial. Of the true carophyllene derivatives, β-caryophyllene chloride[60] (P75(i)) and caryophyllene chlorhydrin[61] (P75(ii)) yielded almost identical conformational detail with ring A in a boat form, ring B a normal chair and ring C buckled. Caryophyllene iodonitrosite[62] (21) is a stable free radical. Ring A adopts a chair form, the ring being slightly flattened to relieve steric hindrance between O(1) and H of C(5). Irradiation of either caryophyllene or *iso*-caryophyllene yields two methyl epimers, both of which have been studied, namely, photocaryophyllene A and D bromoketones[63] (22). The change in configuration at position 4 alters the conformation of ring A considerably. In A, it approximates to a crown while in D, it appears to approximate more to a boat form.

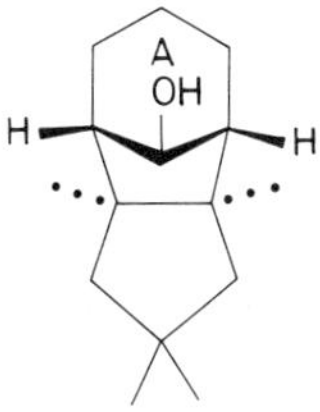

(20) 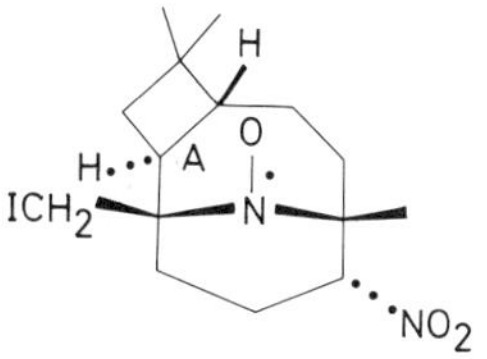 (21) 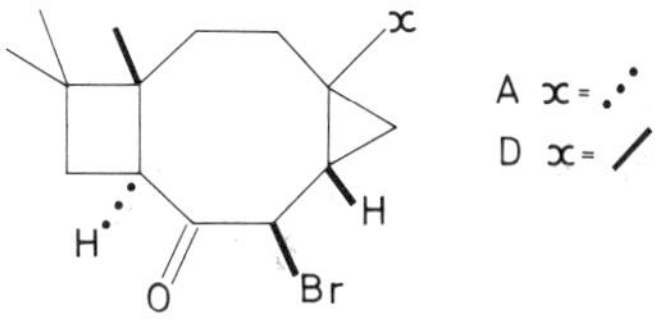 (22)

A number of sesquiterpenes involving a decalin ring system have been studied. In dihydrohydroxyeremophilone[64] (P61), the rings are *cis*-fused; both are in the chair form. X-ray analysis also helped to establish the structure of valeranone[65] (P62). In β-gorgonene[66] (23), both rings are in the chair form as also in bulgarene dihydrobromide[67] (24). The configuration of C(1) in (−)-ε-bulgarene is not established by this analysis. In nor-β-cubebone[68] (25), ring A is modified by 1,5 bond formation. Two analyses of dihalogen dihydrocadinene[69] (P60) established the decalin system as *trans* and the 6-rings as in the chair form. Chamaecynenol[70] (26) is a norsesquiterpenoid, ring A being in a half-chair form, *cis*-fused to ring B which is a normal chair.

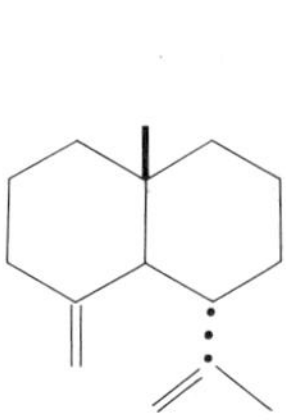

(23) 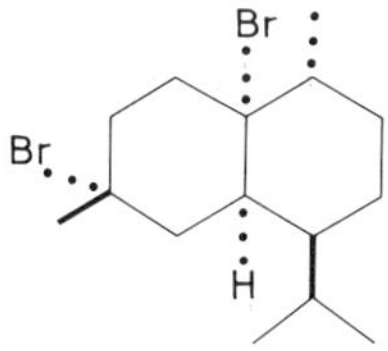 (24)

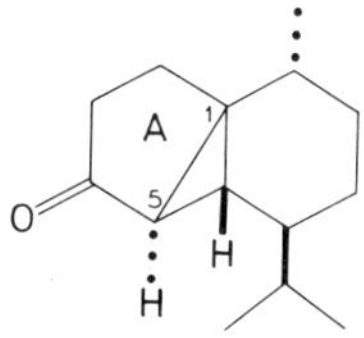

(25) 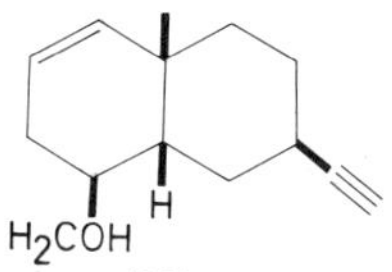 (26)

In himachalene monohydrochloride[71] (27), ring A is in the chair form while ring B approximates to a twist chair. In the case of hirsutic acid *p*-bromophenacyl ester[72] (P81(i)), irradiation during the analysis led to the production of rearranged material (P81(ii)) roughly in a 1 : 1 proportion with the original. The 5-rings are mainly in the envelope form or slightly distorted therefrom.

Iresin di-*p*-bromobenzoate[73] (P63) has a normal chair form for ring A and a half-chair form for ring B. Ring C with C(9) out-of-plane relative to the lactone group has an envelope form. A similar disposition in respect of C(7) relative to the lactone group applies in *O*-(bromoacetyl) tetrahydrodouglanine[74] (28) where rings A and B are in the chair form. In 2-bromo-α-santonin[75] (P64), ring A is essentially planar with C(7) again being out-of-plane relative to the lactone group. A similar situation applies in 2-bromo-β-santonin[76]

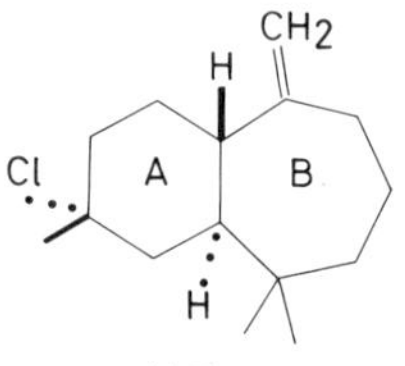

(27) 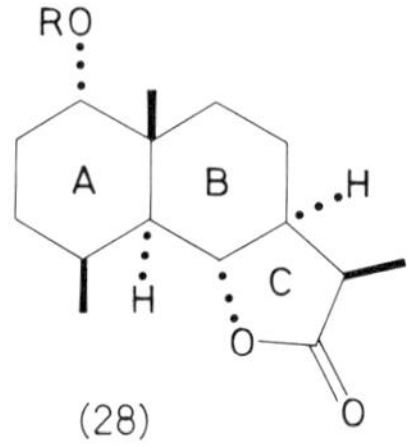 (28) 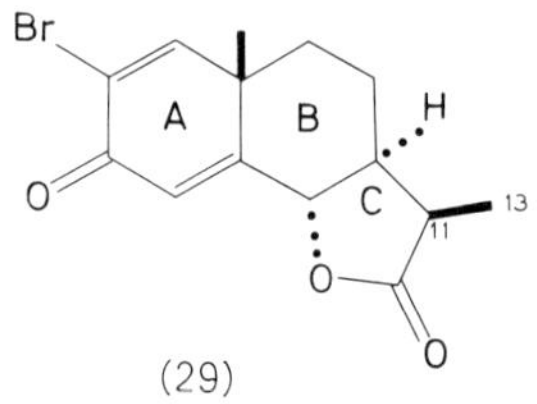 (29)

(29). Ring C is half-way between a half-chair form and envelope with C(8) out-of-plane. The epimeric relationship of C(13) in α- and β-santonin greatly modifies the intra-molecular interactions, C(13)—C(6) and C(13)—C(9), which, in β-santonin are 3.17 and 3.07 Å respectively relative to 3.79 and 3.49 Å in α-santonin, thus influencing the relative stabilities of these two compounds. Other santonin compounds studied have been, 2-bromo-(−)-β-desmotroposantonin[77] (30) in order to confirm the β configuration of the Me at position 11. Ring B is in the sofa form with C(7) out-of-plane by only 0.07 Å but with C(8) out by 0.75 Å. Ring C is again envelope in form with C(8) displaced 0.52 Å from the lactone plane. In 2-bromolumisantonin[78] (31), ring A and its substituents O(3) and Br are coplanar and the cyclopropane ring A deviates C(1) to bring it coplanar with 6,7,9 and 10. Ring B is thus in a sofa form with atom 7, 0.74 Å out-of-plane. In ring C, atom 8 is out-of-plane by 0.61 Å relative to the lactone plane. 2-Bromodihydroiso-photo-α-santonic lactone acetate[79] (P65) involves a rearrangement of rings A and B with ring A

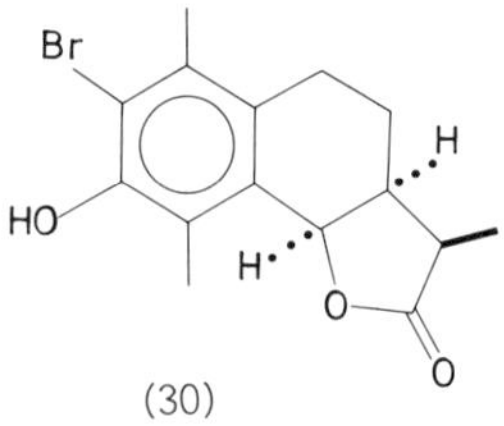

(30) 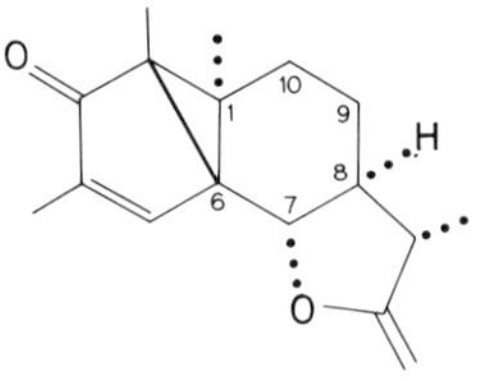 (31) 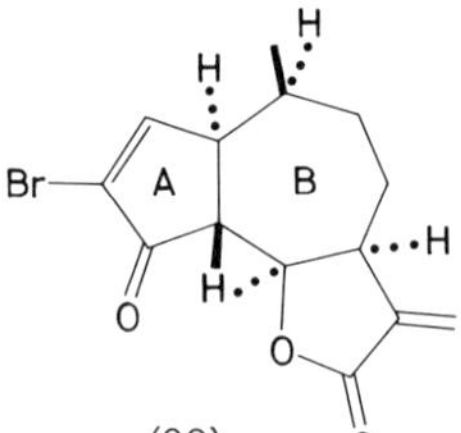 (32)

an envelope with C(4) 0.54 Å out-of-plane. Ring B, now 7-membered, is claimed to adopt a chair form.

A considerable number of sesquiterpene lactones with a pseudoguaianolide carbon skeleton have been analysed. Bromoambrosin[80] (32) and solstitialin[81] (33) have ring A in a C(5)-envelope form, in the former case being rather shallow. Rings B are chairlike. Ring C appears near planar in the former and more C(7)-envelope in the latter. Euparotin bromoacetate[82] (34) possesses an epoxide group and an α,β unsaturated lactone function, features common to other tumour inhibitors which have been analysed, withaferin A[83] and elephantin and elephantopin[54].

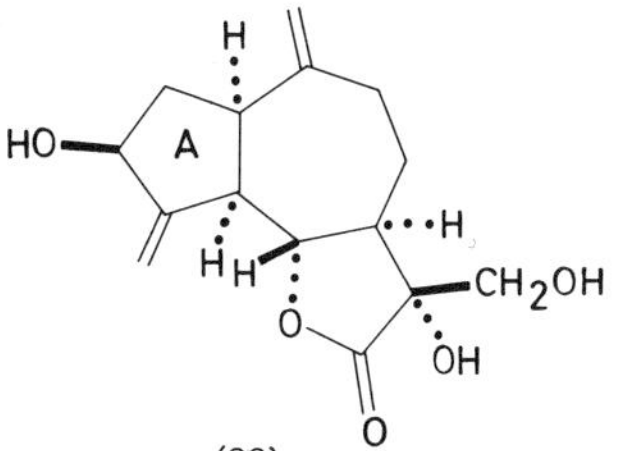

(33) 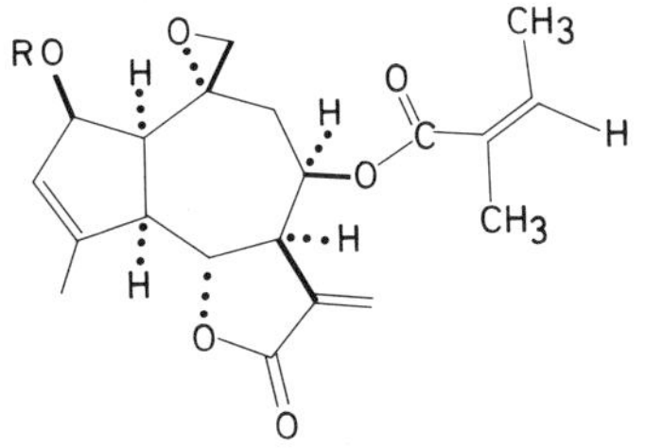 (34)

Structures with the lactone ring closure to C(8) are exemplified by bromogeigerin acetate[84] (P66) with ring A a C(2)-envelope and ring C a C(7)-envelope. Bromoisotenulin[85] (P67) has similar features but interactions between Me (α to C(5)) and the α-hydrogen atoms of C(8) and C(10) cause the molecule to be appreciably folded. In bromohelenalin[86] (P68) the details are similar except that C(11) is apparently out-of-plane. In bromomexicanin-E[87] (P69) which has lost a methyl group, apparently by migration during biogenesis from a regularly constituted guaianolide precursor, the two five-membered rings are *cis*-fused, being *cis-syn-cis*. Ring C appears to involve C(8) being out-of-plane relative to the lactone group. The structure of gaillardin was established by analysis of monogaillardin[88] (35(a)) where the configuration was established at all points except C(1) which was derived from analysis of desacetyldihydrogaillardin *p*-bromobenzoate[88] (35b). Pul-

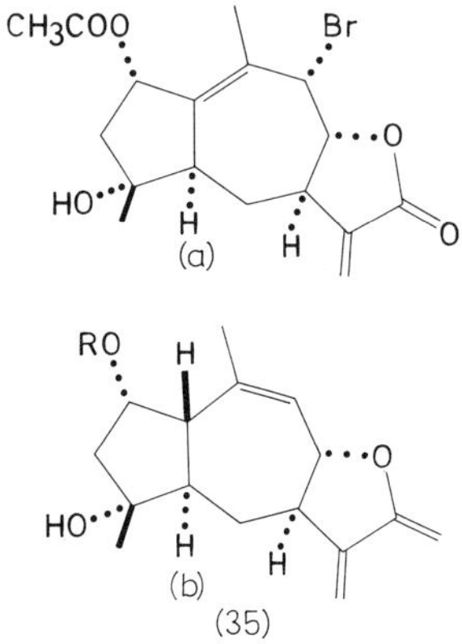

(a) (a)

(b) (b)

(35) 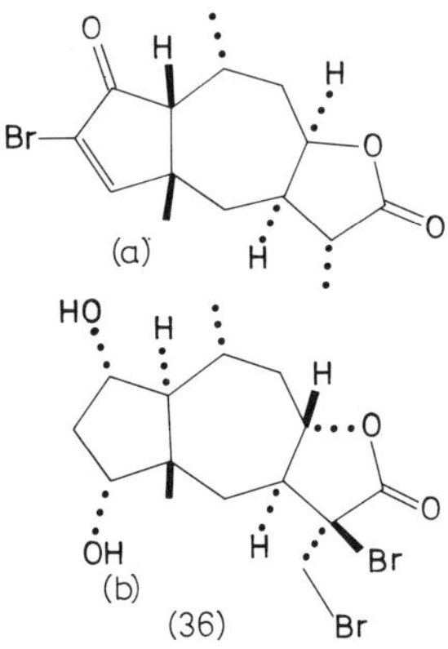 (36)

chellin has been studied as two derivatives, 11,13-dibromopulchellin[89] (36b) with ring A a 5-envelope and ring C a 7-envelope and ring B a distorted flat chair, and bromo-anhydrotetrahydropulchellin[90] (36a). In the preparation of the latter compound, the configuration at C(1) was inverted.

The structure of *p*-bromobenzoyl-laserol[91] (37) shows ring A as a 5-envelope while ring B appears to have a boat-like shape. In its preparation, with phthaloyl-L-alaninyl chloride, epimerisation α to the carbonyl group resulted in daucyl-DL-alaninate hydrobromide[92] (38). Rings A and C are envelope in form while ring B is a chair. In the case of both D and L ester groups the orientation relative to the associated ring B appears to accord with the predicted preferred conformation[8].

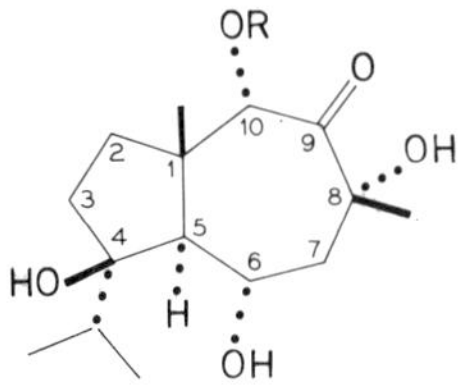

(37)

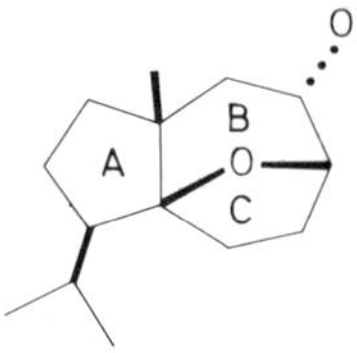

(38)

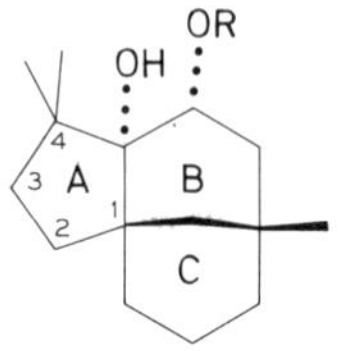

(39)

In isoclovene hydrochloride[93] (P78), ring A is a 5-envelope while rings B and C are in chair conformations. Pseudoclovene, prepared by dehydration with phosphoric acid from caryolan-1-ol, as in isoclovene, has been studied as the diol mono-*p*-bromobenzenesulphonate[94] (39). In this, ring A is a 3-envelope and ring C a chair. Ring B is forced into a boat from the 'bow, sprit' hydrogens at C(6) and C(12) involve severe interaction. The molecule is highly strained, relieved by distortions in bond angles, flattening ring B and also significant twisting of rings B and C.

For a number of longifoline derivatives, similar distributions of strains occur in these tricyclic systems. Longifolene hydrochloride[95] (P76) was an early example while a group, 3-α-bromo-7βH-longifolane (40(a)),3-α-

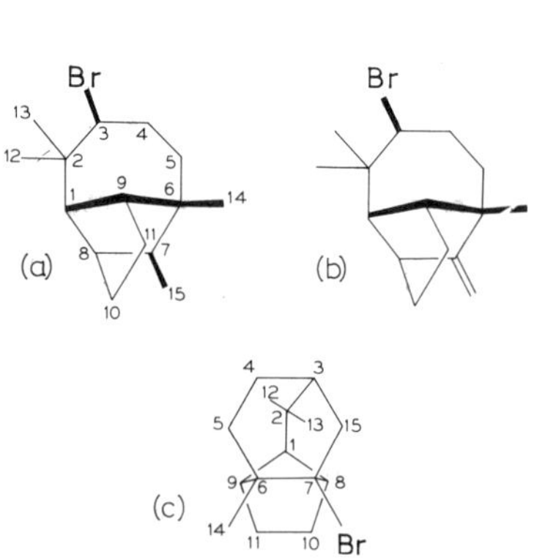

(40)

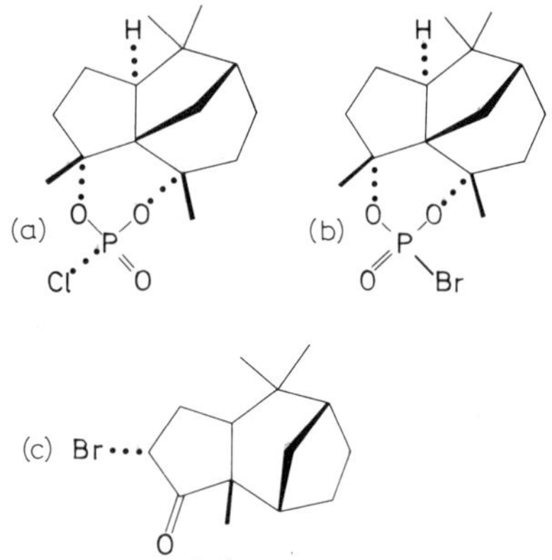

(41)

bromolongifolene (40(b)) and 7-bromocyclo(3.15)longifolane (40c)[96], have recently been studied. In the first two, the BC ring system and substituent atoms 2 and 5 constitute a relatively rigid system which with the dimethyl group at C(2) probably determines the conformational disposition of C(3). The third member involves the bridge 3,15,7 which leads to a complex fused ring system, further distortion is introduced by the interaction Br···C(14). Even so, atoms 3,15,7,8,1,2 form an only slightly distorted chair.

A group of structures studied to elucidate the products of a reaction of a formic acid treatment of cyperene oxide which are the halophosphoridates of cyperene diol (41a and b) and of the ketone (41c)[97]. In the first two, the [321] octane group is deformed significantly by the halophosphoridate group whereas in the third it is undistorted. The anlysis of khusimol *p*-bromobenzoate[98] (42) establishes that it is not stereochemically related to cedrene but is related more closely to eremolactone.

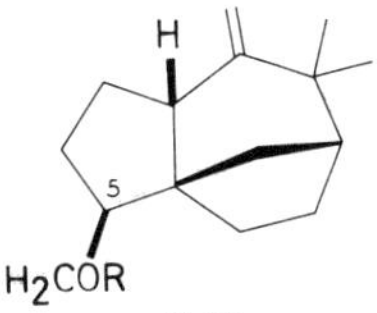

H
5
H_2COR
(42)

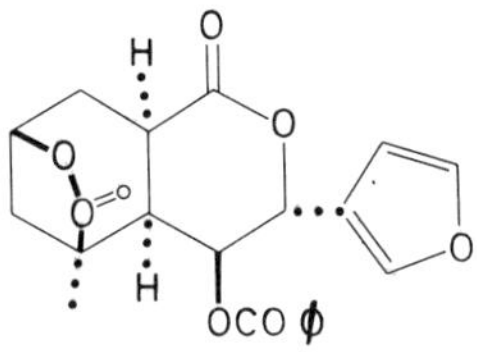

H
O
O
O
O
O
H
OCO ϕ
(43)

In trichodermol *p*-bromobenzoate[99] (P79), ring A is in a half-chair form, ring B is a slightly distorted chair and ring C an 11-envelope. Trichodermin forms a component of the macrocyclic ester verrucarin A[100]. Study of the patchouli alcohol diester of chromic acid[101] (P79) led to a revision of the structure of the alcohol. It involves a bird-cage ring structure with an additional ring fused to it. Shellolic acid bromolactone[102] (P80) has ring A in a chair form, the other rings mainly in the envelope form. In isocollybolide[103] (43), ring A is a chair, B a near half-chair and ring C planar. The molecule is folded on itself in such a way that the γ- and δ-lactones are in virtually parallel planes *c.* 2.7 Å apart.

8.2.3 Diterpenes

In iso-eremolactone[104] (44), ring A is virtually planar while the C ring system and its substituents as far as C(6) are also planar. For enmein, two derivatives have been studied. In acetyl-bromoacetyl-dihydroenmein[105] (P88), ring closure by a lactone group leads to a complex ring system with ring A in a chair form and the *cis*-fused ring D in a 5-envelope form. Ring B is boat-shaped, the lactone group is planar. The molecule is considerably buckled bringing the keto oxygen within 2.97 and 3.08 Å respectively of the two carbon atoms of the bromoacetyl group cf. reference 106. In 16α-methyl-15β-bromacetatoenmein[107] (45), ring A has a chair form slightly distorted by the *cis*-fused lactone group which is 10-envelope. The CD ring system is as in the other enmein derivative, ring C being a high-tilt chair at C(14) and ring D is closer to a half-chair form than an envelope.

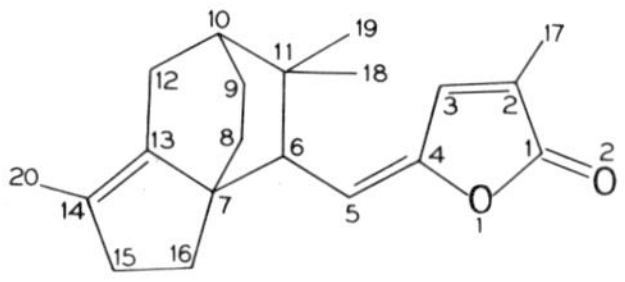

(44)

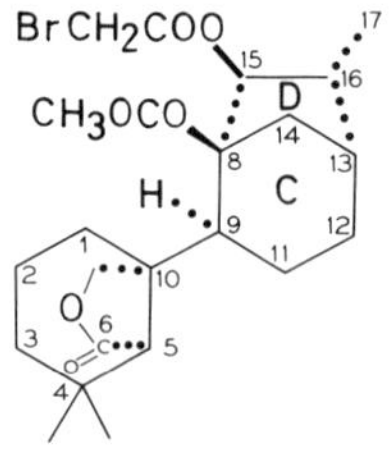

(45)

A group of compounds with similar skeletons comprises aplysin-20 [108] (P87), *p*-bromphenacyl labdanolate[109] (46) and dibromophthyterpol ketone[110] (47), the ring conformations being chair, slightly distorted in some cases. Elaborations of this skeletal structure by ring formation involving the side chain are evident in clerodin bromolactone[111] (P99) (where ring C has an envelope form and ring D is almost planar), isocolumbin 1-*p*-iodophenyl-3-phenylpyrazoline[112] (P100) and deacetylcascarillin[113] (P100a). In the last compound, rings A and B are in the chair form but distorted, with C(2), −0.53 Å and C(5), +0.90 Å, out-of-plane relative to the other four atoms, while for ring B the corresponding displacements are +0.49 and −0.88 Å for C(5) and C(8) respectively. These distortions are attributed partly to the fusion of the five-membered ring across ring B and partly to the pronounced 1,3-axial interaction on the α-face of the molecule. The furan ring in the last two structures is planar.

In rimuene[114] (48), rings A and C in the chair form are slightly distorted and ring B is in a half-chair form while in methyl-6α-bromo-12-methoxy-7-oxopodocarpate[115] (49), ring B is in a boat form with C(7) and C(10) as prow and stern. Rings A and B in beyerol monoethylidene iodoacetate[116] (P97) are chair in form as also is ring C but severely buckled by the C(15)–C(16)

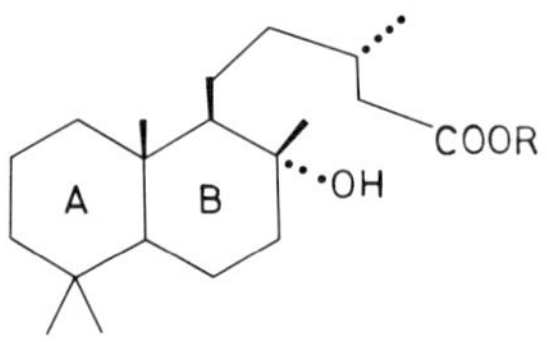

(46)

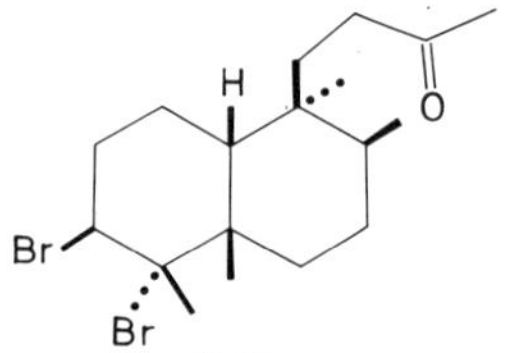

(47)

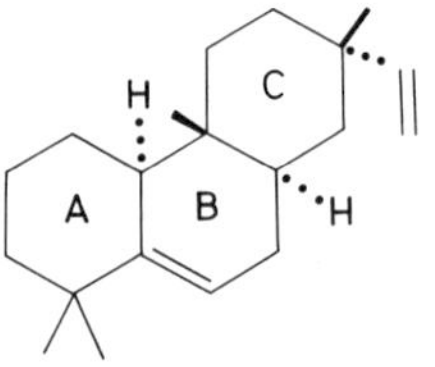

(48)

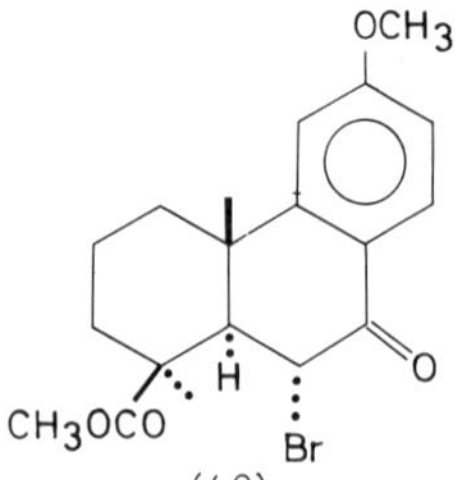

(49)

bridge. C(15) and C(20) interact, twisting C(15)–C(16) from the expected non-strain position. Also plane 8,14,13 is displaced 25 degrees from the plane of atoms 6,7,8. For 1,2-desacetyl-ε-caesalpin-2-*p*-bromobenzoate[117] (50), ring C is a distorted half-chair, approximating to a sofa form, out-of-plane displacements are C(8), −0.58 and C(9), +0.21 Å. Ring D is planar but very slightly tilted (2 degrees) relative to the plane of atoms 11,12,13,14.

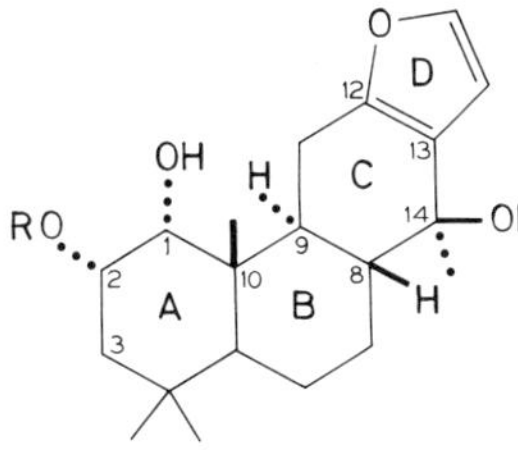

(50)

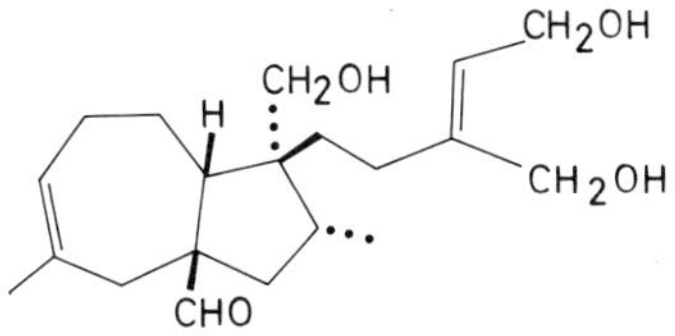

(51)

Use of optical rotatory measurements and x-ray structure analyses of key compounds led to a unified interpretation of the biosyntheses of diterpenes from a geranyl–geraniol precursor. The compounds included bromo-epoxynorcafestanone[118] (P91), dibromo-rosololactone[119] (P92) and gibberellic acid which was studied as two derivatives, methyl gibberellate di-*p*-bromobenzoate[120] (P89) and methyl bromogibberellate[121] (P90). In relation to these last two, formation of a bromo derivative involves an inversion of the two-carbon bridge on ring C. A probable rearrangement from the geranyl–geranyl (or geranyl–linalyl) pyrophosphate precursor is suggested to explain the existence of portulal, analysed as the *p*-bromophenylsulphonylhydrazone[122] (51). In this structure, ring A has a boat form while ring B is a 2-envelope.

A number of phorbol derivatives have been studied. Phorbol itself has been studied as a 20-bromofuroate[123] (52) and simply solvated with ethanol[124] (52, R = H). Ring A is a 4-envelope and ring C is almost sofa in form with

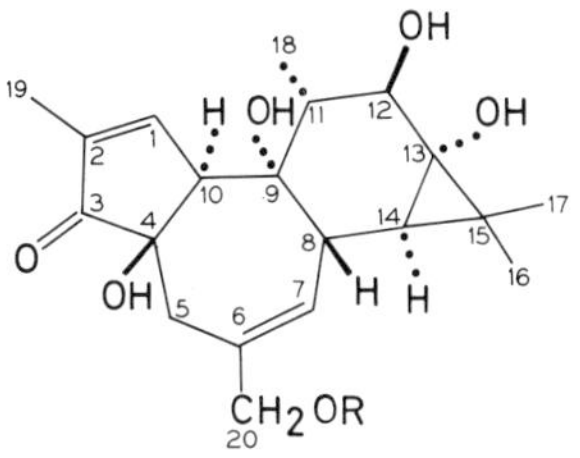

(52)

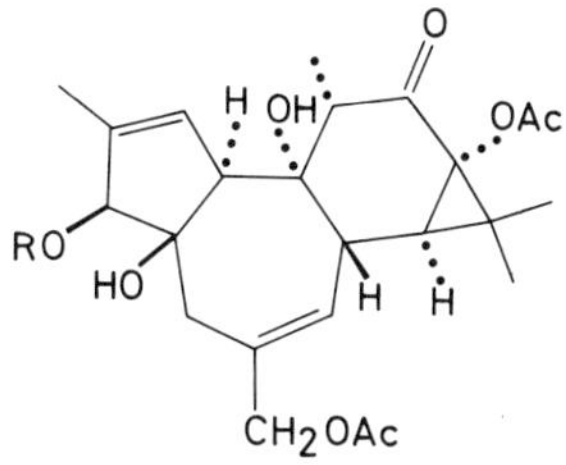

(53)

C(9), −0.72, and C(11), +0.14 Å, out-of-plane relative to atoms 8,14,13,12. Ring B is boat-like in form. In general, the cyclopropane ring effectively imitates a double bond in constraining the associated atoms, here 8,14,13,12, to near coplanarity. In the second analysis of phorbol, the solvate molecule is found 'bonded' to the hydroxyl group at C(9). From this and other studies with phorbol-12,13-diesters, it is thought that this hydroxyl group is essential for interaction with receptor sites in the cell. In neophorbol-13,20-diacetate-3-*p*-bromobenzoate[125] (53), ring A is again a 4-envelope but ring D is described as a slightly skewed chair. An isomer of phorbol, isophorbol, when irradiated as the triacetate, yields 12,13,20-lumiphorbol triacetate which has been analysed as the 4-*p*-bromobenzoate[126] (54). Linkage of the double bonds 1,2 and 7,6, forming a cyclobutane ring, leads to a cage-like structure. Ring C again appears to be in a half-chair form.

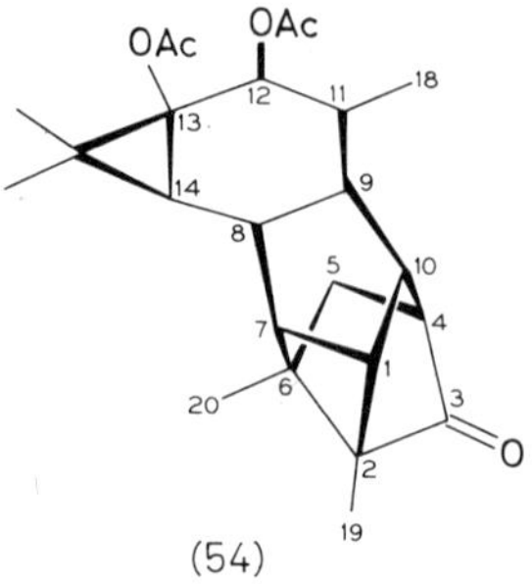

(54)

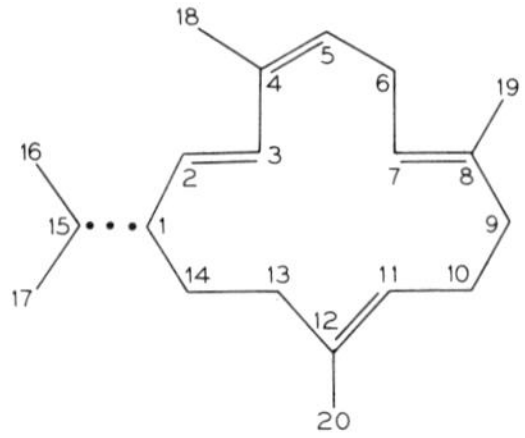

(55)

Cembrene[127] (55) has a *trans* disposition at each double bond except at C(4)—C(5) where it is *cis*. A δ-lactone is fused to a cembrane ring in crassin *p*-iodobenzoate[128] (56) while an oxabridge constitutes an additional feature in eunicin iodoacetate[129] (57). 6,20-Epoxy-lathyrol phenylacetate diacetate[130]

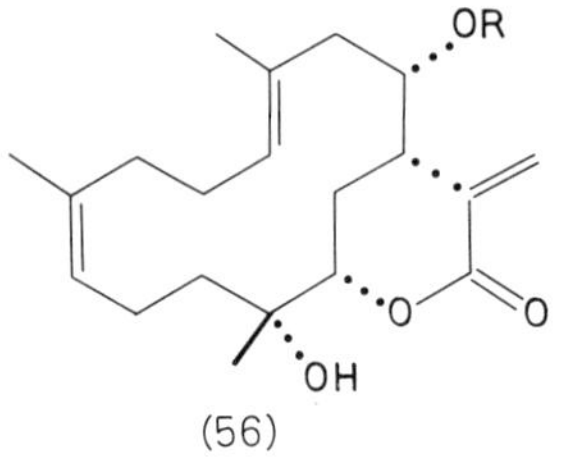

(56)

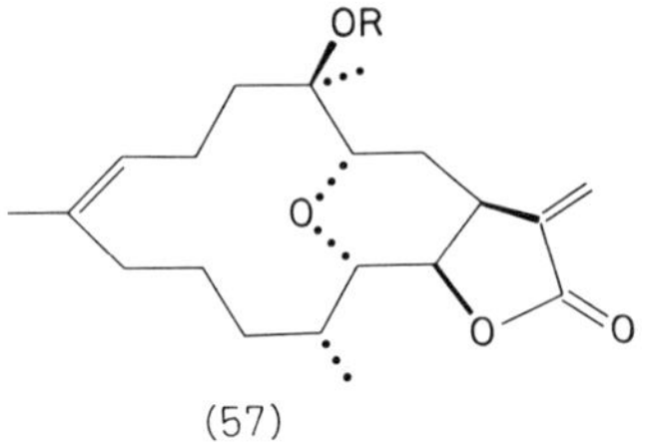

(57)

(58) involves an α,β-unsaturated ketone adjacent to the cyclopropane ring, the combination largely determining the shape of ring B. Ingenol triacetate[131] (59) has a 4-envelope form in ring A while ring B is boat-like in form and ring C chair-like. In this structure, the absolute configuration was determined

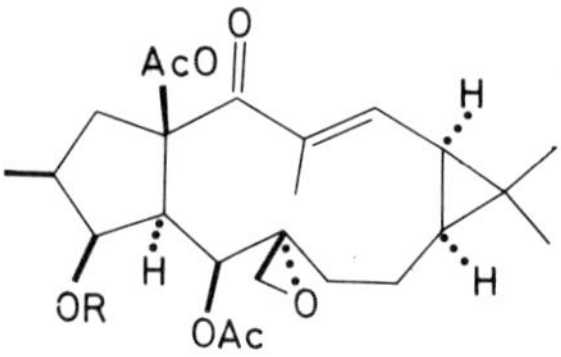

(58)

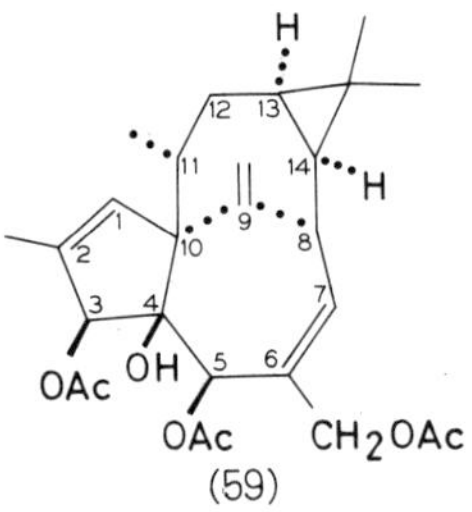

(59)

using the anomalous dispersion of the oxygen atoms, the feasibility of which was established earlier[132]. There is considerable strain in the structure of taxa-4(16),11-diene-5α, 9α, 10β, 13α-tetraol *p*-bromo-benzoate dihydro-anhydroacetonide[133] (60). Ring A is a shallow 1-envelope (0.17 Å out of the plane of atoms 11,12,13 and 14) and repulsive forces between C(18), C(15) and C(18), C(17) are claimed to force C(18), 0.48 Å, below the plane of 11, 12, 13 and 14. C(16) and O(2) are bent away from one another so that ring

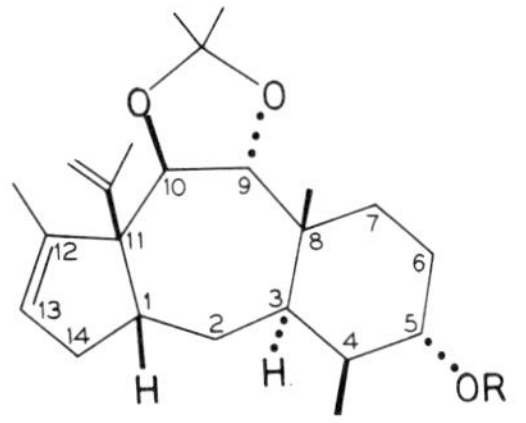

(60)

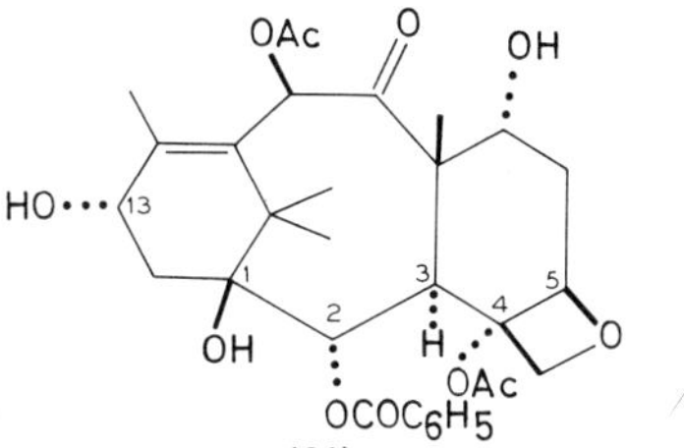

(61)

C is a slightly distorted chair. Ring B adopts a twisted boat form with atoms 1, 2, 8, 10, 11 almost coplanar and 3 and 9 below and above this plane respectively. Ring D is a half-chair with 9 and 10 below the plane of O(3), 22, O(4) by 0.24 and 0.68 Å respectively. In 2,5,9,10-tetra-*O*-acetyl-14-bromotaxinol[134] (P93), ring A has a boat form and C a chair form. Ring B, constrained mainly by ring A has a 'boat-chair' form. A related structure, baccatin-V[135] (61), has an oxetan ring and the configuration of oxygens at C(5) and C(7) (β and α respectively) opposite to those found in other taxane derivatives.

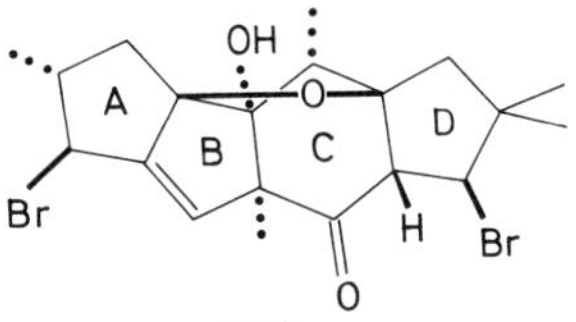

(62)

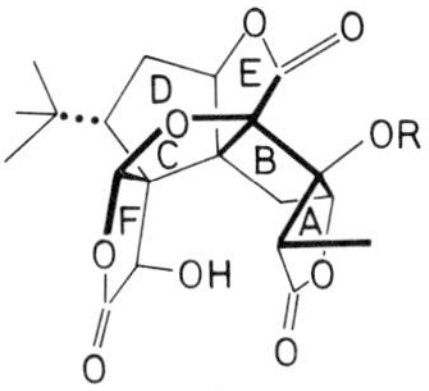

(63)

In jatrophone dihydrobromide[136] (62), ring A is a 1-envelope, B a shallow 14-envelope and ring C has a boat form. Ginkgolide A *p*-bromo-benzoate[137] (63) consists of six 5-membered rings fused into an arc so that the lactone groups of rings A and F are held almost parallel to one another. Ring system

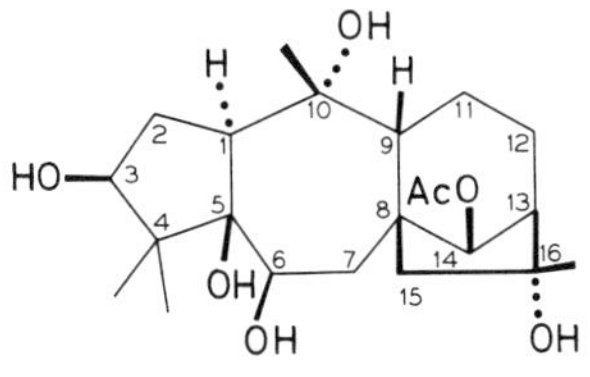

(64)

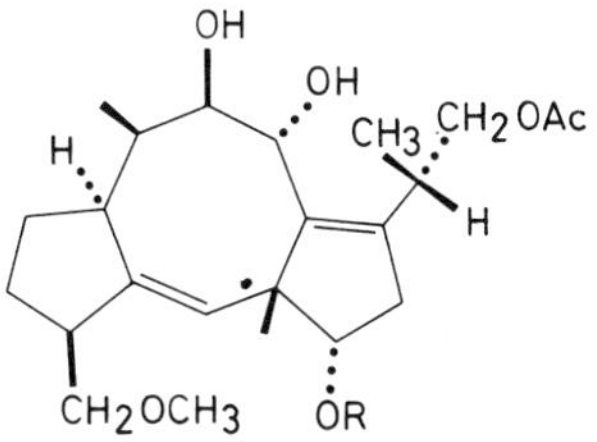

(65)

BD is a spiro[4,4]nonane in which ring B is crosslinked to ring D by a —C—O— and a lactone bridge.

Grayanotoxin-I[138] (64) is tetracyclic with ring A a 4-envelope, ring C a chair distorted by the 15–16 bridge. Ring B is in a chair form. Fusicoccin is a glycoside (with an isoprene unit attached to the glucose component), the aglycone of which, as fusicoccin A *p*-iodobenzenesulphonate[139] (65) is shown to have the ophiobolin ring system.

8.2.4 Sesterterpenes

The structures analysed in this group all have the same ring system. In ophiobolin methoxybromide[140] (P127) rings A, C and D are all half-chair, while B is apparently in a distorted chair form. Ring junctions in ceroplastol-I

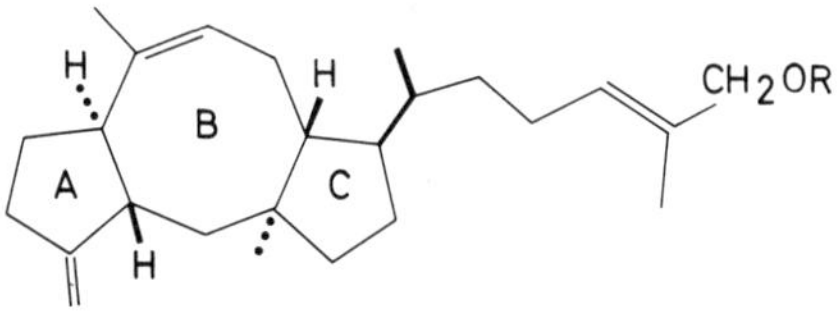

(66)

p-bromobenzoate[141] (66) are *trans-trans*. Here rings A and C are 6- and 11-envelopes respectively while ring B is a distorted chair. Methyl cephalonate bromacetate[142] (67) has ring A planar and ring C a 12-envelope, C(12) being 0.66 Å out of the plane defined by atoms 10, 11, 13, 14. Cephalonic acid has been given the trivial name ophiobolin D.

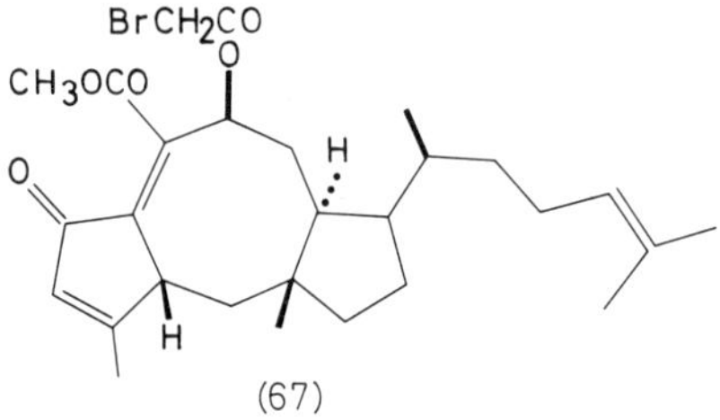

(67)

The distinction between the triterpene and sterol groups is somewhat arbitrary. Allocation of individual compounds in the present case follows, in general, the grouping adopted by the Cambridge Crystallographic Data Centre.

8.2.5 Triterpenes

A number of bitter principles occur in this group. Glaucarubin *p*-bromobenzoate[143] (P105) has ring A and D in the half-chair form. Simarolide 4-iodo-3-nitrobenzoate[144] (P104) has a similar ring skeleton but with a C_5

unit attached to ring C instead of ring D in glaucarubin. Dihydrogedunin-3β-yl iodoacetate[145] (P102) was of interest as intermediate in oxidation pattern between limonin and cedrelone. The former was studied as epilimonol iodoacetate[146] (P101) where ring A is in a boat form. In the CD ring system, ring C is also in a boat form and ring D in a skew form due to the presence of the lactone group and the epoxide group. The latter was investigated as cedrelone iodoacetate[147] (P103) in which ring A is in a sofa form with C(10) out of the plane, B a half-chair and C, a distorted boat arising from the interaction of C(30) and O(4). Ring D is a 17-envelope.

In the previous three structures, formation of the furan ring from a euphol precursor involves also the loss of atoms C(24)–C(27), a feature evident also in the structure of detigloylswietenine *p*-iodobenzoate[148] (P107) in which ring A is in a boat form and ring B is a slightly distorted half-chair; ring D is also a half-chair. A closely related structure, a mexicanolide derivative[149] (P106), with ring A in a slightly skewed boat form, involves a rearrangement of ring A relative to that in gedunin (P102) which is found as 7-deacetoxy-7-oxygedunin together with this derivative.

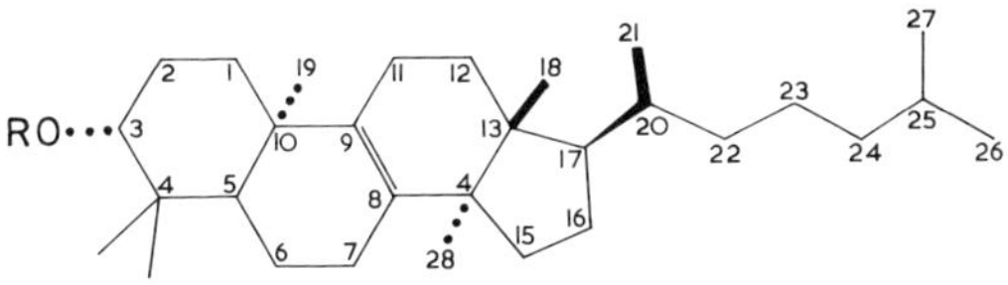

(68)

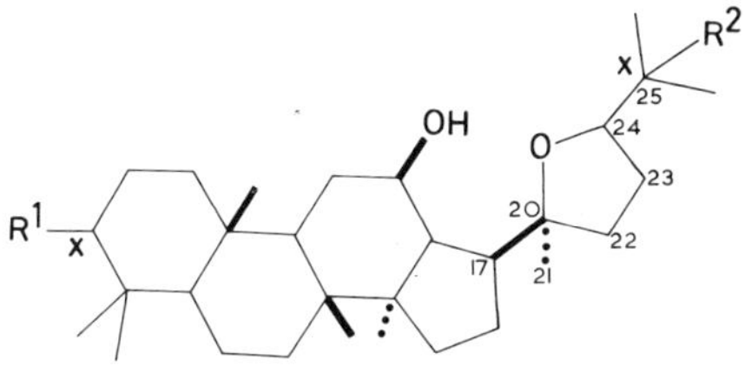

(a) X = ▬ : $R^1 = R^2 = OCOC_6H_4Br$
(b) X ••• : $R^1 = OH$ $R^2 = Br$

(69)

The compounds mentioned above generally derive from euphol, a simple derivative of which, euphenyl iodoacetate[150] (68) has been analysed. Like lanostenol, the rings B and C are in the half-chair form but the side chain is not fully extended. In lanostenol and cholesteral, the C(18) and C(20) are in the β-orientation but the C(21) methyl is α. Here it is also β and hence the side-chain orientation is modified.

Two dammarane-type triterpenes of closely allied structure have been studied. 3β,12β-*O*-Di-*p*-bromobenzoylpyxinol[151] (69a) and 25-bromo-*O*(20), 24-cyclobetulafolianetriol[152] (69b) differ in configuration at atoms 3 and 24. The latter difference markedly affects the orientation of the tetrahydrofuran ring. In the first compound, ring E is nearly coplanar with the remainder of the molecule while in the second compound, rotation about atoms 17–20 is such that ring E is almost at right angles to the main plane of the molecule.

In the structure of methyl melaleucate iodoacetate[153] (P119), attention is drawn to certain long C—C bonds, up to 1.62 Å, associated with fully substituted carbon atoms (cf. diamond). Authentication of error limits in this and similar analyses could very well modify the significance of such bond lengths.

A related group of compounds with a leucotylin skeleton is davallol iodoacetate[154] (P120a) with ring C a half-chair and ring B a skewed boat

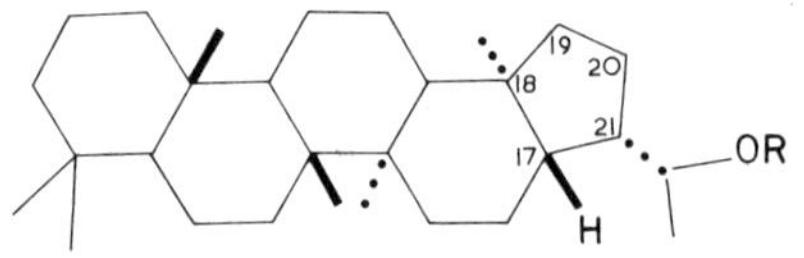

(70)

with atoms 5/8 as prow/stern, adiantol B bromoacetate[155] (70) with ring E an 18-envelope and 6-keto-leucotylin-16*β*-*O*-*p*-bromobenzoate[156] (71) with ring E in a half-chair form. In these three compounds, the individual rings where 1,3 interactions occur, are slightly distorted and the overall molecular

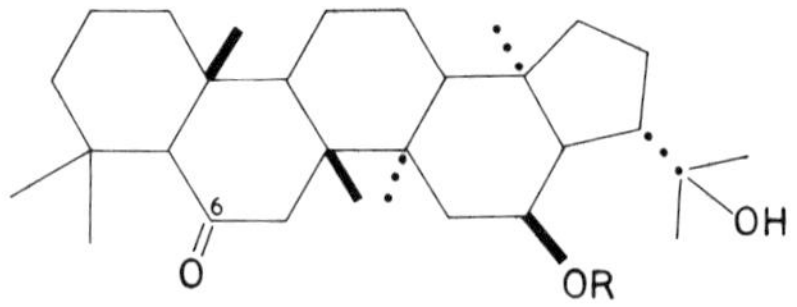

(71)

shape is bow-like. Compounds with the hopane skeleton include 2*α*-bromo-arborinone[157] (P120) with ring C probably a half-chair and ring E a 17-envelope. It is noted that Br is coplanar with 2,3,4 and O in accordance with the expected eclipsing of a single and a double bond. In the analysis of motiol iodoacetate[158] (72), the dimensional evidence appears to suggest the

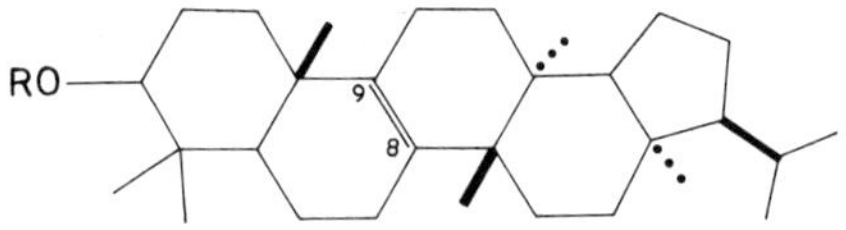

(72)

double bond as being 8,9 instead of 7,8 in which case the structure is *iso*motiol. Rings A and D are in chair form while ring B appears to be in a sofa form with C(5) out of the plane.

Methyl oleanolate iodoacetate[159] (73) has been reported but no accurate conformational information is available. However, the structure of genine

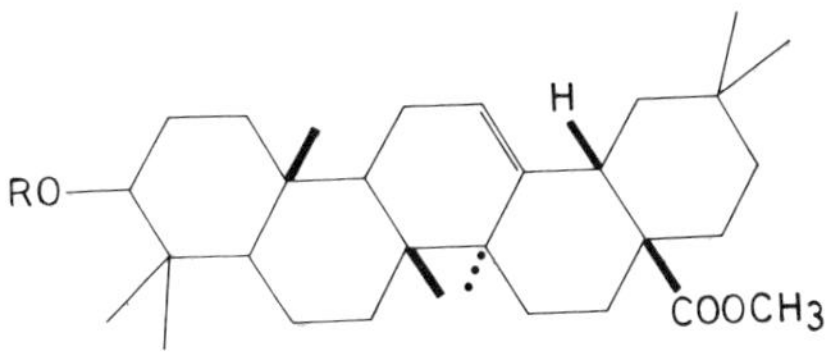

(73)

D[160] (74) is revealed as 3β,16α,28-trihydroxy-21-keto-olean-12,13-ene with the rings in chair form, shallow in the case of ring D and ring C a half-chair. A triterpene isolated from a Nigerian crude oil has been shown to be a modified oleanane structure, 1(10-5)-abeo-3β-methyl-24β-nor-25α,18α-oleanane[161] (75).

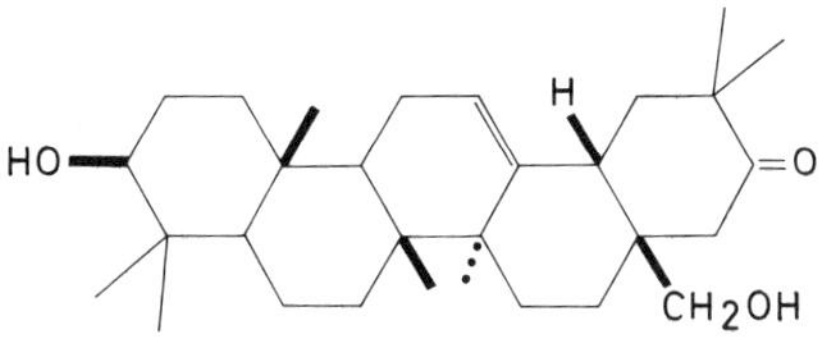

(74)

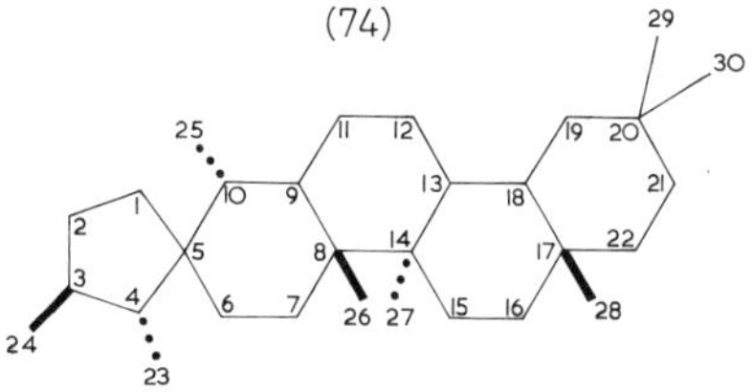

(75)

A further two compounds with an oleanolane-type skeleton are phytolaccagenin 2-oxazoline[162] (P122) and platycodigenin bromolactone[163] (76). In the latter, all the rings are in the chair form but there is considerable distortion due to interaction of axial groups so that the molecule is bowed.

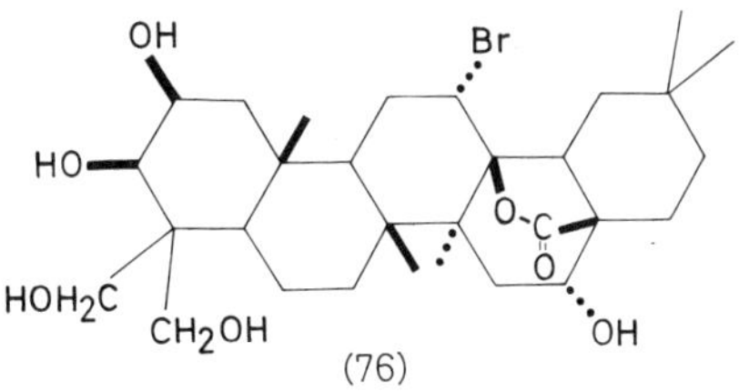

(76)

The high internal symmetry of tetrahymanone[164] (P121) caused problems in the allocation of the keto group to C(3) since an oxygen 'image' appeared adjacent to C(21). The study of methylmicromerol bromacetate[165] (P123), showed it to be ursolic acid and was the first x-ray study of an α-amyrin. This molecule is also bowed due to axial interactions. In eupteleogenin iodoacetate[166] (P124), distortions of the rings from chair form are due to interaction of axial substituents. Ring C is a slightly distorted boat form due to the epoxy group and the interaction between C(26) and O(3).

A compound with an unusual ε-lactone group in ring A, 23-hydroxy-2,3-secours-12-ene-2,3,28-trioic acid (2,23)-lactone 3,28-dimethyl ester[167] (77), has been analysed as also has 3β-methoxy-CΔ(13)-serrateno (20,21-*b*)-(5′-bromoindole)[168] (78), the first pentacyclic triterpenoid with ring C seven-membered. All rings are *trans*-fused. Rings A and B are chair-shaped, B more distorted than A. Ring D is a half-chair while ring E approximates to a sofa form with C(18) out of the plane.

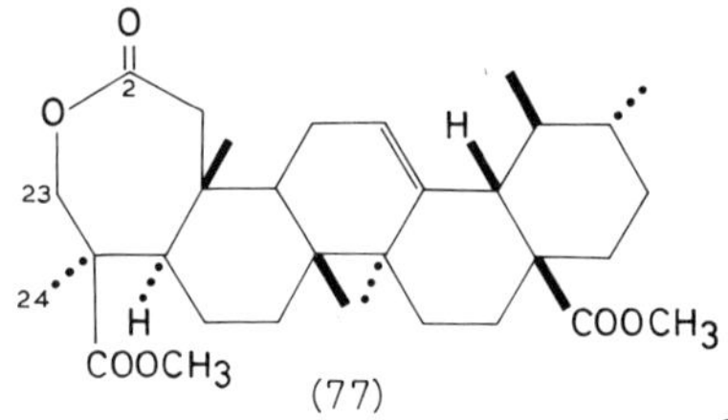

(77)

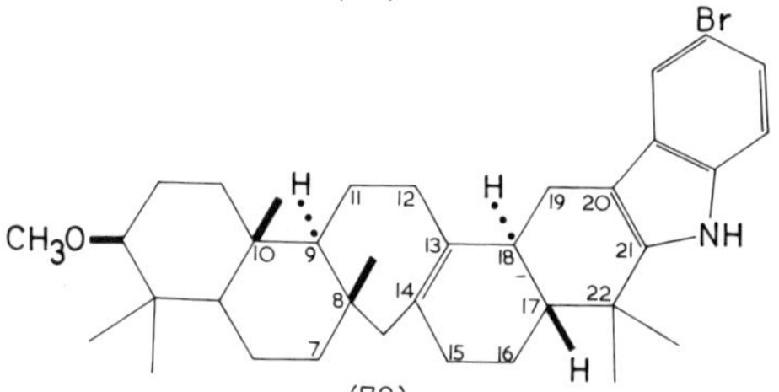

(78)

8.2.6 Tetraterpenes

The compounds listed in this group are closely allied. They are β-carotene[169] (79), 7,7′-dihydro-β-carotene[170] (80), 15,15′-dehydro-β-caratene[171] (81), canthaxanthin[172] (82) and 15,15′-dehydrocanthaxanthin[173] (83). In studies of these structures, some interesting features of conformation, bond angles and intra-molecular interactions have been noted. Thus the polyene chain is practically planar but it is not a straight zig-zag with valence angles of 125

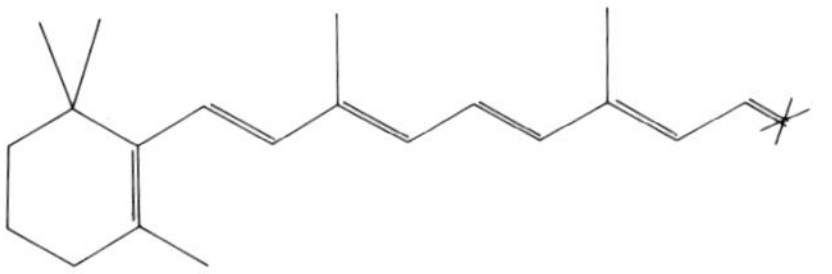

(79)

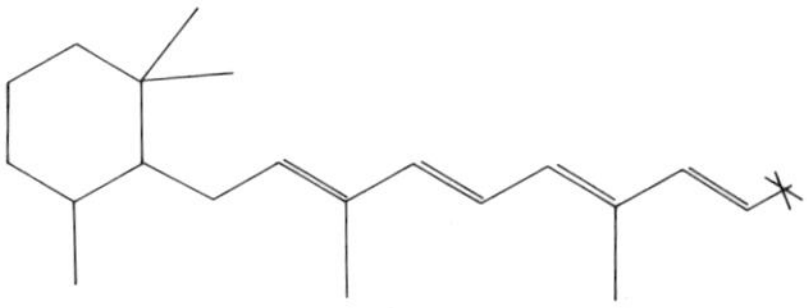

(80)

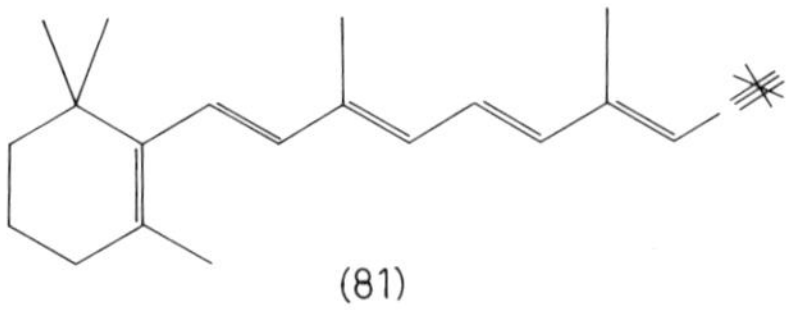

(81)

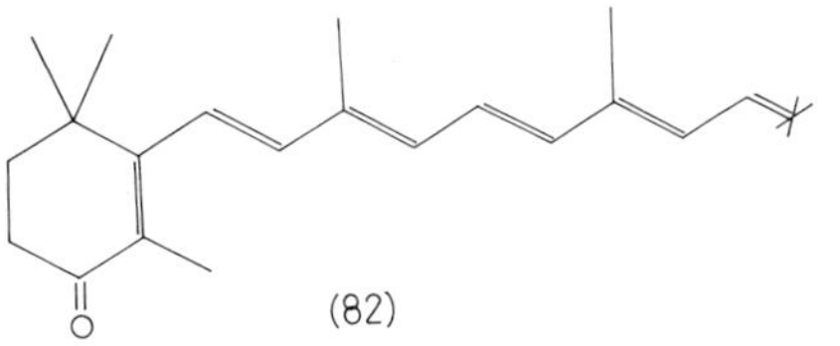

(82)

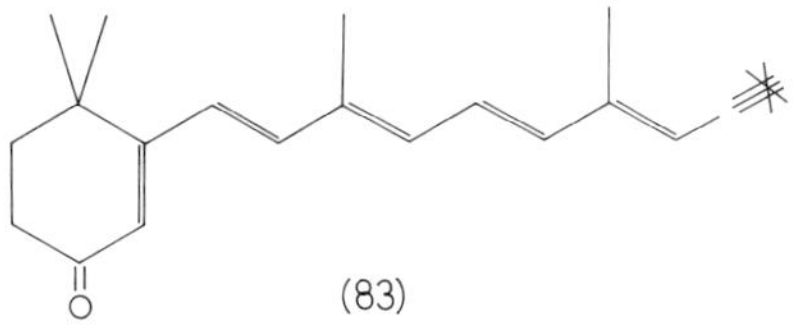

(83)

degrees. Wherever a methyl group is attached, the angle opposite the methyl is <125 degrees whereas the angle opposite the double bond is larger than the expected 110 degrees. The chain is generally curved, due to the interaction of adjacent methyl groups and also to inter-molecular forces. In the β-carotene molecules, the cyclohexene ring is in a half-chair form while in the canthaxanthin molecules, it is in a sofa-form with C(2) out of the plane. The attachment of the ring and chain is generally near *syn-trans* but the angle between the ring double bond and the conjugated chain varies considerably. This is also thought to be due to intra-molecular repulsion between the various methyl groups and adjacent atoms not bonded to the methyl under consideration. The possibility of a systematic decrease in the alternating character of single and double bonds towards the centre of the conjugated chain has been raised but the evidence has not achieved a clear-cut decision although the trend appears to hold in the case of the two 15,15′-dehydro compounds.

8.3 STEROIDS

Of the oestrogens, a number of compounds have been studied, namely 4-bromoestrone[174] (84a, $R^1 = H$, $R^2 = Br$, $R^3 = {=}O$, $R^4 = H$), estradiol (3β, 17β) itself[175] (84b, $R^1 = H$, $R^2 = H$, $R^3 = OH$, $R^4 = H$), two derivatives, the 4-bromo[176] (84c, $R^1 = H$, $R^2 = Br$, $R^3 = {-}OH$, $R^4 = H$), the 3-*p*-bromobenzoate[177] (84d, $R^1 = BrC_6H_5$, $R^2 = H$, $R^3 = {-}OH$, $R^4 = H$) and 3-methoxy-8β-methylestradiol-17-monobromoacetate[178] (84e, $R^1 = CH_3$, $R^2 = H$, $R^3 = {-}OCOCH_2Br$, $R^3 = {-}CH_3$). In the first two estradiol

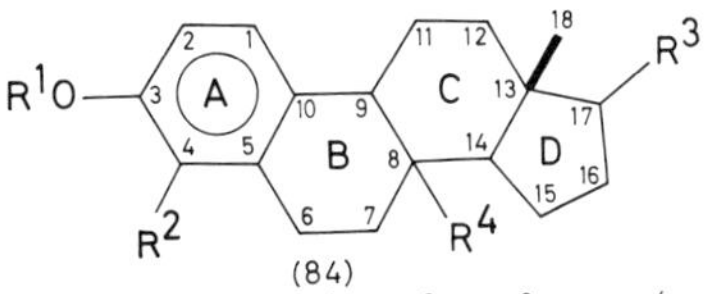

(84)

(a) 4-Bromoestrone $R^1 = H, R^2 = Br, R^3 = {=}O, R^4 = H$

(b) Estradiol $R^1 = H, R^2 = H, R^3 = OH, R^4 = H$

(c) 4-Bromoestradiol $R^1 = H, R^2 = Br, R^3 = {=}OH, R^4 = H$

(d) Estradiol 3-*p* bromobenzoate $R^1 = BrC_6H_4-, R^2 = H, R^3 = {=}OH, R^4 = H$

(e) 3-Methoxy 8β methylestradiol-17-monobromacetate $R^1 = CH_3, R^2 = H, R^3 = OCOCH_2Br, R^4 = CH_3$

structures, ring B appears to be in a half-chair form, while in the third it approximates to a sofa form with C(8) out of the plane. A similar form for ring B occurs in the second (B) form of 2,4-dibromoestradiol[179], examination of the two forms indicating that their differences may be due to 'electronic' (intramolecular) effects more than packing differences. Two other compounds in this group are 3-methoxy-7α,8α,-methylene-1,3,5(10)-estratrien-17β-yl bromoacetate[180] (85) and DL-9β-B-norestrone methyl ester[181]. In the latter, ring B takes the form of a shallow boat due to the A ring and the cyclopropane ring. In estriol[182], with 3β,16α,17β hydroxy groups, although the two molecules in the asymmetric unit are very similar in shape, minor differences are noted namely, ring A in molecle 1 is planar while, in molecule 2, it is slightly folded along C(2)—C(4) with a dihedral angle of 2.6 degrees. Here, the distortion is ascribed to an interaction with C(18) of an adjacent molecule. Distortions in rings B and C are ascribed to a twist about C(9)—C(10) which in molecule 2 leads to a half-chair form of ring B while in molecule 1 it assumes a sofa form with C(8) 0.69 Å above and C(7) 0.06 Å below the plane of the remaining four.

In bromomirestrol[183] (P113), ring conformations are determined by the ring fusions.

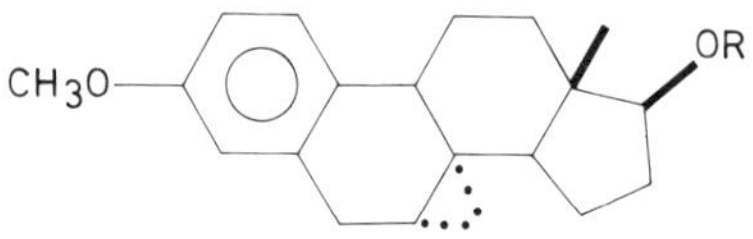

CH_3O— OR

(85)

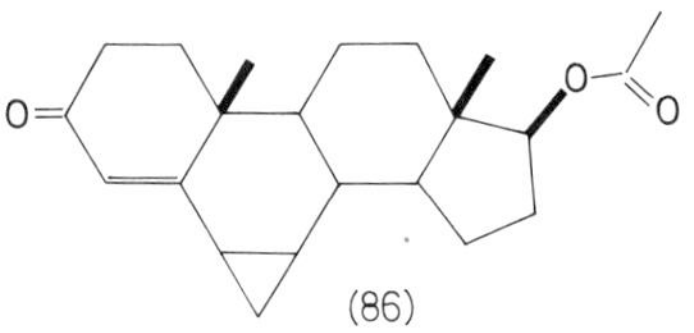

O= O—C=O

(86)

The androgens are represented by a considerable group of androstone derivatives and two testosterone derivatives. The epimers, 5α-androstane-3α (and 3β)-ol-17-one[184,185], have both been studied. With *trans*-fusion of the rings, the general plane of the molecule is bent convex upwards, indicated by the splaying of the bonds, C(10)—C(18) and C(13)—C(18) to the extent of ~14 degrees, a result typical for normal steroids. In 9α-bromo-17β-hydroxy-17α-methylandrost-4-ene-3,11-dione[186], ring A tends to a sofa form with C(2) out of the plane, a feature noted also in 17β-bromoacetoxy-9β,10α-androst-4-ene-3-one[187] but with C(1) out of the plane. In 17*a*β-*p*-brosyloxy-17α-methyl-19-nor-9β,10α-D-homoandrost-4-en-3-one[188], ring A is in a half-chair form. The last two are retro-steroids and hence have a bent shape compared with the relatively flat normal 9α,10β steroids. Study of a group of 7 retro-steroids[189], 9β,10α-androsta-4,6-diene-3,17-dione, 6β-fluoro-6α-methyl- (and 6α-fluoro-6β-methyl)-17β-hydroxy-9β,10α-androst-4-en-3-one-17-acetate, and 9β-hydroxy-10α-androst-4-ene-3,17-dione showed all as having ring A close to a sofa form with C(1) out-of-plane. In 6β,7β-methylene-17β-hydroxy-androst-4-en-3-one-17-acetate[190] (86), the cyclopropane ring with the constraints in ring A make both rings A and B sofa-shaped with C(1) and C(9) out of the plane relative to C(19). The acetate group is oriented approximately normal to ring D with the C=O pointing to the H of C(17) as deduced earlier[8]. Also involving a cyclopropane ring is 3β,6β-dimethoxy-5β,19-cycloandrostan-17-one *N*-acetyl-*p*-brosyl hydrazone[191]. Held in a *cis* relationship by the cyclopropane ring, the conformation of ring A is half-chair with atoms 1,4,5 and 10 coplanar and 2 and 3 out of the plane by +0.36

and −0.42 Å respectively. Atoms 5,6,9 and 10 are also coplanar with 7 and 8 −0.24 and +0.59 Å out of the plane respectively. Other androstane derivatives are 3β-p-bromobenzoyloxy-13α-androst-5-en-17-one[192] with ring B in a half-chair form and ring D a 14-envelope and 17β-iodoacetoxy-4, 4-dimethyl-5α-androstan-3-one and the corresponding 19-nor-5α-androstan-3-one[193]. This last study focused on the conformation of ring A to determine if the presence of C(19) has a significant influence. While not affecting torsion angles near C(10), the results indicate that presence of C(19) causes rotation about C(4)—C(5) so as to reduce the displacement of C(3) from plane 1,2,4,5. The ring D differs significantly in the two compounds. 6α-Bromo-17β-hydroxy-17α-methyl-4-oxa-5α-androstan-3-one[194] is an example of a non-planar δ-lactone in which however atoms 1,2,3,O(4) and O(20) are near planar. The structure[195] arising from the reaction between 17β-acetoxy-1-methyl-1-androsten-3-one with two equivalents of 2,6-dimethylphenyl isocyanide has been studied. In 3-methoxy-5β,19-cyclo-5,10-secoandrosta-1(10),2,4-trien-17β-ol *p*-bromobenzoate[196] (87), ring A has a 'tub' form with atoms 1,4,5 and 10 forming the bottom. The double and single bonds are

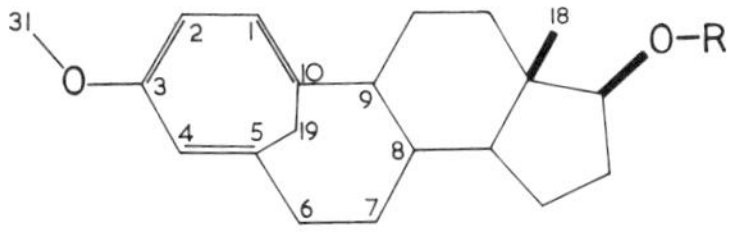

(87)

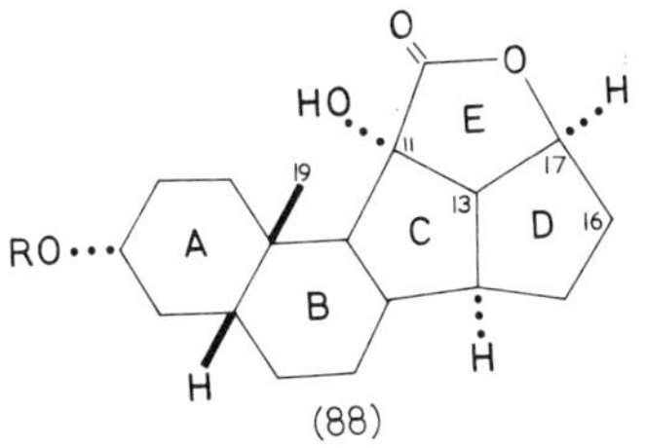

(88)

localised (by contrast with the result for bicyclo[4.4.1]undeca-1,3,5,7,9-pentaene 2-carboxylic acid[197]). Subtle minor deformations are recorded involving displacement of O(30) below the plane of the C(2)—C(3) double bond. In ring B, atoms 5,6,9 and 10 are coplanar with 7 and 8 displaced 0.46 and 0.39 Å on either side of the plane. Benzilic acid rearrangement of 3α, 17β-diacetoxy-11-hydroxy-12-oxo-5β-androst-9(11)-ene leads to 3α,11α,17β-trihydroxy-13α-*C*-nor-5β-androstane-11β-carboxylic acid-11*a*,17-lactone 3-*p*-bromobenzoyl ester[198] (88). The AB ring junction is *cis*, both rings being in the chair form. The three five-membered rings CDE are *cis*-fused. The shape of the molecule forces C(19) and O(12) close together. In a Dreiding model, the distance would be 2.3 Å but it is actually 3.1 Å. This steric repulsion modifies adjacent bond angles, e.g. ∠10,9,11 is increased to 123.5 degrees. In the compact grouping of rings, the lactone group is not coplanar. While C(11), C(12), O(12) and O(17) are coplanar, C(17) and C(13) are 0.27 and 0.21 Å respectively out of the plane.

In testosterone 17β-*p*-bromobenzoate[199] and 8β-methyl-testosterone 17β-monobromacetate[199], ring A is in a sofa form with C(1) out of the plane. The D rings differ marginally with a half-chair form in the first and a slightly distorted 13-envelope in the second. In the testosterone 2:1 complex with mercuric chloride[200], ring A is reported as in a half-chair form. 2β-Hydroxytestosterone-2-acetate-17-chloracetate and 2β-hydroxytestosterone-

2,17-diacetate[201] are described as having ring A in a half-boat form with O(3) 0.3 Å out of the plane of C(3), C(4) and C(5).

A small group of progestogen derivatives has been analysed, 3β,17α-dihydroxy-21-bromo-5α-pregnan-11,20-dione[202] (89a, $R^1 = H$, $R^2 = Br$, $R^3 = H$), 3β-17α-dihydroxy-16β-bromo-5α-pregnan-11,20-dione[203] (89b,

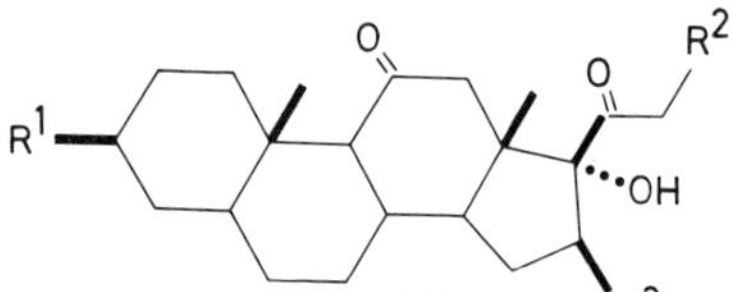

(89)

(a) 3β,17α-Dihydroxy-21-bromo-5α-pregnan-11,20-dione (R^1=H, R^2=Br, R^3=H)

(b) 3β,17-Dihydroxy-16β-bromo-5α-pregnan-11,20-dione (R^1=H, R^2=H, R^3=Br)

(c) 3β-Acetoxy-17α-hydroxy-16β-bromo-5α-pregnan-11,20-dione (R^1=OAC, R^2=H, R^3=Br)

$R^1 = H$, $R^2 = H$, $R^3 = Br$) and 3β-acetoxy-17α-hydroxy-16β-bromo-5α-pregnan-11,20-dione[204] (89c, $R^1 = Ac$, $R^2 = H$, $R^3 = Br$). The rings ABC are all *trans*-fused and have a chair form while ring D is in a distorted half-chair form. In all three compounds, angle 5,10,9 tends to be significantly smaller (av. 105.2 degrees) and angle 8,9,10 larger (av. 114.2 degrees). In the third compound, while the acetate group is planar, it does not lie at right angles to ring A as predicted[8]. Interaction with C(18) of an adjacent molecule is proposed as the factor involved. In 4-bromo-9β,10α-pregna-4,6-diene-3,20-dione[205] (also named 4-bromo-duphaston), the 3-keto-4,6-diene system results in rings A and B being essentially sofa in form, cf. Reference 190, A more nearly than B with C(1) and C(9) out-of-plane. The BC *cis*-junction leads to a strong interaction between C(19) and the axial hydrogens of C(12) and C(14). Three other retro-steroids of this group have been studied, 9β, 10α-pregna-4, 6-diene-3, 20-dione, 6α-methyl-9β, 10α-pregn-4-ene-3, 20-dione and 11α-hydroxy-9β,10α-pregn-4-ene-3,20-dione[189]. In these, ring A is in a sofa form while, in the first, ring B is in a half-chair form. The oxygen bridge in 5α-bromo-6β,19-oxido-pregnan-3β-ol-20-one[206], while it causes certain minor distortions in the structure, does not grossly modify the conformations of rings A, B and C except to torque ring A slightly relative to ring B. Although there are two molecules in the asymmetric unit of this structure they are virtually identical within the error limits of the analyses. Three progesterone derivatives have been studied, namely, 12α-bromo-11β-hydroxy progesterone[207], 6β-bromo-progesterone[208], and 6α,7α-difluoro-methylene-16α-methyl-11β,17α,21-trihydroxypregn-4-en-20-one (3,2–*C*)-2′-phenyl-pyrazole 21-*p*-bromobenzoate[209]. No unusual features appear in the first two; ring A is in a half-chair form while in the third compound ring A is in a skew form due to the 2,4-diene and ring B in a half-chair form due to the cyclopropane ring.

Comparison of the epimers-11β,12α(and 12β)-dibromo-3α,9-oxidocholanic acid methyl ester[210,211] is of interest. The 3,9 oxygen bridge constrains ring A to a boat form and with C(9) forms a regular three-ribbed cage structure.

The two epimers are alike stereochemically except for the region adjacent to C(12). In the *cis* case, the dihedral angle Br, 11,12,13 is 72 degrees whereas in the *trans*, it is 92 degrees almost halfway between staggered and eclipsed. A marked difference in the C(11)—C(12) bond length, 1.46 and 1.64 Å, is noted in the two compounds but this may be partly artefact due to the close proximity of the two Br atoms.

Cholesterol[212] (P108) is of particular significance as it was the first sterol to be studied crystallographically in detail and hence lent considerable stimulus to the application of x-ray techniques to natural product molecules. Rings A, B and C were in the chair form and were *trans*-fused. Subsequent members of the cholestane group investigated were 3β,7α-dibromocholest-5-ene[213] (where the conformational pattern is modified in that ring B is in a half-chair form) and 2β,3α(*a*,*a*)-dichloro-5α-cholestane[214] and 2α,3β(ε,ε)-dichloro-5α-cholestane[215]. These latter two differ only in the disposition of the chlorine atoms. The equatorial halogens do not appear to lead to distortions in ring A whereas the axial substitutions appear to cause some distortion (or artefacts simulating distortion).

Report has been made of a cholic acid derivative undergoing a molecular rearrangement to form methyl-3α-iodoacetoxy-12-methyl-18-nor-5β,17α-chola-8,11,13-trien-24-oate[216] whose structure has been studied. In 4,4-dichloro-2*a*-aza-*A*-homocholestan-3-one[217] the ε-lactam ring is in the chair form; the ring C(2),N,C(3),O(3),C(4) is planar within limits of error. In ecdyson, 2β,3β,14α,22β,25-pentahydroxy-Δ(7)-5β-cholesten-6-one[218] (P118a), rings A and B are *cis*-fused leading to a bent molecule. Ring B is a sofa form with C(10) out-of-plane while rings A and C are chair and D a 13-envelope. In cholestan-4-one-3-spiro (2,5-oxathiolane)[219] (90), both rings A and E are

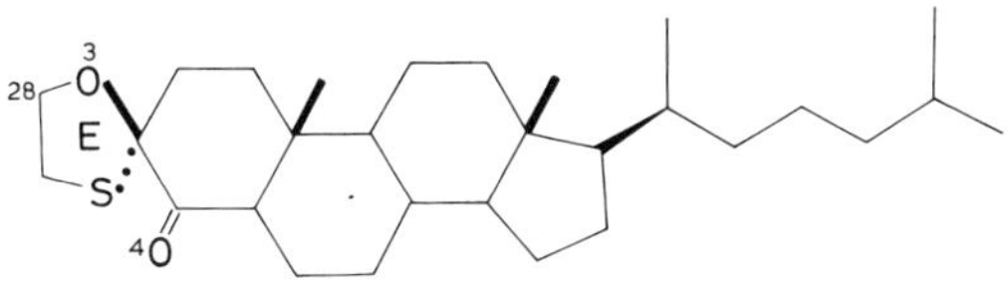

(90)

somewhat distorted. A dipole interaction between O(4) and O(3) (separation 2.74 Å and torsional angle O(4)C(4)C(3)O(3) of +7.5 degrees) is suggested as a factor in ring A being appreciably flattened. Ring E is envelope in form with C(28) 0.51 Å out-of-plane and towards O(4). An interaction between O(4) and the near H of C(28) is regarded as significant. Minor differences between the lengths of the two C—O bonds and the two C—S bonds in ring E is noted cf. the differences noted in carbohydrate structures[220]. The question of the reality of such details is of considerable interest.

22,23-Dibromo-9β-ergost-4-en-2-one[221] was analysed at room temperature and at −180 °C. Only a slight difference in the conformation of ring D was observed. The overall shape of the molecule is remarkably flat with ring A a virtual sofa (dihedral angle O(3),C(3),C(4),C(5) is 13 degrees). As in fusidic acid, rings B and C are twist-boat. A benzenoid steroid, 22,23-dibromo-12-methyl-18-norergostan-8,11,13-trien-3β-yl acetate[222] showed ring B as half-chair and ring A in a chair-form with ring D a 16-envelope. Of the photo- and thermo-isomers of ergosterol, several have been analysed. In calciferol[223] (P110), rings A and C are chair and D half-chair with C(13) and C(14) $+0.47$ and -0.24 Å out-of-plane respectively. The trans-diene system has a dihedral angle of 90 degrees while the *cis* diene has an angle of 54 degrees. The side chain is not a planar zig-zag but adopts a conformation such as to minimise non-bonded repulsive interactions; C(26) and C(27) are, respectively, anti-periplanar and synclinal with respect to C(28). Lumisterol[224] (P111) has a cisoid diene which gives ring B a skew conformation. Suprasterol II [225] (P112) has a three-membered ring B′ *cis*-fused to ring B which is virtually planar. Ring A is in a half-chair form. The side chain has the same conformation as in calciferol and lumisterol, curling at C(24) so that C(24)—C(25) lies almost at right angles to the plane of the double bond cf. in prostaglandin F_2 [226] (P154). Photoisopyrocalciferyl *m*-bromobenzoate[227] (91) contains a novel

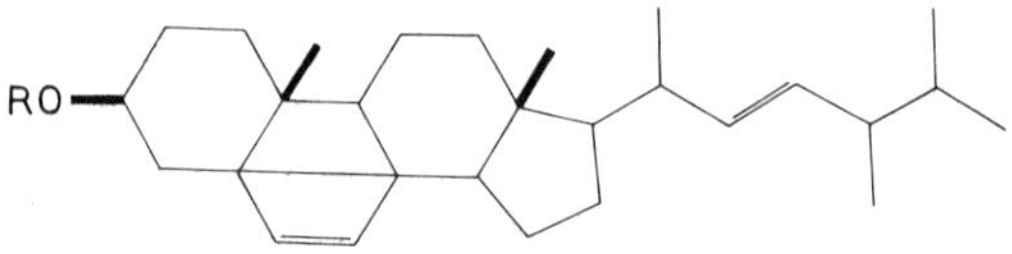

RO

(91)

feature, namely two fused 4-membered rings, one of which contains a double bond. Ring B links rings A and C in a *cis*-configuration, giving a right-angled bend in the overall shape of the molecule. There is considerable distortion from regular dimensions associated with the two 4-membered rings and adjacent bond angles. Ring C is in a sofa form.

In lanostenol iodoacetate[228] (P117), rings B and C are in the half-chair form while in 3β-acetoxy-7α,11α-dibromolanostane-8α,9α-epoxide[229] (P118), the presence of the epoxide ring distorts the usual chair form of rings B and C to the extent that B is in a slightly distorted boat form. Ring C is not readily identified in simple terms.

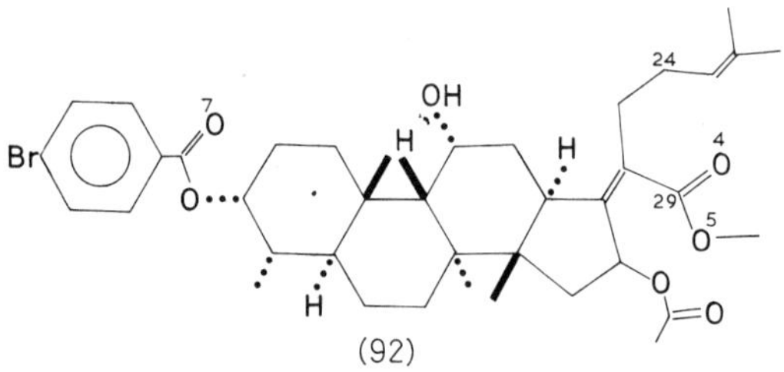

Br O 7 O OH H H H 24 4 O 29 5 O O =O

(92)

Two nitrogen analogue structures have been studied. 8-Azaestrone hydrobromide[230] is stereochemically identical with 4-bromoestrone to which 2,3-dimethoxy-8,13-diaza-18-norestrone[231] also has close resemblances except that the amide character associated with N(13) and C(17)=O is demonstrated by their near coplanarity. Ring D, as a consequence, is a 15-envelope.

In fusidic acid methyl ester[232] (92) and (P118b), all ring junctions are *trans*-fused. Ring B has a boat shape, slightly twisted, with rings A and C *syn* to B. The side-chain at C(17) curls back on itself, partially influenced by a weak interaction between H (of C(24)) and O(4). A similar interaction occurs with respect to O(7) and H of C(3)[8].

Digitoxigenin[233] (93) differs from nearly all other steroids analysed by x-rays in that both the A/B and C/D ring junctions are *cis*, leading to a molecule which is globular rather than flat in shape. Atoms 5,6,8,9,12,13 are nearly coplanar. In ring B, 7 and 10 are 0.90 Å above and below this plane whereas in ring C, 11 and 14 are displaced 0.66 and 0.58 Å from the plane, the corresponding displacements of 3 and 10 relative to the plane 1,2,4,5 are ±0.66 Å. Ring D is a 14-envelope. By contrast, the structure of a cardio-inactive compound, Δ-8,14-anhydro-digitoxigenin[234] showed ring C in a sofa form with C(12) out of the plane and ring D a 17-envelope. A detailed comparison of the two structures is presented. Batrachotoxinin A [235] (94),

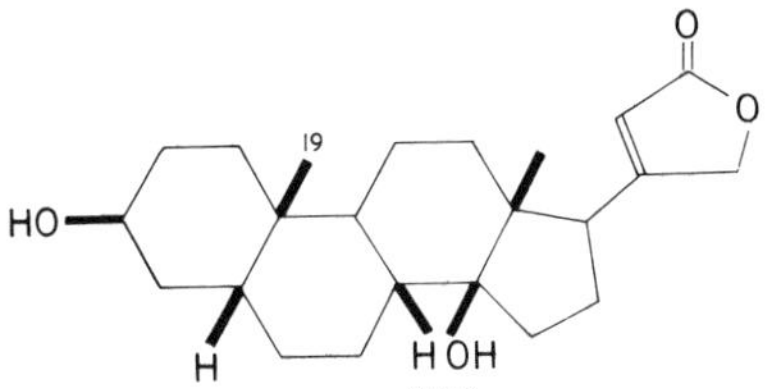

(93)

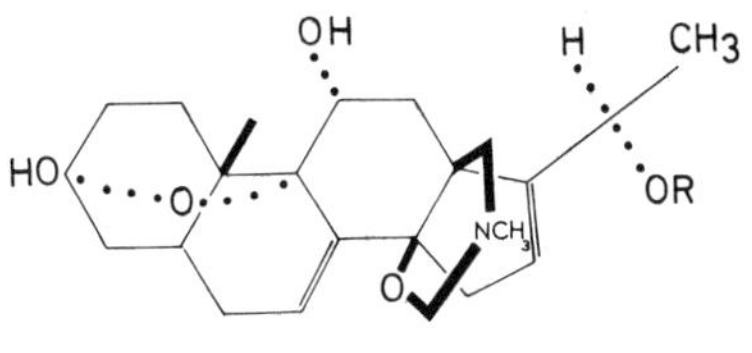

(94)

derived from the Colombian arrow poison frog, has similarities to digitoxigenin in that, in both, A/B and C/D junctions are *cis*. In batrachotoxinin, there is an ether link between atoms 3 and 9 making ring A a boat and forming a trigonal cage. Ring B is a slightly flattened half-chair with atom 5, −0.29 Å, and atom 10, +0.53 Å relative to the plane 6,7,8,9. Ring D is a flattened 14-envelope. Ring E is 7-membered and has a chair form with atoms O(4),23, N and 18 approximately coplanar with atom 22, 0.66 Å above and atoms 13,14, −1.27 and −0.95 Å respectively below plane.

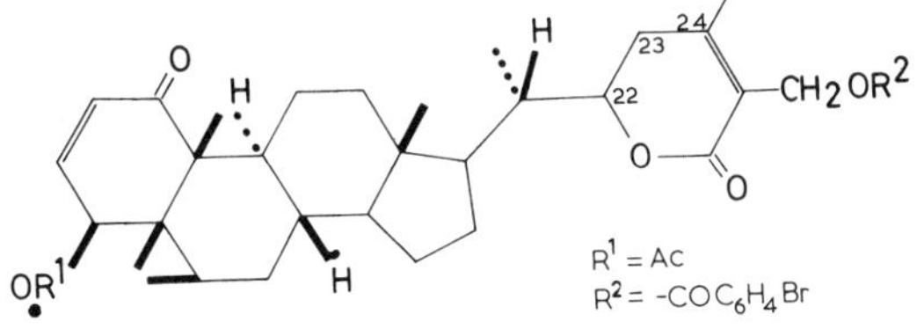

R^1 = Ac
R^2 = $-COC_6H_4Br$

(95)

Withaferin A acetate *p*-bromobenzoate[236] (95) is a novel steroidal lactone. Ring A is boat-like with 5 and 10 on the same side of the plane 1,2,3,4, this being preferred to the more stable half-chair because of possible 1,3 pseudo-axial interaction between C(19) and O(30) substituents. Ring E is not planar but approximates to a sofa form with 22 out of the plane.

Of the steroidal sapogenins, the analysis of diosgenin iodoacetate[237] (P125) involved two molecules in the asymmetric unit. The overall geometry of the two (fully extended) molecules is similar except in the region of the iodine atom of the iodoacetate. Ring junctions B/C and C/D are *trans*-fused with D/E *cis*-fused. Ring D is in a half-chair form. 23-Bromoneotigogenin acetate[238] while not involving a double bond in ring B is conformationally similar. Ring E is in a chair form although certain features are attributed to disorder arising from the presence of ring F in the alternative upright orientation to the extent of 10%.

In the aglycone, protoaescigenin-21-tiglate-22-acetate[239] (96), ring D is

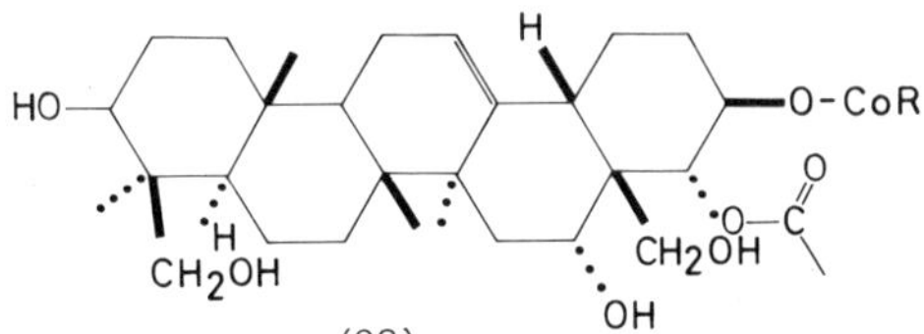

(96)

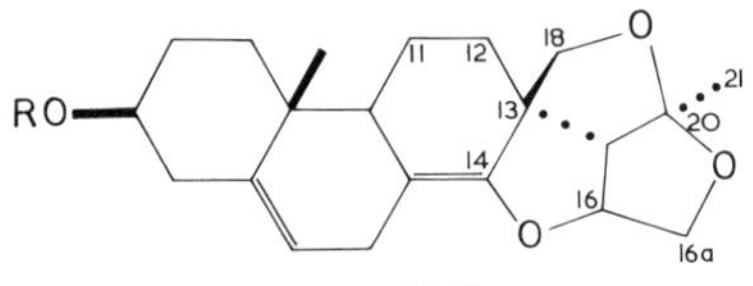

(97)

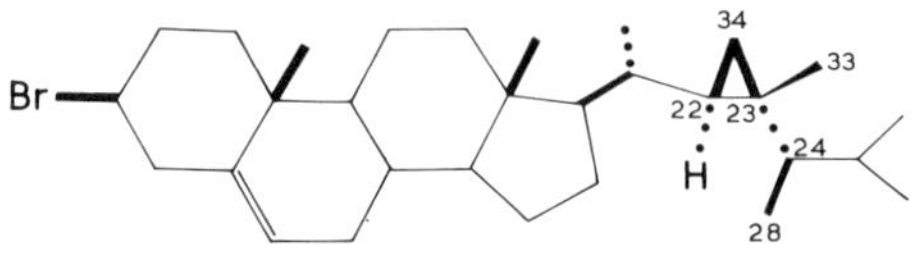

(98)

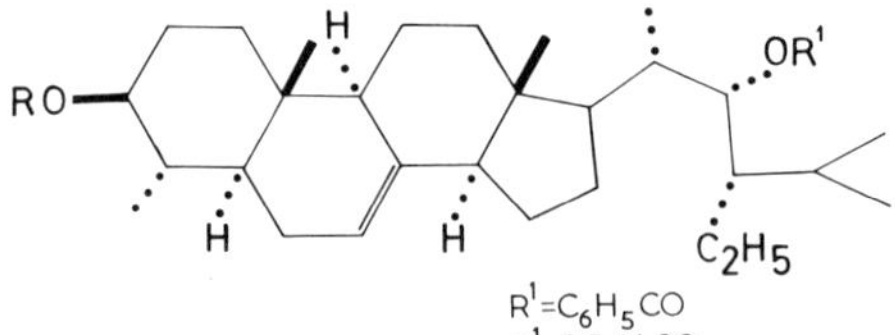

$R^1=C_6H_5CO$
$R^1=IC_6H_4SO_2$—

(99)

flattened by the sp^2 hybridisation at C(13) and by the α-axial substituents at atoms 14, 16 and 18 (which mutually diverge). Rings A and B (distorted chairs) and C (half chair) have conformations to minimise the mutual interaction of atoms 24, 25 and 26. In the aglycone, anhydrohirundigenin *p*-bromobenzoate[240] (97), ring A is in a chair form, while B is intermediate between a sofa and half chair form. Ring C appears also to have a half chair form. The three ring system DEF involve envelope forms with C(18), O(3) and O(4) out of plane in the individual cases.

Some details are available on a number of other steroid structures-3*β*-bromogorgostene[241] (98), carpestol *p*-iodobenzene-sulphonate[242] (99), 11*b*-

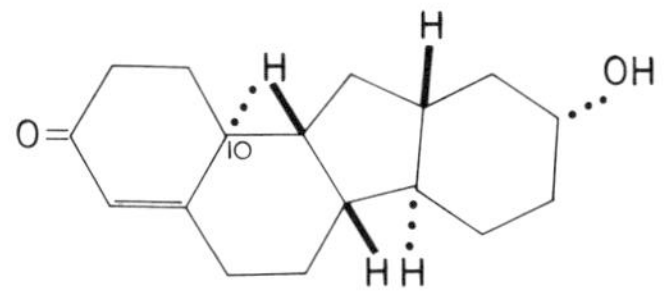

(100)

methyl-9-hydroxy-*C*Δ(4,4*a*)-perhydrobenza-(*A*)-fluorene-3-one[243] (100) and methyl shoreate[244].

8.4 GENERAL REMARKS

It is of interest to note that this survey indicates some changes in emphasis which are likely to influence further developments in this field. Thus, to stress the facility of which present-day techniques are capable, there are cases where two and sometimes even three derivatives have been analysed to establish with certainty all the details of location and identity of substituent atoms or atomic groups necessary to define the complete structure of the original compound and/or modification products. Also, the natural product chemist, with his close concern in this field, has become more directly involved both with the process of analysis and selection of the relevant results reported.

At the level of accuracy normally associated with natural product analyses, the main types of chemical bonds and intermediate ('hybrid') forms can be identified from the dimensions of the bonds and associated angles. In this respect, the data accumulated from x-ray studies has established clearly the constancy of dimensions in frequently occurring chemical groups, a result in accord with the constancy in their characteristic chemical properties and reactions. This information is of value in predicting the probable structure of new compounds. In contrast, there are instances when unexpected chemical groupings are made evident, as in the case of tetrodotoxin.

In terms of reported material, the influence of the natural product chemist is becoming more evident. The details of the structure in which he is interested differ somewhat from those generally regarded as of significance to the crystallographer. He is not so concerned with subtle details of bond lengths and angles but rather with information on conformational aspects, such as torsional angles indicative of the shapes of chemical groups and of ring systems. Indeed, from the viewpoint of the organic chemist, x-ray studies, by their very nature, provide far more numerical detail than is strictly essential for his original intention in requiring the analysis. However, this very detail tends to focus attention on the more subtle aspects of conformation of individual entities, e.g. substituent groups, ring systems, aspects which may be of particular significance for other physico-chemical techniques of structure study.

Most of these techniques operate with the molecules in solution and the question arises as to how far crystallographic evidence may be considered relevant to solution studies. For molecules which are held relatively rigid by bridging bonds (say), the molecular structure in solution is closely similar to that in the crystal. Where this condition does not apply and parts of the molecule may be capable of conformational modification, it has to be recognised that molecules in crystals are 'frozen', i.e. held in specific orientation by the intermolecular forces in that structure. In such circumstances, it may be ill advised and misleading to generalise from an isolated example. However, this objection may be minimised by consideration of a range of crystal structures involving the chemical group concerned, and in which a variety of packing occurs. In this way a distribution (or effective averaging) of forces acting on the group may be considered to imitate the mean buffeting which a molecule experiences in solution. If it is observed that one (or perhaps two) particular conformational pattern(s) for the group persist(s) through the 'statistical' sample of crystal structures, it may be concluded that the intermolecular forces on the group are of less significance than the intramolecular forces. Consequently a generalisation valid for use in relation to isolated molecules may be derived. Such generalisations refer to preferred conformational patterns and do not exclude the possibility that other structural requirements could over-ride or modify them.

For example, the generalisation concerning the planarity of the lactone (and ester) group[245] deduced from x-ray studies indicated that the o.r.d. properties of these groups might be understandable and prompted studies, initially on the lactone group[246]. While most subsequent x-ray analyses have lent support to the planarity generalisation, others have indicated significant deviations, e.g. in 1,5-glucurono-lactone[247]. An explanation for the deviations has not so far emerged but with the increased number of analyses now available, a more satisfactory statistical estimate of the magnitude of the deviations is possible and may give a guide to a suitable explanation.

A further example of the interaction of x-ray on natural-product-molecule studies relates to structures containing a diene group. A rule relating the sign of the Cotton effect for the cisoid diene chromophore to its absolute chirality was proposed earlier[248]. This rule appeared to fit the majority of examples investigated at that time. However, the existence of a number of anomalous examples was revealed, mainly as a result of absolute structure studies with x-rays, an early example being the erythrina bases and another, later, gliotoxin[249]. The accumulation of structures yielding anomalous results in relation to the diene rule forced a careful re-examination of the current evidence and this has now been rationalised in terms of the chirality of the diene-allylic oxygen system[250]. The significance of the x-ray results in stimulating re-assessment of anomalous results and re-activating theoretical investigation should not be under estimated, cf. the disulphide chromophore[251].

In respect of more flexible chemical groupings, study of a number of structures with ester groups attached to six-membered rings indicated a certain preferred conformational pattern[8]. Some use of this generalisation has been made in o.r.d. investigations[252]. Chemical groupings of very much greater biochemical significance in relation to muscle-nerve function are those in the acetylcholine–muscarine–nicotine group of compounds. In order to try to establish the existence of an underlying conformational pattern relevant to the function of such compounds *in vivo*, a range of structural variants has been investigated. Final conclusions have not been presented but comments on the study, which are too extensive to summarise here, are available[253].

It is evident that much valuable information can be acquired from 'statistical' studies of crystal structures. Recent reviews on 5- and 6-membered rings with hetero-atoms[254] and on 10-membered ring systems[255] are indicative of the potential of this approach. Generalisations of this type are valuable in integrating structural evidence, in drawing attention to the underlying physical factors involved and in stimulating further experimental investigation.

References

1. Mathieson, A. McL. (1967). *Perspectives in Structural Chemistry*, Vol. 1, 41. (New York: J. Wiley & Sons)
2. Molecular Structures and Dimensions. (1970). Ed. by O. Kennard and D. G. Watson. Vol. 1 and 2. (Utrecht: Oostoek)
3. Cruickshank, D. W. J. and Jeffery, G. A. (1954). *Acta Crystallogr.*, **7**, 646
4. Smith, H. E., Gray, R. T., Shaffner, T. J. and Lenhert, P. G. (1969). *J. Org. Chem.*, **34**, 136
5. Fleischer, E. B., Axelrod, M., Green, M. and Mislow, K. (1964). *J. Amer. Chem. Soc.*, **86**, 3395
6. Gabe, E. J. and Grant, D. F. (1962). *Acta Crystallogr.*, **15**, 1074
7. Ohrt, J. M. and Parthasarathy, R. (1969). *Acta Crystallogr.*, **A25**, S198
8. Mathieson, A. McL. (1965). *Tetrahedron Lett.*, 4137
9. Blount, J. F., Pawson, B. A. and Saucy, G. (1969). *Chem. Commun.*, 715
10. Wunderlich, J. A. and Lipscomb, W. N. (1960). *Tetrahedron*, **11**, 219
11. Schevitz, R. W., and Rossmann, M. G. (1969). *Chem. Commun.*, 711
12. Sato, T., Shiro, M. and Koyama, H. (1968). *J. Chem. Soc. B*, 935
13. Koyama, H. and Okada, K. (1969). *J. Chem. Soc. B*, 940
14. MacDonald, A. C. and Trotter, J. (1965). *Acta Crystallogr.*, **19**, 456
15. MacDonald, A. C. and Trotter, J. (1965). *Acta Crystallogr.*, **18**, 243
16. Abrahamsson, S. and Nilsson, B. (1966). *J. Amer. Chem. Soc.*, **88**, 3631
17. Fratini, A. V., Britts, K. and Karle, I. L. (1967). *J. Phys. Chem.*, **71**, 2482
18. Johnson, C. K. (1967). *Amer. Cryst. Assoc., Abstr. Papers*, 46
19. Altona, C., and Sundaralingam, M. (1969). *Acta Crystallogr.*, **A25**, S141
20. Soleimani, E. (1966). *Dissert. Abstr. B*, **27**, 1866
21. Cser, F. (1966). *Acta Crystallogr.*, **21**, A115
22. Destro, R., Flilippini, G., Gramaccioli, C. M. and Simonsetta, M. (1969). *Acta Crystallogr.*, **B25**, 2465
23. Destro, R., Gramaccioli, M. and Simonetta, M. (1968). *Chem. Commun.*, 568
24. Williams, J. E., Stang, P. J. and Schleyer, P. V. R. (1968). *Ann. Rev. Phys. Chem.*, **10**, 531
25. Alden, R. A., Kraut, J. and Taylor, T. G. (1968). *J. Amer. Chem. Soc.*, **90**, 74
26. Rerat, C. (1968). *Compt. Rend. Acad. Sci., Fr., C*, **266**, 612
27. Reck, G. (1969). *Z. Chem.*, **9**, 30
28. Williams, P. P. (1969). *Acta Crystallogr.*, **B25**, 409
29. Wiebenga, E. H. and Krom, J. C. (1946). *Rec. Trav. Chim. Pays Bas*, **65**, 663
30. Allen, F. H. and Rogers, D. (1966). *Chem. Commun.*, 837
31. Northolt, M. G. and Palm, J. H. (1966). *Rec. Trav. Chim. Pays Bas*, **85**, 143
32. Oonk, H. A. J. (1965). *Thesis* (University of Utrecht).

33. Ferguson, G., Fritchie, C. J., Robertson, J. M. and G. A. Sim (1961). *J. Chem. Soc.*, 1976
34. Brueckner, D. A., Hamor, T. A., Robertson, J. M. and Sim, G. A. (1962). *J. Chem. Soc.*, 799
35. Wunderlich, J. A. (1967). *Acta Crystallogr.*, **23,** 846
36. Reck, G. and Kutschabsky, L. (1970). *Acta Crystallogr.*, **B26,** 578
37. Kutschabsky, L. (1969). *Z. Chem.*, **9,** 31
38. Brice, M. D., Coxon, J. M., Dansted, E., Hartshorn, M. P. and Robinson, W. T. (1969). *Chem. Commun.*, 356
39. Barrans, Y. (1964). *Compt. Rend. Acad. Sci., Fr., C,* **259,** 796
40. McConnell, J. F., Mathieson, A. McL. and Schoenborn, B. P. (1964). *Acta Crystallogr.*, **17,** 472
41. Schoenborn, B. P. and McConnell, J. F. (1962). *Acta Crystallogr.*, **15,** 779
42. Masaki, N., Hirabayashi, M., Fuji, K., Osaki, K., and Inouye, H. (1967). *Tetrahedron Lett.*, 2367
43. Lentz, P. J. Jnr. and Rossmann, M. G. (1969). *Chem. Commun.*, 1269
44. Karle, I. L. and Karle, J. (1966). *Acta Crystallogr.*, **20,** 555
45. Paul-Roy, S., Schenk, H. and MacGillavry, C. H. (1969). *Chem. Commun.*, 1517
46. Bart, J. C. J. and MacGillavry, C. H. (1968). *Acta Crystallogr.*, **B24,** 1587
47. Cameron, A. F., Ferguson, G. and Robertson, J. M. (1969). *J. Chem. Soc. B,* 692
48. McPhail, A. T. and Sim, G. A. (1968). *J. Chem. Soc. B,* 1104
49. McEachen, C. E., McPhail, A. T. and Sim, G. A. (1966). *J. Chem. Soc. C,* 579
50. McClure, R. J., Schorno, K. S., Bertrand, J. A. and Zalkow, L. H. (1968). *Chem. Commun.*, 1135
51. McPhail, A. T. and Sim, G. A. (1965). *Chem. Commun.*, 276
52. Allen, F. H. and Rogers, D. (1967). *Chem. Commun.*, 588
53a. McClure, R. J., Sim, G. A., Coggon, P. and McPhail, A. T. (1970). *Chem. Commun.*, 128
53b. Coggon, P., McPhail, A. T. and Sim, G. A. (1970). *J. Chem. Soc. B,* 1024
54. Kupchan, S. M., Aynehchi, Y., Cassady, J. M., McPhail, A. T., Sim, G. A., Schnoes, H. K. and Burlingame, A. L. (1966). *J. Amer. Chem. Soc.*, **88,** 3674
55. Sorm, F., Suchy, M., Holub, M., Linek, A., Hadinec, I. and Novak, C. (1970). *Tetra-*
56. Pascard, C. (1970). *Tetrahedron Lett.*, 4131
57. McPhail, A. T. and Sim, G. A. (1966). *J. Chem. Soc. B,* 112
58. Allen, F. H. and Rogers, D. (1968). *J. Chem. Soc. B,* 1047
59. Gemmell, K. W., Parker, W., Roberts, J. S. and Sim, G. A. (1964). *J. Amer. Chem. Soc.*, **86,** 1438
60. Robertson, J. M. and Todd, G. (1955). *J. Chem. Soc.*, 1254
61. Rogers, D. and Ul-Haque, M. (1963). *Proc. Chem. Soc. (London)*, 371
62. Hawley, D. M., Ferguson, G. and Robertson, J. M. (1968). *J. Chem. Soc. B,* 1255
63. Bates, R. B., Forsythe, G. D., Wolfe, G. A., Ohloff, G. and Schulte-Elte, K.-H. (1969). *J. Org. Chem.*, **34,** 1059
64. Grant, D. F. and Rogers, D. (1956). *Chem. and Ind. (London)*, 278
65. Klyne, W., Bhattacharya, S. C., Paknikar, S. K., Narayanau, C. S., Krepinsky, J., Romanuk, M., Herout, V. and Sörm, F. (1964). *Tetrahedron Lett.*, 1443
66. Hossain, M. B. and Van Der Helm, D. (1968). *J. Amer. Chem. Soc.*, **90,** 6607
67. Linek, A., Vlahov, R., Holub, M. and Herout, V. (1968). *Tetrahedron Lett.*, 23
68. Thiessen, W. E. (1969). *Acta Crystallogr.*, **A25,** S144
69a. Hanic, F. (1958). *Collect. Czech. Commun.*, **23,** 1751
69b. Mani, N. V. (1963). *Z. Kristallogr.*, **118,** 103
70. Shimanouchi, H. and Sasada, Y. (1969). *Bull. Chem. Soc. Jap.*, **42,** 334
71. Nilsson, B. (1968). *Ark. Kemi,* **29,** 415
72. Comer, F. W. and Trotter, J. (1966). *J. Chem. Soc. B,* 11
73. Rossmann, M. G. and Lipscomb, W. N. (1958). *Tetrahedron,* **4,** 275
74. Ul-Haque, M., Caughlan, C. N., Emerson, M. T., Geissman, T. A. and Matsueda, S. (1970). *J. Chem. Soc. B,* 598
75. Asher, J. D. and Sim, G. A. (1965). *J. Chem. Soc.*, 6041
76. Coggon, P. and Sim, G. A. (1969). *J. Chem. Soc. B,* 237
77. McPhail, A. T., Rimmer, B., Robertson, J. M. and Sim, G. A. (1967). *J. Chem. Soc. B,* 101

78. Huber, C. P. and Watson, K. J. (1968). *J. Chem. Soc. C*, 2441
79. Asher, J. D. M. and Sim, G. A. (1965). *J. Chem. Soc.*, 1584
80. Emerson, M. T., Herz, W., Caughlan, C. N. and Witters, R. S. (1966). *Tetrahedron Lett.*, 6151
81. Thiessen, W. E. and Hope, H. (1970). *Acta Crystallogr. B*, **26**, 554
82. Kupchan, S. M., Hemingway, J. C., Cassady, J. M., Knox, J. R., McPhail, A. T. and Sim, G. A. (1967). *J. Amer. Chem. Soc.*, **89**, 465
83. Kupchan, S. M., Doskotch, R. W., Bollinger, P., McPhail, A. T., Sim, G. A. and Saenz Renauld, J. A. (1965). *J. Amer. Chem. Soc.*, **87**, 5805
84. Hamilton, J. A., McPhail, A. T. and Sim, G. A. (1962). *J. Chem. Soc.*, 708
85. Rogers, D. and Ul-Haque, M. (1963). *Proc. Chem. Soc. (London)*, 92
86a. Emerson, M. T., Caughlan, C. N. and Herz, W. (1964). *Tetrahedron Lett.*, 621
86b. Ul-Haque, M. and Caughlan, C. N. (1969). *J. Chem. Soc. B*, 956
87. Ul-haque, M. and Caughlan, C. N. (1967). *J. Chem. Soc. B*, 355
88. Dullforce, T. A., Sim, G. A., White, D. N. J., Kelsey, J. E. and Kupchan, S. M. (1969). *Tetrahedron Lett.*, 973
89. Sekita, T., Inayama, S. and Iitaka, Y. (1970). *Tetrahedron Lett.*, 135
90. Ul-Haque, M., Caughlan, C. N. and Emerson, M. T. (1969). *Acta Crystallogr.*, **A25**, S142
91. Linek, A., Novak, C., Vesela, L. and Kupcik, V. (1967). *Collect. Czech. Chem. Commun.*, **32**, 3437
92. Bates, R. B., Green, C. D. and Sneath, T. C. (1969). *Tetrahedron Lett.*, 3461
93. Clunie, J. S. and Robertson, J. M. (1961). *J. Chem. Soc.*, 4382
94. Hawley, D. M., Ferguson, G., McKillop, T. F. W. and Robertson, J. M. (1969). *J. Chem. Soc. B*, 599
95. Cesur, A. F. and Grant, D. F. (1965). *Acta Crystallogr.*, **18**, 55
96. Thierry, J. C. and Weiss, R. (1969). *Tetrahedron Lett.*, 2663
97. Dreyfus, H., Thierry, J. C., Weiss, R. Kennard, O., Motherwell, W. D. S., Coppola, J. C. and Watson, D. G. (1969). *Tetrahedron Lett.*, 3757
98. Coates, R. M., Farney, R. F., Johnson, S. M. and Paul, I. C. (1969). *Chem. Commun.*, 999
99. Abrahamsson, S. and Nilsson, B. (1966). *Acta Chem. Scand.*, **20**, 1044
100. McPhail, A. T. and Sim, G. A. (1965). *Chem. Commun.*, 350
101. Dobler, M., Dunitz, J. D., Gubler, B., Weber, H. P., Buchi, G. and Padilla, J. O. (1963). *Proc. Chem. Soc. (London)*, 383
102. Gabe, E. J. (1962). *Acta Crystallogr.*, **15**, 759
103. Pascard-Billy, C. (1970). *Chem. Commun.*, 1722
104. Oh, Y. L. and Maslen, E. N. (1968). *Acta Crystallogr.*, **B24**, 883
105. Natsume, M. and Iitaka, Y. (1966). *Acta Crystallogr.*, **20**, 197
106. Bolton, W. (1964). *Acta Crystallogr.*, **17**, 147
107. Coggan, P. and Sim, G. A. (1969). *J. Chem. Soc. B*, 413
108. Matsuda, H., Tomiie, Y., Yamamura, S. and Hirata, Y. (1967). *Chem. Commun.*, 898
109. Bjamer, K., Ferguson, G. and Melville, R. D. (1968). *Acta Crystallogr.*, **B24**, 855
110. King, T. J., Rodrigo, S. and Wallwork, S. C. (1969). *Chem. Commun.*, 683
111. Paul, I. C., Sim, G. A., Hamor, T. A. and Robertson, J. M. (1962). *J. Chem. Soc.*, 4133
112. Cheung, K. K., Melville, D., Overton, K. H., Robertson, J. M. and Sim, G. A. (1966). *J. Chem. Soc. B*, 853
113. McEachan, C. E., McPhail, A. T. and Sim, G. A. (1966). *J. Chem. Soc. B*, 633
114. Anderson, B. F., Hall, D. and Waters, T. N. (1970). *Acta Crystallogr.*, **B26**, 882
115a. Clark, G. R. and Waters, T. N. (1970). *J. Chem. Soc. C*, 887
115b. Cambie, R. C., Clark, G. R., Crump, D. R. and Waters, T. N. (1968). *Chem. Commun.*, 183
116. O'Connell, A. M. and Maslen, E. N. (1966). *Acta Crystallogr.*, **21**, 744
117. Birnbaum, K. B. and Ferguson, G. (1969). *Acta Crystallogr.*, **B25**, 720
118. Scott, A. I., Sim, G. A., Ferguson, G., Young, D. W. and McCapra, F. (1962). *J. Amer. Chem. Soc.*, **84**, 3197
119. Scott, A. I., Sutherland, S. A., Young, D. W., Gugielmetti, L., Arigoni, D. and Sim, G. A. (1964). *Proc. Chem. Soc. (London)*, 19
120. Hartsuck, J. A. and Lipscomb, W. N. (1963). *J. Amer. Chem. Soc.*, **85**, 3414
121. McCapra, F., McPhail, A. T., Scott, A. I., Sim, G. A. and Young, D. W. (1966). *J. Chem. Soc. C*, 1577
122. Yamazaki, S., Tamura, S., Marumo, F. and Saito, Y. (1969). *Tetrahedron Lett.*, 359

123. Petterson, R. C., Birnbaum, G. I., Ferguson, G., Islam, K. M. S. and Sime, J. G. (1968). *J. Chem. Soc. B,* 980
124. Hoppe, W., Zechmeister, K., Rohrl, M., Brandl, F., Hecker, E., Kraibich, G. and Burtsch, H. (1969). *Tetrahedron Lett.,* 667
125. Hoppe, W., Brandl, F., Strell, I., Rohrl, M., Gassmann, J., Hecker, E., Bartsch, H., Kreibich, G. and Szczepanski, Ch. V. (1967). *Angew. Chem.,* **79,** 824
126. Hecker, E., Harle, E., Schairer, H. U., Jacobi, P., Hoppe, W., Gassmann, J., Rohrl, M. and Abel, H. (1968). *Angew. Chem.,* **80,** 913
127. Drew, M. G. B., Templeton, D. H., and Zalkin, A. (1969). *Acta Crystallogr.,* **B25,** 261
128. Van der Helm, D. and Hossain, M. B. (1968). *Amer. Cryst. Assoc., Abstr. Papers (Summer Meeting),* 56
129. Hossain, M. B., Nicholas, A. F. and Van der Helm, D. (1968). *Chem. Commun.,* 385
130. Zechmeister, K., Rohrl, M., Brandl, F., Hechtfisher, S., Hoppe, W. Hecker, E., Adolf, W. and Kubinyi, H. (1970). *Tetrahedron Lett.,* 3071
131. Zechmeister, K., Brandl, F., Hoppe, W., Hecker, E., Opferkuch, H. J. and Adolf, W. (1970). *Tetrahedron Lett.,* 4075
132a. Moncrief, J. W. and Sim, S. P. (1969). *Acta Crystallogr.,* **A25,** S74
132b. Hope, H. and de la Camp, U. (1969). *Nature (London),* **221,** 54
133. Bjamer, K., Ferguson, G. and Robertson, J. M. (1967). *J. Chem. Soc. B,* 1272
134. Shiro, M., Sato, T., Koyama, H., Nakanishi, K. and Uyeo, S. (1966). *Chem. Commun.,* 97
135. Della Casa de Marcano, D. P., Halsall, T. G., Castellano, E. and Hodder, O. J. R. (1970). *Chem. Commun.,* 1382
136. Kupchan, S. M., Sigel, C. W., Matz, M. J., Renauld, J. A. S., Haltiwanger, R. C. and Bryan, R. F. (1970). *J. Amer. Chem. Soc.,* **92,** 4476
137. Sakabe, N., Takada, S. and Okabe, K. (1967). *Chem. Commun.,* 259
138. Narayanan, P., Rohrl, M., Zechmeister, K. and Hoppe, W. (1970). *Tetrahedron Lett.,* 3943
139. Brufani, M., Cerrini, S., Fedeli, W. and Vaciago, A. (1969). *Acta Crystallogr.,* **A25,** S202
140. Morisaki, M., Nozoe, S. and Iitaka, Y. (1968) *Acta Crystallogr.,* **B24,** 1293
141. Iitaka, Y., Watanabe, I. and Harrison, I. T. (1969). *Acta Crystallogr.,* **B25,** 1299
142. Itai, A., Nozoe, S., Okuda, S. and Iitaka, Y. (1969). *Acta Crystallogr.,* **B25,** 872
143. Kartha, G. and Haas, D. J. (1964). *J. Amer. Chem. Soc.,* **86,** 3630
144. Brown, W. A. C. and Sim, G. A. (1964). *Proc. Chem. Soc. (London),* 293
145. Sutherland, S. A., Sim, G. A. and Robertson, J. M. (1962). *Proc. Chem. Soc. (London),* 222
146. Arnott, S., Davie, A. W., Robertson, J. M., Sim, G. A. and Watson, D. G. (1961). *J. Chem. Soc.,* 4183
147. Grant, I. J., Hamilton, J. A., Hamor, T. A., Robertson, J. M. and Sim, G. A. (1963). *J. Chem. Soc. B,* 2506
148. McPhail, A. T. and Sim, G. A. (1966). *J. Chem. Soc. B,* 318
149. Adeoye, S. A. and Bekoe, F. A. (1965). *Chem. Commun.,* 301
150. Ladd, M. F. C. and Carlisle, C. H. (1966). *Acta Crystallogr.,* **21,** 689
151. Yamauchi, H., Fujiwara, T. and Tomita, K. (1969). *Tetrahedron Lett.,* 4245
152. Tanaka, O., Tanaka, N., Ohsawa, T., Iitaka, Y. and Shibata, S. (1968). *Tetrahedron Lett.,* 4235
153. Hall, S. R. and Maslen, E. N. (1965). *Acta Crystallogr.,* **18,** 265
154. Oh, Y. L. and Maslen, E. N. (1966). *Acta Crystallogr.,* **20,** 852
155. Koyama, H. and Nakai, H. (1970). *J. Chem. Soc. B,* 546
156. Nakanishi, T., Fujiwara, T. and Tomita, K. (1968). *Tetrahedron Lett.,* 1491
157. Kennard, O., Riva Di Sanseverino, L. and Rollett, J. S. (1967). *Tetrahedron,* **23,** 131
158. Nishi, Y., Ashida, T., Sasada, Y. and Kakudo, M. (1968). *Bull. Chem. Soc. Jap.,* **41,** 1308
159. Abd El Rahim, A. M. and Carlisle, C. H. (1956). *Proc. Math. Phys. Soc. U.A.R.,* **5,** 87
160. Mornon, J. P. (1970). *Compt. Rend. Acad. Sci., Fr., C.,* **270,** 926
161. Smith, G. W. (1970). *Acta Crystallogr.,* **B26,** 1746
162. Stout, G. H., Malofsky, B. M. and Stout, V. F. (1964). *J. Amer. Chem. Soc.,* **86,** 957
163. Akiyama, T., Tanaka, O. and Iitaka, Y. (1970). *Acta Crystallogr.,* **B26,** 163
164. Gordon, J. T. and Doyne, T. H. (1966). *Acta Crystallogr.,* **21,** A113
165. Stout, G. H. and Stevens, K. L. (1963). *J. Org. Chem.,* **28,** 1259
166. Nishikawa, M., Kamiya, K., Murata, T., Tomiie, Y. and Nitta, I. (1965). *Tetrahedron Lett.,* 3223

167. Brewis, S., Halsall, T. G., Harrison, H. R. and Hooder, C. J. R. (1970). *Chem. Commun.*, 891
168. Allen, F. H. and Trotter, J. (1970). *J. Chem. Soc. B*, 721
169. Sterling, C. (1964). *Acta Crystallogr.*, **17**, 1224
170. Sterling, C. (1964). *Acta Crystallogr.*, **17**, 500
171. Sly, W. G. (1964). *Acta Crystallogr.*, **17**, 511
172. Bart, J. C. J. and MacGillavry, C. H. (1968). *Acta Crystallogr.*, **B24**, 1587
173. Bart, J. C. J. and MacGillavry, C. H. (1968). *Acta Crystallogr.*, **B24**, 1569
174. Norton, D. A., Kartha, G. and Lu, C. T. (1963). *Acta Crystallogr.*, **16**, 89
175. Busetta, B. and Hospital, M. (1969). *Compt. Rend. Acad. Sci., Fr. C*, **268**, 1300
176. Norton, D. A., Kartha, G. and Lu, C. T. (1964). *Acta Crystallogr.*, **17**, 77
177. Tsukuda, Y., Sato, T., Shiro, M. and Koyama, H. (1968). *J. Chem. Soc. B*, 1387
178. Tsukuda, Y., Sato, T., Shiro, M. and Koyama, H. (1969). *J. Chem. Soc. B*, 336
179. Cody, V., De Jarnette, F. E., Duax, W. L. and Norton, D. A. (1970). *Amer. Crystallogr. Assoc., Abstr. Papers (Summer Meeting)*, 72
180. Weeks, C. M. and Norton, D. A. (1970). *J. Chem. Soc. B*, 1494
181. Hanson, J. C. and Nordman, C. E. (1970). *Amer. Crystallogr. Assoc., Abstr. Papers (Summer Meeting)*, 73
182. Cooper, A., Norton, D. A. and Hauptman, H. (1969). *Acta Crystallogr.*, **B25**, 814
183. Taylor, N. E., Hodgkin, D. C. and Rollett, J. S. (1960). *J. Chem. Soc.*, 3685
184a. High, D. F. and Kraut, J. (1966). *Acta Crystallogr.*, **21**, 88
184b. Norton, D. A. and Ohrt, J. M. (1964). *Nature (London)*, **203**, 754
185. Hauptman, H., Weeks, C. and Norton, D. A. (1969). *Acta Crystallogr.*, **A25**, S85
186. Cooper, A., Lu, C. T. and Norton, D. A. (1968). *J. Chem. Soc. B*, 1228
187. Oberhansli, W. E. and Robertson, J. M. (1967). *Helv. Chim. Acta*, **50**, 53
188. Puckett, R. T., Sim, G. A., Cross, A. D. and Siddall, J. B. (1967). *J. Chem. Soc. B*, 783
189. Braun, P. B., Hornstra, J. and Leenhouts, J. I. (1969). *Philips Res. Rep., Netherl.*, **24**, 427
190. Braun, P. B., Hornstra, J. and Leenhouts, J. I. (1970). *Acta Crystallogr.*, **B26**, 352
191. Tamura, C. and Sim, G. A. (1968). *J. Chem. Soc. B*, 8
192. Portheine, J. C. and Romers, C. (1970). *Acta Crystallogr.*, **B26**, 1791
193. Ferguson, G., Macaulay, E. W., Midgley, J. M., Robertson, J. M. and Whalley, W. B. (1970). *Chem. Commun.*, 954
194. McKechnie, J. S., Kubina, L. and Paul, I. C. (1970). *J. Chem. Soc. B*, 1476
195. Thiessen, W. E. (1970). *Amer. Cryst. Assoc., Abstr. Papers (Winter Meetings)*, 40
196. Hope, H. and Christensen, A. T. (1968). *Acta Crystallogr.*, **B24**, 375
197. Dobler, M. and Dunitz, J. D. (1965). *Helv. Chim. Acta*, **38**, 1429
198. McKechnie, J. S. and Paul, I. C. (1968). *J. Amer. Chem. Soc.*, **90**, 2144
199. Koyama, H., Shiro, M., Sato, T. and Tsukuda, Y. (1970). *J. Chem. Soc. B*, 443
200. Cooper, A., Gopalakrishna, E. M. and Norton, D. A. (1968). *Acta Crystallogr.*, **B24**, 935
201. Duax, W. L., Osawa, Y., Norton, D. A. and Pokrywiecki, S. (1970). *Amer. Crystallogr. Assoc., Abstr. Papers (Winter Meeting)*, 39
202. Ohrt, J. M., Cooper, A., Kartha, G. and Norton, D. A. (1968). *Acta Crystallogr.*, **B24**, 824
203. Ohrt, J. M., Cooper, A. and Norton, D. A. (1969). *Acta Crystallogr.*, **B25**, 41
204. Ohrt, J. M., Haner, B. A., Cooper, A. and Norton, D. A. (1968). *Acta Crystallogr.*, **B24** 312
205. Romers, C., Hesper, B., van Heijkoop, and Geise, H. J. (1966). *Acta Crystallogr.*, **20**, 363
206. Gopalakrishna, E. M., Cooper, A. and Norton, D. A. (1969). *Acta Crystallogr.*, **B25**, 2473
207. Cooper, A. and Norton, D. A. (1968). *Acta Crystallogr.*, **B24**, 811
208. Gopalakrishna, E. M., Cooper, A. and Norton, D. A. (1969). *Acta Crystallogr.*, **B25**, 639
209. Christensen, A. T. (1970). *Acta Crystallogr.*, **B26**, 1519
210. Gopalakrishna, E. M., Cooper, A. and Norton, D. A. (1969). *Acta Crystallogr.*, **B25**, 143
211. Gopalakrishna, E. M., Cooper, A., and Norton, D. A. (1969). *Acta Crystallogr.*, **B25**, 1601
212. Carlisle, C. H. and Crowfoot, D. (1945). *Proc. Roy. Soc. (London)*, **A184**, 64
213. Burki, H. and Nowacki, W. (1956). *Z. Kristallogr.*, **108**, 206
214. Geise, H. J., Romers, C. and Rutten, E. W. M. (1966). *Acta Crystallogr.*, **20**, 249
215. Geise, H. J. and Romers, C. (1966). *Acta Crystallogr.*, **20**, 257
216. Meney, J., Kim, Y. H., Stevenson, R. and Margulis, T. N. (1970). *Chem. Commun.*, 1706
217. Mootz, D. and Berking, B. (1970). *Acta Crystallogr.*, **B26**, 1362
218. Huber, R. and Hoppe, W. (1965). *Chem. Ber.*, **98**, 2403

219. Cooper, A. and Norton, D. A. (1968). *J. Org. Chem.*, **33**, 3535
220. Berman, H. M., Chu, S. S. C. and Jeffery, G. A. (1967). *Science*, **157**, 1576
221. Hesper, B., Geise, H. J. and Romers, C. (1969). *Rec. Trav. Chim. Pays Bas*, **88**, 871
222. Margulis, T. N., Hammer, C. F. and Stevenson, R. (1964). *J. Chem. Soc.*, 4396
223. Hodgkin, D. C., Rimmer, B. M., Dunitz, J. D. and Trueblood, K. N. (1963). *J. Chem. Soc.*, 4945
224a. Hodgkin, D. C. and Sayre, D. (1952). *J. Chem. Soc.*, 4561
224b. Castells, J., Jones, E. R. H., Williams, R. W. J. and Meakins, G. D. (1958). *Proc. Chem. Soc. (London)*, 7
225. Saunderson, C. P. (1965). *Acta Crystallogr.*, **19**, 187
226. Abrahamsson, S. (1963). *Acta Crystallogr.*, **16**, 409
227. Hardgrove, G. L., Duerst, R. W. and Kispert, L. D. (1968). *J. Org. Chem.*, **33**, 4393
228. Fridrichsons, J. and Mathieson, A. McL. (1953). *J. Chem. Soc.*, 2159
229. Fawcett, J. K. and Trotter, J. (1966). *J. Chem. Soc. B*, 174
230. Majeste, R. and Trefonas, L. M. (1969). *J. Amer. Chem. Soc.*, **91**, 1508
231. Burckhalter, J. H., Abramson, H. N., MacConnell, J. G., Thill, R. J., Olson, A. J., Hanson, J. C. and Nordman, C. E. (1968). *Chem. Commun.*, 1274
232. Cooper, A. and Hodgkin, D. C. (1968). *Tetrahedron*, **24**, 909
233. Karle, I. L. and Karle, J. (1969). *Acta Crystallogr.*, **B25**, 434
234. Gilardi, R. D. and Karle, I. L. (1970). *Acta Crystallogr.*, **B26**, 207
235a. Karle, I. L. and Karle, J. (1969). *Acta Crystallogr.*, **B25**, 428
235b. Gilardi, R. D. (1970). *Acta Crystallogr.*, **B26**, 440
236. McPhail, A. T. and Sim, G. A. (1968). *J. Chem. Soc. B*, 962
237. O'Donnell, E. A. and Ladd, M. F. C. (1967). *Acta Crystallogr.*, **23**, 460
238. Callow, R. K., James, V. H. T., Kennard, O., Page, J. E., Paton, P. N. and Riva Di Sanseverino, L. (1966). *J. Chem. Soc. C*, 288
239. Hoppe, W., Gieren, A., Brodherr, N., Tschesche, R. and Wulff, G. (1968). *Angew. Chem.*, **80**, 563
240. Kennard, O., Fawcett, J. K., Watson, D. G., Kerr, K. A., Stockel, K., Stocklin, W. and Reichstein, T. (1968). *Tetrahedron Lett.*, 3799
241. Ling, N. C., Hale, R. L., and Djerassi, C. (1970). *J. Amer. Chem. Soc.*, **92**, 5281
242. Tsay, Y. H., Silverton, J. V., Beisler, J. A. and Sato, Y. (1970). *J. Amer. Chem. Soc.*, **92**, 7005
243. Green, M. J., Abraham, N. A., Fleischer, E. B., Case, J. and Fried, J. (1970). *Chem. Commun.*, 234
244. Kennard, O., Motherwell, W. D. S., Watson, D. G., Coppola, J. C. and Larson, A. C. (1969). *Acta Crystallogr.*, **A25**, S88
245a. Mathieson, A. McL. and Taylor, J. C. (1961). *Tetrahedron Lett.*, 590
245b. Mathieson, A. McL. (1962). *2nd I.U.P.A.C. International Symposium on the Chemistry of Natural Products, Abstracts*, 73
246a. Jennings, J. P., Klyne, W. and Scopes, P. M. (1965). *J. Chem. Soc.*, 7211
246b. Wolf, H. (1966). *Tetrahedron Lett.*, 5151
246c. Snatzke, G., Ripperger, H., Horstmann, C. and Schreiber, K. (1966). *Tetrahedron*, **22** 3103
247. Hackert, M. L. and Jacobson, R. A. (1969). *Chem. Commun.*, 1179
248. Moscowitz, A., Charney, E., Weiss, U. and Ziffer, H. (1961). *J. Amer. Chem. Soc.*, **83**, 4661
249. Beecham, A. F., Fridrichsons, J. and Mathieson, A. McL. (1966). *Tetrahedron Lett.*, 3131
250. Beecham, A. F., Mathieson, A. McL., Johns, S. R., Lamberton, J. A., Sioumis, A. A., Batterham, T. J. and Young, I. G. (1971). *Tetrahedron*, **27**, 3725
251. Linderberg, J. and Michl, J. (1970). *J. Amer. Chem. Soc.*, **92**, 2619
252. Jennings, J. P., Mose, W. P. and Scopes, P. M. (1967). *J. Chem. Soc. C*, 1102
253a. Chothia, C. H. and Pauling, P. (1970). *Proc. Nat. Acad. Sci.*, **65**, 477
253b. Baker, R. W., Chothia, C. H., Pauling, P. and Petcher, T. J. (1971). *Nature (London)*, **230**, 439
254. Romers, C., Altona, C., Buys, H. R. and Havinga, E. (1969). *Topics in Stereochemistry*, Vol. 4, 39. (New York: Wiley–Interscience).
255. Dunitz, J. D. (1968). *Perspectives in Structural Chemistry*, Vol. 2, 1. (New York: J. Wiley & Sons).